Springer Climate

Series Editor
John Dodson, Institute of Earth Environment, Chinese Academy of Sciences, Xian, Shaanxi, China

Springer Climate is an interdisciplinary book series dedicated to climate research. This includes climatology, climate change impacts, climate change management, climate change policy, regional climate studies, climate monitoring and modeling, palaeoclimatology etc. The series publishes high quality research for scientists, researchers, students and policy makers. An author/editor questionnaire, instructions for authors and a book proposal form can be obtained from the Publishing Editor. **Now indexed in Scopus® !**

More information about this series at http://www.springer.com/series/11741

Song Yang · Renguang Wu · Maoqiu Jian ·
Jian Huang · Xiaoming Hu · Ziqian Wang ·
Xingwen Jiang

Climate Change in Southeast Asia and Surrounding Areas

Science Press
Beijing

Springer

See next page

ISSN 2352-0698 ISSN 2352-0701 (electronic)
Springer Climate
ISBN 978-981-15-8224-0 ISBN 978-981-15-8225-7 (eBook)
https://doi.org/10.1007/978-981-15-8225-7

Jointly published with Science Press.
The print edition is not for sale in China (Mainland). Customers from China (Mainland) please order the print book from Science Press

This Springer imprint is published by the registered company Springer Nature Singapore Pte Ltd.
The registered company address is: 152 Beach Road, #21-01/04 Gateway East, Singapore 189721, Singapore

Song Yang
School of Atmospheric Sciences
Sun Yat-sen University
Zhuhai, Guangdong, China

Guangdong Province Key Laboratory
for Climate Change and Natural
Disaster Studies
Zhuhai, China

Southern Marine Science and Engineering
Guangdong Laboratory (Zhuhai)
Zhuhai, China

Maoqiu Jian
School of Atmospheric Sciences
Sun Yat-sen University
Zhuhai, Guangdong, China

Guangdong Province Key Laboratory
for Climate Change and Natural
Disaster Studies
Zhuhai, China

Southern Marine Science and Engineering
Guangdong Laboratory (Zhuhai)
Zhuhai, China

Xiaoming Hu
School of Atmospheric Sciences
Sun Yat-sen University
Zhuhai, Guangdong, China

Guangdong Province Key Laboratory
for Climate Change and Natural
Disaster Studies
Zhuhai, China

Southern Marine Science and Engineering
Guangdong Laboratory (Zhuhai)
Zhuhai, China

Xingwen Jiang
Institute of Plateau Meteorology
China Meteorological Administration
Chengdu, Sichuan, China

Renguang Wu
School of Earth Sciences
Zhejiang University
Hangzhou, Zhejiang, China

Jian Huang
Institute of Tropical and Marine
Meteorology
China Meteorological Administration
Guangzhou, Guangdong, China

Ziqian Wang
School of Atmospheric Sciences
Sun Yat-sen University
Zhuhai, Guangdong, China

Guangdong Province Key Laboratory
for Climate Change and Natural
Disaster Studies
Zhuhai, China

Southern Marine Science and Engineering
Guangdong Laboratory (Zhuhai)
Zhuhai, China

Contents

1 Introduction 1
References 6

2 Characteristics of the Spring–Summer Atmospheric Circulation Transition Over the South China Sea and Its Surrounding Regions and Their Responses to Global Warming 7
2.1 Characteristics of the Transition Over the SCS and Its Surrounding Regions 8
2.1.1 Characteristics of the Transition 9
2.1.2 Physical Processes of the Transition 14
2.2 Interannual Variation of Net Heat Flux Over the Indian Ocean and the Western Pacific Ocean During the Transition 24
2.3 Impact of Intraseasonal Oscillation on SCSSM Onset 28
2.3.1 Relationship Between SCSSM Onset and MJO Activity 31
2.3.2 Physical Process Behind MJO Impact on SCSSM Onset 32
2.3.3 Synergy Between MJO and Tropical SST 35
2.4 Precursory Signals for SCSSM Onset 38
2.4.1 Multi-time-scale Characteristics of SCSSM Onset Date . . . 39
2.4.2 Precursory SST Signals for Interannual Variation and Interdecadal Change of SCSSM Onset 40
2.4.3 Asymmetry Between the Central–Eastern Equatorial Pacific SSTA and SCSSM Onset 42
2.5 Responses of Subtropical Highs to Global Warming 44
2.5.1 Model, Data, and Method 45
2.5.2 Response of the WNPSH to Global Warming 46
2.5.3 Responses of Subtropical Anticyclones to Global Warming 56

2.6 Response of SCSSM Onset to Global Warming 64
2.6.1 Models, Data, and Method 65
2.6.2 SCSSM Onset in CMIP5 Historical Experiments 65
2.6.3 Response of SCSSM Onset to Global Warming 67
2.7 Summary 72
References 75

3 Air–Sea Interactions and Climate Variability Over the South China Sea and the Adjacent Regions 81
3.1 Air–Sea Interactions on Different Time Scales 82
3.1.1 Intraseasonal Air–Sea Interactions.................. 83
3.1.2 Interannual Air–Sea Interactions 92
3.2 Processes for Interannual Variability of Rainfall Over the South China Sea 97
3.2.1 Interannual Variations of Summer Rainfall in the South China Sea 98
3.2.2 Interannual Variations Of the South China Sea Rainfall During the Spring-To-Summer Transition 102
3.2.3 Influence Of Local Air–Sea Interaction on the South China Sea Climate During the Spring-to-Summer Transition 104
3.3 Interdecadal Variation 109
3.3.1 Interdecadal Variability Of Early Summer Monsoon Rainfall Over South China In Association With the Pacific Decadal Oscillation 109
3.3.2 Interdecadal Modulation of ENSO-Related Spring Rainfall Over South China by the Pacific Decadal Oscillation............................. 112
3.4 Cross-Scale Relation 120
3.4.1 Interannual Variation of the Intensity of the Summer ISO Over the Tropical Western North Pacific 120
3.4.2 The Change of the Intensity of the ISOs Over the Tropical Indo-Pacific Region With the Phase of ENSO 123
3.4.3 Feedback of ISOs on Seasonal Mean SST Anomalies in the Tropical Western North Pacific 125
3.4.4 Influence of ISOs on Interdecadal Change............. 128
3.4.5 Influence of Interannual Variation on Synoptic Disturbances 129
3.5 Summary 131
References 133

4 Land–Atmosphere Interaction and Climate Variability in Southeast Asia and Its Surrounding Area 139
4.1 Field Observations of Typical Underlying Land-Surface Process and Relevant Theoretical Analysis 140
4.1.1 Field Observations of Typical Underlying Surface in the South China 140
4.1.2 Data Processing Method 142
4.1.3 Analysis on Characteristics of Typical Underlying Surface Land–Atmosphere Interaction in South China and Its Surrounding Area 148
4.2 Influence of Land–Atmosphere Interaction on Climate Change of the South China Sea and Its Surrounding Areas as a Response to Global Climate Change 168
4.2.1 Correction, Optimization, and Improvement of Important Land-Surface Process Parameters of the Typical Underlying Surfaces in the South China Sea and the Surrounding Area 169
4.2.2 Influences of Land–Atmosphere Interaction on Spring and Summer Climate Variability in Southeast Asia and Its Surrounding Areas, and Mechanism 175
4.3 Land–Air Interaction and Climate Variation in Southeast Asia and Its Surrounding Areas 184
4.3.1 Relationships Among Surface Energy, Water Balance Characteristics, and Regional Climate Anomalies of Typical Underlying Surface in Southeast Asia 184
4.3.2 Physical Process of Soil Moisture Affecting Summer Precipitation 190
4.3.3 Effect of Soil Moisture on the South China Sea Summer Monsoon 196
References 200

5 Effects of the Tibetan Plateau on Climate 205
5.1 Introduction 205
5.2 Effects of the TP Summer Heating on Climate in Regional Models 206
5.2.1 Models and Experimental Design 206
5.2.2 Two Types of TP Heating and Their Interactions 208
5.2.3 Effects on Meridional Circulation and on Temperature and Circulation in the Upper Troposphere 211
5.2.4 TP Heating and Potential Vorticity Forcing Near the Tropopause 214
5.2.5 Potential Regulation by Tropical Air-Sea Coupling 216

5.3 Key Effect of TP on Eurasian Teleconnection 219
5.3.1 Data and Model 220
5.3.2 The Bridge Effect of TP on Eurasian Teleconnection 220
5.3.3 Direct Impact of TP Heating on East China Summer Rainfall in AGCM Experiments 224
5.4 Thermal Effect of TP on Climate Variation Over the Upstream Regions 225
5.4.1 Data, Model, and Experimental Design 227
5.4.2 Impact of TP Surface Heating on Climate Variation Over West Asia, North Africa, South Europe, and the North Atlantic 228
5.4.3 Conclusions and Discussion 235
5.5 Climate Effect of SAH Variability Over TP 236
5.5.1 Zonal Shift of SAH and Its Relationship with Asian Summer Monsoon Rainfall 236
5.5.2 Relationship with the ISM and EASM Rainfall 237
5.5.3 Impact of ISM Rainfall on the Zonal Shift of SAH 239
5.5.4 Impact of Zonal Displacement of SAH on the Summer Rainfall Over China 242
5.5.5 Impact of YRV Rainfall on the Zonal Shift of SAH 243
5.5.6 Conclusions and Discussion 247
References 248

6 Feedback Attributions of Climate Changes Over the Globe and Over Southeast Asia and Its Adjacent Regions 253
6.1 Overview 253
6.2 Climate Response-Feedback Analysis Method 255
6.3 Attribution of Decadal Climate Difference Between 2002–13 and 1984–95 259
6.4 Feedback Attribution of Interannual Variability 261
6.4.1 Attribution of the Distinct Spatial Patterns of SST Anomalies for Two Types of El Niño 261
6.4.2 Variations of Dominant Modes of the East Asian Winter Monsoon 266
6.5 Feedback Attributions of Climate Changes Over the South China Sea and Its Adjacent Regions 277
6.5.1 Annual Cycles of Surface Temperature 277
6.5.2 Interdecadal Change in Troposphere Temperature 283
6.6 Uncertainty in Tropical Climate Change Under Global Warming 289
6.6.1 Uncertainty in Tropical Rainfall Change 289
6.6.2 Uncertainty in Tropical Pacific SST Change 292
References 297

7 Impact of Climate Change Over Southeast Asia and Its Adjacent Regions on Global Climate 303
7.1 Long-Term Trend and Interannual Variation of Springtime Precipitation 303
7.2 Effect on East Asian Climate 305
7.2.1 Impact of Convection Over the South China Sea and the Philippine Sea on Southern China Precipitation 305
7.2.2 Results from Numerical Model Simulations 307
7.3 Influence of Climate Change Over Southeast Asia on Sahel Summer Monsoon Rainfall 312
7.3.1 Observed Features 314
7.3.2 Numerical Experiments with SAWPSM Latent Heating Anomalies 317
7.3.3 Summary and Discussion 320
7.4 Impact of Precipitation and Associated Latent Heat Release Over South China Sea and Western Pacific on Global Climate During Boreal Spring and Summer 322
7.4.1 Recent Change in Precipitation 322
7.4.2 Teleconnection of Precipitation and Atmospheric Circulation Anomaly 324
7.4.3 Heating Effect During Boreal Summer 330
7.4.4 Heating Effect During Boreal Spring 335
7.4.5 Summary and Discussion 339
7.5 Impacts of Climate Change in the Central-to-Western Tropical Pacific on Arctic Climate 340
7.5.1 Expansion of the Central-Western Pacific Warm Pool Under Global Warming 340
7.5.2 Shifting Mode of El Niño in the Contexts of Global Warming and Warm Pool Expansion 341
7.5.3 Teleconnection Between Shifting El Niño and Arctic Summer Climate Anomalies 343
7.5.4 Convection Differences Associated with Shifting El Niño 346
7.5.5 Summary and Discussion 348
References 350

8 Subseasonal to Seasonal Prediction of Atmospheric Circulation and Rainfall Over Southeast Asia 357
8.1 Introduction 357
8.2 Seasonal Prediction of Atmospheric Circulation and Rainfall Over SEA and Its Adjacent Regions 360
8.2.1 Prediction of the Asian Summer Monsoon 360
8.2.2 Prediction of the Dominant SST Variability in the Pacific and Indian Oceans 362

8.2.3 Prediction of the Western Pacific Subtropical High 367
8.2.4 Prediction of MC Rainfall and Related Atmospheric Circulation 370
8.3 Subseasonal Prediction of Rainfall and Atmospheric Circulation in SEA and Its Adjacent Regions 380
8.3.1 Subseasonal Prediction of Atmospheric Variability Over the Extratropical Northern Hemisphere 380
8.3.2 Subseasonal Prediction of the Tropical Climate Variability 384
8.3.3 Subseasonal Prediction of the Climate Variation Over Southeast Asia 390
8.4 Predictability of Climate Variation Over SEA and Its Adjacent Regions 397
8.4.1 The Most Predictable Patterns of Rainfall and Atmospheric Circulation on Subseasonal and Seasonal Scales 397
8.4.2 Possible Impact Factors of Seasonal Climate Predictability 404
8.4.3 Possible Impact Factors of Subseasonal Climate Predictability 410
8.5 Conclusions and Discussion 413
References 415

Chapter 1
Introduction

Global climate change and its impact are issues concerned gravely by the international community. Under the background of climate change, extreme weather and climate events such as heat waves, droughts, and floods have occurred more and more frequently worldwide, and the impacts of climate change on the human living environment and the development of society and economy have become increasingly significant. Since the United Nations Conference on Environment and Development in 1992, climate change has become the focus of global environmental issues and international political activities. In the past 20 years, the international scientific community has launched several large international research projects such as the World Climate Research Program (WCRP), the International Geosphere-Biosphere Programme (IGBP), the International Human Dimensions Programme on Global Environmental Change (IHDP), the DIVERSITAS projects, and the Global Water System Project (GWSP). The core of these programs is global climate change, especially for the change in physical, chemical, and biological synthesis processes from the interannual, decadal to century time scales and its predictability, as well as the impacts of climate change on the human environment and adaptation policy. In China, the "Twelfth Five-Year Plan for National Economic and Social Development" approved by the 4th Session of the 11th National People's Congress in 2011 clarified the goals and tasks of China's economic and social development during the "Twelfth Five-Year Plan" period, in which climate change response as an important content was formally incorporated into the medium- and long-term planning of national economic and social development.

Southeast Asia and its adjacent regions (SAARs) are located in the core of the Asian monsoon system and in the centers of interactions of the Asian monsoon with El Niño–Southern Oscillation (ENSO), the Pacific and Indian Oceans, as well as the Northern and Southern Hemispheres. The SAAR is one of the regions that experiences most pronounced climate change, and is most adversely affected by global climate change. It consists of seas, land, and a group of islands, and its climate is determined by various atmospheric activities on a range of temporal and spatial scales with frequent severe weather and climate events such as typhoons,

S. Yang et al., *Climate Change in Southeast Asia and Surrounding Areas*,
Springer Climate, https://doi.org/10.1007/978-981-15-8225-7_1

heavy rains, droughts, and floods. Meanwhile, the SAAR is characterized by a large amount of latent heat release and is an important heat source of the global atmospheric circulation. The climate change in the SAAR drives the variations of global atmospheric circulation (Webster and Yang 1992), which exerts profound impacts on regional and global weather and climate (Neale and Slingo 2003).

Located to the north of Southeast Asia, China has experienced an increase in the frequency and intensity of extreme weather and climate events under the background of climate change (Herring et al. 2014). Increases in extreme weather and climate events such as heavy rain and flood disasters, typhoons, droughts, and heat waves have led to unprecedented negative consequences for economic and social development and the safety of human lives and properties. Global and regional climate changes have exerted significant effects on China's economic and social development, a major challenge for sustainable development. Previous studies have also shown that the climate change in the SAAR can directly influence China's weather and climate. Therefore, systematical studies of the characteristics and physical mechanisms of the climate change and a better understanding of the climate predictability in the SAAR are not only of great significance to local social and economic development, but also crucial to improved forecast weather and climate in China.

Previous studies of the SAAR climate variability and changes have been mainly focused on winter and summer seasons, with much less attention to spring, the important transitional season from winter to summer. On the one hand, the tropical monsoon, which obtains its maximum intensity in summer, grows rapidly in Southeast Asia during spring, and ENSO events also develop or terminate in (late) spring. The persistent sea surface temperature (SST) anomalies in the tropical Indian Ocean in spring exert a prominent impact on the climate anomalies in the following summer. The tropical and subtropical land surface also warms rapidly and the convective activity over the South China Sea (SCS) warm pool, and the Maritime Continent (MC) grows abruptly in April and May. Moreover, the high-latitude weather systems in the high latitudes often cause significant influences on Southeast Asia in spring. On the other hand, monsoon, ENSO, Indian Ocean SST, land surface processes, and mid-high latitude systems are highly interactive, importantly determining the surface climate anomalies in tropical Asia and surrounding regions in spring and the following summer. Moreover, the western Pacific, the SCS, Southeast Asia, the Qinghai–Tibet Plateau, and the eastern Indian Ocean are the regions significantly sensitive to these ocean–land–atmosphere interactions. Therefore, a profound understanding of the characteristics and mechanisms of climate change in spring and summer in the SAAR under the background of global climate change is undoubtedly important for climate change adaptation in China, Southeast Asia, and other places of the world, improving the ability of climate prediction and the overall capacity of disaster prevention and reduction and promoting the sustainable social development.

This book comprises the following four parts:

(1) The SAAR climate and its transition between spring and summer under the background of global climate change and impacts on southern China. The SAAR links the Indian Ocean and the western Pacific, and is directly adjacent to the regions of South Asian monsoon, East Asian monsoon, and Australian monsoon. Again, it is located in the center of interaction between the East Asian and the Indian monsoon systems and is one of the most active areas of convective heating in Asia. The onset and the intensity of the SCS monsoon directly control the precipitation, droughts, and floods in eastern China. Under the background of global climate change, the changes in large-scale circulation related to tropical monsoons and marine thermal background lead to changes in the SAAR climate and further influence the variability of climate in southern China and even the whole country. Spring is a very important season in seasonal transition, and the climate variability in springtime has a significant impact on the climate in the following summer. The SAARs are sensitive areas for ocean–land–atmosphere interaction during spring and the transition between spring and summer. Compared with winter and summer seasons, there have been much fewer studies on climate change in spring. Under the background of global warming, the SAAR spring climate shows new characteristics, and the key factors that affect the drought and flood disasters in southern China may also change. These changes have increased the difficulty and uncertainty of the prediction of drought and flood disasters, which urgently requires further studies to better understanding the new characteristics and their underlying mechanisms. That is, it is imperative to explore the features, precursors, and mechanisms of the variability and change of the SAAR climate in spring and summer seasons under the global climate change background.

(2) Impacts of ocean–land–atmosphere interaction over the Tibetan Plateau (TP), Southeast Asia, and the Indian–Pacific regions on the climate in Southeast Asia and China. Southeast Asia is characterized by complex topography and land–sea distribution, including the water domains and islands of the MC where ocean–land–atmosphere interaction is particularly active. The TP, the "roof of the world" adjacent to Southeast Asia, is a major heat source for the establishment, development, and maintenance of the Asian summer monsoon (Yanai and Wu 2006). It also influences the variability of atmospheric circulation and precipitation in the SAAR on interannual and interdecadal time scales; and the SST anomalies in the Indo-Pacific region, the dipole oscillations of SST in the equatorial Indian Ocean, and the SST anomalies in the SCS all exert significant effects on Asian monsoon climate. However, it should also be noted that the TP, the East Asian continent, the tropical ocean, and the Asian monsoon are closely interrelated and can be viewed as a complexly integrated "system". The ocean–land–atmosphere interactions in these areas are specifically important to the climate variability during springtime and the transition between spring and summer, but the physical characteristics and the related dynamic processes, as well as the significance of their role in understanding and predicting climate variability, are still unclear. Therefore, this monograph is aimed at expanding our understanding of the ocean–land–atmosphere interaction among Southeast Asia, the Indo-Pacific region, the East Asian continent, and the TP on multiple time scales, together with their effects on the global climate, especially during the spring and summer seasons.

(3) Attribution of SAAR climate change and its uncertainty, feedback of SAAR climate change on the global climate, and connection between regional and global climate changes. As part of the SAAR, the Indo-Pacific warm pool is the largest heat reservoir in the world, and the subtle changes in this area can lead to an abnormal climate in adjacent regions and even over the globe. As well known, the changes in the western Pacific warm pool are closely linked to ENSO and its variability. The SCS is located on the western edge of the tropical Pacific Ocean, known as "small warm pool," along with the Gulf of Mexico which has become one of the hot spots of climate research in the Americas. The thermodynamic and dynamic variability of the SCS is the key to the onset and evolution of the East Asian monsoon, as the Indian Ocean processes determine the onset and intensity of the southwest monsoon, partially through the Somali jet. The exchanges of water and other physical properties among the western Pacific, the SCS, and the Indian Ocean via the Indonesian through flow and other ocean currents affect the onset and evolution of both the East Asian monsoon and the South Asian monsoon, thereby causing abnormal climate in other regions. By applying a climate feedback-response analysis method (Cai and Lu 2009; Lu and Cai 2009), the climate change in the SAAR is systematically attributed to various physical processes, and the different contributions of these physical processes are quantified. In addition, previous studies are mostly concentrated on the impact of global climate change on regional climate, but the feedback of SAAR climate change on the global climate is seldom addressed. This monograph is intended to fill this gap and demonstrate the impact of climate change in Southeast Asia on the climate of South Asia, East Asia, Africa, and even North America. The results discussed are helpful for enhancing our understanding of the link between regional and global climate changes and arousing more attentions of the international community on SAAR climate change.

(4) Predictability of atmospheric circulation and precipitation in the SAAR on subseasonal-to-seasonal time scales. The weather and climate in Southeast Asia are affected by remote atmospheric circulation systems on the one hand, but exert significant influences on the weather and climate in adjacent and remote regions on the other hand. Therefore, the capability of Southeast Asian weather and climate prediction is not only influential on local social and economic development but also essential to the weather and climate prediction in China and other regions. Currently, two major approaches are generally adopted in operational climate prediction: the traditional statistical method and the dynamic prediction method based on the climate system models. With the continuous improvement of high-performance computers and climate system models, the accuracy of the dynamic prediction method has been constantly improved, e.g., on subseasonal-to-seasonal time scales. Although climate system models can demonstrate their advantages, they also exhibit obvious deficiencies. Therefore, how to extract useful information from model simulations for climate prediction and how to further improve the capabilities of prediction by climate system models are important issues in the areas of climate prediction and model development.

It is worth noticing that most of the contents of this monograph are based on the latest findings from research supported by the National Key Research Program of China under project "Characteristics and mechanisms of climate variability in spring and summer in Southeast Asia and its adjacent regions under the background of global change and its feedback on global climate." With the latest observation and reanalysis data, state-of-the-art numerical models, and advanced climate dynamic diagnostic methods, the variability and change of SAAR climate in spring and the spring–summer transition period were studied systematically, and a series of new features of the physics and dynamics of SAAR climate were presented. It is anticipated that this book will serve as a starting point of relevant studies and attract more attention to the research of SAAR climate to further enhance our understanding of the long-term change and improve the capacity of prediction of the regional weather and climate.

The publication of this book was supported by the National Key Research Program of China (2014CB953900), the National Natural Science Foundation of China (Grants 41690123, 41690120, 91637208, 41975080, 41975074, and 41805050), and the innovation team of "Ocean–Land–Atmosphere Interaction and Its Global Effects" from the Southern Marine Science and Engineering Guangdong Laboratory (Zhuhai).

This book was mainly completed by Song Yang (Sun Yat-sen University), Renguang Wu (Zhejiang University), Maoqiu Jian (Sun Yat-sen University), Jian Huang (Guangzhou Institute of Tropical and Marine Meteorology, China Meteorological Administration), Xiaoming Hu (Sun Yat-sen University), Ziqian Wang (Sun Yat-sen University), and Xingwen Jiang (Chengdu Institute of Plateau Meteorology, China Meteorological Administration).

A list of the names of contributing authors for individual chapters of this book is provided below:

Chapter 1 (this chapter) Song Yang;

Chapter 2 Maoqiu Jian, Zhiping Wen, Ailan Lin, Dejun Gu, Yunting Qiao, and Chao He;

Chapter 3 Renguang Wu, Jiangyu Mao, Guanghua Chen, Liang Wu, and Wenting Hu, Xi Cao;

Chapter 4 Jian Huang, Qian Li, Zhigang Wei, and Jiangnan Li;

Chapter 5 Ziqian Wang, Wei Wei, Song Yang, and Mengmeng Lu;

Chapter 6 Xiaoming Hu, Yana Li, Song Yang, Ping Huang, Ming Cai, Junwen Chen, Wenshi Lin, and Jun Ying;

Chapter 7 Song Yang, Bian He, Chundi Hu, Zhenning Li, Shan He, and Kaiqiang Deng;

Chapter 8 Xingwen Jiang, Song Yang, Xiangwen Liu, Tuantuan Zhang, and Shaorou Dong.

References

Cai M, Lu J (2009) A new framework for isolating individual feedback processes in coupled general circulation climate models. Part II: Method demonstrations and comparisons. Clim Dyn 32:887–900. https://doi.org/10.1007/s00382-008-0424-4

Herring SC, Hoerling MP, Peterson TC et al (2014) Explaining extreme events of 2013 from a climate perspective. Bull Amer Meteor Soc 95(9):S1–S96

Lu J, Cai M (2009) A new framework for isolating individual feedback processes in coupled general circulation climate models. Part I: formulation. Clim Dyn 32:873–885. https://doi.org/10.1007/s00382-008-0425-3

Neale R, Slingo J (2003) The Maritime Continent and its role in the global climate: a GCM study. J Climate 16:834–848

Webster PJ, Yang S (1992) Monsoon and ENSO: Selectively Interactive Systems. Q J R Meteorol Soc 118:877–926. https://doi.org/10.1002/qj.49711850705

Yanai M, Wu GX (2006) Effects of the Tibetan Plateau. In: The Asian Monsoon. Springer Praxis Books. Springer, Berlin, Heidelberg. https://doi.org/10.1007/3-540-37722-0_13

Chapter 2
Characteristics of the Spring–Summer Atmospheric Circulation Transition Over the South China Sea and Its Surrounding Regions and Their Responses to Global Warming

The South China Sea (SCS) connects the Indian Ocean and the Pacific Ocean. It is also a key region where the East Asian monsoon system and the Indian monsoon system interact with each other, and it is a direct source of moisture for the subtropical monsoon system of East Asia. The onset and intensity of SCS summer monsoon (SCSSM) have great influences on precipitation, which may lead to droughts and floods in southern China in summer. In the context of global warming, the changes in the tropical monsoons on large scale and the marine thermal environment may lead to climate change in the SCS and its surrounding areas, which could exert a significant impact on the climate in southern China.

The transition season usually begins in mid-spring and ends in early summer for the circulation system from a winter pattern to a summer pattern in the Asian monsoon region. In the spring–summer transition stage, the SCS and its surrounding areas are highly sensitive to global air–sea and air–land interactions. Compared with the research on winter and summer seasons, there have been fewer studies on climatic characteristics of spring–summer circulation transition (hereafter, the transition), their variation in the region, and their impact on the summer climate in eastern China. In the context of global warming, the transition over the SCS and its surrounding areas may have some new characters, and the key factors for these new changes and the mechanisms behind them are worth investigating. These changes aggravate the difficulty and uncertainty of drought and flood prediction in southern China. We urgently need to study and understand the physical mechanisms behind the present and future climate changes during spring–summer. Therefore, it is of great significance to study the characteristics and mechanisms of the transition over the SCS and its surrounding areas, and their influences on the climate in southern China under global warming.

In this chapter, the characteristics and physical processes of spring–summer atmospheric circulation transition (hereafter, the transition) over the SCS and its surrounding areas are presented using historical observation data and reanalysis outputs. The effects of tropical intraseasonal oscillation on the onset of the SCSSM are discussed. Precursory signals of sea surface temperature (SST) controlling the

S. Yang et al., *Climate Change in Southeast Asia and Surrounding Areas*,
Springer Climate, https://doi.org/10.1007/978-981-15-8225-7_2

early and late onsets of the SCSSM are also analyzed. Finally, the responses of the transition to greenhouse gas (GHG) forcing in Southeast Asia and its surrounding areas, and the large-scale subtropical anticyclones are studied using the model outputs from the Coupled Model Intercomparison Project Phase 5 (CMIP5).

2.1 Characteristics of the Transition Over the SCS and Its Surrounding Regions

The spring–summer transition has always been an important issue for both the academic research community and operational forecasting centers. Yeh et al. (1958) first revealed abrupt changes of the upper atmospheric circulation in the mid- and high latitudes of the Northern Hemisphere in June and October, respectively. Murakami and Nakazawa (1985) investigated the long period of transition from the Southern Hemisphere summer monsoon to Northern Hemisphere summer monsoon, and confirmed the northward migration of low outgoing longwave radiation (OLR) during the transition period. Xie and Sun (1991) emphasized the difference between the transition characteristics of the atmospheric circulation over the Eastern and Western Hemispheres. They considered that the main transition feature of the atmospheric circulation over the Eastern Hemisphere is the establishment of the summer monsoon, which is characterized by three successive phases: moisture slowly increasing, moisture rapidly increasing, and the well-established summer monsoon. Yasunari (1991) found that the interannual variability of the air–ocean–land coupled system over the tropical Pacific and the Asian monsoon regions shows a quasi-biennial nature with the change of sign during the boreal spring to summer, and the concept of "the monsoon year" was proposed as a climate year for the interannual anomalies over the tropics. He et al. (1996) analyzed the seasonal transition features of the Asian summer monsoon (ASM) establishment, and indicated that the transition begins as early as April, followed by abrupt changes in May to June, and the ASM is fully established in June.

The onset of summer monsoon has been considered as an important component of the transition over the monsoon regions. Its mechanism and influence on summer climate have drawn great concerns of researchers domestically and internationally. The transition may have immediate impacts on the summertime occurrence of droughts and floods over Asia (Nitta 1987; Huang 1992; Lau and Yang 1997). An earlier transition may result in an earlier outbreak of southwesterly winds in the low latitudes and more moisture transportation to East Asia, which is conducive to increasing summer precipitation. Therefore, studies on the characteristics of the transition are important for accurate prediction of summer precipitation over East Asia.

In previous studies on the transition, the focus was often on regional summer monsoon. In reality, the onset of summer monsoon occurs in stepwise phases and in different regions (Wang and Lin 2002). The transition has not been adequately understood because of regional distinctions in the summer monsoon regions. In this respect, several questions we attempt to answer in this chapter are as follows. How to define the spring–summer atmospheric circulation transition? What are the characteristics of the transition? What are the factors responsible for the transition?

2.1.1 *Characteristics of the Transition*

(1) Identification of the Transition

In January, a lower-level anticyclone prevails over the Asian monsoon area; meanwhile, the cold air from Mongolia and Siberia moves southward along eastern China into the SCS area, and changes to easterlies over the northern Indian Ocean. In July, the lower-level atmospheric circulation in the Asian monsoon area changes from anticyclonic to cyclonic, with a strong westerly flow over the northern Indian Ocean. In this respect, April–May–June (AMJ) can be considered as a transition season.

In Sect. 2.1, multivariate empirical orthogonal function (MV-EOF) analysis is applied to 5-day running means of OLR and 850-hPa wind from AMJ. The domain of analysis is (30°–150° E, 20° S–40° N) where the ASM is active. The mean state is defined as the climatological AMJ mean, and is subtracted from the data before performing the MV-EOF analysis.

The first MV-EOF mode (Fig. 2.1) shows that the active convections are located over the Arabian Sea (AS), the Bay of Bengal (BOB), and the SCS, while the convections are suppressed over the Arabian Peninsula and the southern Maritime Continent (MC). An anomalous anticyclonic circulation appears over the equatorial Indian Ocean with cross-equatorial flow developing over Somali. The westerlies prevailing in the northern Indian Ocean expand eastward to the SCS, and monsoon troughs develop over the BOB and the SCS.

Figure 2.2 shows a scatter plot of variances explained by the first five modes of the MV-EOF analysis. The first leading mode is statistically distinguishable from

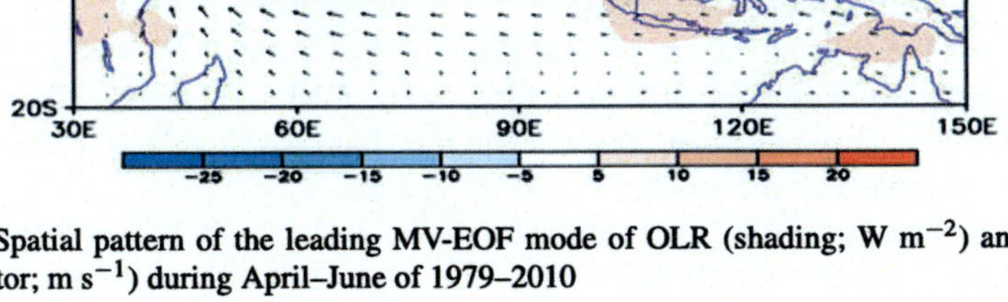

Fig. 2.1 Spatial pattern of the leading MV-EOF mode of OLR (shading; W m^{-2}) and 850-hPa wind (vector; m s^{-1}) during April–June of 1979–2010

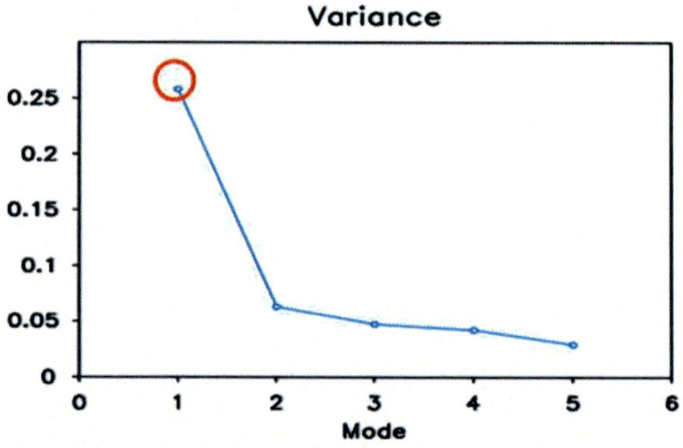

Fig. 2.2 Explained variances of the first five MV-EOF modes

the other four higher modes according to the rule given in North et al. (1982). Furthermore, the spatial pattern of the first MV-EOF mode of OLR in Fig. 2.1 is similar to the standard deviation of daily OLR for AMJ, especially for the areas of active convections (not shown). The first MV-EOF mode exhibits the transition of the spring–summer atmospheric circulation on a large scale, and its principal component (PC1) is thus useful for identifying the timing of the transition.

(2) Timing of the Transition

The corresponding PC1 shown in Fig. 2.3 is mostly negative in the early stage, and turns positive in the later stage from April to June. The PC1 value in some of the years rises sharply and changes sign from negative to positive only once, such as in 1989, reflecting a stable transition of the atmospheric circulation from spring to summer. However, in some years the PC1 fluctuates between negative and positive values, such as in 1997, indicating possible impacts of internal atmospheric processes. To identify the date of the transition, we use the following objective criteria.

The transition date is defined as the first day that satisfies the following two criteria: (1) PC1 changes signs from negative to positive and remains positive for at least five days; (2) in the subsequent 20 days (including the transition date), PC1 value must be positive in for least 10 days.

Table 2.1 lists the dates and corresponding pentads of the transition for each year, based on the above definition. The long-term mean transition date (pentad) is May 16 (Pentad 28). The earliest transition occurred on April 25 (Pentad 23) 2008, while the latest was on June 4 (Pentad 31) 1979 and 1992. The maximum range is about 40 days (8 pentads). The long-term mean transition date is close to that of the SCSSM onset.

Seen from the above analysis, the transition is characterized by strong convection over the AS, the BOB, and the SCS, as well as the establishment of 850-hPa westerlies over the northern Indian Ocean. The transition is thus closely related to the onset of the summer monsoon over the AS, the BOB, and the SCS.

Figure 2.4 shows the time series of the transition dates, the BOB summer monsoon (BOBSM) onset dates, the SCSSM onset dates, and the Indian summer monsoon

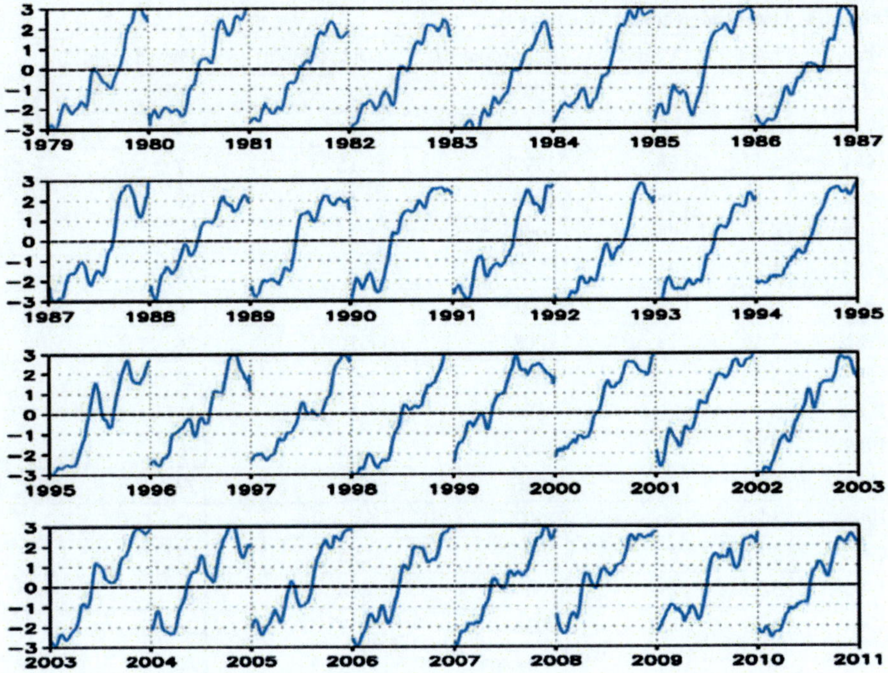

Fig. 2.3 PC1 of the first MV-EOF mode. Only the data from 1 April to 30 June are displayed for each year

(ISM) onset dates. The correlation coefficients of the transition dates with the BOBSM onset dates and the SCSSM onset dates are 0.41 and 0.42, respectively, at the 95% confidence level; and the correlation coefficient between the transition dates and the ISM onset dates is 0.51, at the 99% confidence level. On average, the BOBSM onset is earlier than the transition, the SCSSM onset, and the ISM onset, while the ISM onset is the latest. An interdecadal shift of the SCSSM onset date in the early 1990s is observed. Before the early 1990s, the SCSSM onset is mostly later than the transition date, but it is much closer to the transition date after the early 1990s. These results indicate that the transition, which is characterized by changes in large-scale circulation and convection, is different from the onset of the regional summer monsoon, albeit the two are well correlated.

(3) Evolution of Convection and Atmospheric Circulation during the Transition

To reveal the evolution of convection associated with the transition, we composite the evolution of 10-day-mean OLR based on the transition dates (by centering on the transition day). The transition day is denoted as day 0, while the "-" sign denotes prior to the transition date. The time mean between days −30 and 30 is removed.

During days −30 to −21 (Fig. 2.5, top left panel), three maximum OLR centers (which denote suppressed convection) appear over the AS, the BOB, and the SCS, respectively. The minimum OLR centers (which represent strong convection) cover

Table 2.1 Dates and pentads of the transition

Year	Date	Pentad	Year (cont'd)	Date (cont'd)	Pentad (cont'd)
1979	6.4	31	1995	5.7	26
1980	5.16	28	1996	5.25	29
1981	5.17	28	1997	5.15	27
1982	5.16	28	1998	5.16	28
1983	6.2	31	1999	5.5	25
1984	5.22	29	2000	5.8	26
1985	5.17	28	2001	5.8	26
1986	5.16	28	2002	5.10	26
1987	5.30	30	2003	5.9	26
1988	5.15	27	2004	5.3	25
1989	5.13	27	2005	5.26	30
1990	5.8	26	2006	5.13	27
1991	5.26	30	2007	5.4	25
1992	6.4	31	2008	4.25	23
1993	5.25	29	2009	5.16	28
1994	5.19	28	2010	5.17	28
			Average	**5.16**	**28**

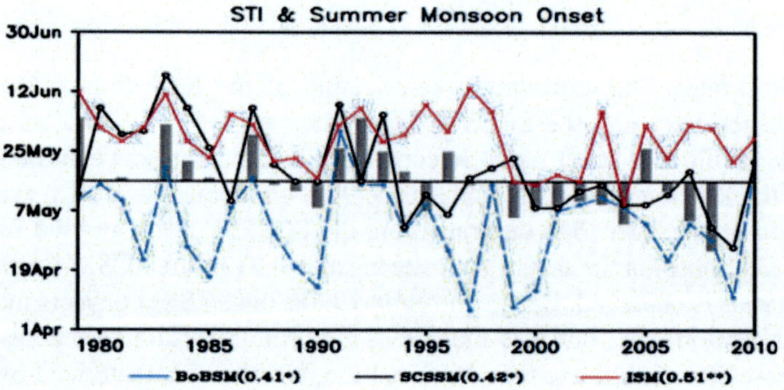

Fig. 2.4 Time series of the spring–summer atmospheric circulation transition dates (gray bars), the BOBSM onset dates (blue dots; 850-hPa zonal wind averaging over 90°–100° E, 5°–15° N; Mao and Wu 2007), the SCSSM onset dates (black circles; 850-hPa zonal wind averaging over 110°–120° E, 5°–15° N; Kajikawa and Wang 2012), and the ISM onset dates (red crosses; 850-hPa zonal wind averaging over 40°–80° E, 5°–15° N; Wang et al. 2009) for the period 1979–2010. The long-term-mean transition date (May 16) is indicated by a black line. The correlation coefficients of transition dates with BOBSM onset dates, with SCSSM onset dates, and with ISM onset dates are 0.41, 0.42, and 0.51, respectively. Symbols * and ** indicate the 95% and 99% confidence levels, respectively

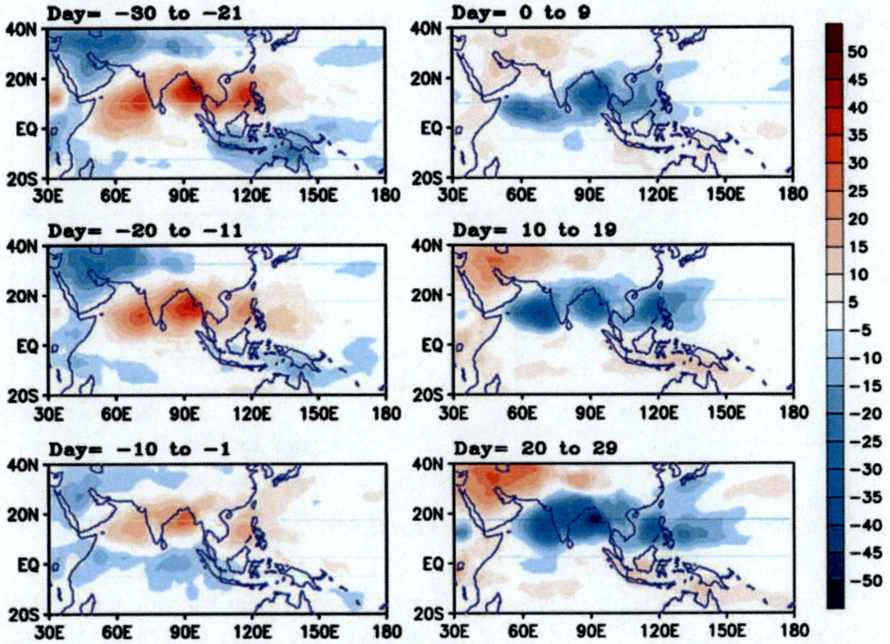

Fig. 2.5 Composites of 10-day-mean OLR anomalies from days –30 to 29. Note that the time mean between days –30 and 30 is removed. Shading interval is 5 W m^{-2}

the MC. For the next 10 days (days −20 to −11, Fig. 2.5, middle left panel), the maximum OLR centers are weakened, so are the minimum OLR centers over the MC, while weak convection develops over the southwestern Indian Ocean and extends northeastward. During the period from days −10 to −1 (Fig. 2.5, bottom left panel), the northeastward convection arrives at the equatorial Indian Ocean. The OLR field changes suddenly after the transition. The active convections are located over the AS, the BOB, and the SCS, while the convections are suppressed over the MC. In the ensuing days (Fig. 2.5, right panel), the three minimum OLR centers intensify, reaching the maximum intensity at days 20–29. Meanwhile, the strong convection over the AS and the SCS extends northward and eastward, respectively.

The characteristics of the wind field during the transition are manifested as the changes of vertical shear of zonal wind. Following Webster and Yang (1992), we calculate the area-mean vertical shear of zonal wind between 850 and 200 hPa in the region of (40°–110° E, 0°–20° N), and show its evolution during the transition in Fig. 2.6. This region has westerly vertical shear before the transition and reverses to easterly vertical shear after the transition, which is conducive to the development of convection. Thus, the convection and vertical shear of zonal wind change notably during the transition, and their transition times are consistent with each other.

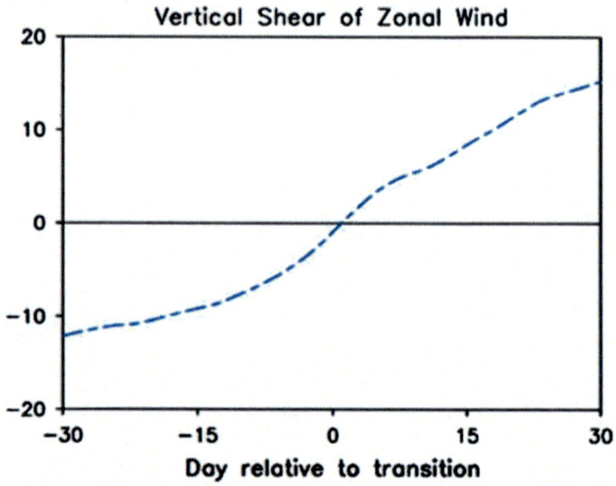

Fig. 2.6 Time series of area-mean vertical shear of zonal wind (U850 minus U200; m s^{-1}) over (40°–110° E, 0°–20° N)

2.1.2 Physical Processes of the Transition

To better understand the physical processes associated with the transition, we define the transition pentad as Pentad 0 and regroup the time series of 224 pentad data (seven pentads centered at the transition pentad each year, for a total of 32 years). An MV-EOF analysis is then reapplied to the 224 pentads of OLR and 850-hPa wind. Figures 2.7 and 2.8 show the spatial patterns and principal components of the first and second MV-EOF modes, respectively. The first MV-EOF mode explains 14.9% of the total variance, and the second MV-EOF mode accounts for 10.3% of the total

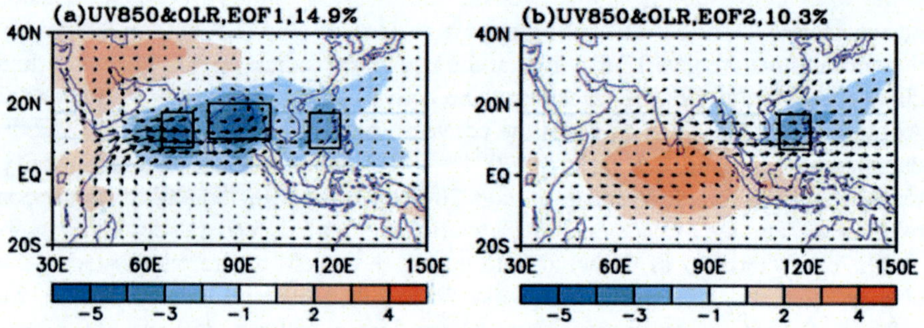

Fig. 2.7 Spatial patterns of **a** the first and **b** the second MV-EOF modes of the 224 pentads (seven pentads centered at the transition pentad × 32 years) OLR (shading; W m^{-2}) and 850-hPa wind (vector; m s^{-1}). The boxes in **a** denote the Arabian Sea region (65°–75° E, 7.5°–17.5° N), the Bay of Bengal region (80°–100° E, 10°–20° N), and the South China Sea region (112.5°–122.5° E, 7.5°–17.5° N), respectively. The box in **b** indicates the South China Sea region only

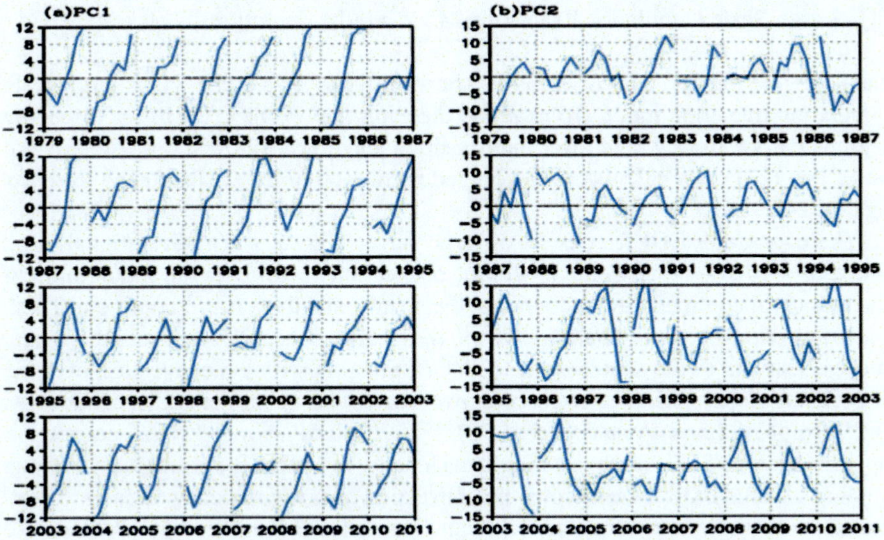

Fig. 2.8 **a** PC1 and **b** PC2 of the MV-EOF modes in Fig. 2.7. Only the data from the seven pentads centered at the onset are displayed for each year

variance. Compared with the daily MV-EOF modes shown in Fig. 2.1, the pentad modes depict similar spatial patterns, except for a weaker OLR center over the SCS.

Table 2.2 presents the correlation coefficients of PC1 and PC2 with OLR in the AS region (65°–75° E, 7.5°–17.5° N), the BOB region (80°–100° E, 10°–20° N), and the SCS region (112.5°–122.5° E, 7.5°–17.5° N), respectively. We can see that PC1 is highly correlated with the OLR in the three key regions. The correlation coefficients are as high as 0.78, 0.86, and 0.54, respectively, and are all significant at the 99% confidence level. In contrast, PC2 has a weak correlation with the OLR in the three key regions. The correlation coefficient between PC2 and OLR in the SCS region only reaches 0.52, comparable to that of PC1 with OLR in the same region. The other two have weak correlations. Therefore, the first MV-EOF mode is indicative of the evolution of the convection over the AS and the BOB during the transition, while both the first and second MV-EOF modes are needed to explain the convection over the SCS.

Table 2.2 Correlation coefficients of PC1 and PC2 with OLR in the three regions shown in Fig. 2.7a. OLR is multiplied by −1

	OLR-AS	OLR-BOB	OLR-SCS
PC1	0.78**	0.86**	0.54**
PC2	−0.27	0.09	0.52**

**Indicates the correlation is confident at the 99% level. AS: the Arabian Sea; BOB: the Bay of Bengal; and SCS: the South China Sea

(1) Large-Scale Circulation Transition Mode and Its Evolution

Let us define the first MV-EOF mode, shown in Figs. 2.7a and 2.8a, as large-scale circulation transition mode. To examine the evolution of the OLR in the three key regions before, during, and after the transition stage, we follow the method in Wu and Chou (2012) by calculating the lagged regression of the OLR of each grid box against PC1.

The evolution of OLR consists of three stages (Fig. 2.9). (a) Gestational stage (Pentads −10 to −7). At Pentad −10, the convection occurs over the southern tip of the Indian subcontinent, the Indochinese Peninsula, and the African continent. Then, the convection over the southern tip of the Indian subcontinent strengthens and moves eastward to the southeast of the BOB, merging with the convection over the Indochinese Peninsula. The convection over the African continent develops locally. (b) Developing stage (Pentads −6 to −3). From Pentads −6 to −3, weak convection appears over the SCS and then strengthens locally, as well as the convection over the southeast of the BOB. Meanwhile, the convection over the African continent moves eastward to the western equatorial Indian Ocean. (c) Booming stage (Pentads −2 to +1). At Pentad −2, the convection over the southeast of the BOB flares up and extends eastward to the southern SCS, and the convection over the western Indian

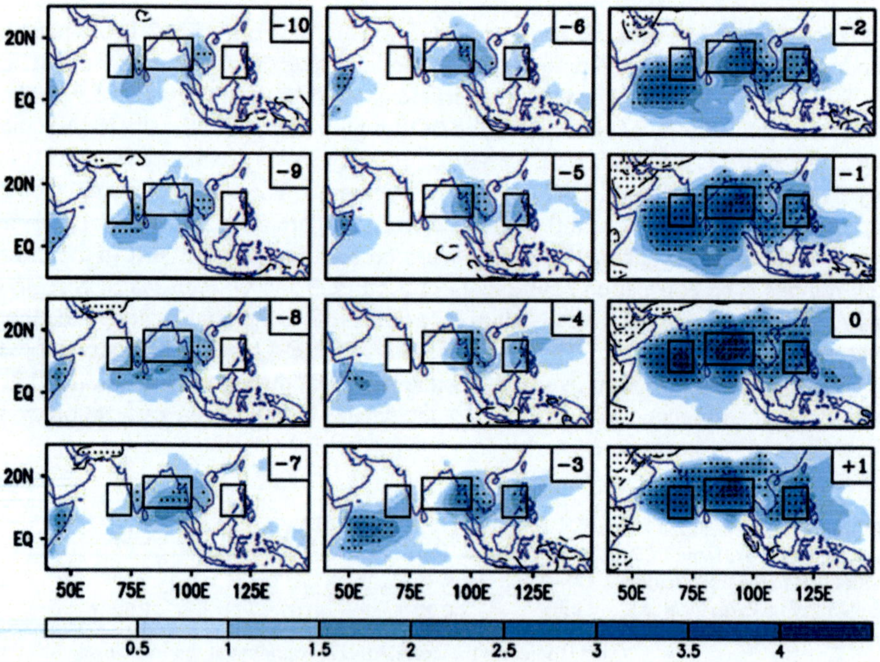

Fig. 2.9 Lagged regression of OLR (W m^{-2}) against PC1. Negative lag denotes the latter lags the former. Stippling indicates the 95% confidence level

Ocean moves northward to the southeastern AS. From Pentads −1 to 0, the convection over the BOB flares up and extends westward, merging with the convection over the AS. Strong convections cover a large area from the BOB to the AS. By Pentad 0, the OLR centers over the AS, the BOB, and the SCS develop into their maxima, with the centers over the northern Indian Ocean being stronger than the one over the SCS. At Pentad +1, the convections are separated by the Indochinese Peninsula into two parts and weakened, and the western convection advances northward.

(a) Gestational Stage

Warm SST has a major effect on the development of convection. The necessary condition for deep convection can be achieved by SST greater than 28 °C in combination with some other factors (Fu et al. 1994). Wu et al. (2012) found that a warm pool (SST > 30.5 °C) often occurs in the eastern part of the central BOB before the formation of the BOB monsoon vortex. To investigate the relationship between convection and warm SST during the gestational stage, we show pentad-mean evolution of SST from Pentads −10 to −7 prior to the transition (Fig. 2.10, left panels). We can see that a warm pool (SST > 29 °C) exists over the Indian Ocean, where SST is the highest near the equator and decreases toward higher latitudes. As time progresses, the area of the warm SST expands and moves slightly northward, which is conducive to the formation of convection over the southern Indian subcontinent (Fig. 2.9, left panels). Figure 2.10 (right panels) shows the differences of SST between consecutive pentads. SST increases in the northern tropical Indian Ocean and decreases in the southern tropical Indian Ocean, which coincides with the northward movement of the Sun during the spring–summer seasons. From Pentads −9 to −7, the increasing SST in the BOB is conducive to the development of convection in this area (Fig. 2.9). Moreover, the SST in the SCS warms up significantly at Pentad −7, which may play an important role in maintaining the convective activities in subsequent pentads. Therefore, warm SST is the main factor for the formation of convection during the gestational stage.

(b) Developing Stage

Figure 2.11 shows the lagged regressions of SST against PC1 from Pentad −6 to −3. There is a significant positive correlation between convection and SST in the SCS and SST in the ocean east of the Philippines (the boxes shown in Fig. 2.11). Anomalous heating due to warm SST anomalies (SSTAs) in the SCS and in the ocean east of the Philippines induces the cyclonic wind anomalies over the BOB and the SCS via a Rossby wave-type response (Gill 1980), which leads to the development of convection in this area (Fig. 2.9).

(c) Booming Stage

High correlations between convection and SST are observed during the gestational stage and the developing stage, but not in the booming stage when the relationship between convection and SST is weakened. Figure 2.12 shows the lagged regressions

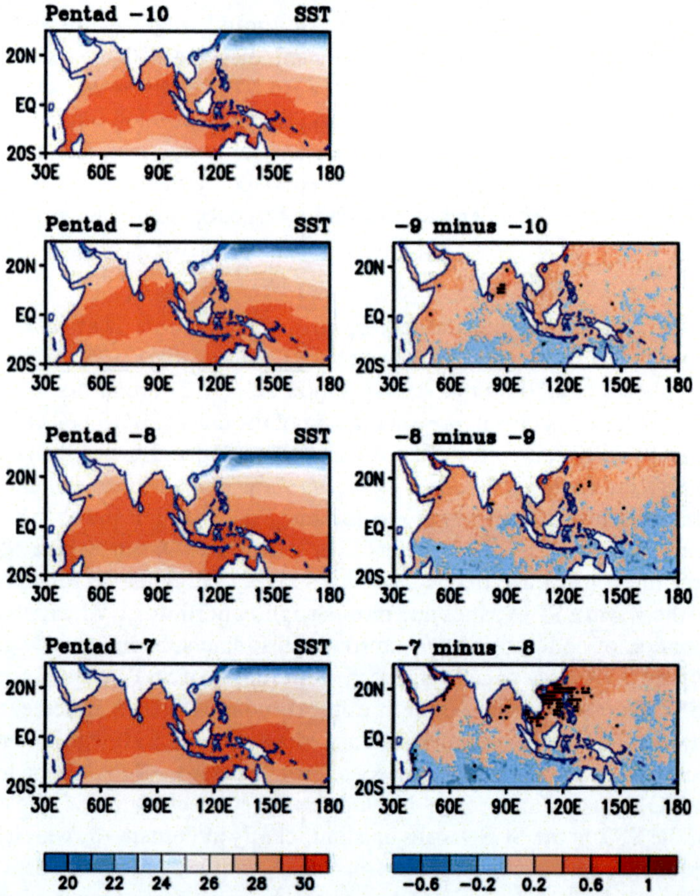

Fig. 2.10 Pentad means of (left) SST (°C) and (right) difference between consecutive pentads. Results are 32-year composites. Stippling indicates the difference is significant at the 95% confidence level

of SST against PC1 from Pentads −2 to +1. Cold SSTAs appear in the western equatorial Indian Ocean and extend northeastward, while warm SSTAs appear in the subtropical Northwest Pacific. At Pentad +1, significant cold SSTAs emerge in the southern AS and southern BOB, while the changes of SSTAs in the SCS slow down. This suggests that no striking warming occurs in the southern AS, the southern BOB, and the southern SCS during the booming stage. Deep convection is active regardless of low SST, which indicates that the SSTAs are not a forcing for the transition, but are a response to deep convection development during the booming stage.

Liu et al. (2014a, b) showed that the sensible heat flux is important for atmospheric circulation. The sensible heat flux from the sea surface can generate available potential energy (APE) efficiently in a region where heating and temperature are positively correlated. It can act as a trigger to release positive convective APE (CAPE) in the

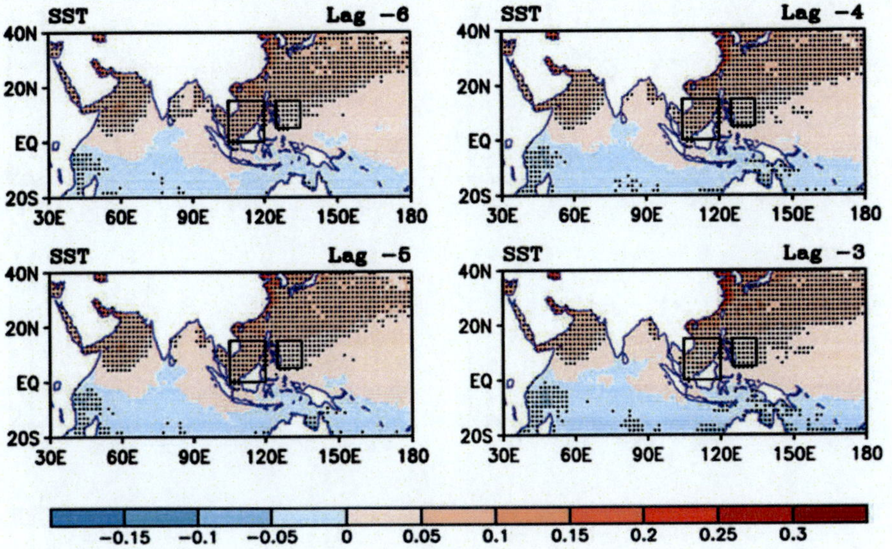

Fig. 2.11 Lagged regressions of SST (°C) against PC1 (Pentads −6 to −3). Negative lag denotes the latter lags the former. Stippling indicates the 95% confidence level

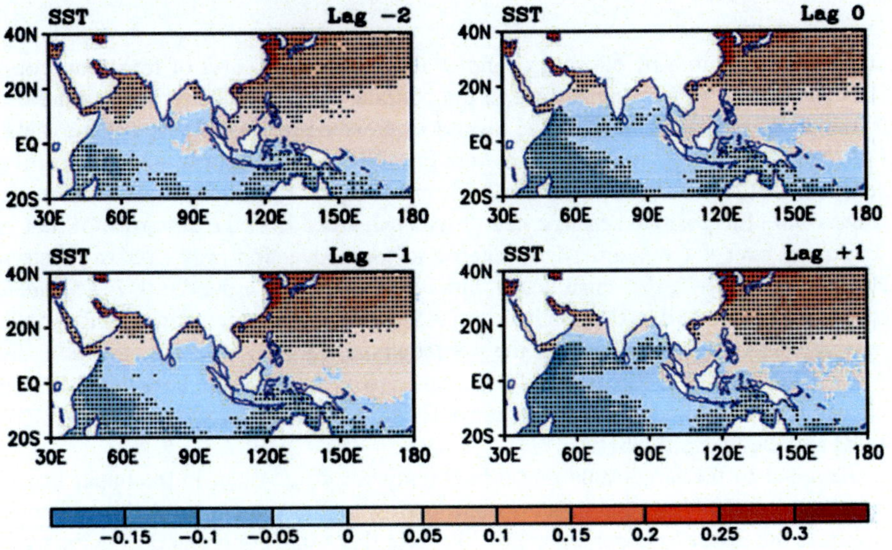

Fig. 2.12 Lagged regressions of SST (°C) against PC1 (Pentads −2 to +1). Negative lag denotes the latter lags the former. Stippling indicates the 95% confidence level

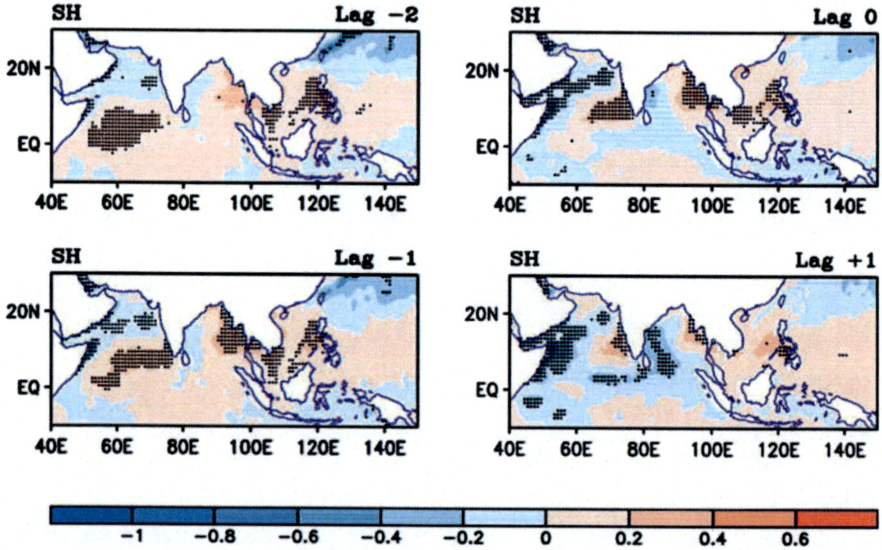

Fig. 2.13 Lagged regressions of sensible heat flux (W m^{-2}) against PC1 (Pentads −2 to +1). Upward is positive. Negative lag denotes the latter lags the former. Stippling indicates the 95% confidence level

atmospheric column by elevating a moist air parcel to the level of free convection (LFC) from the underlying surface in the boundary layer. To better understand the influence of sensible heat flux, the lagged regressions of sensible heat flux against PC1 from Pentads −2 to +1 are shown in Fig. 2.13. From Pentads −2 to 0, positive sensible heat flux increases over the SCS, inducing the development of convection. Meanwhile, the positive sensible heat flux is enhanced over the eastern BOB,but is weakened over the western BOB, leading to the convection only over the eastern BOB. At Pentad −2, the sensible heat flux center is located around (60° E, 0°) and it moves northeastward to the southeastern AS at Pentad 0, which corresponds to the movement of the convection over the AS shown in Fig. 2.9. After the transition, the positive sensible heat flux is clearly decreased over the SCS, the eastern BOB, and the eastern AS. Moreover, strong negative sensible heat flux occurs over the western BOB and the western AS.

Reversal of the large-scale meridional temperature gradient in the upper troposphere is another important factor contributing to the transition. A warm center occupies the upper troposphere at 300 hPa over the Asian continent during the transition (not shown). Figure 2.14 shows the lagged regressions of 300-hPa temperature and geopotential height against PC1 from Pentads −2 to +1. At Pentad −2, a warm center is located over northeastern China and extends southwestward to the northern Indian subcontinent. An anticyclonic circulation prevails over the northern Arabian Peninsula and northeastern China. At Pentad −1, a warm center moves westward over the Tibetan Plateau. In contrast, the high-pressure region located over the northern Arabian Peninsula is connected with the high-pressure area in northeastern China,

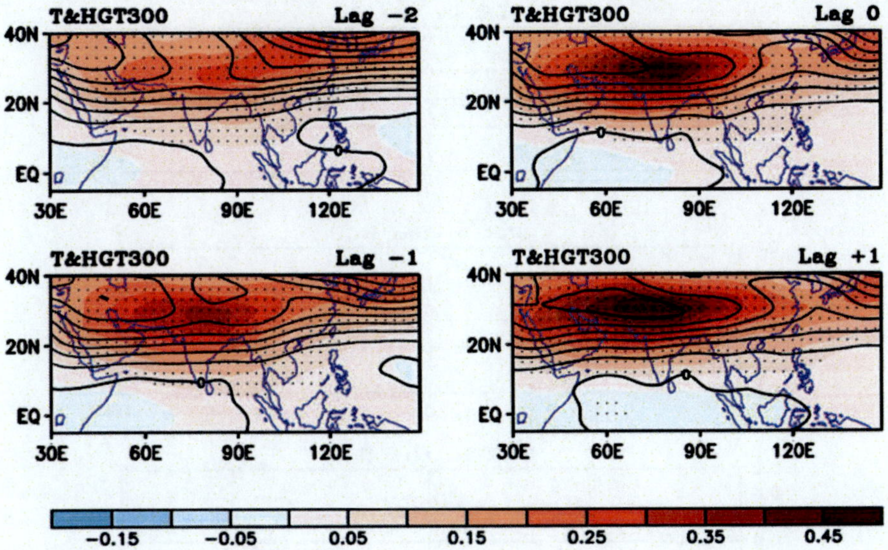

Fig. 2.14 Lagged regressions of 300-hPa temperature (shading; K) and geopotential height (contour; gpm) against PC1 (Pentads −2 to +1). Negative lag denotes the latter lagging the former. Stippling indicates the 95% confidence level

and the warm zone is continuously strengthened, which makes the high-pressure area coincide with the warm zone, and forms a stable warm South Asian high (SAH) over the Tibetan Plateau. Warming over the northern parts of the AS and the BOB makes it easier to form the summer pattern of temperature field in the upper troposphere with warm air in the north and cold air in the south, which induces strong easterly (westerly) at the upper (lower) level, over the AS, the BOB, and the SCS. Such strong easterly vertical shear strengthens the ascending motion, which is favorable to the development of convection over the AS and the BOB.

(2) Intraseasonal Oscillation Mode and Its Relationship to the Transition

Let us define the second MV-EOF mode as intraseasonal oscillation (ISO) mode, and investigate the evolution of ISO during the transition. Figure 2.15 displays the lagged correlation coefficients of OLR in the regions with PC1 and PC2 as shown in Fig. 2.7. The lagged correlation coefficients between PC1 and OLR over the AS and BOB peak at the transition pentad, while the lagged correlation coefficient between PC1 and OLR over the SCS reaches its maximum one pentad after the transition. Compared with PC1, the lagged correlation coefficients of PC2 with OLR in the three key regions show clear periodic fluctuations, indicating a possible connection of the transition and the propagation of the ISO with a period of eight pentads. The lagged correlation coefficient between PC2 and OLR over the AS reaches a local maximum at Pentads −3 and +5, with the largest correlation coefficient at Pentad + 5. The lagged correlation coefficient between PC2 and OLR over the BOB reaches a local maximum at Pentads −2 and +6, with the largest correlation coefficient at

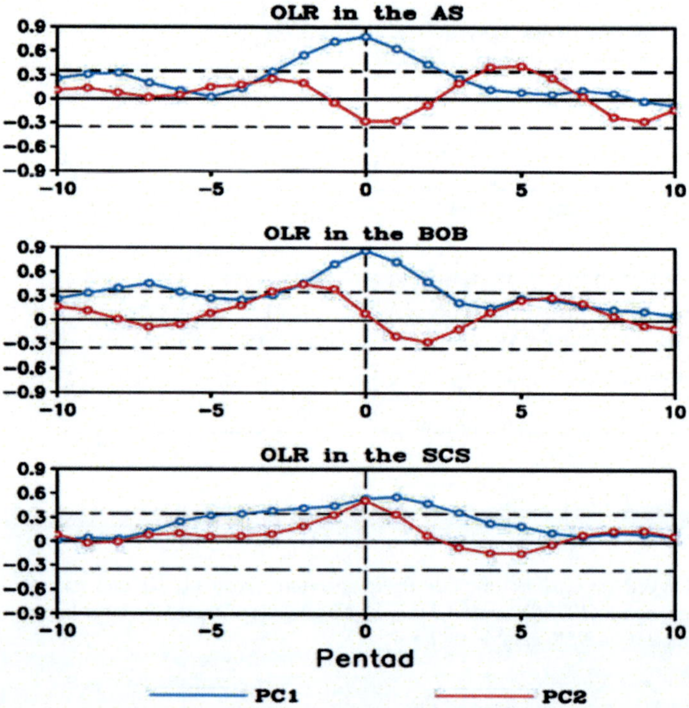

Fig. 2.15 Lagged correlation coefficients of OLR in the regions shown in Fig. 2.7 with PC1 (blue line) and PC2 (red line). Negative lag denotes the latter lags the former. The dashed lines indicate the 95% confidence level

Pentad −2. The lagged correlation coefficient between PC2 and OLR over the SCS reaches a local maximum at Pentads 0 and +9, with the largest correlation coefficient at Pentad 0. The three largest correlation coefficients are all at the 95% confidential level.

There are two propagation processes of convection during the transition (Fig. 2.16). During the first propagation process (Pentads −6 to +1), the wet phase first appears over the western equatorial Indian Ocean (Pentad −6), and then advances eastward to the southern BOB, connecting to the convection over the western Pacific (Pentads −5 to −3). At Pentad −2, enhanced convection covers most of the BOB, indicating the BOBSM onset. At Pentad −1, the convection over the BOB becomes weaker, while the convection over the SCS is enhanced and peaks at Pentad 0, corresponding to the SCSSM onset. Besides, a dry phase occurs over the western equatorial Indian Ocean at Pentad −2 and follows the pathway of the previous wet phase to move eastward, covering most of the equatorial Indian Ocean at Pentad 0. During the second propagation process (Pentads +2 to +9), a wet phase appears over the western equatorial Indian Ocean once more, strengthens and expands northeastward to the southeastern AS at Pentad +5, indicating the ISM onset. After this onset, the enhanced convection continues to move northeastward and reaches the northern

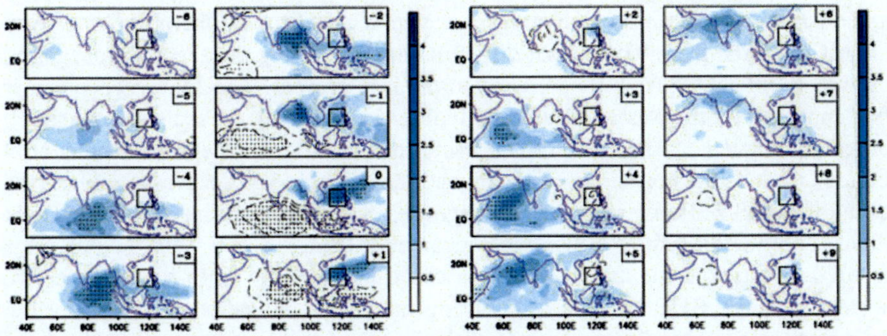

Fig. 2.16 Lagged regressions of OLR (W m^{-2}) against PC2. Shadings indicate wet phase, and dashed contours indicate dry phase. Negative lag denotes the latter lagging the former. Stippling indicates the 95% confidence level

Indian subcontinent at Pentad +6; it then gradually disappears. The first and second propagation processes have different pathways, and the convection intensity during the first propagation process is stronger than that during the second one.

To investigate the change in the large-scale circulation associated with the ISO during the transition, we show the lagged regressions of 850-hPa velocity potential and divergent wind against PC2 (Fig. 2.17). During the first propagation process (Pentads −6 to +1), a large-scale lower-level convergence center is located over the western equatorial Indian Ocean at Pentad −5, and propagates eastward along the equator to the eastern equatorial Indian Ocean at Pentad −4, covering most of the northern Indian Ocean. The intensity of the convergence center peaks at Pentad −3 with its center over the MC. At Pentad −2, the convergence center moves to the western Pacific Ocean; the Indochinese Peninsula and the BOB are under the influence of lower-level convergence, corresponding to the onset of BOBSM. At the transition pentad, the convergence center moves off the western Pacific Ocean,

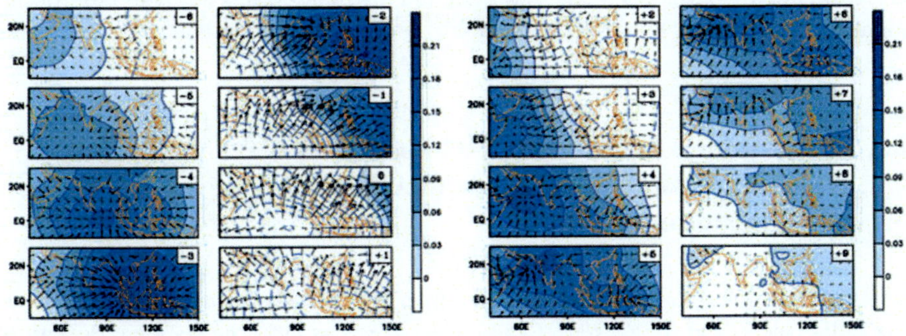

Fig. 2.17 Lagged regressions of 850-hPa velocity potential (shading; 10^6 m^2 s^{-1}) and divergent wind (vector; m s^{-1}) against PC2. Shadings indicate convergnce, and dashed contours indicate divergence. Negative lag denotes the latter lagging the former

and a large-scale lower-level divergence covers most of the northern Indian Ocean, the MC, and the SCS. There exist different changes during the second propagation process (Pentads +2 to +9) compared to the first process. The convergence center first appears over the western equatorial Indian Ocean during the second propagation process (Pentad +3), and then propagates northeastward from the western equatorial Indian Ocean to the northern Indian subcontinent at Pentad +5, corresponding to the onset of the ISM. Furthermore, a convergence center also appears over the MC. At Pentad +6, the intensities of the two convergence centers reach their maxima. This result suggests that the arrival of the eastward-propagating large-scale convective system over the BOB and the SCS during the first propagation process plays an important role in inducing the onsets of the BOBSM and SCSSM, while the arrival of the northeastward-propagating large-scale convective system over the AS during the second propagation process has a greater influence on the onset of the ISM.

Figures 2.15, 2.16, and 2.17 show that the lagged correlation coefficient of OLR over the SCS with PC1 strengthens and the lagged correlation coefficient with PC2 peaks at Pentad 0. As a result, the SCS is under the influence of lower-level southwesterly divergence wind. At Pentad −2, the lagged correlation coefficient of OLR over the BOB with PC2 reaches its maximum and the lagged correlation coefficient with PC1 strengthens. The BOB is covered by large-scale lower-level convergence. At Pentad +5, the relationship between OLR over the southeastern AS and PC2 is most significant, while the relationship between OLR over the Indian subcontinent and PC1 is the strongest (not shown). Lower-level convergence covers the AS and the Indian subcontinent, with its center over the southeastern AS. As shown in Table 2.1, the long-term-mean transition pentad is Pentad 28 (May 16–20), therefore, convection is enhanced at Pentad 28 (Pentad 0) over the SCS, at Pentad 26 (Pentad −2) over the BOB, and at Pentad 33 (Pentad +5) over the AS. These timings are close to the onset dates of the summer monsoon in the three key regions, respectively, and the superposition characteristics of the two modes are similar to the booming features of the summer monsoon in each region.

In conclusion, the staged onsets of the ASM components, i.e., SCSSM, BOBSM, and ISM, are under the combined influence of both large-scale circulation transition mode and ISO mode.

2.2 Interannual Variation of Net Heat Flux Over the Indian Ocean and the Western Pacific Ocean During the Transition

To investigate the maximum regions of the interannual variation of net heat flux over the India Ocean and the western Pacific Ocean, we analyze the standard deviation of the net heat flux in March, April, May, and June, respectively, during 1985–2009 (Fig. 2.18).

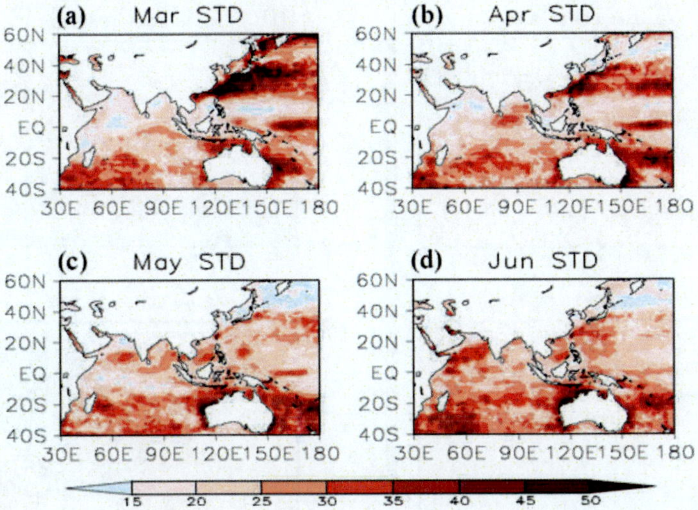

Fig. 2.18 Standard deviation of net heat flux (W m^{-2}) in **a** March, **b** April, **c** May, and **d** June

The maximum center of standard deviation in Fig. 2.18 indicates strong interannual variation. In March and April, the maximum center is located over the SCS and Kuroshio area (Fig. 2.18a, b). In May and June, the distribution of the standard deviation is a bit more scattered. However, we can still identify the maximum center over the tropical region (Fig. 2.18c and d). From Fig. 2.18a–d, the maximum center of the net heat flux is generally confined within the region (30°–160° E, 30° S–30° N) over the Indian Ocean and the western Pacific Ocean.

The net heat flux is the total effect of latent heat flux, sensible heat flux, longwave radiation, and shortwave radiation. The first three flux terms are positive when they are upward, while the shortwave radiation and the net heat flux are positive when they are downward. To find out the most important component of the net heat flux, an EOF analysis is applied to the pentad data of the climatological mean heat fluxes to reveal the major characteristics of the transition. Figure 2.19 shows the EOF1 and PC1 of the climatology of the net heat flux, latent heat flux, sensible heat flux, shortwave radiation, and longwave radiation during Pentad 1 (January 1–5) to Pentad 42 (July 25–29).

The percentage variances of the first EOF mode of the net heat flux, latent heat flux, sensible heat flux, shortwave radiation, and longwave radiation are 93.8%, 82.58%, 92.74%, 90.98%, and 78.04%, respectively. In the EOF1 of the net heat flux, the zero contour is almost along the equator with the positive value in the Southern Hemisphere and negative value in the Northern Hemisphere. For the corresponding PC1, it is positive before Pentad 18, but is negative after Pentad 19. The positive-to-negative transition time of the net heat flux is around Pentad 19 (Fig. 2.19f). The first EOF patterns of latent heat flux and sensible heat flux are similar. In Fig. 2.19b, the positive values of latent heat flux are mainly in the Northern Hemisphere with

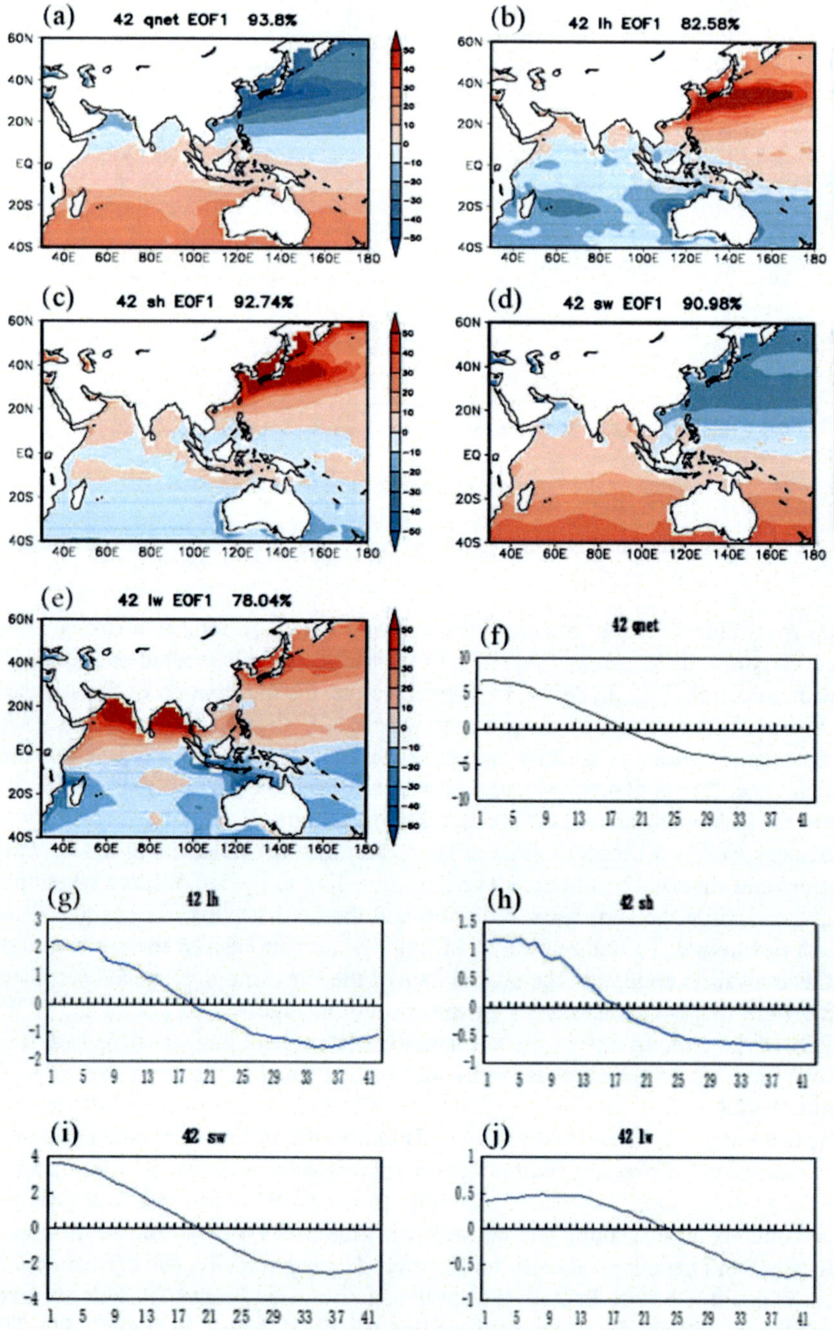

Fig. 2.19 EOF1 and PC1 of the climatology of the net heat flux (**a**; qnet), latent heat flux (**b**; lh), sensible heat flux (**c**; sh), shortwave radiation (**d**; sw), and longwave radiation (**e**; lw) during Pentads 1 to 42

the maximum center over the SCS and Kuroshio area, while the negative values are mainly in the Southern Hemisphere, which is similar to the EOF1 of the net heat flux (Fig. 2.19a). Furthermore, the transition time of latent heat flux is also at Pentad 19, the same as that of the net heat flux, indicating the important contribution of latent heat flux to the variation of the net heat flux. In Fig. 2.19c, the positive center of sensible heat flux is also located over the SCS and Kuroshio area; however, the absolute value in the Southern Hemisphere is very small. The distribution of shortwave radiation is also similar to that of the net heat flux: there are obvious negative values over the SCS and Kuroshio area, and positive values in the Southern Hemisphere (Fig. 2.19d). Both the PC1s of sensible heat flux and shortwave radiation show decreasing trends. In Fig. 2.19e, the most outstanding positive center is located over the northern Indian Ocean. The corresponding PC1 shows a small increasing trend and then a decreasing trend (Fig. 2.19j). Comparing Fig. 2.19a–j, both latent heat flux and shortwave radiation are highly consistent with the net heat flux for both spatial distribution (EOF1) and the principal component (PC1). In other words, the transition of the net heat flux is mainly contributed by the variation of latent heat flux and shortwave radiation.

To examine the transition time in each year, the EOF analysis is applied to the pentad data of the net heat flux for each year individually from 1985 to 2009. The PC1 is used to determine the pentad of the transition time of the net heat flux. Because there are 25 years, we cannot show the results for all individual years. We select 1985, 1986, and 1988 to represent moderate year, early year, and late year of the transition. Figure 2.20 shows the EOF1s and PC1s in 1985, 1986, and 1988.

The percentage variances are 71.63%, 73.57%, and 68.01% in 1985, 1986, and 1988, respectively. Figure 2.20a–c show they all have similar patterns. All the EOF1s show negative values in the Northern Hemisphere but positive values in the Southern Hemisphere. The EOF1s for the other years are also similar to those of the years 1985, 1986, and 1988 (not shown). The positive center is located near the Kuroshio area. The PC1s show apparent decreasing trends. The approximate transition of PC1 from positive to negative occurs around Pentad 19. After the transition time, the net heat flux becomes positive in the Northern Hemisphere but negative in the Southern Hemisphere.

Similarly, we define the transition time based on the PC1 for each year. The definition is given as follows: when the PC1 of the net heat flux becomes negative and lasts for more than four pentads, the first pentad of the negative values of PC1 is the transition time (T1) for each year. We call it the spring–summer transition time. Thus, the transition times of 1985, 1986, and 1988 were Pentad 19, Pentad 17, and Pentad 20, respectively. The time series of the transition time during 1985–2009 is shown in Fig. 2.21.

The average transition time was Pentad 18.48 during 1985–2009, which is close to the transition time for the climatological mean of the net heat flux. The early years, defined by the standardized value less than -1, include the years 1986, 2002, and 2005. The late years, defined by the standardized value more than 1, include the years 1988, 1993, 1996, and 1999.

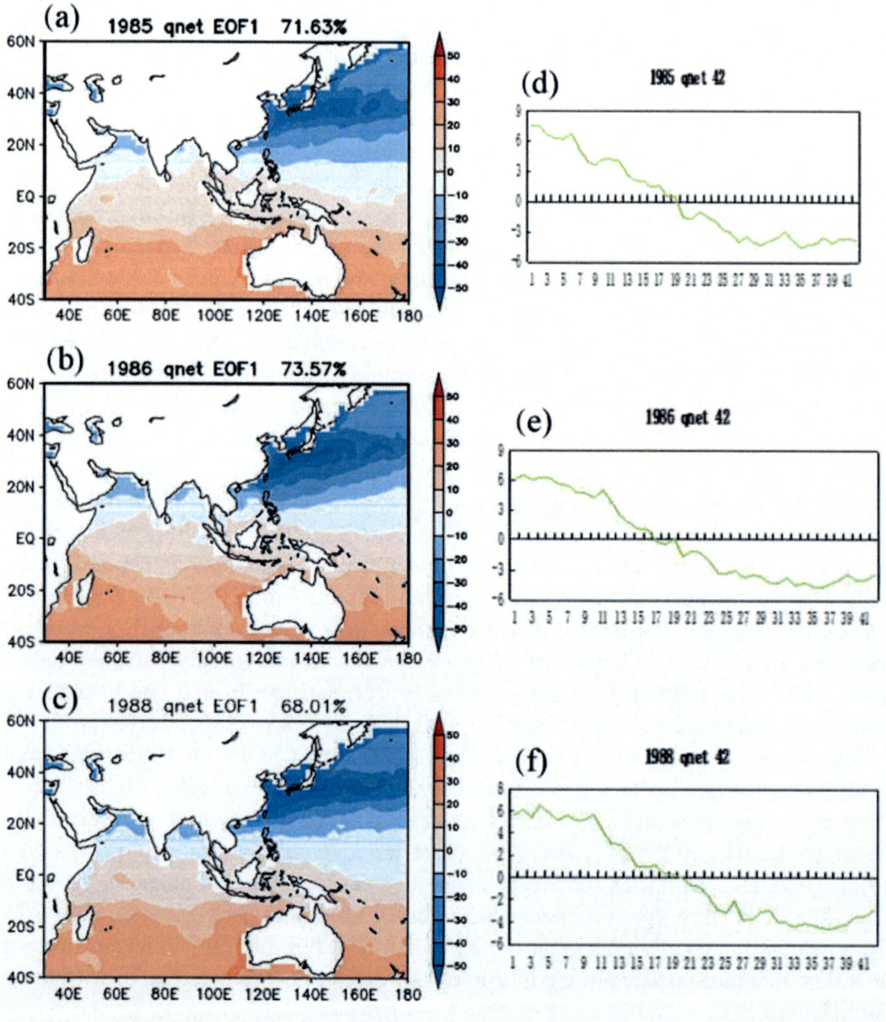

Fig. 2.20 EOF1s (left) and PC1s (right) of the pentad data of the net heat flux in 1985, 1986, and 1988

2.3 Impact of Intraseasonal Oscillation on SCSSM Onset

The onset of the ASM indicates the rapid strengthening of atmospheric energy and water cycle (Wu et al. 2013). Monsoon onset and activity have important influences on weather and climate in Asia, even globally. The East Asian summer monsoon (EASM) starts over the SCS. The SCSSM onset is the basis for the northward movement of the summer rain belt. So, the SCSSM onset is one of the important factors for predicting the rainy season in China. Studies on the dynamics and interannual variation of SCSSM onset have attracted a great deal of attention from domestic as

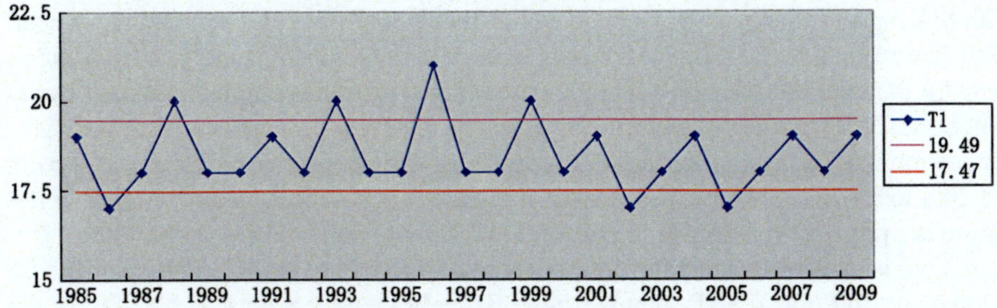

Fig. 2.21 Time series of the spring–summer transition of the net heat flux (T1; pentads). The purple and red lines represent the range of one times the standard deviation centered at the mean value

well as international researchers. Studies by Liu et al. (2002) and Wu et al. (2013) showed the ASM break out first over the BOB, and an environment favorable for summer monsoon onset is set up over the eastern BOB and the SCS. The Rossby wave excited by monsoon precipitation over the BOB propagates eastward, resulting in the onset of the SCSSM. Much has been studied regarding the impact of air–sea interaction on SCSSM onset (Chen et al. 1999; Ding and Li 1999; Zhao et al. 2000). It was shown that the SSTAs represent the El Niño (La Niña) pattern in the years of early (late) onset of the SCSSM (Mao et al. 2000). Huang et al. (2005) indicated that convection over the Philippines is enhanced during a warmer state in the western tropical Pacific in spring, resulting in an early onset of SCSSM, and vice versa. Besides the impact of the tropical Pacific, SCSSM onset also is related to the SST in the Indian Ocean (Liang et al. 2006) and even in the Southern Ocean (Lin et al. 2013a, b). Some techniques based on SST for short-term climate prediction of SCSSM onset were established, which achieved certain good results (Zhou et al. 2002; Gu et al. 2011). However, the prediction based on SST in the spring showed that the SCSSM onset date would be a little earlier than normal in 2014. But the actual date was June 5, which was abnormally late. It is worth exploring why the forecast error of SCSSM onset in some years was so big.

Atmospheric ISO is related to the SCSSM onset to a certain extent (Lin 1998). The ISO developing and extending to the SCS from the eastern Philippines shows great effects on the ISO over the SCS, and triggers the onset of SCSSM finally from the view of 850-hPa low-frequency wind field distribution in 1998 (Mu and Li 2000). Using multiyear data, Wen et al. (2006) indicated that when low-frequency convection over the southern Philippines develops and moves westward, low-frequency convection over southern China moves southward, and low-frequency convection over Kalimantan Island is active and moves northward during the period May 1–15, and the SCSSM tends to start early. When the convections over the above regions are inactive, the SCSSM tends to start late. The ISO moving eastward along the equator, i.e., Madden–Julian Oscillation (MJO), is one of the dominant modes of the tropical ISO. A great deal of analyses showed that the MJO impacts the weather and climate not only in the tropics but also outside the tropics (e.g., Mo and Higgins 1998; Higgins and Shi 2001; Jones et al. 2004; Lorenz and Dennis 2006; Jeong et al.

2008; Ding and Liang 2010; Jia et al. 2011; Zhang et al. 2011; Lü et al. 2012; Li et al. 2013a, b; Lin et al. 2013b, 2015; Hung et al. 2014). Composite analysis by Hendon and Liebmann (1990) showed that the onset of the Australia summer monsoon corresponds to MJO passing over northern Australia. Our results here show that both SST in interannual and even interdecadal time scales and atmospheric ISO are impacting factors to the SCSSM onset. However, there are a few references on whether or not there is a physical relationship between MJO activity and SCSSM onset. Our objective here is to explore the possible impact of MJO activity on SCSSM onset and its synergy with tropical SST signals using relatively long-term data.

The data used in this section include (i) daily mean atmospheric variables from the National Center for Environmental Prediction/Department of Energy (NCEP/DOE) Reanalysis 2 (Kanamitsu et al. 2002); (ii) daily mean OLR (Liebmann and Smith 1996) provided by the National Oceanic and Atmospheric Administration (NOAA); and (iii) monthly extended reconstructed SST (ERSST.v2) from the NOAA (Smith and Reynolds 2004). The period of these data is from 1979 to 2013 with a global coverage at a resolution of 2.5° × 2.5° in (i) and (ii), and 2° × 2° in (iii). The MJO index used in this section is the real-time multivariate MJO index advanced by Wheeler and Hendon (2004).

The definition of SCSSM onset in Liang and Wu (2002) is considered to be more reasonable among various definitions (Huang et al. 2008). For this definition, two conditions need to be satisfied: one is the "zonal wind averaged over the SCS greater than zero," and the other is "westerlies over the SCS coming from the southern BOB primarily." The second condition avoids the situation that westerlies are brought about by a southern branch of the westerly trough, but this condition is determined by subject analysis. To facilitate automatic calculation, a new definition of SCSSM onset was proposed by Lin et al. (2013a) based on the definition by Liang and Wu (2002). The new definition is as follows: when both southwesterlies at 850 hPa and easterlies at 200 hPa prevail continuously over the monitored domain of the SCS (105°–120° E, 5°–20° N) for five days or longer, then the first day is defined as SCSSM onset date (Lin et al. 2013a). The analysis by Lin et al. (2013a) showed that the criterion that "200-hPa flow is easterlies over the SCS" achieves the same result with the criterion of "westerlies over the SCS coming from the southern BOB" in previous studies (Liang and Wu 2002). Both of the abovementioned criteria indicate the retreat of the westerlies and the northward push of the tropical circulation. Here, the definition of SCSSM onset by Lin et al. (2013a) is used. The date of SCSSM onset in each year during 1979–2013 is given in Table 2.1. The mean date averaged over the 35 years is May 16, which agrees with that using surface wind to define the onset date by Wu et al. (2013). The standard deviation of the SCSSM onset date is 11.5 days. According to 1.0 (1.5) times of standard deviation to select abnormal (very abnormal) years, there are three late years (very late years) and three early years (very early years), respectively.

2.3.1 Relationship Between SCSSM Onset and MJO Activity

Figure 2.22 shows the phase space of MJO on the date of the SCSSM onset during 1979–2013. There are 31 years when MJO is active in phases 4, 5, 6, or 7 on the onset date among the 35 years (about 89%), and the vast majority of MJO is strong (strength greater than or equal to 1). There are very few years when MJO is active in phase 8, 1, 2, or 3 on the onset date. MJO in phases 4 to 7 indicates that convection is active over the MC and the western Pacific (hereinafter referred to as the western Pacific phase), and MJO in phase 8 and phases 1 to 3 indicates convection is active over Africa and the Indian Ocean (hereinafter referred to as the Indian Ocean phase). It is necessary to declare that the probability of MJO in the western Pacific phase is close to that in the Indian Ocean phase in AMJ in terms of statistics based on several-decade data. It is documented that MJO activity is likely to impact on monsoon onset date. MJO in the western Pacific phase (Indian Ocean phase) does (does not) favor SCSSM onset.

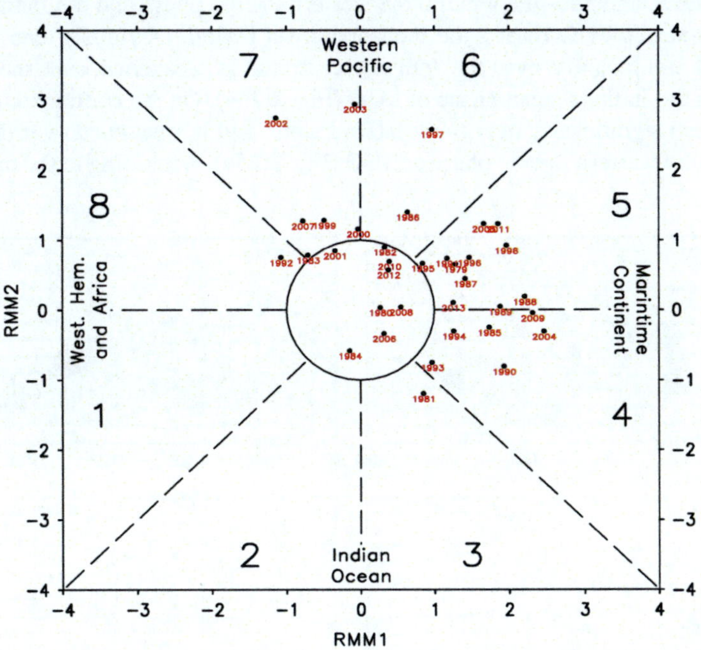

Fig. 2.22 Phase space of MJO on the date of SCSSM onset during 1979–2013. The year is given in red

2.3.2 Physical Process Behind MJO Impact on SCSSM Onset

The primary atmospheric circulation variation characteristics around SCSSM onset are discussed next. The western North Pacific subtropical high (WNPSH) in the lower-level troposphere retreats eastward, the western ridge of the subtropical high withdraws eastward from the SCS, tropical westerlies expand eastward, and westerlies prevail over the SCS (Fig. 2.23). The SAH in the upper-level troposphere is established over the Indochinese Peninsula and northern SCS, and easterlies prevail over most of the SCS region (central and southern parts) (Fig. 2.24). Then, the question becomes whether or not the MJO activity influences the abovementioned circulation variation.

The earliest and latest onset dates are April 17 and June 11, respectively. So, the period from the fourth pentad in April to the second pentad in June is considered as the monsoon onset period (56 days each year). Statistics show that there are 163, 152, 150, 154, 149, 152, 158, and 163 days in phases 1–8 of strong MJO, respectively, during the monsoon onset period in the 35 years. Figure 2.25 shows the composite fields of OLR and 850-hPa wind in the western Pacific phase and the Indian Ocean phase for strong MJO during the monsoon onset period. Obviously, convection is enhanced significantly over the Indian Ocean and is weakened over the western Pacific in the Indian Ocean phase of MJO (Fig. 2.25a). On the contrary, convection is enhanced significantly over the western Pacific and is weakened over the Indian Ocean in the western Pacific phase of MJO (Fig. 2.25b). According to the theoretical

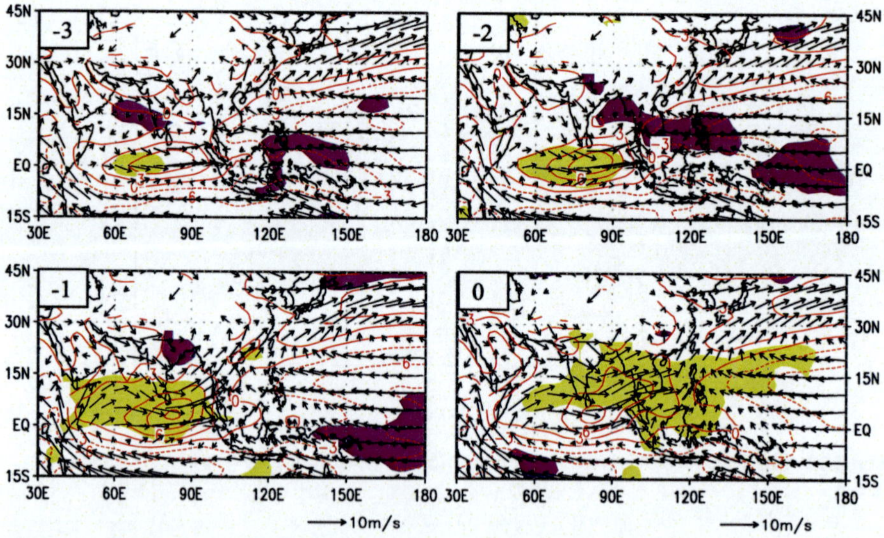

Fig. 2.23 Composite 850-hPa wind (vector; units: m s^{-1}) and zonal wind (contour; units: m s^{-1}) in three pentads, two pentads, and one pentad prior to SCSSM onset and in the onset pentad. Yellow or purple shading indicates the 95% confidence level of zonal wind greater or less than the climatological mean, respectively. Pentad 0 is the onset date of SCSSM

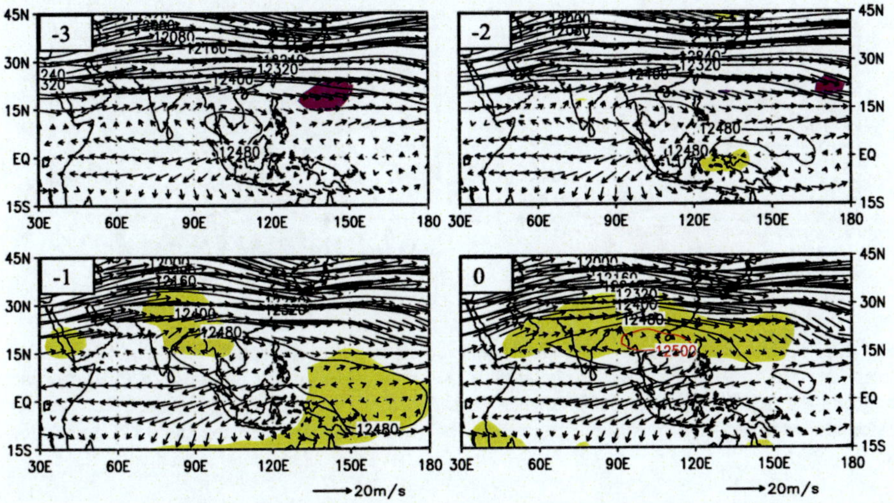

Fig. 2.24 Same as Fig. 2.23, except for composite 200-hPa wind (vector; m s^{-1}) and geopotential height (contour; gpm)

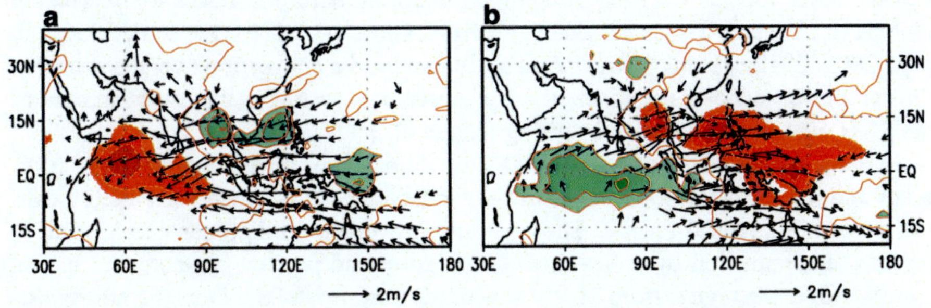

Fig. 2.25 Composite 850-hPa wind anomalies (vector; m s^{-1}) and OLR anomalies (contour; W m^{-2}) for strong MJO in the Indian Ocean phase **a** and in the western Pacific phase **b** during the SCSSM onset period (between Pentad 4 in April and Pentad 2 in June). Red (green) color indicates stronger (weaker) convection with OLR anomalies less (greater) than −6 W m^{-2} (6 W m^{-2}). Only the 850-hPa wind anomaly vectors exceeding the 95% confidence level are shown

model of Gill (1980) about atmosphere responses to equatorial heating (cooling), a pair of anomalous cyclones (anticyclones) appears in the northwest and southwest of the heating (cooling) because of Rossby wave responses, and anomalous easterlies (westerlies) emerge near the equator east of the heating (cooling) source because of Kelvin wave responses. Therefore, in the western Pacific phase of MJO there is an anomalous cyclonic circulation over the SCS and western North Pacific (WNP), which is located on the northwest side of the heating source because of the latent heat released by convection. The cyclonic anomaly favors the retreat of the WNPSH. In the Indian Ocean phase of MJO, the situation is basically the opposite. There is

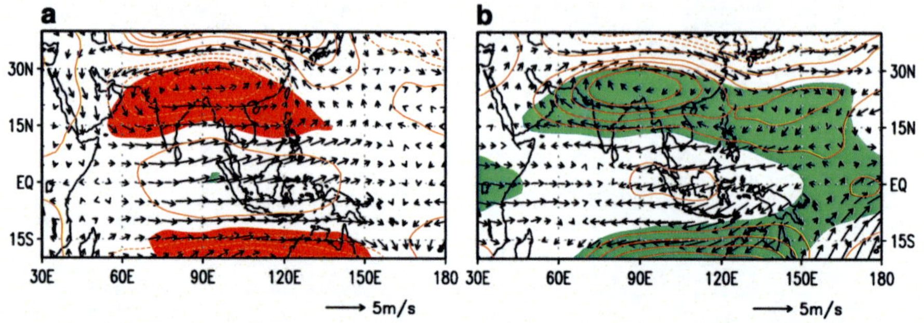

Fig. 2.26 Composite 200-hPa wind anomalies (vector; m s^{-1}) and geopotential height anomalies (contour; gpm) for strong MJO in the Indian Ocean phase **a** and in the western Pacific phase **b** during the SCSSM onset period (between Pentad 4 in April and Pentad 2 in June). Red (green) color indicates negative (positive) geopotential height anomalies exceeding the 95% confidence level

an anticyclonic anomaly over the SCS and the WNP, which favors the enhancement and elongation toward the west of the subtropical high over the WNP.

MJO activity is closely related to the SAH in the upper troposphere. It was shown that the SAH appears over the SCS region in spring before it moves to the Tibetan Plateau (Reiter and Gao 1982; Zhu et al. 1986; Chen et al. 1991; He et al. 2006; Liu et al. 2009, 2012). From the viewpoint of dynamics, the convection over the southern Philippines produces a negative vorticity source, and results in the establishment of the SAH over the SCS region in spring (Liu et al. 1999, 2013).

In the western Pacific phase of MJO (Fig. 2.26b), the convection over the Philippines and its nearby regions is also enhanced besides the enhanced convection over the western equatorial Pacific. The convection over the Philippines and its nearby regions, accompanied by anticyclonic circulation and positive geopotential height anomaly over the region from South Asia to the northern SCS, favors the northwestward movement from the western Pacific and enhancement for the SAH. Then, the SAH is established over the Indochinese Peninsula and northern SCS. In the Indian Ocean phase of MJO (Fig. 2.26a), the convection over the Philippines and its nearby regions is also decreased, which brings cyclonic circulation and negative geopotential height anomaly over the region from South Asia to the northern SCS. These conditions make it difficult to establish the SAH over the Indochinese Peninsula and northern SCS.

The above analyses document that the primary physical process for the impact of MJO on SCSSM is atmospheric response to heating source. When the MJO is in the western Pacific phase, more latent heat is released from enhanced convection, resulting in an anomalous cyclonic circulation in the low-level troposphere of the tropical Pacific as a Rossby wave response. The anomalous cyclonic circulation is favorable for the eastward retreat of the subtropical high, and is conducive to the eastward expansion of tropical southwesterlies to the SCS. On the other hand, the heating source over the Philippines and its vicinity promote the establishment of the SAH over the Indochinese Peninsula and northern SCS, and the wind in the upper-level troposphere changes to the easterly over the SCS. The corporation between lower and upper levels of the troposphere is conducive to the establishment of SCSSM. When the MJO is in the Indian Ocean phase, the situation is basically opposite, which is not conducive to the SCSSM onset.

2.3.3 Synergy Between MJO and Tropical SST

There exist significant correlations between SCSSM onset and the SST in both western and central tropical Pacific (Fig. 2.27). Figure 2.28 shows that the onset date of SCSSM varies inversely with the SSTA in the western tropical Pacific (130°–160° E, 0°–15° N) but varies congruously with the SSTA in the central tropical Pacific (180°–140° W, 5° S–5° N). Considering the opposite correlation signs of the two regional SSTAs, we composite a tropical Pacific SST synthetic signal using the SST

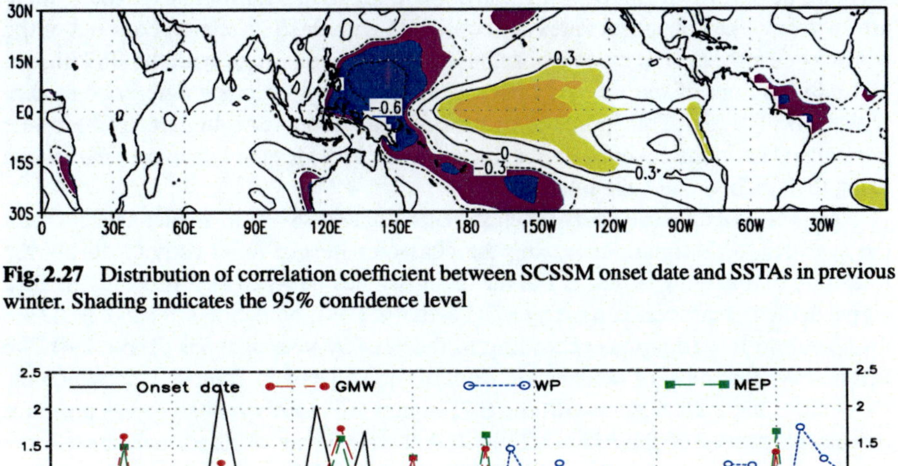

Fig. 2.27 Distribution of correlation coefficient between SCSSM onset date and SSTAs in previous winter. Shading indicates the 95% confidence level

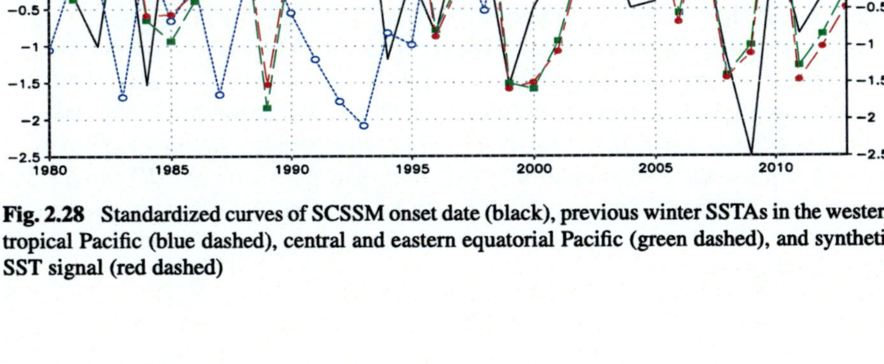

Fig. 2.28 Standardized curves of SCSSM onset date (black), previous winter SSTAs in the western tropical Pacific (blue dashed), central and eastern equatorial Pacific (green dashed), and synthetic SST signal (red dashed)

difference between the central tropical Pacific and the western tropical Pacific. We can see that the onset date of SCSSM varies congruously with the SST synthetic signal (red curve in Fig. 2.28). The correlation coefficient between them reaches 0.56, passing the 99.9% confidence level. We conclude that SSTAs in the tropical Pacific are important factors for SCSSM onset.

By comparing the black solid curve and the red dashed curve in Fig. 2.28, we can see the difference between the onset date and the synthetic SST signal can be quite large and can even be out of phase in some years, despite a high correlation between them. Here, the difference between the standardized value of the onset date of SCSSM and the synthetic SST signal is used to determine whether or not the onset date is consistent with the SST signal. When the absolute value of the difference between the two standardized values is within 1 (up to or above 1), it is considered that the onset date of SCSSM is consistent (inconsistent) with the synthetic SST signal. We found that the SCSSM onset dates are consistent with the SST signal for 21 years and are inconsistent for 14 years during the 35-year study period (Table 2.3). It is shown that the forecast errors for 60% of the years are within one standard deviation (or less than 12 days), but the forecast errors for 40% of the years are up to or above one standard deviation (or more than 12 days). Among the 23 normal onset years, there are 16 years when the onset dates are consistent with the SST signal. Four out of the six early onset years are those whose onset dates are consistent with the SST signal, and only one out of the six late onset years is that whose onset date is consistent with the SST signal. This indicates that the forecast accuracy is higher (69%) for the normal and early onset years, and is much lower (17%) for the late onset years when using the synthetic SST signal.

Next, we examine the years when the onset date of SCSSM is inconsistent with the synthetic SST signal, analyzing the characteristics of MJO activity during the transition to find out whether or not the inconsistency between the onset and the SST signal in these years was caused by MJO activity. The modulation effects of MJO are grouped into five situations according to the year-by-year analysis (Table 2.4). We can observe that most of the years (12 out of the 14 years in Table 2.3) in which the onset dates are inconsistent with the SST signal are closely related to the presence of various phases of the MJO activity. Of the six years of abnormal onset (over 1.5 times standard deviation in Fig. 2.28) in the past 35 years, four abnormal years (1987, 1991, 1993, and 2009) are obviously modulated by the MJO. Regarding the abnormal late onset case of 2014 mentioned at the beginning of Sect. 2.3, the MJO activity in May is quite unusual. The MJO moves eastward from phase 8 to phase 1, phase 2, and phase 3 during the first half of May. However, the MJO does not move eastward into the western Pacific as usual; instead, it wanes and stays stagnant or even retreats westward in the Indian Ocean and does not reinvigorate until the end of May (not shown). Therefore, the characteristics of MJO activity are favorable for a later SCSSM onset in 2014. Clearly, it is important to consider the modulation of MJO to improve the prediction of SCSSM onset. The interseasonal MJO activity and interannual SST variability jointly influence SCSSM onset. Of course, the year in which the SCSSM onset date coincides with the SST signal is not entirely immune from MJO activity.

Table 2.3 Onset dates of SCSSM, consistency with the SST signal, MJO intensity, and phase on onset date during 1979–2009

Year	Sequence number of onset date	Onset date	Consistency with SST signal	MJO intensity and phase
1979	133	13 May	Yes	Strong, 5
1980	136	15 May	Yes	weak
1981	132	12 May	Yes	Strong, 3
1982	124*	4 May	Yes	weak
1983	148+	28 May	Yes	Strong, 7
1984	118**	27 April	Yes	weak
1985	147+	27 May	No	Strong, 4
1986	136	16 May	Yes	Strong, 6
1987	162++	11 June	No	Strong, 5
1988	141	20 May	Yes	Strong, 5
1989	139	19 May	No	Strong, 5
1990	137	17 May	Yes	Strong, 4
1991	159++	8 June	No	Strong, 5
1992	142	21 May	No	Strong, 8
1993	155++	4 June	No	Strong, 4
1994	122*	2 May	No	Strong, 4
1995	135	15 May	No	Strong, 5
1996	126	5 May	Yes	Strong, 5
1997	141	21 May	Yes	Strong, 6
1998	140	20 May	No	Strong, 5
1999	118**	28 April	Yes	Strong, 7
2000	130	9 May	No	Strong, 7
2001	137	17 May	No	weak
2002	139	19 May	Yes	Strong, 7
2003	141	21 May	Yes	Strong, 7
2004	130	9 May	Yes	Strong, 4
2005	131	11 May	No	Strong, 5
2006	147+	27 May	No	weak
2007	140	20 May	Yes	Strong, 7
2008	122	1 May	Yes	weak
2009	107**	17 April	No	Strong, 4
2010	141	21 May	Yes	weak
2011	126	6 May	Yes	Strong, 5
2012	132	11 May	Yes	weak
2013	133	13 May	Yes	Strong, 4

Note *, **, +, and ++ indicate early onset, very early onset, late onset, and very late onset, respectively

Table 2.4 Characteristics of SST signal and MJO activity for five situations of SCSSM onset

Situation	SST signal anomaly	SCSSM Onset	MJO	Years
First	Negative	Late	MJO is basically in the Indian Ocean phase during the first and second 10-day period in May	1985, 1989, 2006
Second	Positive	Very late	MJO is in the Indian Ocean phase from the end of the second 10-day period in May to early June	1987, 1991, 1993
Third	Positive	Normal or early	MJO activities in the western Pacific during the first 10-day in May	1994, 2005, 1995
Forth	Negative	Very early	MJO is very strong persistently in April and moves into the western Pacific in mid-April	2009
Fifth	Strong Positive	Normal	MJO moves into the western Pacific in mid-May and remains strong	1992, 1998

2.4 Precursory Signals for SCSSM Onset

The SCSSM is an important part of the EASM. The onset of the EASM and the beginning of the rainy season in China are marked by the onset of the SCSSM. Its activities have an important impact not only on the Asian monsoon system, but also on the circulation and weather in the Northern Hemisphere (Li and Zhang 1999). Therefore, researchers have paid much attention to the attribution analysis of the early and late onset of the SCS summer. The prediction of SCSSM onset has also become one of the main concerns of scientific meetings on climate trends, as well as of consultations on flood seasons in China.

In this section, we adopt the definition of the onset date of the SCSSM proposed by Liang and Wu (2002). That is, the average zonal wind speed at 850 hPa over the SCS is greater than zero (an indication that most part of the WNPSH moves out of the SCS), and the westerly wind in the SCS mainly comes from the south of the BOB. When the above two conditions are met simultaneously for more than five days, and a subsequent continuous interruption (the average zonal wind in the SCS area is less than zero) is not longer than three times the number of days in the early southwest monsoon, then the first day when these conditions are met, is the onset date of the SCSSM.

Observational studies indicate that the onset of SCSSM may be affected by SSTAs. For example, the onset of the SCSSM in the El Niño year tends to be late. The onset of SCSSM has significant interannual and interdecadal variations, and the physical causes of evolution at different time scales can be different. We will analyze the precursory signals in interannual and interdecadal variations of SST and possible mechanisms.

In carrying out the analysis of this section, we used NCEP/NCAR reanalysis data and the third edition of monthly extended reconstructed SST data (ERSST.v3) from NOAA (Smith and Reynolds 2004).

2.4.1 Multi-time-scale Characteristics of SCSSM Onset Date

The average onset time of the SCSSM is May 18 for the study period from 1958 to 2016, with a standard deviation of about 10.5 days. Figure 2.29 shows the anomalous onset dates of the SCSSM, the wavelet transform coefficient, and the total mean wavelet power spectrum. The linear trend of the onset of the SCSSM is −0.1 days per year, or a weak trend of early onset (Fig. 2.29a). The wavelet coefficient before the end of the 1980s shows a periodicity of 4-year to 7-year cycles, and most of which pass the 90% confidence level; after the 1980s, it mainly shows the interdecadal

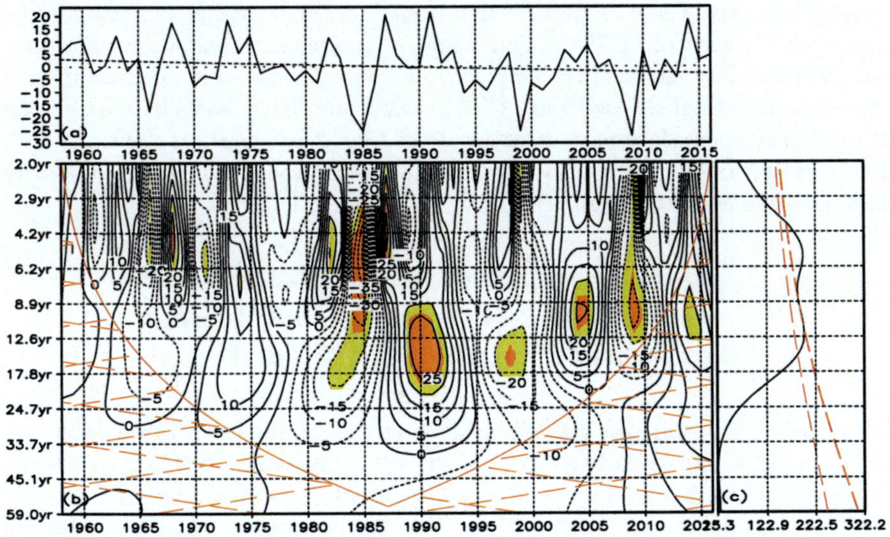

Fig. 2.29 The anomalous sequence (solid curve) and its trend (dashed line) (**a**; days) of SCSSM onset date, wavelet transform coefficient (**b**), and averaged total wavelet power spectrum (**c**). In (**b**), the light and dark shadings represent 90% and 95% confidence levels, respectively. The thick and thin dashed lines in (**c**) are the 95% and 90% confidence level lines s, respectively. The Monte Carlo method is used for the significance level test

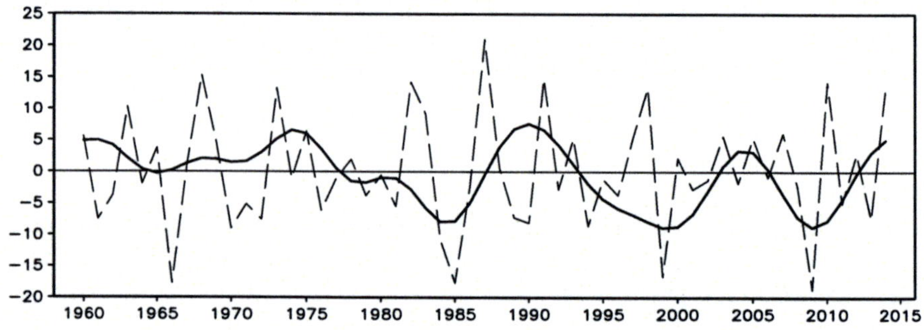

Fig. 2.30 Changes in SCSSM onset date on the interannual (dashed) and interdecadal (solid) time scales

change of a period of 10–16 year, and all pass the 90% confidence level (Fig. 2.29b). The averaged global wavelet power spectrum shows that there are two major periods at 12.4 and 5.5 years, passing the significant level test of the $\alpha = 0.05$ (Fig. 2.29c).

Figure 2.29c shows that the oscillation with an eight-year period is the weakest among the periods of 5–12 years. Therefore, the time scale of eight years can be regarded as the separation point of the interannual and interdecadal changes. We use the Lanczos time filter to decompose the SCSSM date sequence into the interannual and interdecadal components on the two time scales. The standard deviation of the interannual component is 8.87 days, and the standard deviation of the interdecadal component is 4.76 days, which is slightly more than half of the former. Therefore, the onset of SCSSM is dominated by interannual variability, but the interdecadal variability should not be ignored. A large amplitude of interannual variation appears in the middle of the 1960s, 1980s, 1990s, and 2010s, and a large amplitude of interdecadal change appears after the 1980s (Fig. 2.30), which is consistent with the characteristics of interannual variation and interdecadal change in the wavelet transform coefficient diagram.

2.4.2 Precursory SST Signals for Interannual Variation and Interdecadal Change of SCSSM Onset

The correlation coefficients between the interannual variation of SCSSM onset date and the mean SST in the early winter show that the regions with significant correlation are in the tropical Pacific and tropical India Ocean, or to be more precisely, the central–eastern equatorial Pacific (180°–80° W, 10° S–14° N, named region 1), the western tropical Pacific (126°–144° E, 0°–14° N, named region 2), and the southwestern India Ocean (58°–72° E, 16° S–0°, named region 3) (upper panel in Fig. 2.31). The interannual variation of SCSSM onset is positively correlated with the SST in regions 1 and 3, and negatively correlated with the SST in region 2. By averaging the winter SSTs in these three regions, we can obtain three precursory signals of

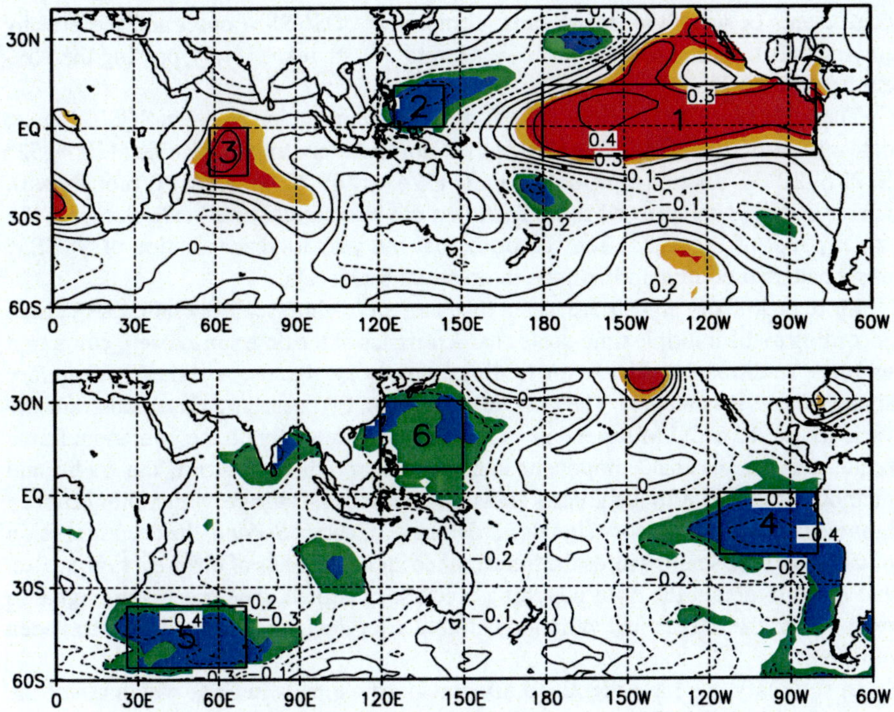

Fig. 2.31 Correlation coefficients between interannual variation of SCSSM onset date and previous winter SST (upper), and between interdecadal changes of SCSSM onset date and averaged SST in the previous year (bottom). Yellow and green shadings indicate the 90% confidence level; red and blue shadings indicate the 95% confidence level

the interannual variability of SCSSM onset. Corresponding to regional numbers, the correlation coefficients between SCSSM onset date and precursory signals in regions 1, 2, and 3 are 0.37, −0.3, and 0.3 on the interannual time scale, respectively, passing the 95% confidence level.

The correlation coefficients between the interdecadal component of SCSSM onset date and the mean SST in the previous year (from March to the following February) show that there are three significantly correlated regions: the southeastern Pacific (80°–124° W, 20° S–0°, named region 4), the southwestern India Ocean (26°–70° E, 56°–36° S, named region 5), and the Northwest Pacific (120°–150° E, 8°–30° N, named region 6). The interdecadal component of SCSSM onset date and the SSTs in regions 4, 5, and 6 are significantly negatively correlated; and the largest correlation coefficients in these three regions are −0.47, −0.46, and −0.39, respectively. Therefore, the yearly mean SSTs from March to the following February in the above three regions are regarded as the precursory signals associated with the interdecadal change of SCSSM onset. Corresponding to the regional numbers, the correlation coefficients between the interdecadal component of SCSSM onset and the precursory signals 4, 5, and 6 are −0.37, −0.41, and −0.38, respectively, passing the 95% confidence level.

There are two major regions with a significant correlation between SCSSM onset date and the SST in the previous winter: the northwestern tropical Pacific (120°–152° E, 2° S–20° N) and the equatorial Pacific (124°–178° W, 8° S–14° N) (not shown), which is quite similar to what we found on the interannual time scale shown in the upper panel of Fig. 2.31. The reason is that the year-to-year variation of SCSSM onset date is predominantly on the interannual time scale.

More precursory signals are found by using multi-time-scale evolution sequences according to the multiple time-scale characteristics of the time series itself, compared with the traditional methods for finding precursory signals. Are these precursory signals more reasonable? The answer is yes for two reasons. First, although the variation of the SCSSM onset date is mainly predominantly on the interannual time scale, the interdecadal component cannot be ignored. In contrast, the traditional methods usually overlook the early signal of interdecadal change, which should not be ignored, namely, precursory signals in regions 4, 5, and 6. Second, the decomposition of different time scales avoids the interference among signals of different time scales, and helps to obtain the right matching precursory signals. The precursory signal in region 3 for the interannual variation is found after the interdecadal change has been removed.

In summary, we have identified six precursory signals in SST, which affect the onset of the SCSSM. They are the winter averaged SST in the central–eastern tropical Pacific (180°–80° W, 10° S–14° N), the northwestern tropical Pacific (126°–144° E, 0°–14° N), the southwestern tropical India Ocean (58°–72° E, 16° S–0°), the averaged SST from previous March to the following February in the southeastern tropical Pacific (80°–116° W, 20° S–0°), the southwestern India Ocean (26°–70° E, 56°–36° S), and the Northwest Pacific (120°–150° E, 8°–30° N).

2.4.3 Asymmetry Between the Central–Eastern Equatorial Pacific SSTA and SCSSM Onset

There is an asymmetry in the relationship between the central–eastern equatorial Pacific SSTA and SCSSM onset. Next, we describe this asymmetry in more detail. Based on a criterion of 0.75 times of the standard deviation of SCSSM onset date and Fig. 2.29a, we find 10 anomalous early onsets of SCSSM for the period 1958–2014; they are 1966, 1984, 1985, 1994, 1996, 1999, 2001, 2008, 2009, and 2011. We also find 11 anomalous late onsets of SCSSM; they are 1959, 1960, 1963, 1968, 1973, 1975, 1982, 1987, 1991, 2005, and 2014. Our composite analysis shows that significant winter SSTAs occur mainly in the central–eastern equatorial Pacific and the northwestern tropical Pacific, the southwestern subtropical Pacific, and the southern India Ocean for anomalous early onset cases (Fig. 2.32, upper panel). However, for

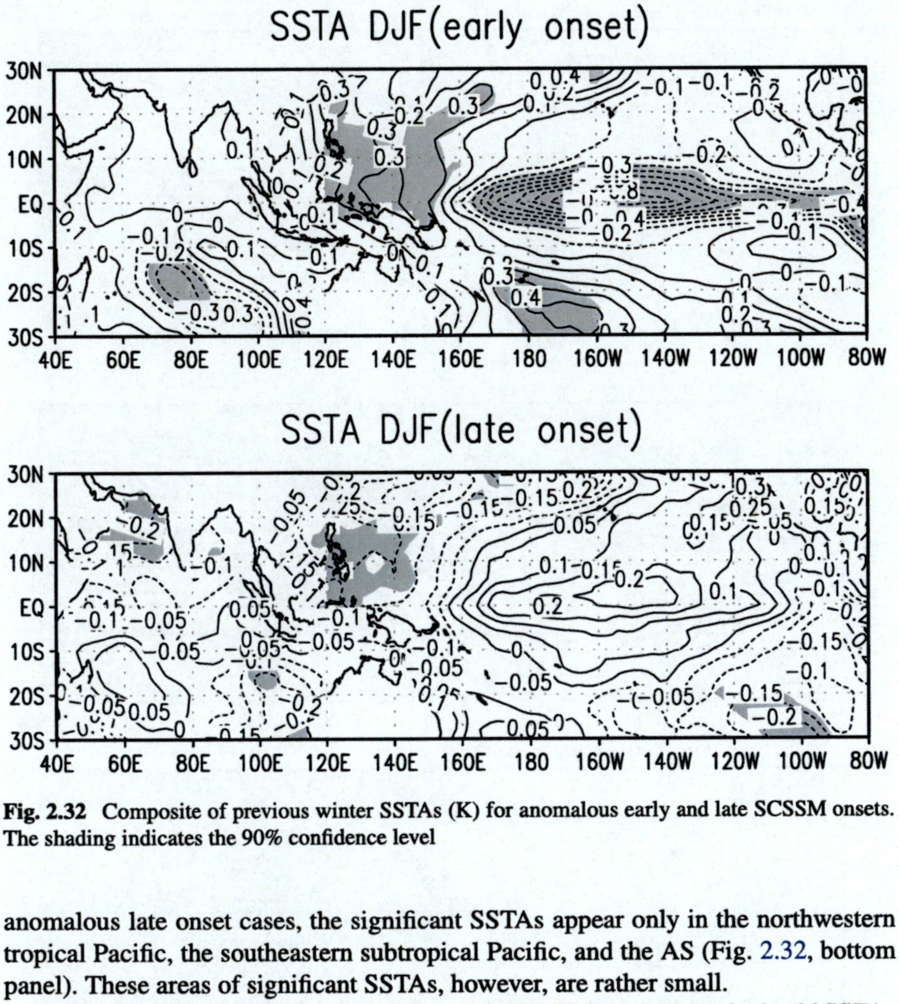

Fig. 2.32 Composite of previous winter SSTAs (K) for anomalous early and late SCSSM onsets. The shading indicates the 90% confidence level

anomalous late onset cases, the significant SSTAs appear only in the northwestern tropical Pacific, the southeastern subtropical Pacific, and the AS (Fig. 2.32, bottom panel). These areas of significant SSTAs, however, are rather small.

Among the ten anomalous early onset cases of SCSSM, six cases had cold SSTAs in the central–eastern Pacific in winter, i.e., 1985, 1996, 1999, 2008, 2009, and 2011. Composite analysis shows that the SSTAs in the tropical Pacific exhibit a La Niña-like pattern in winter, and the response of the lower atmosphere to the SSTAs in the tropical Pacific exhibits an anomalous cyclonic circulation over the SCS and its vicinity (Fig. 2.33, upper panel), which is maintained until spring and then intensified (Fig. 2.33, middle panel). In May, there are obvious anomalous westerlies in the central and southern SCS, which corresponds to the early onset of SCSSM (Fig. 2.33, bottom panel). The distributions of SST and atmospheric circulation anomalies conform to the East Asia-Pacific teleconnection (Wang et al. 2000); thus, the lower SSTs in the central–eastern equatorial Pacific Ocean in winter affect the early SCSSM onset via East Asia-Pacific teleconnection.

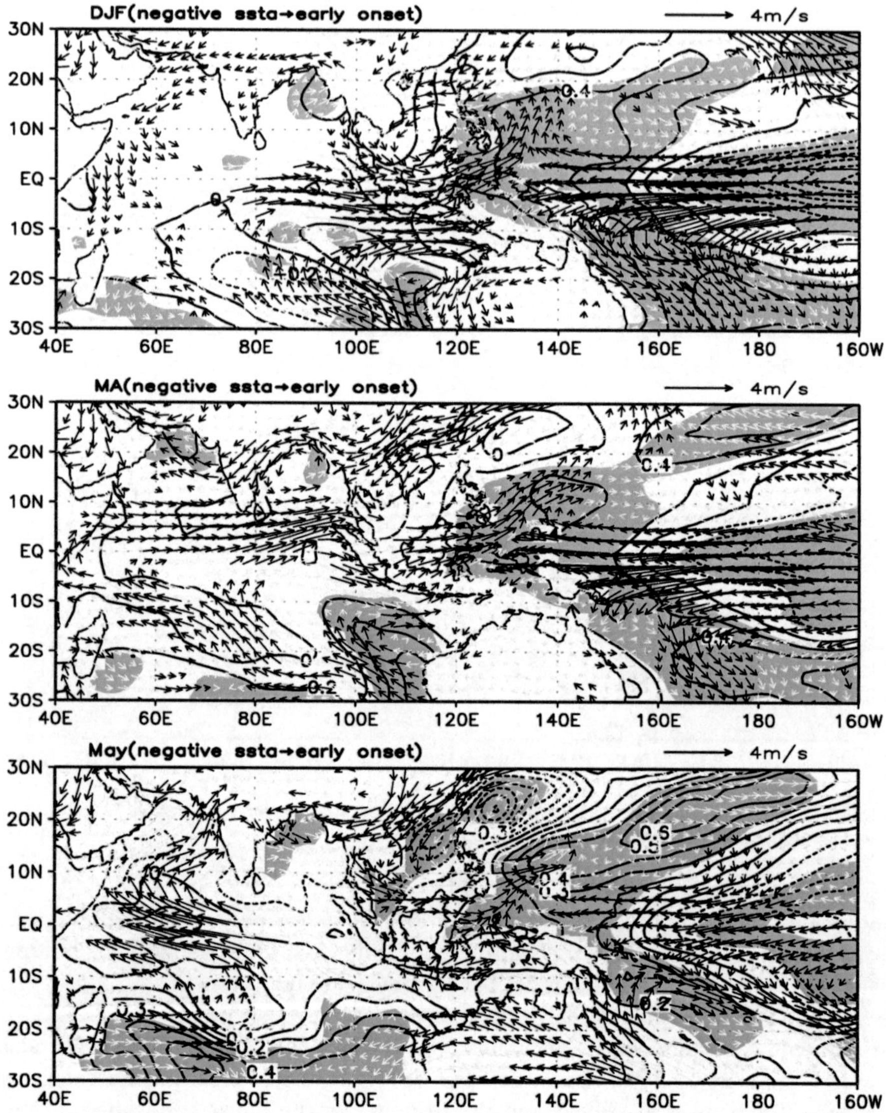

Fig. 2.33 Composite SST (contour; K) and 850-hPa circulation anomalies caused by strong negative anomalies in the central–eastern equatorial Pacific in December–February (upper panel), March–April (middle panel), and May (bottom panel). Shading indicates the 90% confidence level

2.5 Responses of Subtropical Highs to Global Warming

With the increase of GHG concentration in the atmosphere due to anthropogenic emission, global air temperature rises and the atmospheric circulation changes accordingly. As an important large-scale atmospheric circulation system over the

WNP, the WNPSH has a profound impact on the East Asian climate, especially in summer. Abnormal activity of the WNPSH can lead to great floods or droughts over large areas within East Asia. However, possible future change of the WNPSH under global warming is not well understood. Here, the responses of the WNPSH and the other subtropical anticyclones over the global subtropical oceans to the ongoing global warming are investigated, and the mechanisms for these responses are proposed (He and Zhou 2015; He et al. 2015, 2017). Given the possibility that the rising trend of global geopotential height under global warming might mislead the real-time monitoring of the WNPSH, new indices for measuring the WNPSH under a warming climate are proposed (He et al. 2018).

2.5.1 Model, Data, and Method

Many factors contribute to the climate change at the decadal time scale and beyond, including anthropogenic forcing (e.g., anthropogenic GHG, anthropogenic aerosols), natural forcing (e.g., solar insolation, orbital precession), and internal variability of the climate system (e.g., Pacific decadal oscillation, Atlantic multi-decadal oscillation). The observational and reanalysis data are too short to extract the forced response of the climate system from the internal variability, and the observational data in earlier decades are subject to large uncertainty. Coordinated by the World Climate Research Programme (WCRP), a large number of world modeling groups performed the same experiments using different climate system models, to investigate the response of the climate to anthropogenic forcing. Using a large number of models helps to suppress the internal variability of the climate system and random biases, and to extract the response of the climate system to anthropogenic forcing. For this study, we adopted a total of 33 models (Table 2.5) participating in the Coupled Model Intercomparison Project Phase 5 (CMIP5) for conducting the historical experiments, and the Representative Concentration Pathway 4.5 (RCP4.5) and RCP8.5 experiments to investigate the responses of the subtropical highs to global warming. The NCEP/NCAR reanalysis data set (Kalnay et al. 1996) is used for model evaluation.

Table 2.5 List of 33 CMIP5 models

Names of the CMIP5 models used for analysis				
ACCESS1.0	ACCESS1.3	BCC-CSM1.1	BNU-ESM	CanESM2
CCSM4	CESM1-BGC	CESM1-CAM5	CESM1-CAM5.1-FV2	CMCC-CM
CNRM-CM5	CSIRO-Mk3.6.0	FGOALS-g2	FGOALS-s2	GFDL-CM3
GFDL-ESM2G	GISS-E2-H	GISS-E2-R	HadGEM2-AO	HadGEM2-CC
HadGEM2-ES	INM-CM4	IPSl-CM5A-LR	IPSL-CM5A-MR	IPSL-CM5B-LR
MIROC-ESM	MIROC-ESM-CHEM	MIROC5	MPI-ESM-LR	MPI-ESM-MR
MRI-CGCM3	NorESM1-M	NorESM1-ME		

The RCP4.5 is a mitigation pathway toward a radiative forcing of 4.5 Wm^{-2} in the year 2100, equivalent to 650 ppm CO_2 concentration, and RCP8.5 is a business-as-usual pathway toward a radiative forcing of 8.5 Wm^{-2} in the year 2100, equivalent to 1370 ppm CO_2 concentration (Vuuren et al. 2011). In this section, the late twenty-first century outputs (referred to as the 21C) in RCP4.5 or RCP8.5 experiment are compared with the late twentieth century outputs (referred to as the 20C) in the historical experiment, and the multi-model ensemble of the difference between 21C and 20C is considered as the GHG-forced response. The supplementary analysis shows that the forced response of the WNPSH per degree of surface warming does not depend on the choice of scenarios, so we mainly focus on the RCP8.5 scenario. The statistical significance of the forced response relative to internal variability and stochastic model bias is evaluated by the inter-model consistency, which is defined as the percentage of models that agree in sign with the multi-model mean (MME). According to Power et al. (2012), an inter-model consistency of 68% is approximately equivalent to the 5% significance level based on Student's *t*-test.

2.5.2 Response of the WNPSH to Global Warming

The WNPSH is located in the lower to mid-troposphere, and its center tilts westward with height. The WNPSH is usually measured by geopotential height (H), and an anomalous high (low) geopotential height over the WNP is interpreted as anomalously strong (weak) WNPSH. The operational metric for monitoring the WNPSH is based on H at 500 hPa, and the 5880-m contour of H at 500 hPa is usually considered as the boundary of the WNPSH. As shown in Fig. 2.34a, the contour of $H =$ 5880 m is located near the eastern coast of East Asia. The Chinese Meteorological Administration defines the area index, intensity index, and western boundary index for the WNPSH based on the contour of $H = 5880$ m, and these indices are adopted in the real-time monitoring of WNPSH (Zhao 1999; Liu et al. 2014a, 2014b).

The WNPSH indices based on geopotential height can well describe the area, intensity, and western boundary if there were no climate change. However, Yang and Sun (2003) noted that the geopotential height is subject to a substantial long-term increasing trend under the warming climate, but the strength of the anticyclone circulation associated with the WNPSH does not show an enhancement. As consistently shown by multiple reanalysis data sets, the anticyclone circulation in the lower troposphere associated with WNPSH is weakened after the late 1970s, despite an increase of the geopotential height itself (Huang et al. 2015). The decadal weakening of WNPSH is consistent with the weakening of the EASM (Huang et al. 2015). At 500 hPa of the mid-troposphere, the contour of $H = 5880$ m expands westward substantially associated with a systematic increase of geopotential height at 500 hPa, but the anticyclone circulation associated with the WNPSH in the mid-troposphere shows a moderate weakening trend (Wu and Wang 2015). Therefore, the WNPSH in both lower and mid-troposphere was weakened in the past few decades, despite the increase of geopotential height at 500 hPa.

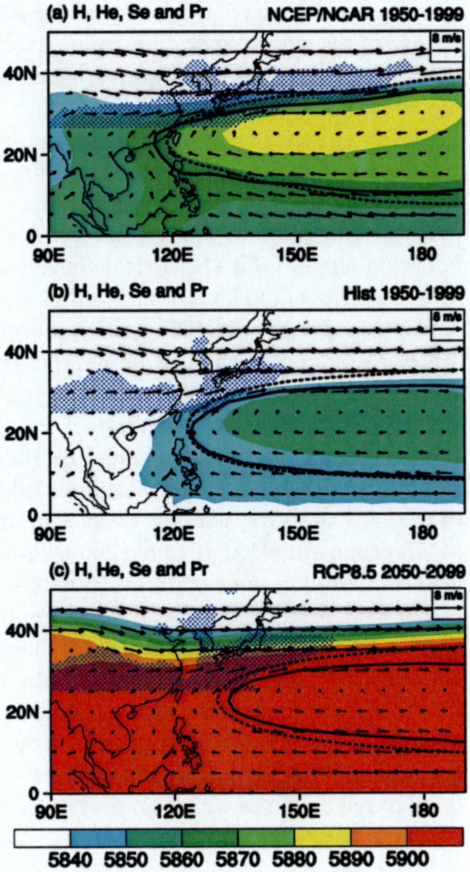

Fig. 2.34 The geopotential height (m) and wind (m s^{-1}) at 500 hPa over the western North Pacific in summer. **a** The mean state for 1950–1999 in NCEP/NCAR reanalysis data. **b** The mean state for the 20C based on the MME of the historical experiments by the CMIP5 models. **c** The mean state for the 21C based on the MME of RCP8.5 experiments by the CMIP5 models. The regions with a precipitation rate above 5 mm per day north of 25° N are stippled. The solid and dashed black contours are the zero contour for eddy geopotential height and eddy stream function, respectively. After He et al. (2015)

The above results show that the metrics based on geopotential height are not suitable for evaluating the WNPSH at decadal time scale and beyond. The theoretical basis for this fact is based on the hypsometric equation (the integration form of the hydrostatic equation) for the atmosphere. According to the hypsometric equation,

$$H(p) \infty T \ln(ps/p), \tag{2.1}$$

which states that the geopotential height (H) for a specific pressure level p is determined by two factors: one is the surface pressure (ps) and the other is the averaged temperature (T) between the surface and the pressure level p. As globally averaged ps is determined by the total mass of global atmosphere that is not impacted by global warming, geopotential height (H) is proportional to temperature (T) and will increase with T in a warming climate. However, the atmospheric circulation is determined by the horizontal gradient of H and has no relation to the absolute value of H; therefore, a uniform increase of H under global warming does not mean that high-pressure systems are enhanced and low-pressure systems are weakened. To obtain a robust change of the atmospheric circulation associated with the WNPSH, it is necessary to compare the changes in multiple variables and be cautious in interpreting the increase of geopotential height.

Wind is a direct measure of atmospheric circulation, but it is a vector and thus not suitable for quantitatively assessing changes in atmospheric circulation. Some recent works proposed to use eddy geopotential height (*He*) as a measure of the WNPSH in a warming climate (Zhou et al. 2009; Huang et al. 2015; Wu and Wang 2015). Focusing on the WNPSH, *He* is defined as the deviation of *H* from the synchronous averaged *H* over the latitudes 0°–40° N. For comparison, eddy stream function (*Se*) is also defined by subtracting the synchronous averaged stream function over 0°–40° N. We compare changes in multiple variables associated with the atmospheric circulation and changes in precipitation, to obtain a robust conclusion on the response of the WNPSH to global warming.

The climatological geopotential height and wind at 500 hPa over the WNP in summer from the observation and the MME of the coupled models are shown in Fig. 2.34. Compared with the mean state for 1950–1999 in the reanalysis data, the MME geopotential height of the 20C is lower, and the contour *of* $H = 5880$ m cannot be identified since none of the grid points has an *H* value higher than 5880 m. But the MME wind field for the 20C is consistent with the reanalysis result (Fig. 2.34a, b), and all the individual models capture the anticyclone wind pattern associated with the WNPSH (not shown). Therefore, the negative bias in the geopotential height for the models does not mean the models cannot capture the WNPSH, since the atmospheric circulation associated with the WNPSH is determined by the horizontal gradient of *H,* rather than by the absolute value of *H*. If the boundary of the WNPSH is measured by the zero contour of eddy geopotential height or eddy stream function, the simulated WNPSH by the MME is very close to that of the reanalysis (contours in Fig. 2.34a, b).

Compared with the 20C, the geopotential height rises substantially in the 21C of the RCP8.5 experiments (Fig. 2.34c). Despite a negative bias of the mean state geopotential height in the 20C of the historical simulations, the mean geopotential height in the 21C based on the RCP8.5 experiments is systematically above 5900 m south of 30° N. However, the zero contours for the eddy geopotential height and the eddy stream function retreat eastward by about 10° in longitude instead of extending westward. The subtropical rain belt over East Asia does not shift northwestward following the westward shift of the contour of $H = 5880$ m, but its location remains generally unchanged and it slightly expands eastward by about 10° in longitude, consistent with the eastward retreat of the zero contours for the eddy geopotential height and eddy stream function.

To demonstrate the response of the WNPSH under global warming, Fig. 2.35a shows the difference of the geopotential height and wind between the 21C in the RCP8.5 experiments and the 20C in the historical experiments. It is clear that the change in wind is characterized by a cyclonic wind anomaly surrounding the WNPSH center, suggesting the anticyclonic circulation associated with WNPSH is weakened. In addition, the westerly wind on the northern flank of WNPSH is substantially more

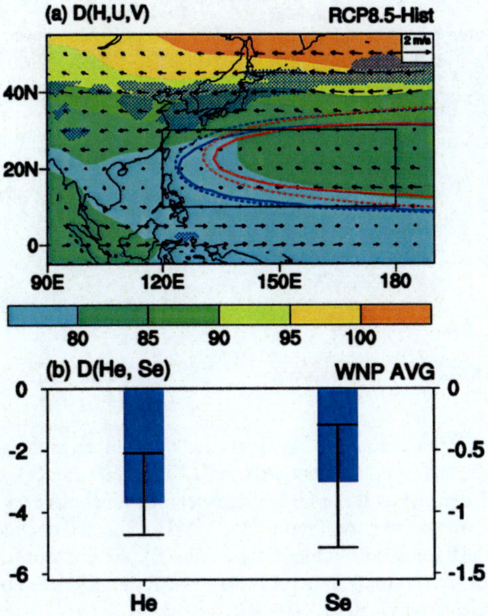

Fig. 2.35 **a** MME-projected change of geopotential height (shading; m) and wind (vector; m s^{-1}) at 500 hPa in the 21C relative to those in the 20C. Blue (red) solid and dashed contours are the zero contour for the eddy geopotential height and eddy stream function in the 20C (21C), respectively. **b** Projected changes in the WNPSH index by the MME (blue bar) and the thin black whiskers show the range of the 25th and 75th percentiles of individual models. The two WNPSH indices are the regionally averaged eddy geopotential height (*He*; m, left *y*-axis) and eddy stream function (*Se*; m^2 s^{-1}, right *y*-axis) over the black box in (**a**). After He et al. (2015)

weakened compared to the reduction of the easterly wind on the southern flank of WNPSH, making a major contribution to the weakened WNPSH. On the northern flank of WNPSH, the increase of geopotential height is not spatially uniform but features a meridional gradient, the magnitude of the rise in *H* increases with latitude, suggesting the mean state meridional gradient of geopotential height decreases under global warming. Therefore, the projected change in wind is consistent with the change of geopotential height gradient.

To quantitatively evaluate the change of WNPSH, we describe the intensity of WNPSH using the regionally averaged eddy geopotential height or eddy stream function within the box shown in Fig. 2.35a. The changes in these two variables (or indices) in 21C relative to 20C are shown in Fig. 2.35b. These two indices consistently show that the WNPSH is weakened under global warming based on the coupled models, and more than 75% of the individual models consistently show a tendency toward weakened WNPSH, corroborating the significance of the MME-projected change. The above results based on multiple variables suggest that the WNPSH at 500 hPa in the mid-troposphere is weakened under anthropogenic GHG forcing.

To investigate the mechanism for the weakened WNPSH under global warming, Fig. 2.36 shows the latitude-height section of the change in the zonal wind and

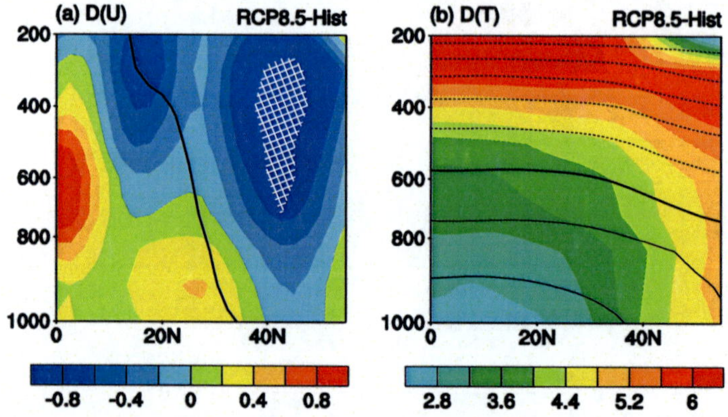

Fig. 2.36 Longitude-height section of the MME-projected changes in the zonal wind (**a**; shading, m s^{-1}) and temperature (b; shading, K) averaged over 120° E–180°. In (**a**), the black contour is the ridgeline of the WNPSH defined as the interface between tropical easterly wind and mid-latitude westerly wind in the 20C historical experiment. The MME-projected changes agreed by over 75% of the individual models are filled with white crosses. In (**b**), the contours are the isotherms in the 20C historical experiment; the contour interval is 10 K with the negative contours dashed, and the thick black contour is zero. After He et al. (2015)

temperature averaged within 120° E–180°. It is clear from that the weakened westerly wind on the northern flank of WNPSH is confined to the mid- to upper troposphere but insignificant in the lower troposphere (Fig. 2.36a). The amplitude of warming is horizontally uniform on the southern flank of WNPSH, possibly due to the mixing effect of equatorial waves. But the amplitude of warming shows a strong meridional gradient on the northern flank of WNPSH, with stronger warming in the higher latitudes, consistent with the so-called "polar amplification." As the mean state temperature decreases with latitude on the northern flank of WNPSH, the change in temperature suggests a weakened meridional temperature gradient (Fig. 2.36b). According to the thermal wind relationship, this weakened meridional temperature gradient can lead to the weakened westerly wind in the mid-troposphere.

The inter-model spread of the projected changes also supports the importance of meridional temperature gradient on the northern flank of WNPSH. As shown in Fig. 2.37, the models with a larger decrease of meridional temperature gradient project a larger decrease of westerly wind on the northern flank of WNPSH, with an inter-model correlation coefficient of 0.92. There is still a high inter-model correlation coefficient of 0.75 if the four outliers on the lower-left corner of Fig. 2.37 are removed, still significant at the 99% confidence level. The above findings based on MME and the inter-model spread confirm that the weakened WNPSH in the mid-troposphere under global warming is a result of the weakened meridional temperature gradient on the northern flank of WNPSH, constrained by the thermal wind relationship.

Under the anthropogenic GHG forcing, the response of WNPSH in the lower troposphere to global warming is different from that in the mid-troposphere. The change in the 850-hPa wind does not show any cyclonic or anticyclonic anomaly

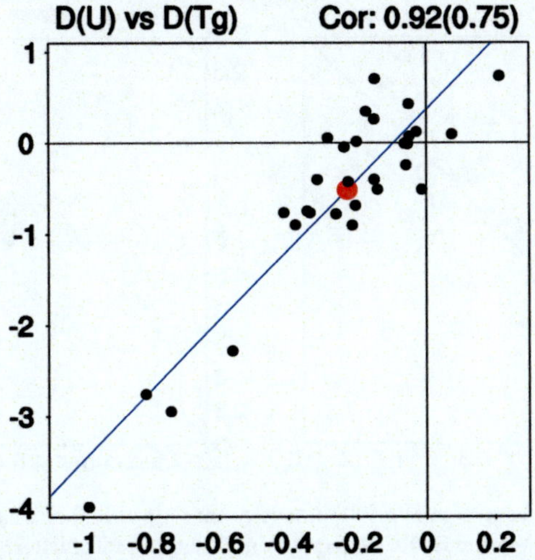

Fig. 2.37 Scatter plot for the projected changes in the zonal wind (m s^{-1}) on the northern flank of WNPSH (*y*-axis) as a function of the change in the meridional temperature gradient (K; *x*-axis) from the individual models. The zonal wind is averaged over (120° E–180°, 25°–40° N), and the meridional temperature gradient is defined as the difference in the regionally averaged temperature between (120° E–180°, 40°–50° N) and (120° E–180°, 15°–25° N). After He et al. (2015)

straddling the ridgeline of the WNPSH (Fig. 2.38), suggesting that the intensity of WNPSH at 850 hPa is generally unchanged in the MME projection. The geopotential height at 850 hPa also rises systematically, similar to that at 500 hPa. This again

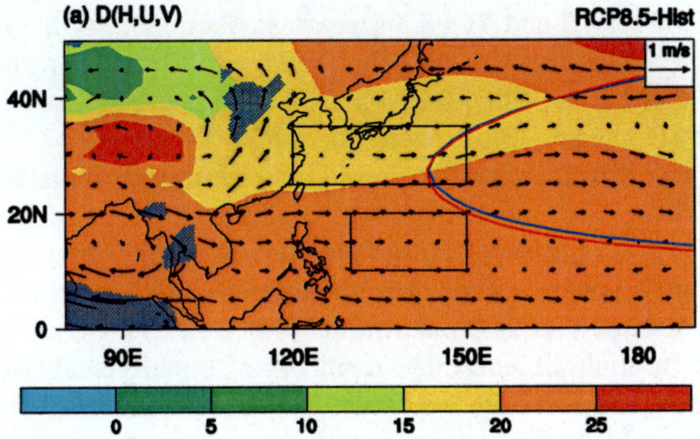

Fig. 2.38 MME-projected changes in the geopotential height (shading; m) and wind (vectors) at 850 hPa. Stippling indicates the changes in wind agreed by more than 70% of the individual models. Blue and red contours are the zero contour of eddy geopotential height for the 20C and the 21C, respectively. The difference in the regionally averaged zonal wind between the northern box (120°–150° E, 25°–35° N) and the southern box (130°–150° E, 15°–20° N) is used to measure the WNPSH intensity at 850 hPa, shown in Fig. 2.39

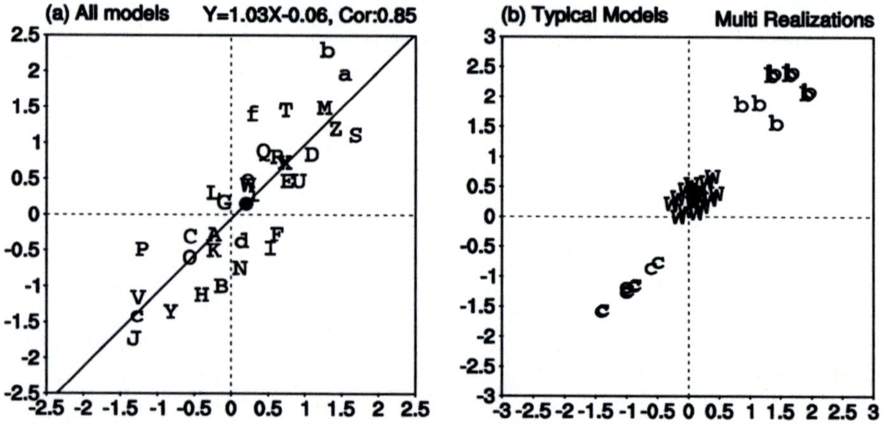

Fig. 2.39 **a** Projected changes of WNPSH intensity (m s^{-1}) at 850 hPa by the 33 models under RCP8.5 (*y*-axis) versus those under RCP4.5 (*x*-axis). The alphabetical letters indicate the individual models, and the black dot indicates the MME. **b** Projected changes in WNPSH intensity by the multiple realizations of three typical models, including IPSL-CM5A-LR (W), MIROC5 (**b**), and MPI-ESM-LR (**c**). After He and Zhou (2015)

suggests that it is not suitable to use geopotential height to measure the long-term change of the WNPSH.

To assess the inter-model and inter-scenario uncertainty in the projected change of the WNPSH at 850 hPa, the intensity of WNPSH at 850 hPa is measured by the anticyclonic zonal wind shear over the WNP, i.e., the difference in the regionally averaged zonal wind between the northern box and the southern box in Fig. 2.38. Figure 2.39a shows the projected change in the WNPSH at 850 hPa among individual models under the RCP4.5 and RCP8.5 scenarios. About 50% of the models project an increase of WNPSH intensity, and about 50% of the models project a decrease of WNPSH intensity, under either RCP4.5 or RCP8.5 scenario. The MME-projected change in WNPSH intensity is near zero under both scenarios.

Since only one ensemble member is used for each model, the inter-model differences in Fig. 2.39a could originate from two possible sources, i.e., different responses of different models to the same prescribed forcing, or the internal variability of the air–sea coupled system. Given that multiple ensemble members are available only for a few models, Fig. 2.39b shows the projected changes of WNPSH intensity at 850 hPa by multiple ensemble members of three typical models in which multiple ensemble members are available. These three models are IPSL-CM5A-LR, MIROC5, and MPI-ESM-LR, which show unchanged, enhanced, and weakened WNPSH intensity, respectively. It is evident that the diversity (or spread) is small among different ensemble members of a given model, but the results are quite different among different models (Fig. 2.39b). This suggests that the inter-model diversity in Fig. 2.39a is mainly due to different responses of different models to the same prescribed forcing, rather than due to internal variability of the system.

To investigate why different models respond differently to the same prescribed external forcing, inter-model regressions of the projected changes in SST, precipitation, and tropospheric temperature (TT) onto the projected changes in WNPSH intensity are shown in Fig. 2.40a and b. The models with an increase of 850-hPa WNPSH intensity are usually associated with stronger warming in the tropical Indian Ocean and weaker warming in the western tropical Pacific Ocean. The stronger warming in the tropical Indian Ocean is associated with a local positive precipitation anomaly and a warm tropospheric temperature anomaly emanating from the tropical Indian Ocean and extending into the Pacific Ocean (Fig. 2.40b), suggesting that the atmospheric anomalies are forced by warm SSTA in the tropical Indian Ocean. Previous studies revealed that warm (cold) SSTA in the tropical Indian Ocean favors an anomalous strong (weak) WNPSH at 850 hPa (Wu et al. 2009; Xie et al. 2009; Zhou et al. 2009), consistent with our results shown here that a stronger amplitude of SST warming in the tropical Indian Ocean favors an increase of WNPSH intensity under global warming, and vice versa.

To further quantify how the changes in 850-hPa WNPSH intensity are associated with the change in tropical SST, Fig. 2.40c–e shows the inter-model relation between

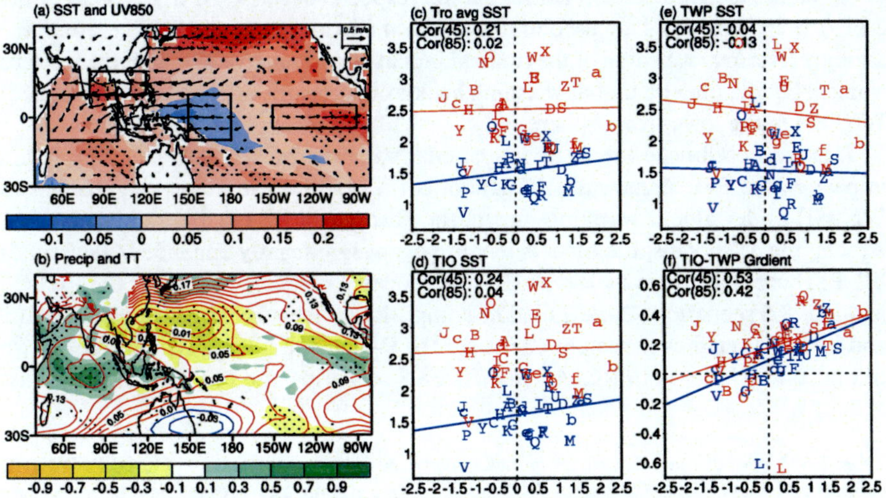

Fig. 2.40 **a** Inter-model regression of projected changes in SST (shading; K) and 850-hPa wind onto projected changes in the intensity of 850-hPa WNPSH index. **b** Inter-model regression of projected changes in precipitation (shading; mm day^{-1}) and tropospheric temperature (contour) onto projected changes in the intensity of 850-hPa WNPSH index. Stippling in **a** and **b** indicates that the regressed SST and precipitation, respectively, exceed the 95% confidence level based on the Student's *t*-test. **c–f** Scatter diagrams for the inter-model relationship between changes in the intensity 850-hPa WNPSH (*x*-axis) and changes in the SST over key oceanic regions (*y*-axis). The *x*-axis in **c–f** is the changes in averaged SST in the tropics (**c**), averaged SST in the tropical Indian Ocean (**d**), averaged SST in the western tropical Pacific (**e**), and the zonal SST gradient between tropical Indian Ocean and western tropical Pacific (**f**). Dots and least square regression line in blue (red) are for the RCP4.5 (RCP8.5) scenario. After He and Zhou (2015)

the changes in WNPSH intensity with the changes in SST in the entire tropics, the tropical Indian Ocean, and the western tropical Pacific. None of the inter-model correlation coefficients are significant for these three key regions under either RCP4.5 or RCP8.5 scenario (Fig. 2.40c–e). However, the change in WNPSH intensity is significantly correlated with the change in zonal SST gradient between the tropical Indian Ocean and western tropical Pacific (Fig. 2.40f), with an inter-model correlation coefficient of 0.42 under RCP4.5 and 0.53 under RCP8.5, above the 95% confidence level. If the outlier model of CSIRO-Mk3.6.0 (represented as "L" in Fig. 2.40f) is removed, the above correlation coefficients rise to 0.64 and 0.57, respectively, both exceeding the 99% confidence level. These results further suggest that the change in the low-level WNPSH is associated with the change in the zonal SST contrast between the tropical Indian Ocean and the western tropical Pacific Ocean.

The modulation of the Indo-Pacific zonal SST gradient on WNPSH intensity can be confirmed by the idealized experiments of Community Atmospheric Model version 4 (CAM4). If CAM4 is forced by both positive SSTAs in the tropical Indian Ocean and in the western tropical Pacific, with the SSTA in the tropical Indian Ocean being stronger, the model results show that the WNPSH is enhanced. In the inter-model regressed SSTA shown in Fig. 2.40a, positive SSTAs are seen from the BOB to SCS and eastern tropical Pacific, but the response of the WNPSH is weak if the SSTAs in these regions are used to force CAM4 (He and Zhou 2015). Therefore, the stronger (weaker) warming of the tropical Indian Ocean relative to that of the western tropical Pacific under a global warming background favors an enhanced (weakened) WNPSH in the lower troposphere.

It is well established that the interannual variation of WNPSH has a profound impact on the East Asian rainfall pattern, but it is still unclear how the changes in WNPSH under global warming impact the changes in East Asian rainfall pattern. Among the 33 models, the models that project a significantly enhanced (weakened) WNPSH are referred to as P-type (N-type models), respectively, under the RCP4.5 and RCP8.5 scenarios (Table 2.6). The composite projected changes in precipitation and 850-hPa wind over eastern China for the P-type and N-type models are shown in Fig. 2.41. Under both RCP4.5 and RCP8.5 scenarios, the P-type models consis-

Table 2.6 List of the models with a significant increased (decreased) intensity of 850-hPa WNPSH under RCP4.5 and RCP8.5 scenarios. The models with a significant increase (P-type) or decrease (N-type) in WNPSH intensity under both RCP4.5 and RCP8.5 scenarios are marked in boldface

	P-type	N-type
RCP4.5	BNU-ESM, **FGOALS-g2**, HadGEM2-AO, **IPSL-CM5A-MR, MIROC-ESM, MIROC-ESM-CHEM, MIROC5**	**CMCC-CM, inmcm4, IPSL-CM5B-LR, MPI-ESM-LR**
RCP8.5	**FGOALS-g2**, GISS-E2-H, GISS-E2-R, HadGEM2-CC, **IPSL-CM5A-MR, MIROC-ESM, MIROC-ESM-CHEM, MIROC5**, NorESM1-M	**CMCC-CM, inmcm4, IPSL-CM5B-LR, MPI-ESM-LR**

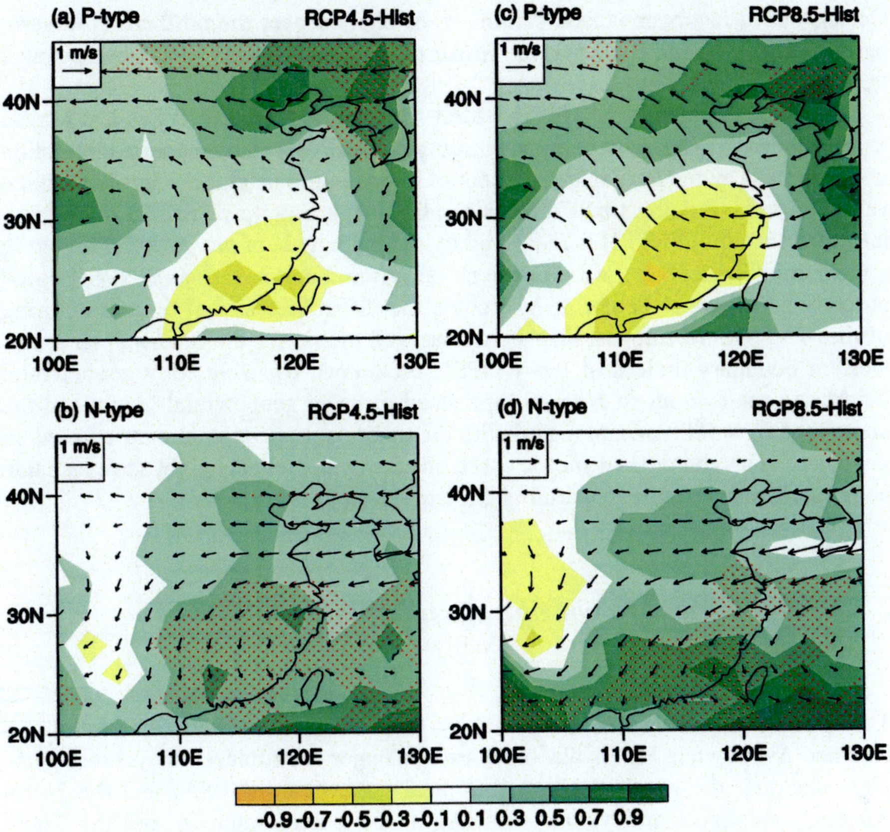

Fig. 2.41 Composite changes in summer rainfall (shading; mm day^{-1}) and 850-hPa wind (vectors; m s^{-1}) based on the models with a significant enhancement of 850-hPa WNPSH under RCP4.5 **a** and RCP8.5 **c** scenarios. **b, d** Same as (**a, c**), except for the composite of the models with a significant reduction of 850-hPa WNPSH intensity. Stippling indicates all of the P-type (or N-type) models agree in the sign of changes in the precipitation. After He and Zhou (2015)

tently project enhanced precipitation over the northern part of eastern China, associated with strengthened monsoon southerly wind, but the N-type models consistently project enhanced precipitation over the southern part of eastern China, associated with weakened monsoon southerly wind.

Based on the above analyses, the WNPSH is weakened in the mid-troposphere but shows great inter-model uncertainty in the lower troposphere under anthropogenic GHG forcing. The weakened WNPSH in the mid-troposphere results from the weakened meridional temperature gradient constrained by the thermal wind relationship, and the change in WNPSH in the lower troposphere depends on the changes in the zonal SST gradient between the tropical Indian Ocean and tropical WNP. The WNPSH tends to be enhanced (weakened) in the models with a stronger (weaker) warming over the tropical Indian Ocean than that over the western tropical Pacific

Ocean, which favors enhanced summer precipitation over the northern (southern) part of eastern China. However, the MME of the changes in WNPSH in the lower troposphere is nearly zero.

Although geopotential height is a widely used metric for the WNPSH, it is subject to a systematic rising trend under a warming climate. Since atmospheric circulation is determined by the horizontal gradient of geopotential height, not by its absolute value, the strong rising trend of geopotential height may give rise to misleading information if the WNPSH is measured by the magnitude of geopotential height in a warming climate. Here, we propose an operational metric for the WNPSH based on eddy geopotential height, to overcome the difficulties in real-time monitoring of the WNPSH. We suggest that the operational metric for the intensity, area, and western boundary indices of the WNPSH be defined by using eddy geopotential height and the contour of $He = 0$ m, instead of using geopotential height and the contour of $H = 5880$ m. Compared with the traditional operational metric based on geopotential height, the new metric based on eddy geopotential height shows a much better performance under a warming climate (He et al. 2018).

2.5.3 Responses of Subtropical Anticyclones to Global Warming

The subtropical oceans are controlled by large-scale anticyclonic circulations in summer. As shown in Fig. 2.42a, there are five major subtropical anticyclones globally, including the North Pacific subtropical anticyclone (NPSA) and the North Atlantic subtropical anticyclone (NASA) in the boreal summer, and the South Pacific subtropical anticyclone (SPSA), the South Atlantic subtropical anticyclone (SASA), and the South Indian Ocean subtropical anticyclone (SISA) during the austral summer. The WNPSH is the western section of the NPSA. As consistently shown by reanalysis data and model simulations, the anticyclonic vorticity spreads over the entire subtropical oceans, but the subsidence and low-level divergence are confined to the eastern part of the subtropical ocean (Fig. 2.42).

The projected changes in wind, relative vorticity, divergence at 925 hPa, and the vertical velocity at 700 hPa are shown in Fig. 2.43. Over the North Pacific Ocean, the projected change in wind is characterized by westerly wind anomaly south of 30° N, easterly wind anomaly near 40° N, and southerly wind anomaly along the western coast of North America (Fig. 2.43a). These wind anomalies constitute an anomalous cyclone around the center of NPSA, suggesting that the anticyclone circulation associated with NPSA is weakened under global warming. The projected change in relative vorticity at 925 hPa is characterized by positive anomalies along 30° N over the subtropical North Pacific Ocean, opposing the mean state negative relative vorticity (Fig. 2.43b), consistent with the weakened anticyclonic circulation. The projected changes in subsidence and low-level divergence are also opposite to their

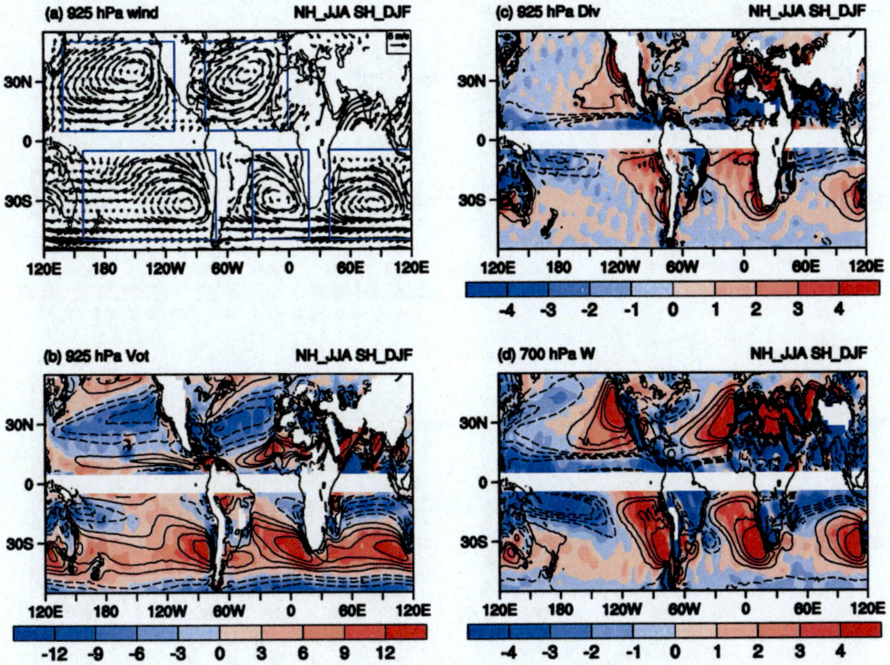

Fig. 2.42 Observed and simulated MME mean state for the 20C, showing the boreal summer (JJA) for the Northern Hemisphere and the austral summer (DJF) for the Southern Hemisphere. The variables include **a** wind (m s^{-1}) at 925 hPa, **b** relative vorticity (10^{-6} s^{-1}) at 925 hPa, **c** divergence (10^{-6} s^{-1}) at 925 hPa, and **d** vertical velocity (10^{-2} Pa s^{-1}) at 700 hPa. The wind in **a** and the shading in **b–d** are based on the NCEP/NCAR reanalysis data, and the contours in **b–d** are based on the MME of the simulations. After He et al. (2017)

mean states (Fig. 2.43c, d). All these results suggest that the intensity of NPSA is weakened under global warming.

The projected change in the wind of the subtropical North Atlantic is characterized by an anomalous cyclonic curl off the western coast of Africa, and an anticyclone anomaly in the central to the western part of the subtropical North Atlantic Ocean. The projected changes in relative vorticity, divergence at 925 hPa, and vertical velocity at 700 hPa consistently show that the subsidence, anticyclone wind curl, and low-level divergence are weakened at the southeastern part of the subtropical North Atlantic Ocean, but enhanced over the northwestern part. These changes suggest that NASA shifts northwestward under global warming, consistent with the finding of Shaw and Voigt (2015). It is difficult to say, however, if NASA is enhanced or weakened based on Fig. 2.43.

In the subtropical South Pacific during the austral summer, the projected changes in relative vorticity, divergence, and vertical velocity consistently suggest that the NPSA is weakened in the eastern part of the ocean but enhanced in the central part of the ocean. These results suggest that the SPSA shifts westward under global warming.

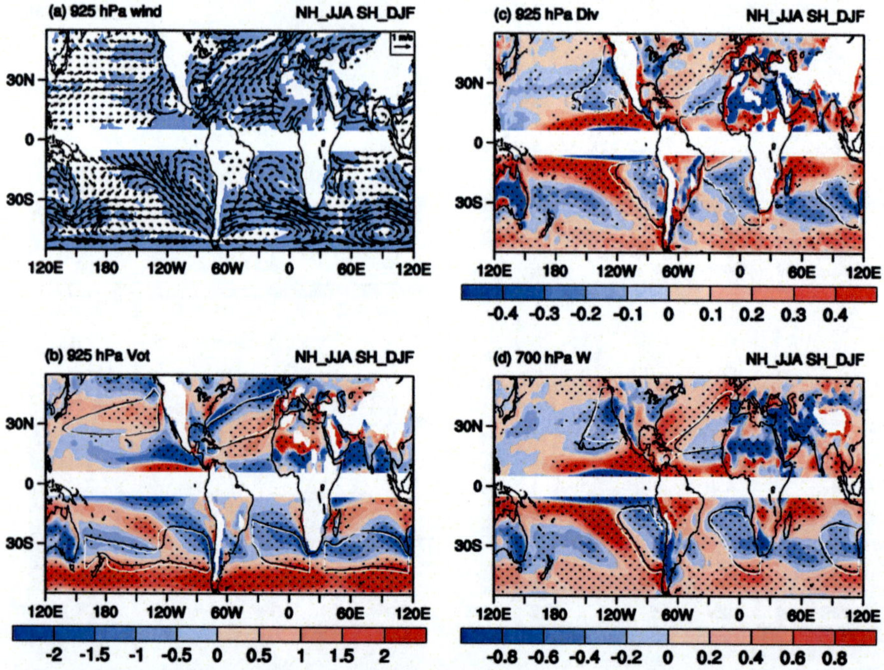

Fig. 2.43 Differences between 21C of RCP8.5 experiments and 20C of the historical experiments, for wind (m s^{-1}) at 925 hPa (**a**), relative vorticity (10^{-6} s^{-1}) at 925 hPa (**b**), divergence (10^{-6} s^{-1}) at 925 hPa (**c**), and vertical velocity (10^{-2} Pa s^{-1}) at 700 hPa (**d**). The MME-projected changes agreed by more than 70% of the individual models are indicated by shading in (**a**) and by stippling in (**b–d**). Black and white contours in **b–d** are the characteristic contours for the boundary of the subtropical anticyclones in the 20C and 21C, respectively, including the $6 \times 10^{-6} s^{-1}$ contour for relative vorticity ($-6 \times 10^{-6} s^{-1}$ for the Northern Hemisphere), the $1 \times 10^{-6} s^{-1}$ contour for divergence, and the 2×10^{-2} Pa s^{-1} for vertical velocity. After He et al. (2017)

In the subtropical South Atlantic Ocean, the signs of the projected changes in relative vorticity, divergence, and vertical velocity are all opposite to their mean states, suggesting that both SASA and SISA are weakened under global warming. Meanwhile, the projected changes in relative vorticity, divergence, and vertical velocity consistently have the same sign as their mean states over a small area on the southern tip of the subtropical anticyclone (near 40° S), suggesting a poleward shift of the subtropical anticyclone over the Southern Hemisphere under global warming.

To quantitatively measure the intensity changes for the five subtropical anticyclones, we introduce an intensity index. According to the Stokes law of circulation, the strength of the rotational flow along the boundary of a domain equals the integrated vorticity within the domain, and the strength of the divergent flow out of the boundary of a domain equals the integrated divergence within the domain. The location and boundary for each subtropical anticyclone are identified over each basin for the 20C and 21C, respectively, shown as the black (white) contours for the 20C (21C) in Fig. 2.43b–d. Different domains are used for the 20C and 21C in a contour-following perspective, to extract the change of intensity of the subtropical anticyclones independent of their location changes. The characteristic contours taken as the boundary of the subtropical anticyclones are the $6 \times 10^{-6}\ s^{-1}$ contour for relative vorticity ($-6 \times 10^{-6}\ s^{-1}$ contour in the Northern Hemisphere), the $1 \times 10^{-6}\ s^{-1}$ contour for divergence, and the 2×10^{-2} Pa s^{-1} contour for vertical velocity. The Vot925 index for the subtropical anticyclone over each basin is defined as the integrated relative vorticity within the domain enclosed by the black (white) contour in Fig. 2.43b for the 20C (21C) over each basin. The Div925 index for the subtropical anticyclone over each basin is defined as the integrated divergence within the domain enclosed by the black (white) contour in Fig. 2.43c for the 20C (21C) over each basin. The W700 index is defined in a similar fashion.

The percentage changes in the intensity of the five subtropical anticyclones based on the Vot925 index, the Div295 index, and the W700 index are shown in Fig. 2.44, scaled and unscaled by the amplitude of global warming. These three indices consistently show that the NPSA, SASA, and SISA are weakened, whereas NASA and

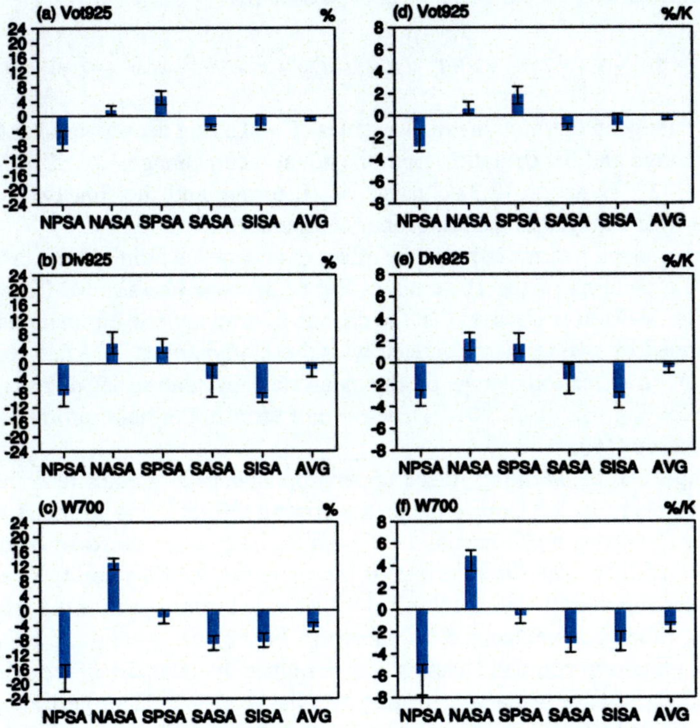

Fig. 2.44 Percentage (%) of intensity change for the five subtropical anticyclones, based on the Vot925 index (**a**), Div925 index (**b**), and W700 index (**c**). **d–f** Same as (**a–c**), except the projected changes are scaled by the amplitude of tropical warming (30° S–30° N; % K^{-1}). The blue bars indicate the MME, and the thin black whiskers indicate 30% and 70% percentiles of the individual models, respectively. After He et al. (2017)

SPSA are enhanced under global warming. The averaged intensities for the five subtropical anticyclones are reduced based on all three indices. There is some uncertainty for the SPSA, in which the change in the W700 index is inconsistent with the changes in the Vot925 and Div925 indices. Among the five subtropical anticyclones, the reduction of the NPSA intensity is the largest, reaching about 5% per degree of the mean state warming.

To investigate the mechanism for the changes in the intensity of the subtropical anticyclones, diagnoses are performed based on the thermodynamic equation of the atmosphere. Since we only focus on the mean state, the temporal tendency terms can be omitted. The resulting thermodynamic equation can be written as follows:

$$-\mathbf{V}\cdot\nabla T + Sw + Q = 0 \tag{2.2}$$

where $\mathbf{V}, T, S, \omega$, and Q represent horizontal wind vector, temperature, static stability, vertical velocity, and diabatic heating, respectively. Designating the climatology in 20C with a bar and the difference between 21C and 20C with a prime, the difference between 21C under RCP8.5 and 20C is expressed by

$$-\mathbf{V}'\bullet\nabla\bar{T} + \bar{S}\omega' - \bar{\mathbf{V}}\bullet\nabla T'_m - \bar{\mathbf{V}}\bullet\nabla T'_e + S'_m\bar{\omega} + S'_e\bar{\omega} - \mathbf{V}'\bullet\nabla T' + S'\omega' + Q' + \mathrm{Res} = 0 \tag{2.3}$$

where the change in thermodynamic variables (T and S) are decomposed as the zonal mean of change and the deviation from the zonal mean change, i.e., $T' = T'_m + T'_e$, $S' = S'_m + S'_e$. Equation (2.3) consists of 10 terms, and the first two terms are associated with changes in the atmospheric circulation.

To examine which terms balance the effect of changes in atmospheric circulation, we show the patterns of the 10 terms in Eq. (2.3), based on the MME of CMIP5 models (Fig. 2.45). It is obvious that the change of atmospheric circulation is essentially balanced by two terms associated with thermodynamics. The first term is the advection of stratification change by the mean state circulation ($S'_m\bar{\omega}$, referred to as MASC following Ma et al. 2012). The second term is the nonuniform change of diabatic heating (Q').

The amplitude of warming in the upper troposphere is greater than that in the lower troposphere in the tropics under a warming climate, due to moist adiabatic adjustment (Knutson and Manabe 1995), giving rise to an increase of the mean state static stability. The direct effect of the increased static stability is enhanced adiabatic heating over the climatological subsidence region, even if the intensity of descending motion is unchanged, as shown in Fig. 2.45e. According to Ma et al. (2012), the enhanced adiabatic heating acts to reduce the intensity of the subsidence. Meanwhile, the change in diabatic heating over the subtropical Northern Hemisphere is characterized by a zonal wavenumber-one pattern, with enhanced Q' over the central–western subtropical Pacific and weakened Q' over the subtropical Atlantic, and Q' is substantially weakened over the central subtropical South Pacific Ocean (Fig. 2.45i).

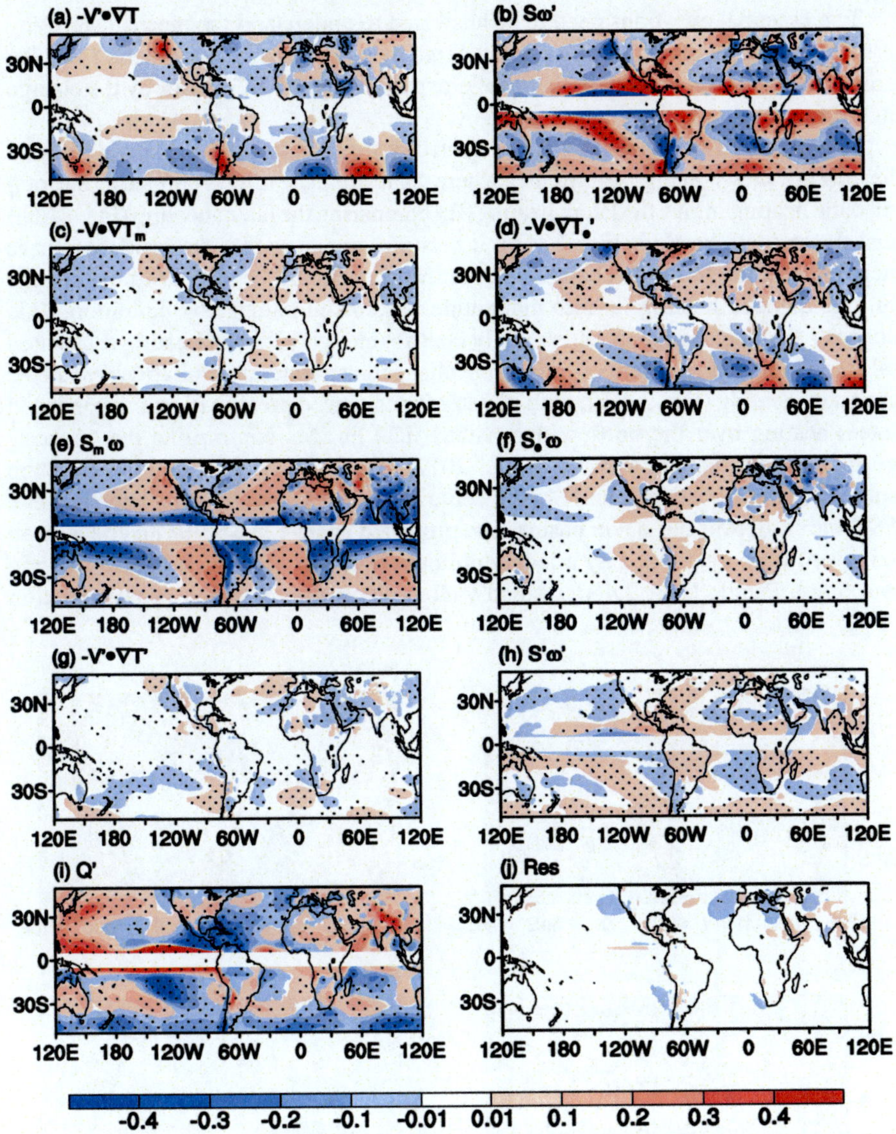

Fig. 2.45 Patterns for the vertical integration of each of the 10 terms in Eq. (2.3) (K day^{-1}), as diagnosed by the MME of the differences between the 21C in RCP8.5 experiments and the 20C in the historical experiments. Stippling indicates the projected sign of change is agreed upon by more than 70% of the models. After He et al. (2017)

Two essential questions need to be answered to understand the mechanism. Why is the change in diabatic heating characterized by a strong zonal asymmetry? Can the change of the subtropical anticyclones be explained by MASC and/or by the change in diabatic heating?

To answer the first question, the projected changes in latent heating and residual heating are shown in Fig. 2.46a and b, where the residual heating is defined as the total diabatic heating minus the latent heating. By comparing the latent heating (Fig. 2.46a) and the residual heating (Fig. 2.46b), it becomes apparent that the residual term is negligible. Therefore, the change in latent heating should be similar to the change in total diabatic heating in both magnitude and spatial pattern, by definition. This suggests that the change in Q' is mainly contributed by the latent heating associated with changes in precipitation. In fact, the change in summer precipitation resembles the latent heating (Fig. 2.46c), with a center for enhanced precipitation and enhanced latent heating over the northwestern subtropical Pacific, conforming the "richest-get-richer" mechanism (Zhang and Li 2017). The reduced precipitation over the subtropical North Atlantic and South Pacific can be explained by the relatively weak SST warming over these two basins. The projected change in SST (with zonal mean removed) is characterized by a local minimum over the subtropical North Atlantic and South Pacific (Fig. 2.46d), which well explains the reduction of precipitation

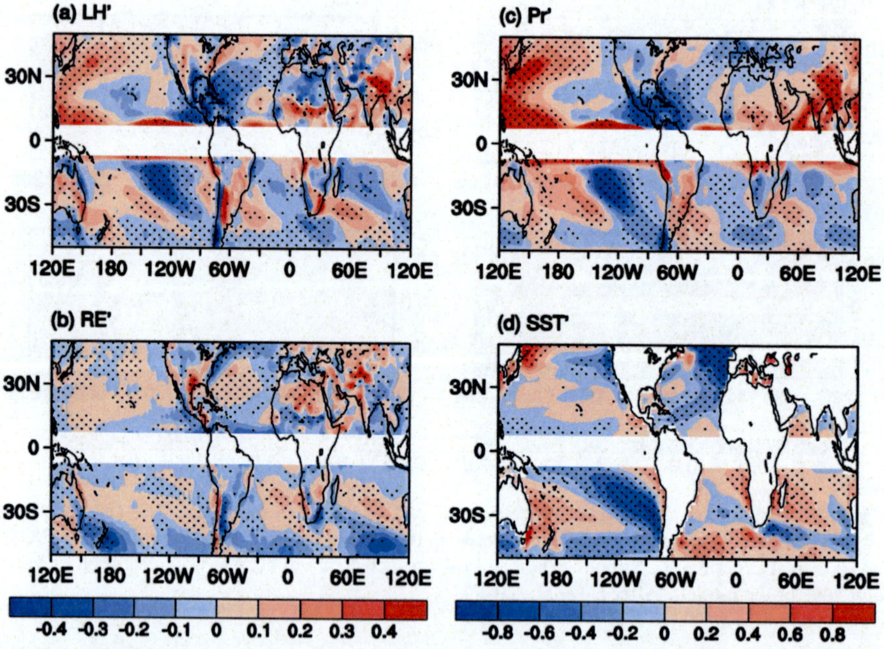

Fig. 2.46 MME-projected changes in latent heating (a; K day^{-1}), residual heating (b; K day^{-1}), precipitation (c; mm day^{-1}), and SST (d; K) of 21C with respect to 20C. The contemporary zonal mean value is removed for SST before calculating its change. Stippling indicates the projected sign of change is agreed upon by more than 70% of the models. After He et al. (2017)

over these two ocean basins, conforming the "warmer-get-wetter, colder-get-drier" mechanism (Xie et al. 2010).

To answer the second question, we look at the relative contributions of MASC and change in diabatic heating (*Q*') on the changes in the subtropical anticyclone over each ocean basin. To proceed, a linear baroclinic model (LBM) is adopted to examine the response of atmospheric circulation to prescribed heating. The 3-dimensional anomalies of MASC and *Q*' (whose vertical integrated patterns are shown in Fig. 2.44) are prescribed to force the LBM. The LBM is integrated for 50 days to reach its equilibrium. The equilibrium responses of wind and relative vorticity at 925 hPa are shown in Fig. 2.47a–c, and the quantitative change of the intensity for the subtropical anticyclone is shown in Fig. 2.47d. Under the forcing of MASC, the intensities of all five subtropical anticyclones are weakened. Under the forcing of *Q*', the NASA and SPSA are enhanced, the NPSA is weakened, and the changes for the other two subtropical anticyclones are unclear. NPSA shows the strongest tendency toward weakening because both MASC and *Q*' act to weaken it. Over the North Atlantic and the South Pacific, the effect of *Q*' overwhelms MASC, and the subtropical anticyclones over these two ocean basins are enhanced. The SASA and SISA are weakened because MASC dominates over these two ocean basins whereas the contribution from *Q*' is small.

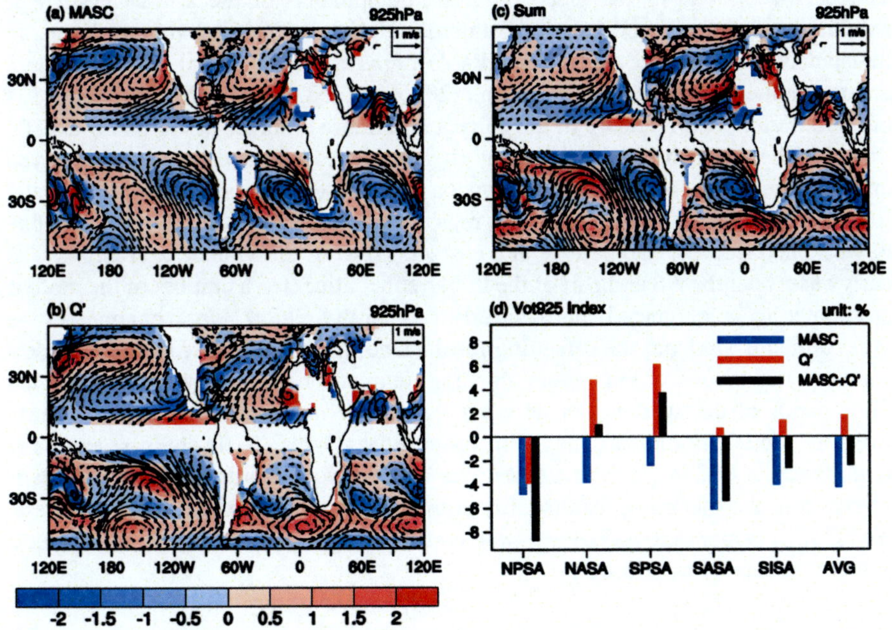

Fig. 2.47 Responses of wind and relative vorticity (10^{-6} s^{-1}) at 925 hPa to MASC **a** and *Q*' **b** simulated by the LBM. **c** The sum of the responses shown in **(a)** and **(b)**. **d** The simulated responses of the intensity of each subtropical anticyclone to the forcing of MASC and Q' and their sum. After He et al. (2017)

In summary, the responses of the subtropical anticyclones to global warming under anthropogenic GHG forcing are controlled by two factors: the enhanced static stability of the troposphere and the change in diabatic heating. The enhanced static stability of the tropical troposphere acts to induce stronger adiabatic heating over the subsidence region, and it acts to weaken all five subtropical anticyclones. The nonuniform change of diabatic heating is dominated by latent heating associated with the nonuniform changes in precipitation. The diabatic heating is enhanced over the northwestern subtropical Pacific possibly due to the "richest-get-richer" mechanism, and it further weakens the NPSA. The amplitude of SST warming is characterized by a local minimum over the subtropical North Atlantic and the subtropical South Pacific, which favors reduced local precipitation and latent heating and further enhances the subtropical anticyclones over these two ocean basins. The combined effect of increased static stability and the change in diabatic heating is the enhanced NASA and SPSA and the weakened NPSA, SASA, and SISA, with the largest reduction in the intensity of NPSA.

2.6 Response of SCSSM Onset to Global Warming

The SCSSM is one of the most important components of the EASM system. It exhibits pronounced interannual and interdecadal variation. Global warming is an indisputable fact, and it exerts remarkable impacts on the global climate system and atmospheric circulation. Dong et al. (2016) studied the CMIP5-simulated ASM onset, duration, and intensity in the current and future climates based on 26 CMIP5 models. Their results show that a majority of the models tend to suggest delayed onset for the South Asian summer monsoon in the eastern tropical Indian Ocean and Indochinese Peninsula and its nearby region under global warming, primarily due to weakened tropical circulations and eastward shift of the Walker circulation. The early onset over the AS and part of the Indian subcontinent in a number of the models are related to an enhanced southwesterly flow in the region. Weak changes in the other domains are from the offsetting results among the models, with some models showing earlier onsets but others showing delayed onsets. There is a significant discrepancy of the averaged onset dates among the models, and large uncertainty exists in model-simulated changes in onset/retreat dates. Their study stresses the importance of SST warming patterns over both tropical Pacific and Indian Oceans in affecting the modeling results. However, as a component of the EASM system, the SCSSM response to global warming is still not well understood in many respects and is the focus of this section.

Table 2.7 List of 22 CMIP5 models

BCC-CSM1.1	BCC-CSM1.1 M	BNU-ESM	CanESM2	CCSM4
CNRM-CM5	CSIRO-Mk3.6.0	FGOALS-g2	GFDL-CM3	GFDL-ESM2G
HadGEM2-CC	INM-CM4	IPSl-CM5A-LR	IPSL-CM5A-MR	IPSL-CM5B-LR
MIROC5	MIROC-ESM	MPI-ESM-LR	MPI-ESM-MR	MRI-CGCM3
MRI-ESM1	NorESM1-M			

2.6.1 Models, Data, and Method

To analyze the response of SCSSM onset to global warming, we adopt a total of 22 CMIP5 models (Table 2.7). We use their historical experiments and RCP8.5 experiments (Taylor et al. 2012). Detailed descriptions of models and data are available at http://cmippcmdi.llnl.gov/cmip5/guide_to_cmip5.html. The late twenty-first century (2050–2099) outputs of RCP8.5 experiments are compared with the late twentieth century (1950–1999) outputs of historical experiments, and the multi-model ensemble of the difference between RCP8.5 and historical experiments is considered as the response to global warming. The ERA-40 reanalysis data for 1958–2001 (Uppala 2002) provided by the European Centre for Medium-Range Weather Forecasts (ECMWF) are adopted for a brief evaluation of the model results.

In the present study, starting from the definition of SCSSM onset by Lin et al. (2013a), we first revise the criteria for the SCSSM onset as follows: based on the zonal wind, when both 850-hPa westerlies and 250-hPa easterlies prevail simultaneously and continuously over the domain of the SCS (110°–120° E, 5°–17.5° N) for five days or longer, the first day is defined as SCSSM onset date.

2.6.2 SCSSM Onset in CMIP5 Historical Experiments

The SCSSM is marked by the southwesterly and northeasterly prevailing at lower and upper levels over the SCS, respectively. Figure 2.48 shows the model-simulated mean seasonal evolution of 850- and 250-hPa zonal winds (hereafter referred to as U850 and U250, respectively) over the SCSSM domain. The ERA-40 reanalysis data (marked with "Obs") show that U850 turns from easterly to westerly at Pentad 28 (May 16–20) climatologically, while U250 changes from westerly to easterly at Pentad 24 in spring. This seasonal evolution of zonal winds at low and upper levels from observations is mostly reproduced by MME average of the 22 CMIP5 models, although there are differences in the timing of zonal wind reversal and the amplitude of zonal wind itself. There are some discrepancies among these models. For example, some models, including MRI-CGM3, MRI-ESM1, BCC-CSM1-1-M, CSIRO-MK3.6.0, HADGEM2-CC, and IPSL-CM5B-LR have a large simulated amplitude of U850.

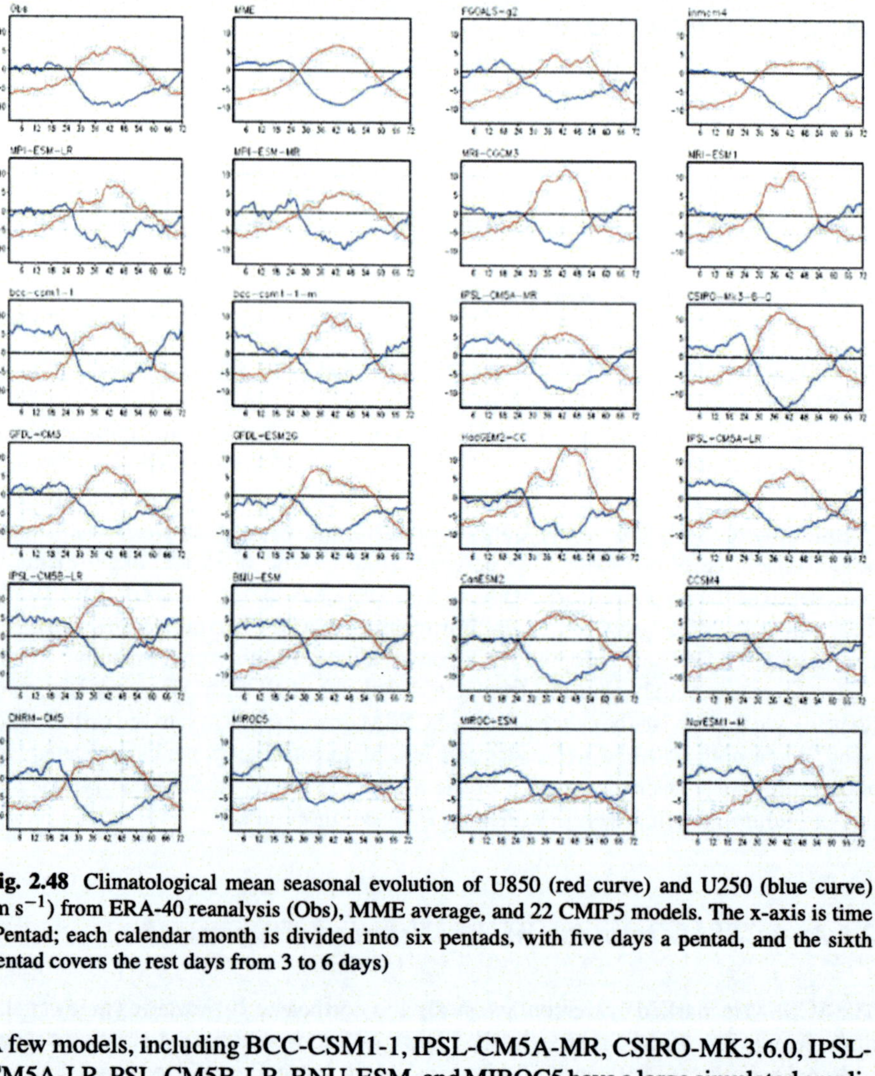

Fig. 2.48 Climatological mean seasonal evolution of U850 (red curve) and U250 (blue curve) (m s^{-1}) from ERA-40 reanalysis (Obs), MME average, and 22 CMIP5 models. The x-axis is time (Pentad; each calendar month is divided into six pentads, with five days a pentad, and the sixth pentad covers the rest days from 3 to 6 days)

A few models, including BCC-CSM1-1, IPSL-CM5A-MR, CSIRO-MK3.6.0, IPSL-CM5A-LR, PSL-CM5B-LR, BNU-ESM, and MIROC5 have a large simulated amplitude of U250 before the reversal from westerly to easterly. MIROC-ESM presents poor simulation of both U850 and U250 seasonal evolution. Therefore, this model is removed from the 22 CMIP5 models for the rest of this study. Because the RCP8.5 daily data from the CCSM4 is unavailable, the CCSM4 is also removed from further discussion. As a result, we use 20 CMIP5 models in this section.

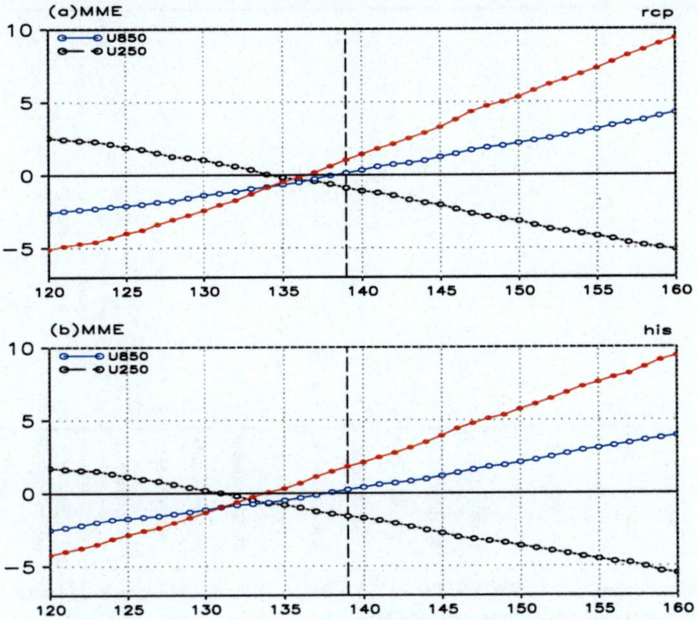

Fig. 2.49 MME mean seasonal evolutions of U850 (blue line), U250 (black line), and U850 minus U250 (m s^{-1}) from 20 CMIP5 models of RCP8.5 and historical runs. The x-axis is calendar date (days from January 1)

2.6.3 Response of SCSSM Onset to Global Warming

(1) MME Mean Changes in SCSSM Onset

Figure 2.49 shows the temporal evolution of MME U850 and U250 over the SCS in the onset stage of SCSSM from the historical runs and RCP8.5 runs. Although U250 shows a delayed reversal from westerly to easterly over the SCS in the RCP8.5 runs compared with that in the historical runs, there is almost no change in the timing of U850 reversal. It implies that there is no significant change in the onset date of SCSSM under the global warming of the RCP8.5 Scenario. This is confirmed by the MME onset dates estimated based on the RCP8.5 runs and historical runs (Fig. 2.50), respectively. The MME average displays a delayed onset by only one day between RCP8.5 and the historical runs, indicating no distinct change of SCSSM onset under global warming. Twelve of the 20 models showed delayed onset, while five models display early onset. The FGOALS-g2 model displays the largest delayed onset by 27 days among the 20 models, and this change is so exceptional that we also exclude the FGOALS-g2 model for the rest of the discussion.

Although the aforementioned results indicate no significant change of SCSSM onset in the MME average of the 20 CMIP5 models, a large discrepancy exists in model-simulated changes in the SCSSM onset date. Therefore, it is necessary to

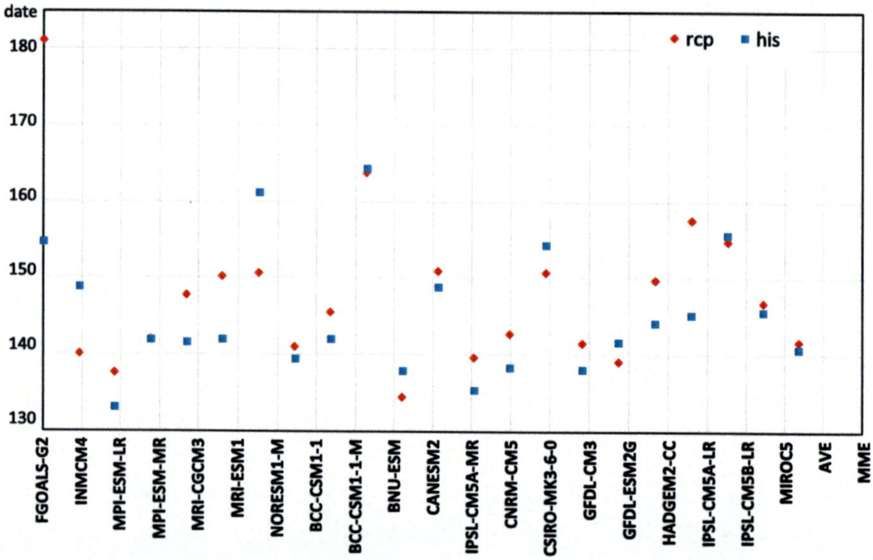

Fig. 2.50 Model onset dates of SCSSM in RCP8.5 (red diamond) and historical (blue square) runs. The y-axis is onset date (days from January 1)

examine the performance of circulation simulation by different model groups closely, to provide some guidance for model improvement. For this purpose, we screen the models of the delayed onset and early onset groups according to the criterion of onset change amplitude greater than half of the standard deviation from the change of onset dates of all 19 models (Fig. 2.51a). The early onset group includes five models (inmcm4, NorESM1-M, CanESM2, GFDL-CM3, and MIROC5) with an averaged 6-day change, and the delayed onset group includes six models (MRI-CGCM3, MRI-ESM1, CNRM-CM5, CSIRO-Mk3.6.0, IPSL-CM5A-LR, and IPSL-CM5B-LR) with an averaged 7-day change. The SCSSM onset process is closely associated with the reversal of the meridional gradient of temperature (MGT) in the mid- and upper levels of the troposphere over the SCS (Jian and Luo 2001; Wu and Wang 2001). The changes in the reversal date of MGT in the RCP8.5 runs compared to those in the historical runs from all 19 models are presented in Fig. 2.51b. The two curves in Fig. 2.51a and b are consistent with each other, indicating the reversal of MGT is crucial to the SCSSM onset.

(2) Simulated Changes in Circulation for Early Onset Cases of SCSSM

Figure 2.52a shows the evolutions of U850 and U250 for the early onset of SCSSM from the five selected models. In May (days 121–151), there is a distinct earlier reversal of U850 from easterly to westerly in RCP8.5 than in the historical experiments, while no change is observed in U250. However, the U850 of the early onset group has a later reversal from easterly to westerly than the MME average in the historical experiments (Fig. 2.52c), while the U250 of the early onset group

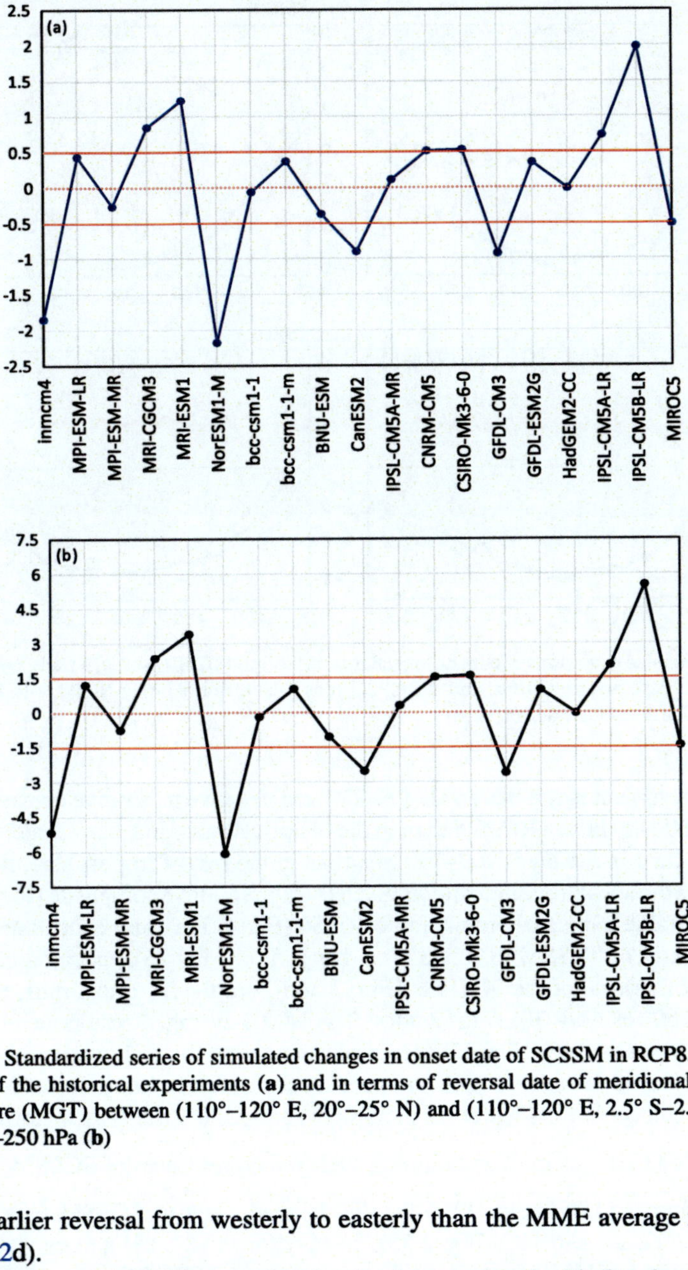

Fig. 2.51 Standardized series of simulated changes in onset date of SCSSM in RCP8.5 compared to those of the historical experiments (**a**) and in terms of reversal date of meridional gradient of temperature (MGT) between (110°–120° E, 20°–25° N) and (110°–120° E, 2.5° S–2.5° N) in the layer 500–250 hPa (**b**)

has an earlier reversal from westerly to easterly than the MME average in RCP8.5 (Fig. 2.52d).

The change of SCSSM onset date is closely associated with the change of sea–land thermal contrast, which exerts influence on the thermal state of the troposphere. Figure 2.53 displays the seasonal evolution of MGT over the SCSSM region in the historical runs and RCP8.5 runs for the simulated early onset cases, together with the MME results of 19 CMIP5 models. First of all, the results from the early onset cases

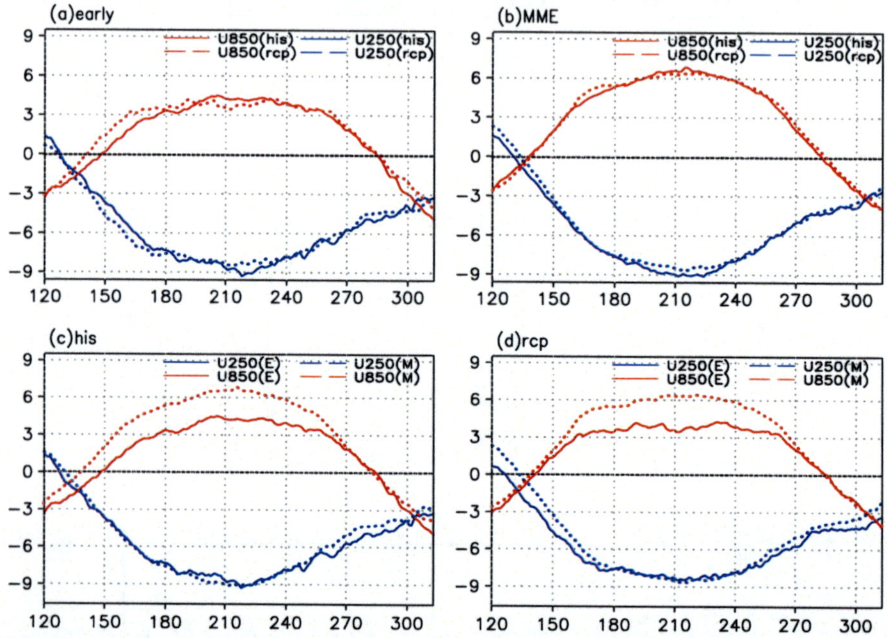

Fig. 2.52 Seasonal evolutions of climatological mean U850 and U250 (m s^{-1}). **a** Early onset group; **b** MME average; **c** historical runs; and **d** RCP8.5 runs (his: historical; rcp: RCP8.5: E: early; and M: MME). The x-axis is calendar date

display a prominent earlier reversal of MGT from negative to positive consistently in the layer 900–400 hPa in RCP8.5 than in the historical runs, and show larger changes of reversal time in the layer 900–700 hPa than in the rest of troposphere, no matter whether it is the absolute composite (Fig. 2.53a) or the relative composite (Fig. 2.53b). This is favorable to an early transition of U850 (Fig. 2.54a), and contributes to a 5-day early onset of SCSSM in RCP8.5 on day 145 (or May 25), accompanied by an eastward withdrawal of the WNPSH (Fig. 2.54b). For the historical runs, the onset is on day 150 (or May 30) (Fig. 2.54c). The MME-averaged results in Fig. 2.53c and d shows weak changes in MGT reversal timing with the earlier reversal in lower levels but delayed reversal in mid- to upper levels, which is the reason why there is no distinct change in SCSSM onset dates between RCP8.5 runs and historical runs.

(3) Simulated Changes in Circulation for Delayed Onset Cases of SCSSM

The evolutions of U850 and U250 for the delayed onset of SCSSM from the six selected models are shown in Fig. 2.55a. In May (days 121–151), there is a distinct delayed reversal of U250 from westerly to easterly in RCP8.5 than in the historical runs, while a weak delayed change is observed U850. However, the U850 of the delayed onset group has an earlier reversal from easterly to westerly than the MME average in the historical experiments (Fig. 2.55c), while the U250 of the delayed

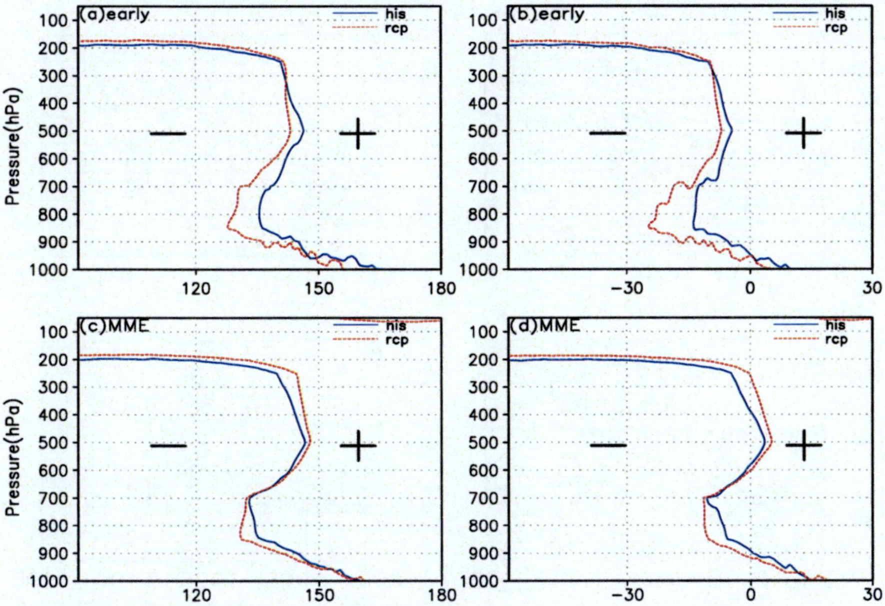

Fig. 2.53 Height-time sections of MGT between (110°–120° E, 20°–25° N) and (110°–120° E, 2.5° S–2.5° N) in the historical runs and RCP8.5 runs. **a** Early onset group with calendar date; **b** early onset group with relative composite date; **c** MME average with calendar date; and **d** MME average with relative composite date. Blue and red lines are the zero line of MGT, and minus and plus signs denote negative and positive MGTs, respectively (his: historical; rcp: RCP8.5). The x-axis in the left panels denotes calendar date from January 1. The x-axis in the right panels denotes composite date with "0" referring to the onset date of SCSSM

onset group has a later reversal from westerly to easterly than the MME average in RCP8.5 (Fig. 2.55d), which is different from the results of the early onset group.

Figure 2.56 displays the seasonal evolution of MGT over the SCSSM region in the historical runs and RCP8.5 runs for the simulated delayed onset cases. First of all, the results from the delayed onset cases display a prominent later reversal of MGT from negative to positive consistently in the layer 750–250 hPa in RCP8.5 than in the historical runs, and show large changes of reversal time in the upper layer 400–250 hPa, while there exists an early reversal in the lower layer, no matter whether it is the absolute composite (Fig. 2.56a) or the relative composite (Fig. 2.56b). This is favorable to a delayed transition of U250 (Fig. 2.55a), and contributes to a 7-day delayed onset of SCSSM in RCP8.5 on day 147 (May 27), accompanied by the eastward withdraw of the WNPSH (Fig. 2.57b). For the historical runs, the onset is on day140 (May 20) (Fig. 2.57c).

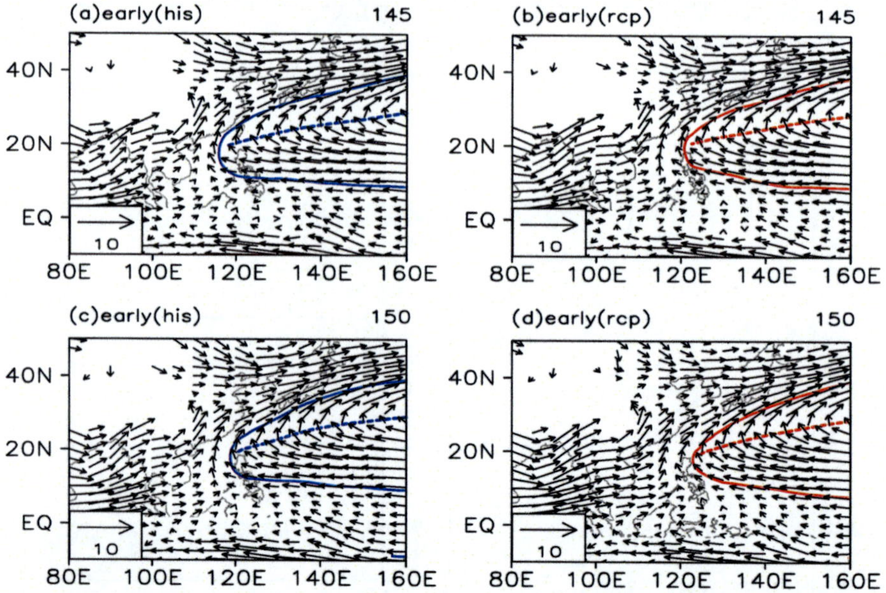

Fig. 2.54 Composite 850-hPa circulation simulated by five early onset models. **a** Day 145 (May 25) in the historical runs; **b** day 145 (May 25) in RCP8.5 runs; **c** day 150 (May 30) in the historical runs; **d** day 150 (May 30) in RCP8.5 runs. Solid contour and dashed line indicate the subtropical high and its ridge, respectively

2.7 Summary

In this chapter, we discussed issues associated with the characteristics of the spring–summer atmospheric circulation transition over the SCS and its surrounding regions and their responses to global warming. The main findings are summarized as follows.

The transition over the SCS and its surrounding regions mainly consists of two leading modes. The first leading mode is a large-scale circulation transition mode, which indicates the seasonal variation of large-scale atmospheric circulation and the evolution of the convection in the AS and the BOB during the transition. The second leading mode is an ISO mode. The onset of the ASM is under the combined influence of both large-scale circulation transition mode and ISO mode.

The MJO in the western Pacific phase is favorable to the onset of SCSSM, while the MJO in the Indian Ocean phase is unfavorable to the onset of SCSSM. When the MJO is in the western Pacific phase, more latent heat is released from enhanced convection, resulting in an anomalous cyclonic circulation in the low-level troposphere over the SCS–Northwest Pacific as a Rossby wave response. The anomalous cyclonic circulation is favorable for the eastward retreat of the subtropical high. On the other hand, the heat source over the Philippines promotes the establishment of the SAH over the Indochinese Peninsula and the northern SCS, and causes the wind to change to the easterly over the SCS in the upper-level troposphere, which is also conducive

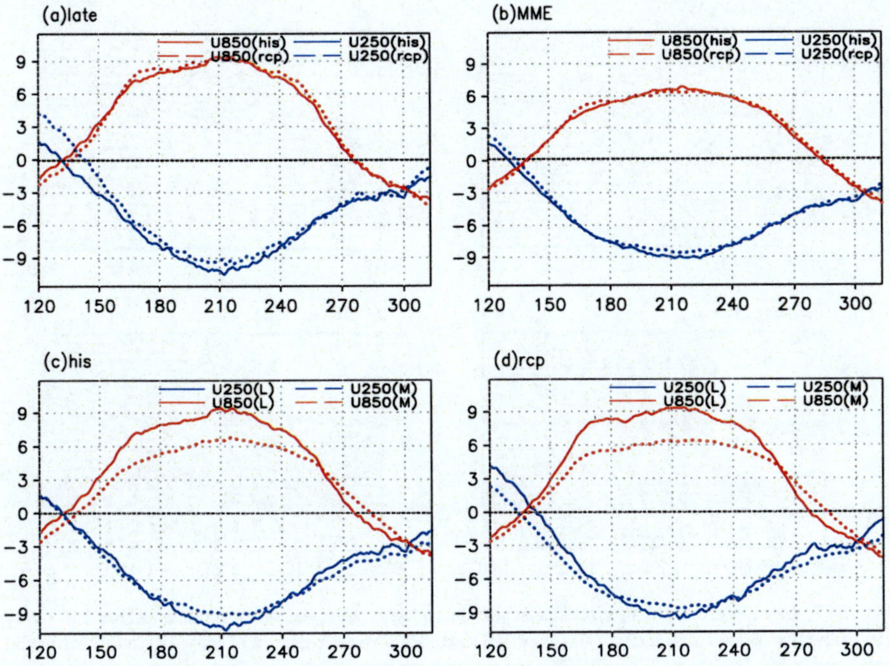

Fig. 2.55 Seasonal evolutions of climate mean U850 and U250 (m s^{-1}). **a** Delayed (late) onset group; **b** MME average; **c** historical runs; and **d** RCP8.5 runs (his: historical; rcp: RCP8.5; L: late; and M: MME). The x-axis is calendar date (days from January 1)

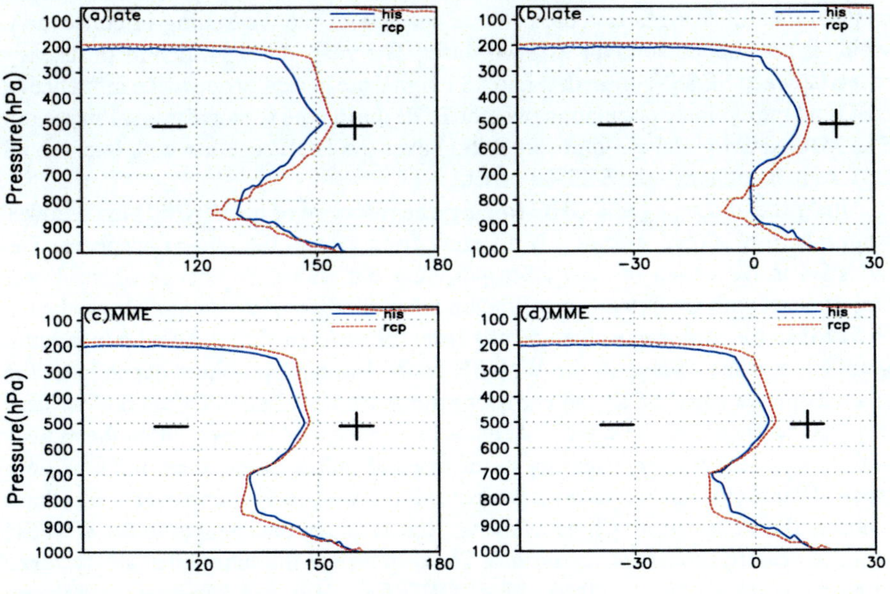

Fig. 2.56 Same as Fig. 2.53, except for the delayed onset group

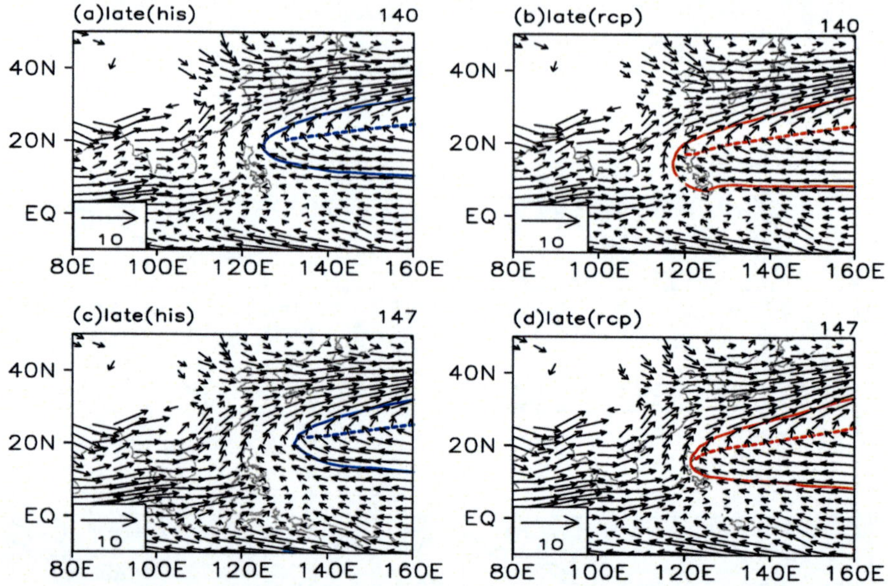

Fig. 2.57 Composite 850-hPa circulation simulated by six delayed onset models. **a** Day 140 (May 20) in the historical runs; **b** day 140 (May 20) in RCP8.5 runs; **c** day 147 (May 27) in the historical runs; and **d** day 147 (May 27) in RCP8.5 runs. Solid contour and dashed line indicate the subtropical high and its ridge, respectively

to the onset of SCSSM. When the MJO is in the Indian Ocean phase, the western tropical Pacific changes to a cold source because of the weakening of convection, which is not conducive to the SCSSM onset. The years during which inconsistency exists between SCSSM onset date and SST signal are basically due to the influence of MJO and the cyclonic circulation over the BOB during the seasonal transition period. The MJO activity and cyclonic circulation over the BOB together with interannual SST signals influence the SCSSM onset.

Under the anthropogenic GHG forcing, the response of the WNPSH in the lower troposphere to global warming is different from that in the mid-troposphere. The WNPSH is weakened in the mid-troposphere but shows no change in the lower troposphere, with great inter-model uncertainty under global warming. The weakened WNPSH in the mid-troposphere results from the weakened meridional temperature gradient, and the change in the WNPSH in the lower troposphere depends on the changes in the zonal SST gradient between the tropical Indian Ocean and western tropical Pacific Ocean. The WNPSH tends to be enhanced (weakened) in the models with a stronger (weaker) warming in the tropical Indian Ocean than in the western tropical Pacific Ocean, which favors enhanced summer precipitation over the northern (southern) part of eastern China. However, the MME of the changes in the WNPSH in the lower troposphere is nearly zero. The response of the subtropical anticyclones to global warming under anthropogenic GHG forcing is controlled by two factors: the enhanced static stability of the troposphere and the change in diabatic heating.

The combined effect of increased static stability and the change in diabatic heating results in the enhanced NASA and SPSA and the weakened NPSA, SASA, and SISA, with the strongest reduction in the intensity of NPSA.

The onset date of SCSSM shows no significant response to global warming. Large inter-model discrepancies exist in the model-simulated SCSSM onset date.

References

Chen LX, Liu HQ, Wang W et al (1999) Preliminary study on the characteristics and mechanism of summer monsoon onset over South China Sea and region adjacent to it. Acta Meteor Sinica (in Chinese) 57(1):16–29

Chen LX, Zhu QG, Luo HB et al (1991) East Asian Monsoon (in Chinese). Beijing China Meteorological Press, Beijing, p 362

Ding YH, Li CY (1999) Onset and evolution of the South China Sea monsoon and its interaction with the ocean (in Chinese). Beijing China Meteorological Press, Beijing, p p423

Ding YH, Liang P (2010) Extended range forecast basing on MJO. Meteorological Monthly (in Chinese) 36(7):111–122

Dong G, Zhang H, Moise A, Hanson L, Liang P, Ye H (2016) CMIP5 model-simulated onset, duration and intensity of the Asian summer monsoon in current and future climate. Clim Dyn 46(1–2):355–382. https://doi.org/10.1007/s00382-015-2588-z

Fu R, Del Genio AD, Rossow WB (1994) Influence of ocean surface conditions on atmospheric vertical thermodynamic structure and deep convection. J Climate 7(7):1092–1108. https://doi.org/10.1175/1520-0442(1994)007%3c1092:IOOSCO%3e2.0.CO;2

Gill AE (1980) Some simple solutions for heat-induced tropical circulation. Quart J Roy Meteor Soc 106(449):447–662. https://doi.org/10.1002/qj.49710644905

Gu DJ, Ji ZP, Li CH (2011) Multi-scale correlation of onset date of South China Sea summer monsoon with sea surface temperature and optimal subset regression prediction. Acta Oceanologyca Sinica (in Chinese) 33(6):55–63

He C, Zhou T (2015) Responses of the Western North Pacific Subtropical High to Global Warming under RCP4.5 and RCP8.5 Scenarios Projected by 33 CMIP5 Models: The Dominance of Tropical Indian Ocean–Tropical Western Pacific SST Gradient. J Climate 28(1):365–380. https://doi.org/10.1175/jcli-d-13-00494.1

He C, Zhou T, Lin A, Wu B, Gu D, Li C, Zheng B (2015) Enhanced or Weakened Western North Pacific Subtropical High under Global Warming? Scientific Reports 5:16771. https://doi.org/10.1038/srep16771

He C, Wu B, Zou L, Zhou T (2017) Responses of the Summertime Subtropical Anticyclones to Global Warming. J Climate 30(16):6465–6479. https://doi.org/10.1175/jcli-d-16-0529.1

He C, Lin A, Gu D, Li C, Zheng B, Wu B, Zhou T (2018) Using eddy geopotential height to measure the western North Pacific subtropical high in a warming climate. Theor Appl Climatol. https://doi.org/10.1007/s00704-016-2001-9

He JH, Wen M, Wang L et al (2006) Characteristics of the onset of the Asian summer monsoon and the importance of Asian–Australian "Land Bridge". Adv Atmos Sci 23:951–963

He JH, Zhu QG Murakami M (1996) T_{BB} data-revealed features of Asian-Australian monsoon seasonal transition and Asian summer monsoon establishment. J Tropical Meteorol 12(1):34–42. (in Chinese)

Hendon HH, Liebmann B (1990) A composite study of the onset of the Australian summer monsoon. J Atmos Sci 47:2227–2240

Higgins RW, Shi W (2001) Intercomparison of the principal modes of interannual and intraseasonal variability of the North American monsoon system. J Climate 14:403–417

Huang RH (1992) The East Asia/Pacific pattern teleconnection of summer circulation and climate anomaly in East Asia. Acta Meteor Sinica 6(1):25–37

Huang RH, Gu L, Xu YH et al (2005) Characteristics of the interannual variations of onset and advance of the East Asian summer monsoon and their associations with thermal states of the tropical western Pacific. Chinese J Atmos Sci (in Chinese) 29:20–36

Huang RH, Gu L, Chen JL et al (2008) Recent progresses in studies of the temporal-spatial variations of the East Asian monsoon system and their impacts on climate anomalies in China. Chinese J Atmospheric Sci (in Chinese) 32(4):691–719

Huang Y, Wang H, Fan K, Gao Y (2015) The western Pacific subtropical high after the 1970s: westward or eastward shift? Clim Dynam 44(7–8):2035–2047. https://doi.org/10.1007/s00382-014-2194-5

Hung CW, Lin HJ, Hsu HH (2014) Madden–Julian Oscillation and the Winter Rainfall in Taiwan. J Climate 27:4521–4530. https://doi.org/10.1175/JCLI-D-13-00435

Jeong JH, Kim BM, Ho CH et al (2008) Systematic variation in wintertime precipitation in East Asia by MJO-induced extratropical vertical motion. J Climate 21:788–801

Jia X, Chen L, Ren F et al (2011) Impacts of the MJO on winter rainfall and circulation in China. Adv Atmospheric Sci 28(3):521–533

Jian M, Luo H (2001) Heat sources over Qinghai-xizang Plateau and surrounding areas and their relationships to onset of SCS summer monsoon. (in Chinese). Plateau Meteorol 20:381–387

Jones C, Waliser DE, Lau KM et al (2004) Global occurrences of extreme precipitation events and the Madden–Julian oscillation: Observations and predictability. J Climate 17:4575–4589

Kajikawa Y, Wang B (2012) Interdecadal change of the South China Sea summer monsoon onset. J Climate 25(9):3207–3218. https://doi.org/10.1175/JCLI-D-11-00207.1

Kalnay E, Kanamitsu M, Kistler R, Collins W, Deaven D, Gandin L, Iredell M, Saha S, White G, Woollen J, Zhu Y, Chelliah M, Ebisuzaki W, Higgins W, Janowiak J, Mo KC, Ropelewski C, Wang J, Leetmaa A, Reynolds R, Jenne R, Joseph D (1996) The NCEP/NCAR 40-year reanalysis project. B Am Meteorol Soc 77(3):437–471. https://doi.org/10.1175/1520-0477(1996)077%3c0437:tnyrp%3e2.0.co;2

Kanamitsu M, Ebisuząki W, Woollen J et al (2002) NCEP–DOE AMIP-II Reanalysis (R-2). Bull Amer Meteor Soc 83:1631–1643

Knutson TR, Manabe S (1995) Time-Mean response over the tropical pacific to increased CO2 in a coupled ocean-atmosphere Model. J Climate 8(9):2181–2199

Lau KM, Yang S (1997) Climatology and Interannual Variability of the Southeast Asian Summer Monsoon. Adv Atmos Sci 14(2):141–162. https://doi.org/10.1007/s00376-997-0016-y

Li CY, Zhang LP (1999) SCS summer monsoon activity and its impacts. Chinese J Atmospheric Sci (Chinese) 23(3):257–266

Li CY, Pan J, Song J (2013a) Progress on the MJO research in recent years. Chin J Atmospheric Sci (Chin) 37(2):229–252. https://doi.org/10.3878/j.issn.1006-9895.2012.12318

Li KP, Yu WD, Li T (2013b) Structures and mechanisms of the first-branch northward-propagating intraseasonal oscillation over the tropical Indian Ocean. Clim Dyn 40:1707–1720

Liang JY, Wu SS (2002) A study of southwest monsoon onset date over the South China Sea and its impact factors. Chin J Atmospheric Sci (Chin) 26(6):829–844

Liang ZN, Wen ZP, Wu LJ (2006) The relationship between the Indian Ocean sea surface temperature anomaly and the onset of South China Sea summer monsoon. I. Coupling analysis. Chin J Atmospheric Sci (Chin) 30(4):619–634

Liebmann B, Smith CA (1996) Description of a complete (interpolated) outgoing longwave radiation dataset. Bull Amer Meteor Soc 77:1275–1277

Lin AL, Gu DJ, Zheng B et al (2013a) Relationship between South China Sea summer monsoon onset and Southern Ocean sea surface temperature variation. Chin J Geophys (Chin) 56(2):383–391. https://doi.org/10.6038/cjg20130203

Lin AL, Li CH, Zheng B et al (2013b) Modulation effect of MJO on the precipitation over Guangdong and its relation with the low latitude circulation system in June. J Appl Meteorol Sci (Chinese) 24(3):397–406

Lin AL, Li CH, Gu DJ et al (2015) Impact of tropical intraseasonal oscillations on the precipitation of Guangdong in Junes. J Tropical Meteorol 21(4):326–336

Lin AL (1998) The characteristics of low-frequency oscillation over South China Sea. J Tropical Meteorol (Chin) 14(2):113–118

Liu YM, Wu GX, Liu H et al (1999) The effect of spatially non-uniform heating on the formation and variation of subtropical high part III: CondensatiOn heating and South Asia high and western Pacific subtropical high. Acta Meteor Sinica (Chin) 57(5):525–538

Liu YM, Chan JCL, Mao JY et al (2002) The role of bay of Bengal convection in the onset of the 1998 South China Sea summer monsoon. Mon Wea Rev 130:2731–2744

Liu BQ, He JH, Wang L (2009) Characteristics of the South Asia high establishment processes above the Indo-China Peninsula from April to May and their possible mechanism [J]. Chin J Atmospheric Sci (in Chinese) 33(6):1319–1332

Liu BQ, He JH, Wang L (2012) On a possible mechanism for southern Asian convection influencing the South Asian High establishment during winter to summer transition. J Trop Meteor 18:473–484

Liu BQ, Wu GX, Mao JY, He JH (2013) Genesis of the South Asian High and its impact on the Asian Summer Monsoon Onset. J Clim 26(9):2976–2991. https://doi.org/10.1175/JCLI-D-12-00286.1

Liu BQ, Liu YM, Wu GX, Yan JH, He JH, Ren SL (2014a) Asian summer monsoon onset barrier and its formation mechanism. Clim Dyn 45(3–4):1–16. https://doi.org/10.1007/s00382-014-2296-0

Liu Y, Li W, Zuo J, Hu Z-Z (2014b) Simulation and projection of the western pacific subtropical high in CMIP5 models. J Meteorol Res 28(3):327–340. https://doi.org/10.1007/s13351-014-3151-2

Lorenz DJ, Dennis LH (2006) The Effect of the MJO on the North American Monsoon. J Climate 19:333–343

Lü JM, Ju JH, Ren JZ et al (2012) The influence of the Madden-Julian Oscillation activity anomalies on Yunnan's extreme drought of 2009–2010. Sci China Earth Sci (in Chinese) 55:98–112. https://doi.org/10.1007/s11430-011-4348-1

Ma J, Xie SP, Kosaka Y (2012) Mechanisms for tropical tropospheric circulation change in response to global warming. J Climate 25(8):2979–2994

Mao JY, Xie A, Song YY (2000) Impact of sea surface temperature and its variations on the onset of summer monsoon over South China Sea. Acta Meteorologica Sinica (Chin) 58(5):556–569

Mao JY, Wu GX (2007) Interannual variability in the onset of the summer monsoon over the Eastern Bay of Bengal. Theor Appl Climatol 89(3–4):155–170. https://doi.org/10.1007/s00704-006-0265-1

Mo KC, Higgins RW (1998) Tropical convection and precipitation regimes in the western United States. J Climate 11:2404–2423

Mu MQ, Li CY (2000) On the outbreak of South China Sea summer monsoon in 1998 and activity of atmospheric intraseasonal oscillation. Climatic Environ Res (Chin) 5(4):375–387. https://doi.org/10.3878/j.issn.1006-9585.2000.04.05

Murakami T, Nakazawa T (1985) Transition from Southern Hemisphere to Northern Hemisphere summer monsoon. Mon Weather Rev 113(9):1470–1486. https://doi.org/10.1175/1520-0493(1985)113%3c1470:TFTSTN%3e2.0.CO;2

Nitta T (1987) Convective activities in the tropical western Pacific and their impact on the northern hemisphere summer circulation. J Meteor Soc Japan 65(3):373–390. https://doi.org/10.2151/jmsj1965.65.3_373

North GR, Thomas LB, Cahalan RF (1982) Sampling errors in the estimation of empirical orthogonal functions. Mon Weather Rev 110(7):699–706. https://doi.org/10.1175/1520-0493(1982)110%3c0699:SEITEO%3e2.0.CO;2

Power SB, Delage F, Colman R et al (2012) Consensus on twenty-first-century rainfall projections in climate models more widespread than previously thought. J Climate 25(11):3792–3809

Reiter ER, Gao D (1982) Heating of the Tibet Plateau and movements of the South Asian High during Spring. Mon Wea Rev 110:1694–1711

Shaw TA, Voigt A (2015) Tug of war on summertime circulation between radiative forcing and sea surface warming. Nature Geosci 8(7):560–566. https://doi.org/10.1038/ngeo2449

Smith TM, Reynolds RW (2004) Improved Extended Reconstruction of SST (1854–1997). J Clim 17:2466–2477

Taylor KE, Stouffer RJ, Meehl GA (2012) An overview of CMIP5 and the experiment design. Bull Am Meteorol Soc 93:485–498. https://doi.org/10.1175/BAMS-D-11-00094.1

Uppala S (2002) ECMWF reanalysis, 1957–2001, ERA-40[J]. ERA-40 Project Report Series 3: 1–10

Vuuren D, Edmonds J, Kainuma M, Riahi K, Thomson A, Hibbard K, Hurtt G, Kram T, Krey V, Lamarque J-F, Masui T, Meinshausen M, Nakicenovic N, Smith S, Rose S (2011) The representative concentration pathways: an overview. Clim Change 109(1–2):5–31. https://doi.org/10.1007/s10584-011-0148-z

Wang B, Wu R, Fu X (2000) Pacific-East Asia teleconnection: How does ENSO affect East Asian climate? J Climate 13(9):1517–1536

Wang B, Lin H (2002) Rainy season of the Asian-Pacific summer monsoon. J Climate 15(4):386–398. https://doi.org/10.1175/1520-0442(2002)015%3c0386:RSOTAP%3e2.0.CO;2

Wang B, Ding Q, Joseph P (2009) Objective definition of the Indian summer monsoon onset. J Climate 22(12):3303–3316. https://doi.org/10.1175/2008JCLI2675.1

Webster PJ, Yang S (1992) Monsoon and Enso: Selectively Interactive Systems. Q J R Meteorol Soc 118(507):877–926. https://doi.org/10.1002/qj.49711850705

Wen ZP, Huang RH, He HY et al (2006) The influences of anomalous atmospheric circulation over mid-high latitudes and the activities of 30-60 d low frequency convection over low latitudes on the onset of the South China Sea summer monsoon. Chin J Atmospheric Sci (Chin) 30(5):952–964. https://doi.org/10.3878/j.issn.1006-9895.2006.05.23

Wheeler M, Hendon HH (2004) An all-season real-time multivariate MJO index: Development of an index for monitoring and prediction. Mon Wea Rev 132(8):1917–1932

Wu R, Wang B (2001) Multi-stage onset of the summer monsoon over the western North Pacific. Clim Dyn 17:277–289. https://doi.org/10.1007/s003820000118

Wu B, Zhou TJ, Li T (2009) Seasonally evolving dominant interannual variability modes of East Asian climate. J Climate 22(11):2992–3005. https://doi.org/10.1175/2008jcli2710.1

Wu CH, Chou MD (2012) Upper-Tropospheric Forcing on Late July Monsoon Transition in East Asia and the Western North Pacific. J Climate 25(11):3929–3941. https://doi.org/10.1175/JCLI-D-11-00343.1

Wu GX, Guan Y, Liu YM, Yan JH, Mao JY (2012) Air-sea interaction and formation of the Asian summer monsoon onset vortex over the Bay of Bengal. Clim Dyn 38(1–2):261–279. https://doi.org/10.1007/s00382-010-0978-9

Wu GX, Duan AM, Liu YM, Yan JH et al (2013) Recent advances in the study on the dynamics of the Asian summer monsoon onset. Chin J Atmospheric Sci (Chin) 37(2):211–228. https://doi.org/10.3878/j.issn.1006-9895.2012.12312

Wu L, Wang C (2015) Has the Western Pacific Subtropical High Extended Westward since the Late 1970s? J Climate 28(13):5406–5413. https://doi.org/10.1175/jcli-d-14-00618.1

Xie SP, Deser C, Vecchi GA, Ma J, Teng HY, Wittenberg AT (2010) Global Warming Pattern Formation: Sea Surface Temperature and Rainfall. J Climate 23(4):966–986

Xie SP, Hu KM, Hafner J, Tokinaga H, Du Y, Huang G, Sampe T (2009) Indian Ocean Capacitor Effect on Indo-Western Pacific Climate during the Summer following El Nino. J Climate 22(3):730–747. https://doi.org/10.1175/2008jcli2544.1

Xie A, Sun LQ (1991) Seasonal transition features in the tropical region. Acta Oceanol Sin 13(6):786–796 (in Chinese)

Yang H, Sun SQ (2003) Longitudinal displacement of the subtropical high in the western Pacific in summer and its influence. Adv Atmos Sci 20(6):921–933

Yasunari T (1991) The monsoon year-A new concept of the climate year in the Tropics. Bull Amer Meteor Soc 72(9):1331–1338. https://doi.org/10.1175/1520-0477(1991)072%3c1331:TMYNCO%3e2.0.CO;2

Yeh TC, Dao SY, Li MT (1958) The abrupt change of circulation over northern hemisphere during June and October. Acta Meteor Sinica 29(4):249–263 (in Chinese)

Zhang LN, Lin PF, Xiong Z et al (2011) Impact of the Madden-Julian oscillation on pre-flood season precipitation in South China. Chin J Atmospheric Sci (Chin) 35(3):560–570. https://doi.org/10.3878/j.issn.1006-9895.2011.03.15

Zhang L, Li T (2017) Relative roles of differential SST warming, uniform SST warming and land surface warming in determining the Walker circulation changes under global warming. Clim Dynam 48(3):987–997. https://doi.org/10.1007/s00382-016-3123-6

Zhao ZG (1999) Droughts and floods over China in summer and their environmental factors. Chinese Meteorological Press, Beijing (In Chinese)

Zhao YP, Chen YL, Bai XZ et al (2000) The relations between the SST anomalies in South China Sea-Tropical Eastern Indian Ocean and the South China Sea monsoon. J Tropical Meteorol (Chin) 16(2):115–123

Zhou W, Wen ZP, Chen CM (2002) Forecast of the South China Sea southwest monsoon onset. Acta Scientiarum Naturalium Universitatis Sunyatseni (in Chinese) 41(3):95–98

Zhou TJ, Yu RC, Zhang J, Drange H, Cassou C, Deser C, Hodson DLR, Sanchez-Gomez E, Li J, Keenlyside N, Xin XG, Okumura Y (2009) Why the Western Pacific Subtropical High Has Extended Westward since the Late 1970s. J Climate 22(8):2199–2215. https://doi.org/10.1175/2008jcli2527.1

Zhu QG, He J, Wang P (1986) A study of circulation differences between East Asian and Indian summer monsoon with their interaction. Adv Atmos Sci 3:466–477

Chapter 3
Air–Sea Interactions and Climate Variability Over the South China Sea and the Adjacent Regions

The South China Sea is a part of the tropical Indo-western Pacific warm pool. Air–sea interactions on various time scales have been identified in the South China Sea region (Wang et al. 1997; Liang and Wu 2000; Wu and Wang 2001; Wu 2002; Xie et al. 2007a; Wu 2010). On an interannual time scale, the air–sea interaction processes over the South China Sea and tropical western North Pacific play an important role in the maintenance of an anomalous anticyclone over the Philippine Sea during the decaying phase of El Niño events, which forms initially due to the remote El Niño–Southern Oscillation (ENSO) forcing (Wang et al. 2000, 2003; Wu et al. 2003). The East Asian winter monsoon variability may induce sea surface temperature (SST) anomalies in the South China Sea region in boreal winter, which may maintain to the succeeding summer through local air–sea coupled processes and affect the East Asian summer monsoon variability (Chen et al. 2000; Chen 2002; Wu et al. 2014). Thus, SST anomalies in the South China Sea act as a connection between the East Asian winter and summer monsoons. On the intraseasonal time scales, local air–sea interaction processes in the South China Sea may affect the phase transition and propagation of the intraseasonal oscillations (ISOs) and the summer monsoon onsets (Wu and Wang 2000; Wu 2002, 2010; Roxy and Tanimoto 2012; Ye and Wu 2015).

Being located between the Pacific and Indian Oceans, the South China Sea acts as a channel connecting climate systems in the surrounding regions. On the one hand, the climate variability in the South China Sea is affected by the thermal status of the Pacific and Indian Oceans (Wang et al. 2000; Xie et al. 2009; He and Wu 2014; He et al. 2016). On the other hand, the thermal status of the South China Sea may induce anomalous heating in the overlying atmosphere that may in turn affect the climate in the surrounding countries and over East Asia through atmospheric teleconnections (Nitta 1987; Huang and Sun 1992; Wang et al. 2001). So, the South China Sea is a medium in the influence of the tropical central–eastern Pacific and the tropical Indian Ocean on the East Asian climate variability. The South China Sea is a junction of the East Asian monsoon, the South Asian monsoon, the western North Pacific monsoon, and the Australian monsoon. On the one hand, the climate variability over the South China Sea is connected to the variability of those monsoons. On the other hand, the

S. Yang et al., *Climate Change in Southeast Asia and Surrounding Areas*, Springer Climate, https://doi.org/10.1007/978-981-15-8225-7_3

South China Sea climate variability may play a role of bridging the climate among those monsoon systems. Tao and Chen (1987) indicated that the South Asian and Australian monsoons are linked to the East Asian monsoon through the climate over the South China Sea. In boreal winter, the cold surges associated with the East Asian winter monsoon activity may be channeled through the South China Sea and affect the convection over the Maritime Continent and the Australian summer monsoon variability (Chang and Chen 1992). In boreal summer, the ISOs originated from the tropical western Pacific may move westward through the South China Sea to the Indochinese Peninsula, the Bay of Bengal, and even India (Annamalai and Slingo 2001; Kemball-Cook and Wang 2001).

There exist interconnections among variations on different time scales over the South China Sea and tropical western North Pacific. The interannual anomalies in this region may affect the background fields for the formation and movement of tropical cyclones over the western North Pacific (Chen et al. 1998; Wang and Chan 2002; Camargo et al. 2007; Wu et al. 2012a, b; Cao et al. 2014). The tropical cyclones tend to form during the wet phase of the ISOs, and the wet intraseasonal phase guides the movement of tropical cyclones (Zhu et al. 2004; Kim et al. 2008; Chen and Huang 2009; Sun et al. 2009; Mao and Wu 2010; Pan et al. 2010; Huang et al. 2011; Feng et al. 2013; Li et al. 2013). The seasonal mean change induced by interannual anomalies modulates the intensity of ISOs (Kajikawa and Yausnari 2005; Yang et al. 2008; Wu and Cao 2017; Wu and Song 2018). The intensity of ISOs may change the signal-to-noise ratio of seasonal mean climate anomalies, which is related to the predictability of seasonal mean climate.

This chapter documents systematically the air–sea relationship on intraseasonal and interannual time scales in the South China Sea and the tropical western North Pacific, the influence of ENSO and tropical Indian Ocean SST anomalies on the climate variability in the South China Sea region and associated processes, the interdecadal variability and its modulation of the relationship between interannual variations, and the connections among different time-scale variations.

3.1 Air–Sea Interactions on Different Time Scales

There exist air–sea interactions on different time scales in the South China Sea and the tropical western North Pacific. Intraseasonal air–sea interactions affect the phase transition and propagation of ISOs. Interannual air–sea interactions modulate the maintenance of remote forcing induced anomalies, providing delayed effects of remote forcing on regional climate. This section presents the air–sea relationship on intraseasonal and interannual time scales.

3.1.1 *Intraseasonal Air–Sea Interactions*

The ISO is an important component of regional climate and monsoon variability in the tropical western North Pacific and the South China Sea. The ISOs affect the time and suddenness of the summer monsoon onset in this region (Chen and Chen 1995; Mu and Li 2000; Wu and Wang 2001; Chan et al. 2002; Wu 2002; Mao and Chan 2005; Zhou and Chan 2005; Wen et al. 2006; Wu 2010). The ISOs influence directly the formation and movement of tropical cyclones and typhoons in this region (Chen and Huang 2009; Pan et al. 2010; Huang et al. 2011; Feng et al. 2013). The northward and northwestward propagating ISOs from the western North Pacific and the South China Sea lead to more or less summer rainfall in southern China (Li and Li 1997; Feng et al. 2013; Wang et al. 2017). The changes in convection and atmospheric heating induced by ISOs in this region modulate the climate in the neighboring regions through atmospheric teleconnections, imposing an important impact on the East Asian climate anomalies, such as the drought and flood in eastern China (Yang and Li 2003; Zhu et al. 2003; Mao et al. 2010; Ren et al. 2013; Li et al. 2013).

Intraseasonal variations are identified in tropical oceans (Lau and Sui 1997; Sengupta and Ravichandran 2001; Sengupta et al. 2001; Gao ad Zhou 2002; Xie et al. 2007b; Wu 2010). The coexistence of intraseasonal variations in both atmosphere and ocean suggests that there may be interactions between ISOs in the atmosphere and ocean (Woolnough et al. 2000; Kemball-Cook and Wang 2001; Wang et al. 2009; Wu 2010). There are two prominent ISOs: one on the 10–20-day time period and the other on the 30–60-day time period (Chen and Chen 1995; Fukutomi and Yausnari 1999; Annamalai and Slingo 2001; Mao and Chan 2005; Kikuchi and Wang 2009; Ye and Wu 2015). The two ISOs display different horizontal structures and propagations (Kajikawa and Yasunari 2005; Mao and Chan 2005; Wang et al. 2009; Ye and Wu 2015; Cao et al. 2017). The ISOs display obvious different characteristics in winter and summer (Ye and Wu 2015).

This subsection distinguishes the 10–20-day and 30–60-day ISOs and compares the difference of air–sea relationship and the seasonal change of the relationship between the two ISOs. The focus is on the spatial dependence of the consistency between atmospheric and oceanic intraseasonal variations and the influence of the East Asian winter monsoon variability on intraseasonal SST variations in the South China Sea.

1. Time-scale dependence of boreal summer intraseasonal air–sea relationship

During boreal summer (May through September), the intraseasonal air–sea relationship displays different spatial features in the South China Sea and tropical western North Pacific between the 10–20-day (quasi-biweekly) and 30–60-day ISOs (Ye and Wu 2015; Wu et al. 2015a, b, c; Cao et al. 2017). On the quasi-biweekly time scale, the correlation of SST with precipitation, surface shortwave radiation, and latent heat flux displays a southwest–northeast-oriented distribution, whereas on the 30–60-day time scale, the correlation is characterized by a zonal distribution (Ye and Wu 2015).

A common feature in the air–sea relationship between the two ISOs is that the atmospheric ISOs influence intraseasonal variations in SST through cloud-radiation and wind-evaporation effects. In the off-equatorial region, the intraseasonal SST variations affect the atmosphere by modulating the atmospheric stability. In comparison, the influence of SST on the atmosphere is larger on the 30–60-day time scale than on the quasi-biweekly time scale (Ye and Wu 2015).

The difference of intraseasonal air–sea relationship between the 10–20-day and 30–60-day ISOs is illustrated in Fig. 3.1 that presents the maximum and minimum correlation coefficients and the corresponding lead and lag times between intraseasonal SST and precipitation variations on the two time scales during northern summer (May through September). The lead–lag correlation is calculated based on intraseasonal variations of SST and rain rate at the same grid on the 10–20-day and 30–60-day time scales separately. The SST and rain rate are from the Tropical Rainfall Measuring Mission (TRMM) Microwave Imaging (TMI) data for the period 1998–2012 (Wentz et al. 2000). The 10–20-day time-scale variations are derived by 9-day running mean minus 21-day running mean. The 30–60-day time-scale variations are derived by 29-day running mean minus 61-day running mean. The maximum (positive) and minimum (negative) correlation coefficients and the corresponding lead and lag times are determined based on the temporal evolution of the lead–lag correlation at each grid point and they are plotted in Fig. 3.1. According to Fig. 3.1, on the 10–20-day time scale, the maximum and minimum correlation when SST leads and lags precipitation is observed along a southwest–northeast-oriented zone extending from the South China Sea to the subtropical western North Pacific (Fig. 3.1a, 3.1c). The SST leading time is about 5 days (Fig. 3.1b) and the SST lagging time is about 3–5 days (Fig. 3.1d). In the lower latitudes of the western North Pacific, the SST lagging time is longer. On the 30–60-day time scale, the maximum and minimum correlation is observed in a zonal band from the South China Sea to the Philippine Sea (Fig. 3.1e, 3.1g). The SST leading and lagging times are asymmetric. The SST leading time reaches about two weeks (Fig. 3.1f) and the SST lagging time is about 8 days (Fig. 3.1h). In the subtropics, the SST leading time is longer, whereas the SST lagging time is shorter. In the lower latitudes, the SST lagging time is longer. The regional difference in the lead and lag times may be related to the spatial change in the ocean mixed-layer depth (Wu et al. 2015a, b, c). In the lower latitudes, the ocean mixed layer is deeper and it takes a longer time for the SST to respond to the atmospheric change.

The influence of the atmosphere on intraseasonal SST variations includes surface heat flux, wind-driven ocean mixing, and upwelling. Figure 3.2 shows the local correlation of the SST tendency with net surface heat flux, surface wind stress curl, and surface wind speed. The wind stress curl indicates the change in ocean upwelling. The wind speed indicates the change in ocean mixing. The SST data are from TMI (Wentz et al. 2000), surface heat flux data are from the NCEP-DOE reanalysis 2 (Kanamitsu et al. 2002), and surface wind stress data are from TropFlux (Kumar et al. 2012). On the 10–20-day time scale, the SST tendency and surface heat flux display the most consistent coupled variations in the South China Sea and subtropical

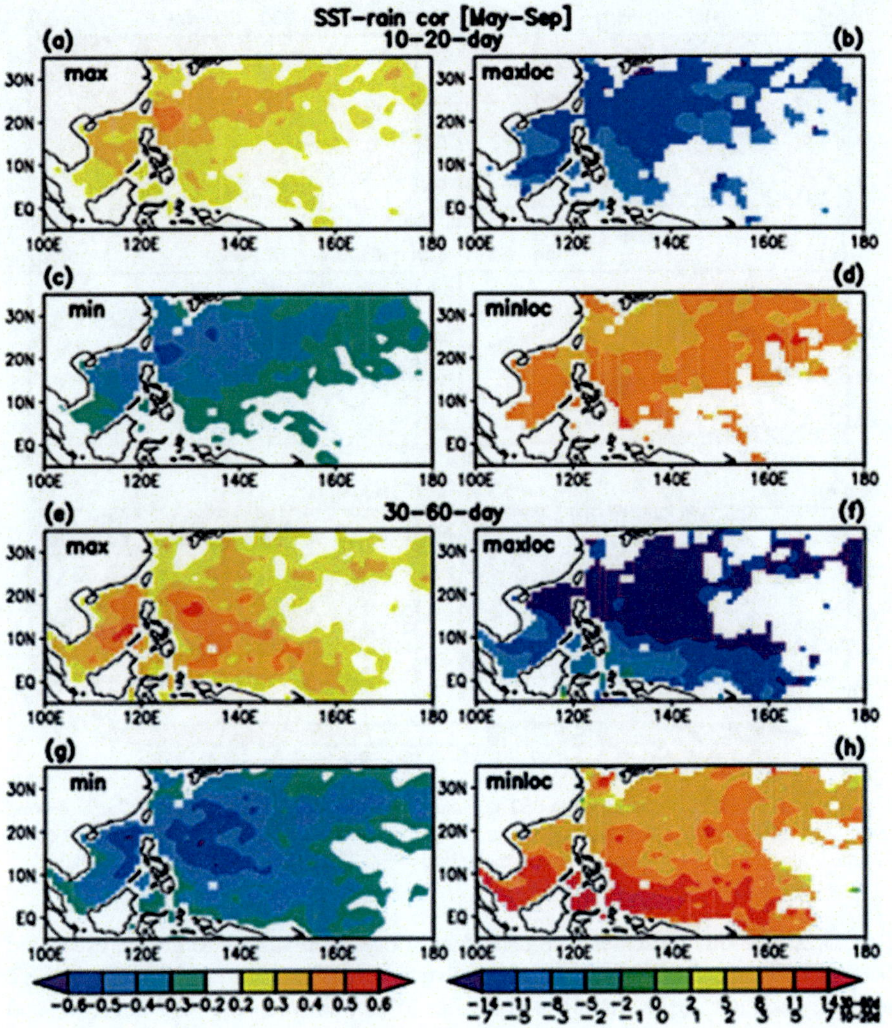

Fig. 3.1 The maximum (**a**, **e**) and minimum (**c**,) lag–lead correlations of SST with regard to rain rate and their corresponding days by which SST leads (**b**, **f**) and lags (**d**, **h**) rain rate during May–September for the period 1998–2012 on the 10–20-day (**a**–**d**) and 30–60-day (**e**–**h**) time scales. Only regions with the magnitude of correlation coefficient exceeding 0.2 are shown. SST and rain rate data are from TMI (Wentz et al. 2000). (From Ye and Wu 2015)

western North Pacific (Fig. 3.2a). This is the region where the leading mode of 10–20-day SST variations has the largest loading (Cao et al. 2017). On the 30–60-day time scale, the region of consistent coupled variations in the SST tendency and surface heat flux extends eastward from the South China Sea to the Philippine Sea (Fig. 3.2b). The wind stress curl and the SST tendency have an obvious negative correlation in the region from the northern South China Sea to the subtropical western North Pacific

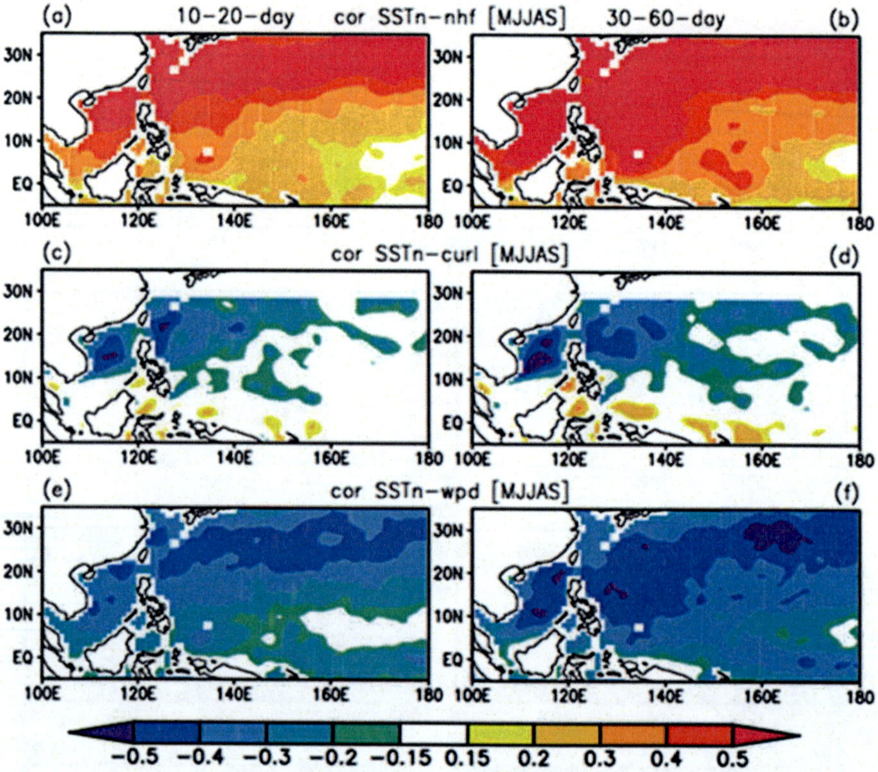

Fig. 3.2 Point-wise correlation of the SST tendency with surface net heat flux (**a**, **b**), surface wind stress curl (**c**, **d**), and surface wind speed (**e**, **f**) on the 10–20-day (**a**, **c**, **e**) and 30–60-day (**b**, **d**, **f**) time scales during May–September for the period 1998–2012. (From Wu et al. 2015a, b, c)

(Fig. 3.2c–d). This indicates a contribution of ocean upwelling to the intraseasonal SST variations in the above region. The distribution of correlation between the SST tendency and surface wind speed is similar to that between the SST tendency and surface heat flux. The high correlation region on the 10–20-day time scale extends northeastward from the South China Sea, and the high correlation region on the 30–60-day time scale extends eastward from the South China Sea (Fig. 3.2e–f). This indicates a contribution of oceanic mixing to the SST tendency consistent with that of surface heat flux.

The above comparison shows that the horizontal structure of coupled variations is different between the 10–20-day and 30–60-day time scales. The discrepancy in the spatial distribution of consistent coupled variations is related to the difference in the horizontal structure of the two ISOs. The 10–20-day precipitation and lower-level wind variations display a southwest–northeast distribution, whereas the 30–60-day variations display a zonal distribution (Cao et al. 2017). Wu et al. (2015a, b, c) estimated quantitatively the contribution of surface heat flux to the SST tendency.

The results showed a larger contribution of surface heat flux on the 30–60-day time scale than on the 10–20-day time scale, with the former being about two times or more of the latter. The contribution of latent heat flux is larger than shortwave radiation, in particular, on the 10–20-day time scale.

Intraseasonal SST variations may feedback on the atmospheric ISOs through modifying the lower-level moisture convergence and near-surface atmospheric stability (Stephens et al. 2004; Fu et al. 2006; Tao et al. 2009a, b; Wu 2010; Roxy and Tanimoto 2012; Roxy et al. 2013; Ye and Wu 2015). Before the South China Sea summer monsoon onset, anomalously high SST increases surface air temperature and humidity, reduces the atmospheric stability, which provides a favorable thermodynamic condition for the development of convection (Wu 2010). Both surface evaporation and lower-level moisture convergence contribute to the increase of moisture in the atmospheric boundary layer. The effect of SST on the atmospheric stability is present in the off-equatorial region on both 10–20-day and 30–60-day variations, but more prominent on the 30–60-day variations (Ye and Wu 2015). In the equatorial western Pacific, the feedback of intraseasonal SST variations on the atmosphere is mainly through modulating the lower-level convergence and divergence (Ye and Wu 2015), but this feedback appears weak in the South China Sea and the Philippine Sea (Wu 2010).

2. Coupled intraseasonal air–sea variations during boreal winter

During boreal winter (November–March), the relationship between intraseasonal variations of SST and surface wind speed displays a pronounced regional difference, with a prominent correlation in the South China Sea (Wu and Chen 2015). Figure 3.3 presents the maximum and minimum correlation coefficients and the corresponding lead and lag times between intraseasonal SST and surface wind speed variations on the 10–20-day and 30–60-day time scales from November through March for the period 1998/99–2012/13. Different from boreal summer, precipitation and its variations are small in boreal winter, whereas surface wind variations are large in the off-equatorial regions of the South China Sea and the western North Pacific. So, the relationship between intraseasonal SST and surface wind variations is analyzed in boreal winter. The correlation is calculated using SST data from TMI (Wentz et al. 2000) and surface wind data from the NCEP-DOE reanalysis 2 (Kanamitsu et al. 2002). The surface wind has been interpolated to the SST grid when calculating the grid-point lead–lag correlation.

The largest positive and negative correlation is observed in the South China Sea (Fig. 3.3a, c, e, g). For the 10–20-day variations, high SST leads large wind speed by about 7 days (Fig. 3.3b). Low SST appears about 2 days after large wind speed (Fig. 3.3d). This indicates a quick response of SST to atmospheric wind change. The correlation between SST and wind speed variations is relatively small, the SST leading time is relatively short, and the SST lagging time is relatively long in the tropical western North Pacific. These features may be related to the spatial distribution of the ocean mixed-layer depth. The mixed layer is shallower in the South China Sea than in the western North Pacific. For the 30–60-day variations, the time lag relationship between SST and wind speed variations is basically similar to that for

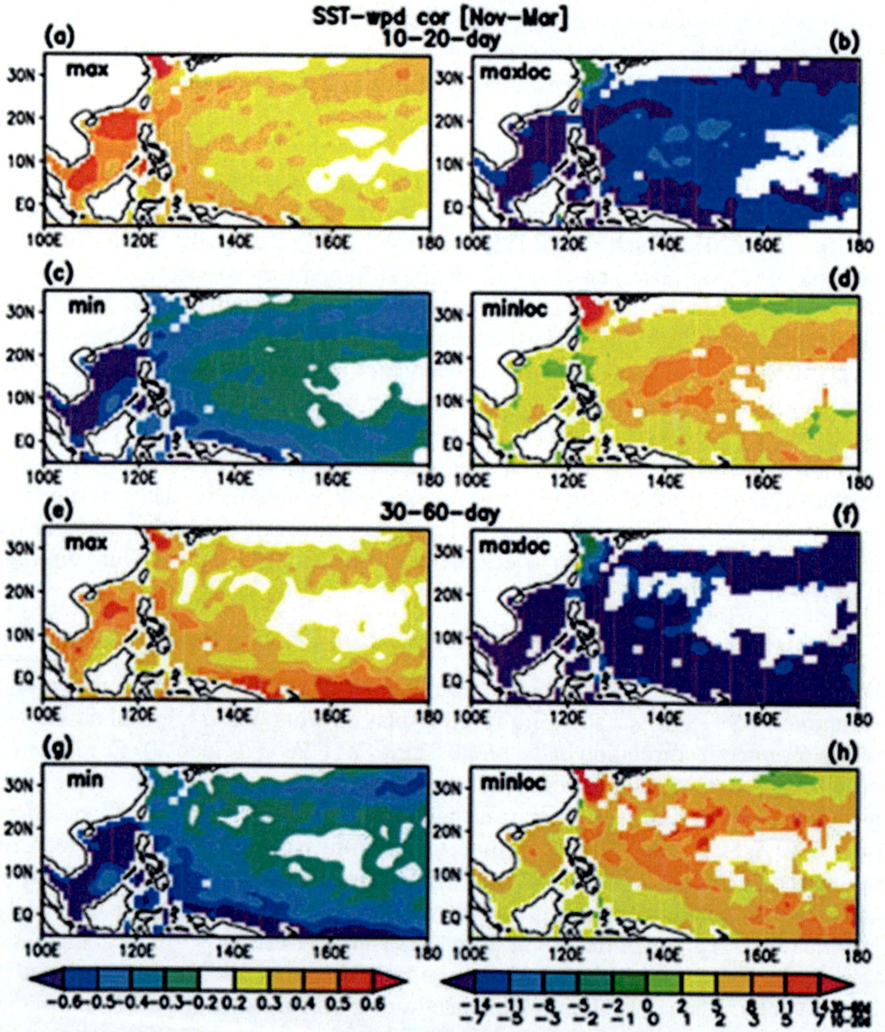

Fig. 3.3 The maximum (**a**, **e**) and minimum (**c**, **g**) lag–lead correlations of SST with regard to 10-m wind speed and their corresponding days by which SST leads (**b**, **f**) and lags (**d**, **h**) wind speed during November–March for the period 1998/99–2012/13 on the 10–20-day (**a**–**d**) and 30–60-day (**e**–**h**) time scales. Only regions with the magnitude of correlation coefficient exceeding 0.2 are shown. Data are from TMI (Wentz et al. 2000) and NCEP-DOE reanalysis (Kanamitsu et al. 2002)

the 10–20-day variations, but the SST leading time does not show a big difference in the South China Sea and the western North Pacific (Fig. 3.3f, h).

The coupled intraseasonal ocean–atmosphere variations in boreal winter may be illustrated from the consistency of SST variations with surface heat flux and wind speed variations. Figure 3.4 shows the local correlation of the SST tendency with net surface heat flux, surface wind stress curl, and surface wind speed during boreal

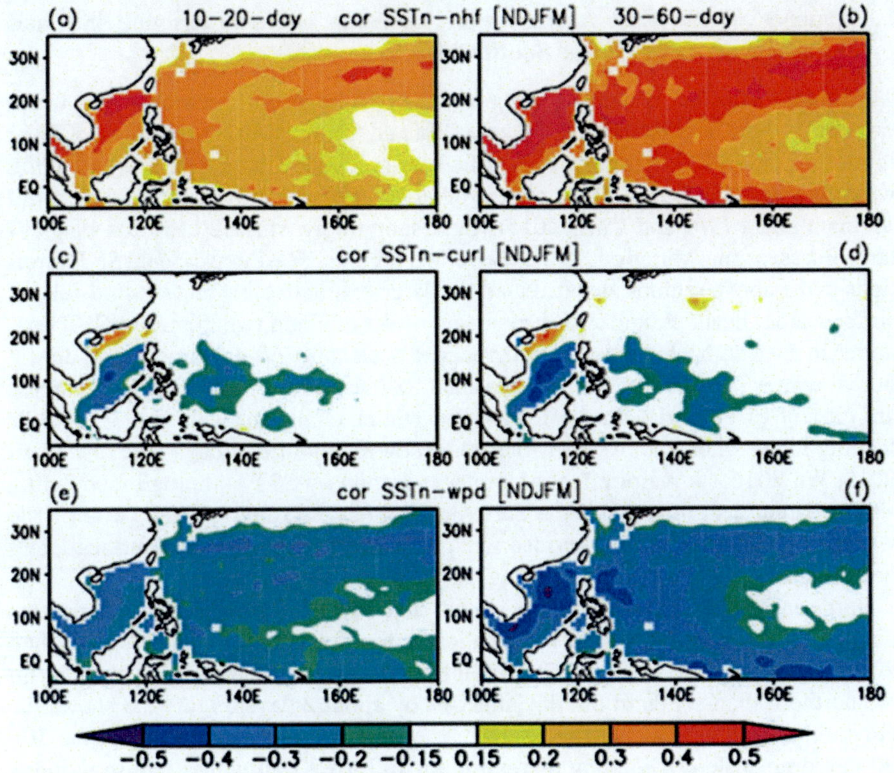

Fig. 3.4 Point-wise correlation of the SST tendency with surface net heat flux (**a**, **b**), surface wind stress curl (**c**, **d**), and surface wind speed (**e**, **f**) on the 10–20-day (**a**, **c**, **e**) and 30–60-day (**b**, **d**, **f**) time scales during November–March for the period 1998/99–2012/13. (From Wu et al. 2015a, b, c)

winter. The data used are the same as those in Fig. 3.2. The SST tendency and net surface heat flux variations display a high consistency in most of the South China Sea and subtropical western North Pacific (Fig. 3.4a–b). The SST tendency and surface wind speed variations display a similar spatial pattern of consistency (Fig. 3.4e–f). The correlation between the SST tendency and wind stress curl is opposite between the northern and central–southern South China Sea (Fig. 3.4c–d). This indicates that the contributions of surface heat flux, oceanic mixing, and oceanic upwelling to the SST tendency are consistent in the central–southern South China Sea, whereas the SST tendency is mainly determined by surface heat flux and oceanic mixing in the northern South China Sea. The comparison showed a larger contribution of surface heat flux to the SST change on the 30–60-day time scale than on the 10–20-day time scale, and latent heat flux has a much larger contribution to the SST change than shortwave radiation (Wu et al. 2015a, b, c).

3. Influence of the East Asian winter monsoon on intraseasonal SST and precipitation variations in the South China Sea

The South China Sea is one of the regions with the largest intraseasonal SST variations in the Northern Hemisphere winter (Wu and Chen 2015). Two regions of large intraseasonal SST variations are identified in the South China Sea, one extending westward from the Luzon Strait and the other extending southward from the central Vietnam coast (Wu and Chen 2015). Correspondingly, surface heat flux displays large intraseasonal variations in the above two regions. The intraseasonal SST variations in the above regions are attributed to a large extent to the effect of wind-related surface latent heat flux and a supplementary effect of cloud radiation. The SST variations in the South China Sea have a spectrum peak at 30–60-day time period during boreal winter (Wu 2016). The intraseasonal SST variations in the South China Sea are closely associated with the East Asian winter monsoon, with the higher SST lagging the weaker East Asian winter monsoon by about 3–5 days (Wu and Chen 2015; Wu 2016). A major effect of winter monsoon on SST is through modulating surface latent heat flux. When the East Asian winter monsoon is weak (strong), the weakening (intensifying) of surface wind speed reduces (enhances) surface latent heat flux, leading to warming (cooling) of the ocean surface water.

Intraseasonal SST, surface wind speed, and surface latent heat flux variations display a consistent southward propagation in the South China Sea (Wu and Chen 2015; Wu 2016; Cao et al. 2017). The intraseasonal SST signals appear earlier in the northern than southern South China Sea by about 2 days. Figure 3.5 shows the Hovmöller diagrams along 107.5°–120° E of intraseasonal anomalies on the 10–30-day time scale obtained by regression on an East Asian winter monsoon index (EAWMI). The data used are the same as in Fig. 3.3. The EAWMI is defined using meridional wind averaged over the region of 0–25° N and 105°–135° E. At about 2–3 days after the East Asian winter monsoon intensifies, negative SST anomalies appear in the South China Sea (Fig. 3.5a). The southward intensification of the winter monsoon is accompanied by the southward movement of a region of increase in surface wind speed and evaporation. Correspondingly, the low SST region moves from the northern to the southern South China Sea. The above relationship may be explained using the wind-evaporation mechanism. The climatological mean winds are northerly during boreal winter over the South China Sea. Anomalous northerly winds enhance surface wind speed (Fig. 3.5b) and induce more evaporation (Fig. 3.5c). The ocean loses more heat, lowering the ocean surface temperature. The wind speed anomaly is larger north of the SST anomaly region. So, surface evaporation anomaly is larger north of the SST anomaly, leading to southward movement of the region of SST anomalies. The region of wind speed and evaporation anomalies moves southward accordingly. The southward movement of intraseasonal cloud and shortwave radiation anomalies is limited to the northern part of the South China Sea (Wu and Chen 2015; Wu 2016).

The southward propagating intraseasonal wind anomalies associated with the East Asian winter monsoon activity may induce convergence when they reach the Maritime Continent, and thus lead to intraseasonal precipitation changes in the

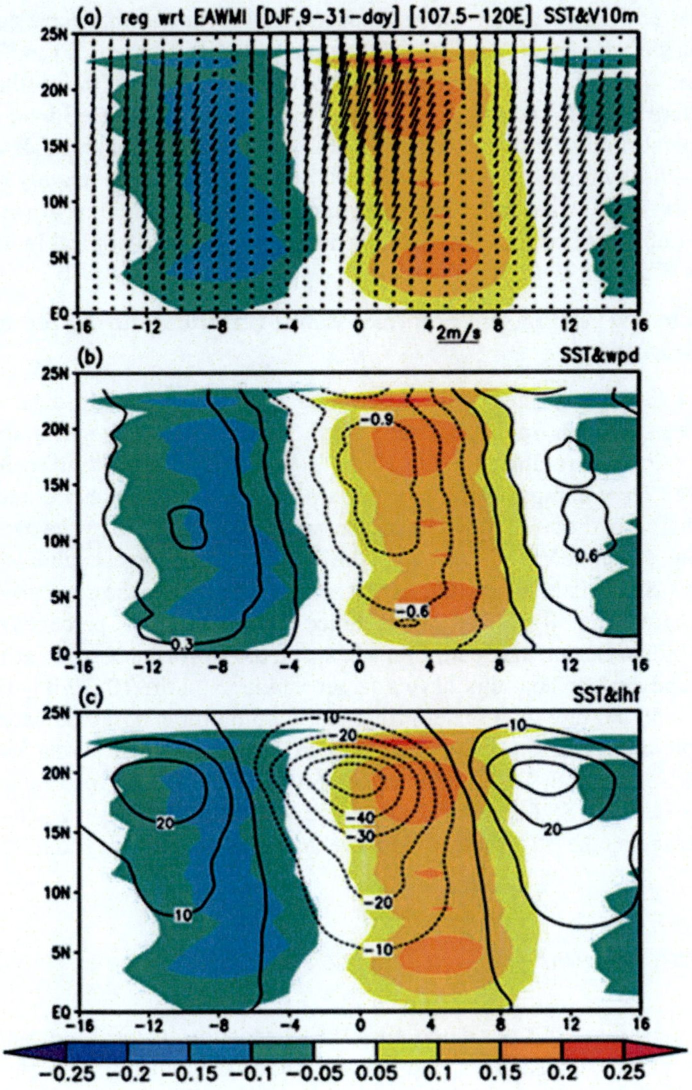

Fig. 3.5 Hovmöller diagrams of anomalies of SST (shading, °C) and 10-m wind (vector, m/s, scale at bottom) (**a**), surface wind speed (contour, 0.3 m/s) (**b**), and surface latent heat flux (contour, W/m^2) (**c**) along 107.5°–120° E from 16 days before to 16 days after the EAWMI obtained by regression with respect to the normalized EAWMI. (From Wu 2016)

southern South China Sea (Wu and Chen 2015; Wu 2016; Cao and Wu 2017, 2018). On the other hand, the eastward propagating ISOs from the tropical Indian Ocean may influence the intraseasonal variations over the Maritime Continent. The 10–20-day intraseasonal variations of precipitation in the southern South China Sea are closely related to the East Asian winter monsoon, with the maximum correlation at about

1–2 days after the East Asian winter monsoon peak, whereas they have no obvious connection with the 10–20-day ISOs of the tropical Indian Ocean (Cao and Wu 2017, 2018). The 30–60-day intraseasonal variations of precipitation in the southern South China Sea are related to both the East Asian winter monsoon and the 30–60-day ISOs of the tropical Indian Ocean. This indicates that the intraseasonal rainfall variations on the 10–20-day time scale in the southern South China Sea are mainly due to the impact of the East Asian winter monsoon, and those on the 30–60-day time scale are subject to impacts of both the East Asian winter monsoon and tropical Indian Ocean variability (Cao and Wu 2017, 2018).

4. Local air–sea relationship of intraseasonal variations during the spring-to-summer transition

Hu and Li (2017) analyzed the intraseasonal air–sea relationship over the North Indian Ocean through tropical western North Pacific during the spring-to-summer transition (April–May–June) on the 10–20-day and 30–60-day time scales during 1998–2014. They compared the relationship of various atmospheric and oceanic variables between the two ISOs. The percent variances explained by the two ISOs are comparable, but there are differences in the location of the largest value region. The 10–20-day SST variations have the largest percent variance in the equatorial western Pacific, whereas the 30–60-day SST variations have the largest percent variance in the North Indian Ocean, the South China Sea, and the North Pacific. The atmospheric variables and surface heat flux have a larger variance on the 10–20-day time scale than on the 30–60-day time scale. Prominent correlation is observed between SST and precipitation variations in the North Indian Ocean, the South China Sea, and the Philippine Sea on both 10–20-day and 30–60-day time scales. In comparison, the correlation between SST and surface heat flux variations is stronger and has a larger areal coverage on the 30–60-day time scale than on the 10–20-day time scale.

3.1.2 Interannual Air–Sea Interactions

Air–sea interaction not only determines the connection of local atmospheric and oceanic variations, it also plays an important role in the influence of remote forcing on regional climate. Remote forcing-induced atmospheric anomalies may be maintained through regional air–sea coupled processes, which provide a delayed impact of remote forcing on regional climate variability. In the El Niño decaying years, the air–sea interaction in the western North Pacific sustains the El Niño-induced anomalous anticyclone over the Philippine Sea from winter to the following summer (Wang et al. 2000, 2003; Wu et al. 2003) so that the El Niño's impact may be conveyed to the following East Asian summer monsoon variability. The East Asian winter monsoon-induced SST anomalies in the South China Sea may be sustained to the following summer through air–sea interaction processes and thus affect the East Asian summer monsoon (Chen et al. 2000; Wu et al. 2014). Thus, air–sea interaction provides a pathway for connecting the East Asian winter and summer monsoons.

This section introduces the relationship between interannual variations of precipitation and SST over the South China Sea and tropical western North Pacific in different seasons. The emphasis is on the unique feature of air–sea relationship over the South China Sea during the spring-to-summer transition.

1. Seasonality of interannual air–sea relationship in the South China Sea

The air–sea relationship on the interannual time scale displays prominent seasonality and regional feature in the South China Sea. Using the approach developed by Wu et al. (2006), He and Wu (2013) analyzed the nature of air–sea relationship in the South China Sea during the four seasons. It is found that the atmospheric forcing is dominant in the central and southern South China Sea during the warm season and the oceanic forcing is present in the northern South China Sea during the cold season (He and Wu 2013). According to a heat budget analysis of the ocean mixed layer during April–June, the atmospheric influence on SST is characterized by obvious cloud-radiation effect in the central South China Sea, by wind-evaporation effect in the central and southern South China Sea, and by wind-driven oceanic effect along the west coast (He and Wu 2013). During November–February, SST anomalies induce local convection through modulating lower-level convergence and atmospheric stability.

The seasonal change in the air–sea relationship is indicated by local precipitation-SST and precipitation-SST tendency correlation. A positive precipitation-SST correlation denotes an oceanic forcing of the atmosphere, whereas a negative precipitation-SST tendency correlation signifies an atmospheric forcing of the ocean (Wu et al. 2006). Figure 3.6 presents the distribution of correlation between precipitation and SST and between precipitation and SST tendency during the four seasons. The monthly data used to calculate the correlation are from the NOAA Optimal Interpolation SST version 2 (Reynolds et al. 2002) and the NASA Global Precipitation Climatology Project (GPCP) version 2 (Adler et al. 2003). In the South China Sea region, the positive precipitation-SST correlation is limited to the northern part during boreal winter (Fig. 3.6a). Negative precipitation-SST tendency correlation dominates in the other seasons (Fig. 3.6f, g, h). This indicates an oceanic influence on the atmosphere in the northern part during winter and mainly atmospheric impact on the ocean during the other seasons, consistent with He and Wu (2013). In the lower latitudes of the tropical western Pacific, positive precipitation-SST correlation is obvious during winter and spring, indicative of the important influence of local oceanic status on atmospheric variability. Different features are observed during summer and fall when the precipitation-SST correlation is relatively weak, whereas the precipitation-SST tendency correlation is negative, signifying an important role of the atmosphere in local SST variations. These results are in agreement with Wu and Kirtman (2007).

He and Wu (2013) evaluated the local air–sea relationship in the South China Sea in 24 climate models that participate in the Coupled Model Inter-comparison Project Phase 5 (CMIP5) based on the precipitation-SST and precipitation-SST tendency correlation. They obtained a large difference in local correlation in the South China Sea region among climate models. Most models have the worst performance in

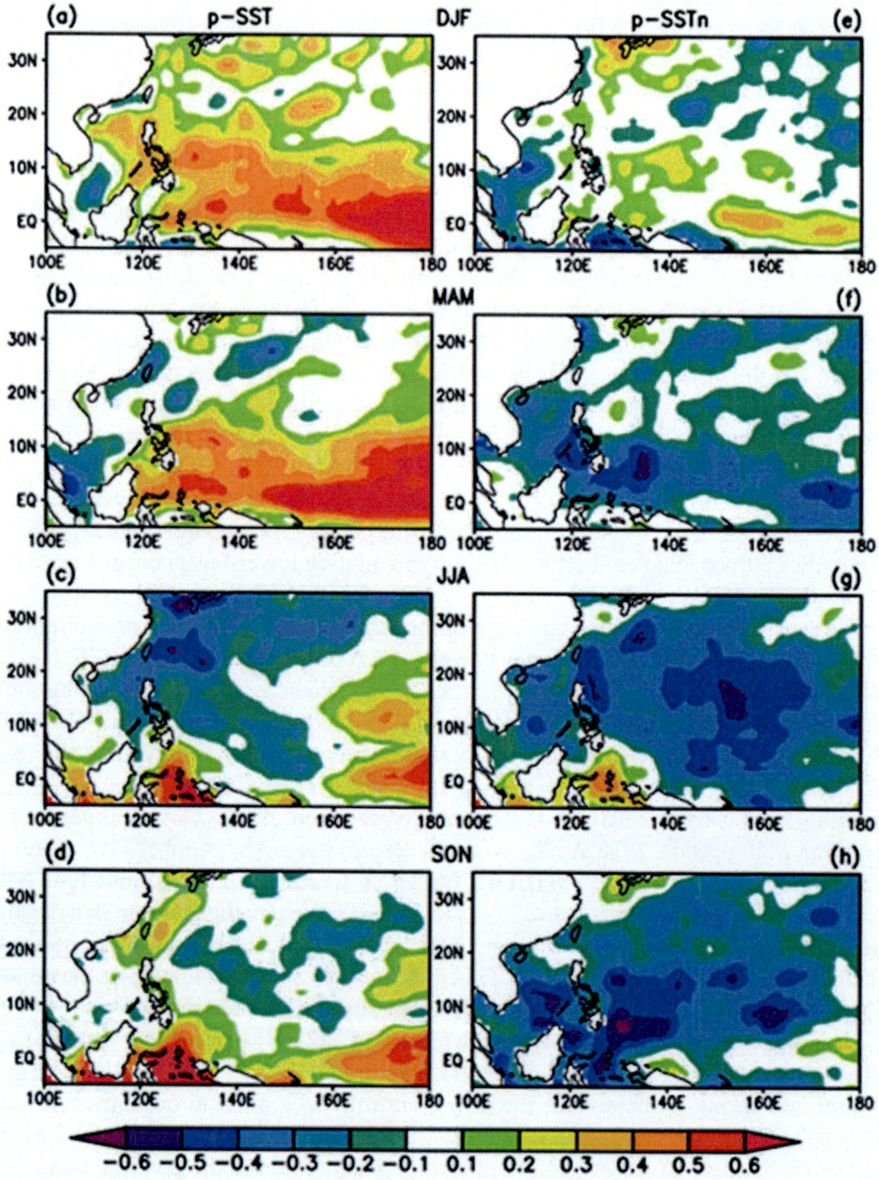

Fig. 3.6 Point-wise correlation of precipitation-SST and precipitation-SST tendency during 1982–2010 in winter (**a, e**), spring (**b, f**), summer (**c, g**), and fall (**d, h**). Shading denotes correlation coefficient reaching the 90% confidence level. The contour interval is 0.1. The data are from Reynolds et al. (2002) and Adler et al. (2003)

spring, and the atmospheric forcing is better captured than the oceanic forcing in the models (He and Wu 2013).

2. Unique feature of air–sea relationship during the spring-to-summer transition season in the South China Sea

The South China Sea summer monsoon onset is usually around mid-May. Thus, April–June signifies the spring-to-summer transition in the South China Sea and the surrounding regions. The air–sea relationship displays a unique feature during the spring-to-summer transition in the South China Sea, which manifests in the local relationship between precipitation and SST variations. Figure 3.7 displays the point-wise correlation between precipitation and SST, between precipitation and SST tendency, and between SST and precipitation tendency during April–June. The data used to calculate the correlation are the same as those in Fig. 3.6. There are obvious negative precipitation-SST tendency correlation and positive SST-precipitation tendency correlation in the South China Sea (Fig. 3.7b–c). This implies an air–sea interaction in the South China Sea. On the one hand, precipitation increase lowers SST. On the other hand, higher SST favors a subsequent increase in precipitation. This feature differs from the relationship in the equatorial central–eastern Pacific and tropical southwest Indian Ocean where there is a large positive precipitation–SST correlation (Fig. 3.7a), indicating an impact of SST anomalies on precipitation there (Wu and Kirtman 2007).

Hu and Wu (2016) analyzed the local correlation between monthly precipitation and SST anomaly departure (defined as the difference of monthly anomaly minus the seasonal mean anomaly during the spring-to-summer transition). The local correlation shows obvious different distributions among April, May, and June. In April, the delayed impact of SST on precipitation is observed in the Arabian Sea, the equatorial Indian Ocean, and the South China Sea (Fig. 3.8c), whereas the quick response of SST to atmosphere appears in the equatorial central–eastern Pacific (Fig. 3.8a). In May, there is a prominent negative correlation between precipitation and the SST tendency in the tropical Indian Ocean (Fig. 3.8e), indicative of more (less) precipitation followed by lower (higher) SST. In the South China Sea and the Philippine Sea, there is a positive correlation between SST and precipitation tendency (Fig. 3.8f), indicating a delayed impact of SST on precipitation. In June, positive precipitation-SST correlation is observed in the equatorial central–eastern Pacific, tropical Indian Ocean, and part of the Arabian Sea (Fig. 3.8g). There is positive correlation between SST and the precipitation tendency in the tropical Indian Ocean and the South China Sea (Fig. 3.8i) and a negative correlation between precipitation and the SST tendency in the North Indian Ocean and the South China Sea (Fig. 3.8h). This indicates a coupled feature of precipitation and SST change in the North Indian Ocean and the South China Sea.

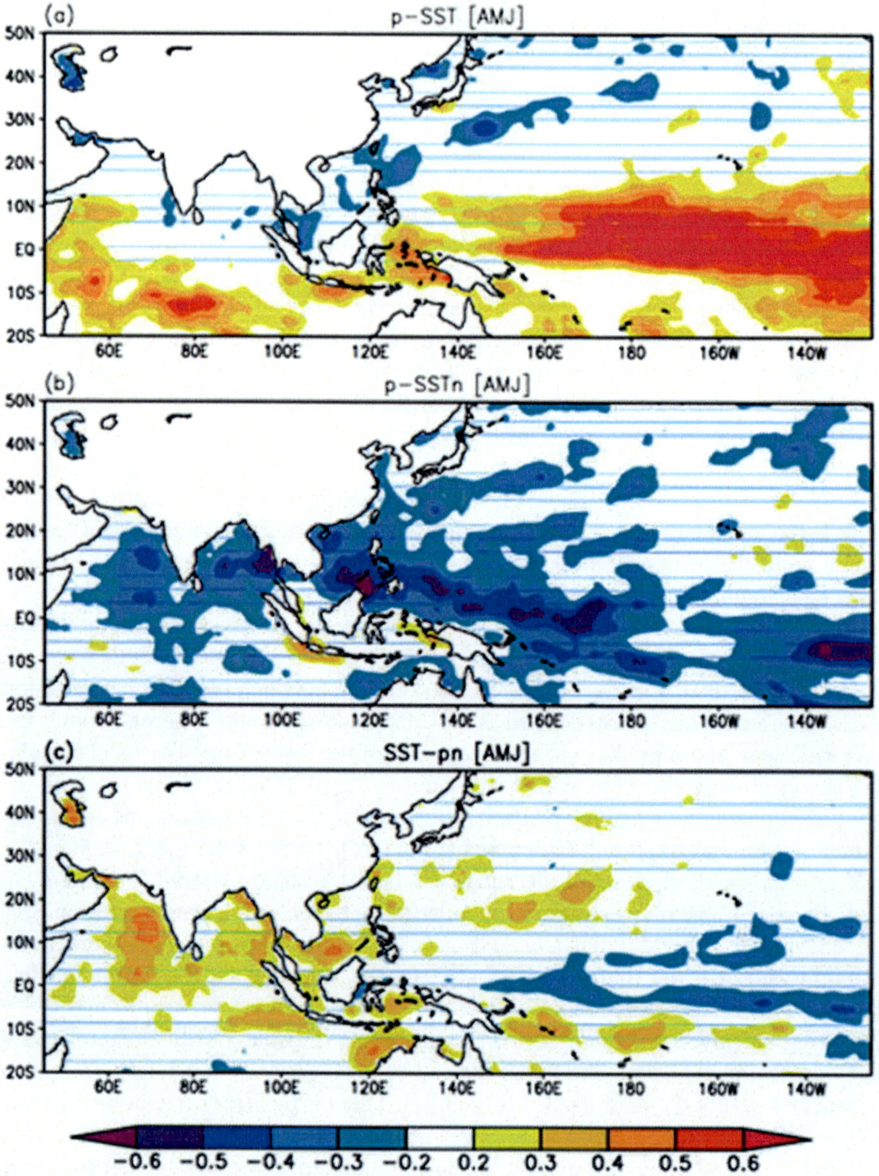

Fig. 3.7 Point-wise correlation of precipitation-SST (**a**), precipitation-SST tendency (**b**), and SST-precipitation tendency (**c**) during April–June of 1982–2010. (From Wu and Hu 2015)

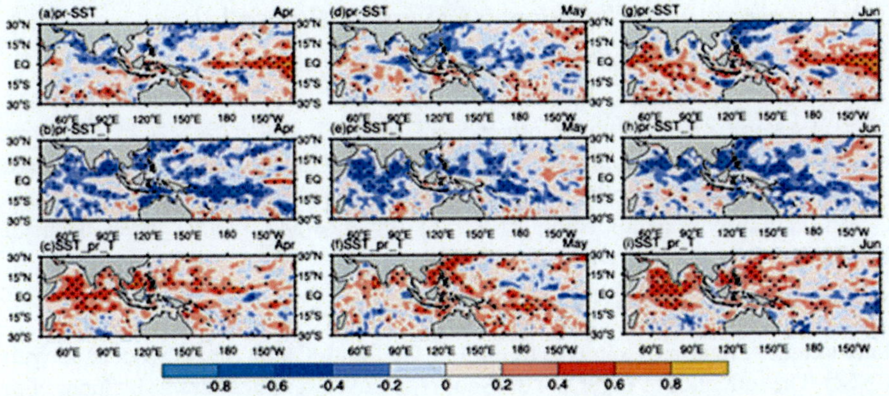

Fig. 3.8 Point-wise correlation of precipitation and SST, precipitation and SST tendency, and SST and precipitation tendency derived from monthly anomaly departures in **a–c** April, **d–f** May, and **g–i** June for the period 1979–2014. Dotted region denotes correlation coefficient significant at the 90% confidence level. (From Hu and Wu 2016)

3.2 Processes for Interannual Variability of Rainfall Over the South China Sea

Climate variability in the South China Sea is influenced by the thermal status in the tropical Pacific and Indian Oceans. The resultant change in the thermal condition in turn may induce anomalous heating in the overlying atmosphere and affect the climate over East Asia through atmospheric teleconnection. Thus, the South China Sea acts as a medium in the influence of tropical central–eastern Pacific and the Indian Ocean on the East Asian climate variability. The South China Sea is located in the junction of the East Asian monsoon, the South Asian monsoon, the western North Pacific monsoon, and the Australian monsoon. On the one hand, the climate variability in the South China Sea is related to the changes in those monsoons. On the other hand, it acts as a bridge connecting the changes among those monsoon systems.

This section first documents the factors of and processes for interannual climate variability over the South China Sea during boreal summer. Particular attention is paid to the influences of SST anomalies in the tropical Pacific and Indian Oceans. Then, this section investigates the occurrence of precipitation anomalies in the South China Sea during the spring-to-summer transition, including the seasonal mean precipitation anomalies and the departure of monthly mean anomalies from the seasonal mean anomalies.

3.2.1 *Interannual Variations of Summer Rainfall in the South China Sea*

This subsection is focused on the influences of tropical Indo-Pacific SST anomalies on the summer rainfall variability in the South China Sea. Influences of both concurrent and preceding tropical Pacific SST anomalies are considered. The roles of the Indian Ocean include both as a medium of ENSO influence and a factor independent of ENSO.

1. Roles of tropical Pacific SST anomalies

Interannual variation of summer rainfall in the South China Sea is associated with ENSO. The influence of ENSO on the South China Sea summer climate includes both direct and indirect impacts (He and Wu 2014). Concurrent positive SST anomalies in the equatorial central–eastern Pacific induce an anomalous lower-level cyclone over the South China Sea–western North Pacific through a Rossby wave-type response, leading to more rainfall over the northern South China Sea. Preceding winter negative SST anomalies in the equatorial central–eastern Pacific induce negative SST anomalies in the tropical Indian Ocean through the atmospheric bridge (Klein et al. 1999; Alexander et al. 2002). The Indian Ocean's negative SST anomalies maintain to the following summer and then lead to the formation of an anomalous cyclone over the western North Pacific through an anomalous west–east overturning circulation (Wu et al. 2014; He and Wu 2014), favoring more rainfall over the northern South China Sea. During this process, the North Indian Ocean SST anomalies act as a medium for the indirect influence of preceding equatorial Pacific SST anomalies on the succeeding summer rainfall over the South China Sea. Summer rainfall anomalies over the South China Sea are dominated by the direct influence of ENSO in some years, by the indirect influence of ENSO in some other years, and by both the direct and indirect influences of ENSO in some other years. When the direct and indirect influences of ENSO coexist, the resultant summer rainfall anomalies are more prominent over the South China Sea (He and Wu 2014). This points to the importance of synthesizing the direct and indirect influences of ENSO.

The influence of the Pacific and Indian Ocean SST anomalis on the South China Sea summer rainfall is verified in numerical experiments of the atmospheric general circulation model. In observations, SST anomalies are obtained in both the equatorial central–eastern Pacific and the North Indian Ocean corresponding to summer rainfall anomalies in the South China Sea (Fig. 3.9a). He and Wu (2014) conducted numerical experiments with the atmospheric model of the Community Earth System Model (CESM) with SST anomalies specified in the equatorial central–eastern Pacific, in the North Indian Ocean, and in both the above two regions. The model experiments show that the SST anomalies in the above regions induce abnormal rainfall in the South China Sea individually (Fig. 3.9c–f). When opposite SST anomalies are present in the two regions (Fig. 3.9g), the simulated rainfall and wind anomalies (Fig. 3.9h) are closer to observations (Fig. 3.9b). This confirms the importance of a combined effect of SST anomalies in different regions.

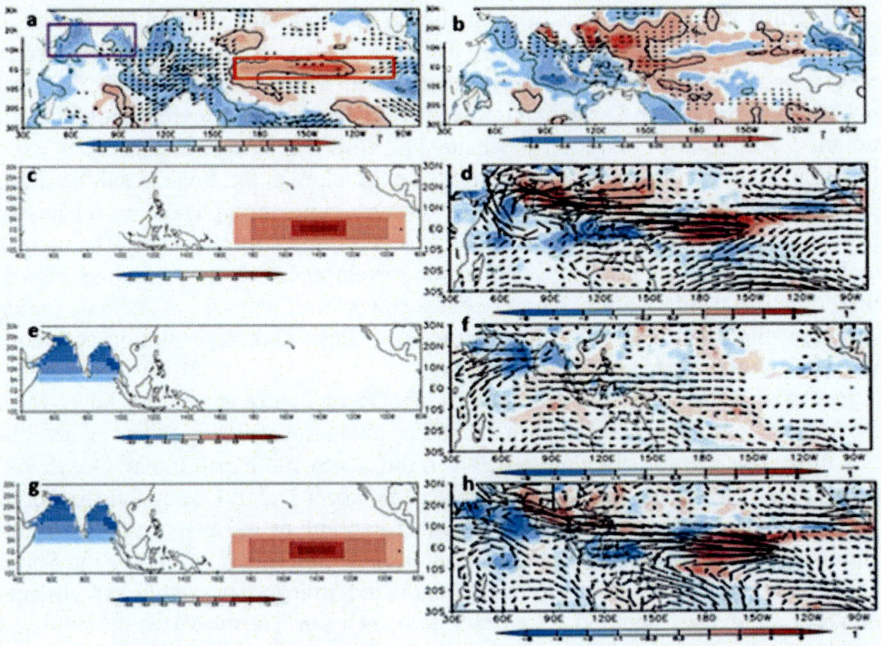

Fig. 3.9 Anomalies of **a** SST and 200 hPa wind, and **b** precipitation and 850 hPa wind corresponding to above-normal summer rainfall in the South China Sea in observations. SST anomalies specified in **c** the equatorial central–eastern Pacific and **d** the response of precipitation and 850 hPa wind, **e** the North Indian Ocean and **f** the response of precipitation and 850 hPa wind, and **g** both regions and **h** the response of precipitation and 850 hPa wind in the atmospheric model of CESM. (From He and Wu 2014)

The South China Sea summer rainfall has a negative correlation with concurrent SST anomalies in the tropical southeastern Indian Ocean. The latter induces a cross-equatorial north–south overturning circulation with the north branch over the South China Sea and thus affects the South China Sea rainfall variability (He and Wu 2014). This is validated by a numerical experiment with the atmospheric component of CESM with SST anomalies specified in the tropical southeastern Indian Ocean (He and Wu 2014).

2. Signal of the South China Sea summer rainfall variability in the tropical Indian Ocean

The Indian Ocean SST anomalies not only act as a medium in the influence of ENSO on the South China Sea summer rainfall variability, but also affect the South China Sea rainfall variability independent of ENSO. After removing the ENSO signal, the summer rainfall in the South China Sea has a negative correlation with the preceding winter SST anomalies in the equatorial western Indian Ocean and a positive correlation with concurrent summer SST anomalies in the tropical Indian Ocean (He et al. 2016).

Preceding winter negative SST anomalies in the equatorial western Indian Ocean and the associated zonal SST gradient form an anomalous west–east overturning circulation. The resultant lower-level westerlies and ascending motion over the tropical southeastern Indian Ocean induce negative SST anomalies there through cloud-radiation and wind-evaporation mechanisms. With the change in the mean wind direction when spring arrives, negative SST anomalies in the tropical southeastern Indian Ocean are maintained from spring to early summer through positive feedbacks of wind-ocean upwelling-SST and wind-evaporation-SST mechanisms. The associated south–north SST gradient leads to the development of an anomalous meridional overturning circulation. Consequently, lower-level anomalous cyclonic winds form over the South China Sea, leading to anomalous ascending motion and more precipitation there (He et al. 2016).

Positive SST anomalies in the South Indian Ocean induce anomalous lower-level convergence and more precipitation there. The anomalous heating drives an anomalous direct overturning circulation between the South and North Indian Ocean and an indirect overturning circulation between the North Indian Ocean and the South China Sea. This leads to the development of anomalous cyclones over the South China Sea–western North Pacific, resulting in more summer rainfall over the South China Sea (Fig. 3.10). He et al. (2016) conducted atmospheric model experiments with SST anomalies specified in the South Indian Ocean. The model results validated the observational analysis.

3. Combined effects of the Indian Ocean and western North Pacific SST anomalies

Wu et al. (2014) identified that the SST anomalies tend to be opposite in the North Indian Ocean and the tropical central North Pacific during the decaying years of El

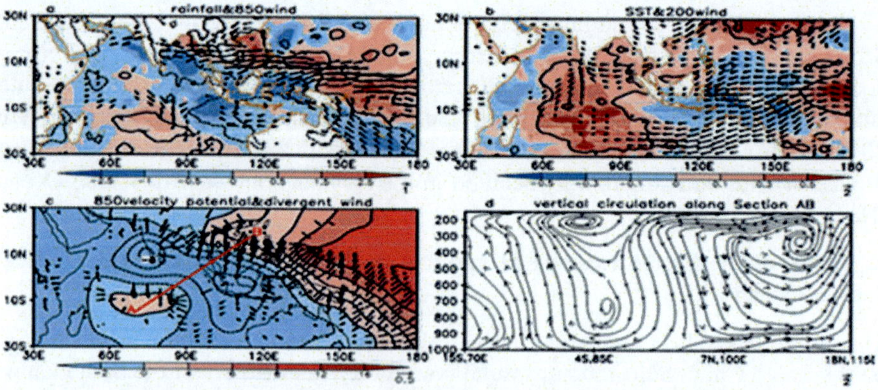

Fig. 3.10 Anomalies corresponding to more summer rainfall in the South China Sea after removing the ENSO signal. **a** Precipitation (mm/day) and 850 hPa wind (m/s), **b** SST (°C) and 200 hPa wind (m/s), **c** 850 hPa velocity potential (10^{-7} s^{-1}) and divergent wind (m/s), and **d** a cross section of vertical circulation along AB in (**c**). (From He et al. 2016)

Niño and La Niña events. This forms a west–east SST gradient between the North Indian and Pacific Oceans. The SST gradient is favorable to an in-phase relationship of rainfall in the South China Sea between preceding winter when El Niño/La Niña matures and decaying summer. Wu et al. (2014) demonstrated the impacts of the zonal SST gradient through atmospheric model experiments with specified SST anomalies. When positive SST anomalies are specified in the central North Pacific or negative SST anomalies specified in the North Indian Ocean, anomalous cyclone and more rainfall is induced over the South China Sea (Fig. 3.11a–b). When the SST anomalies are specified in both regions with opposite signs, the simulated wind and rainfall anomalies over the South China Sea are more prominent. This indicates the importance of the combined effects of the west–east contrasting SST anomalies in the South China Sea summer rainfall variability.

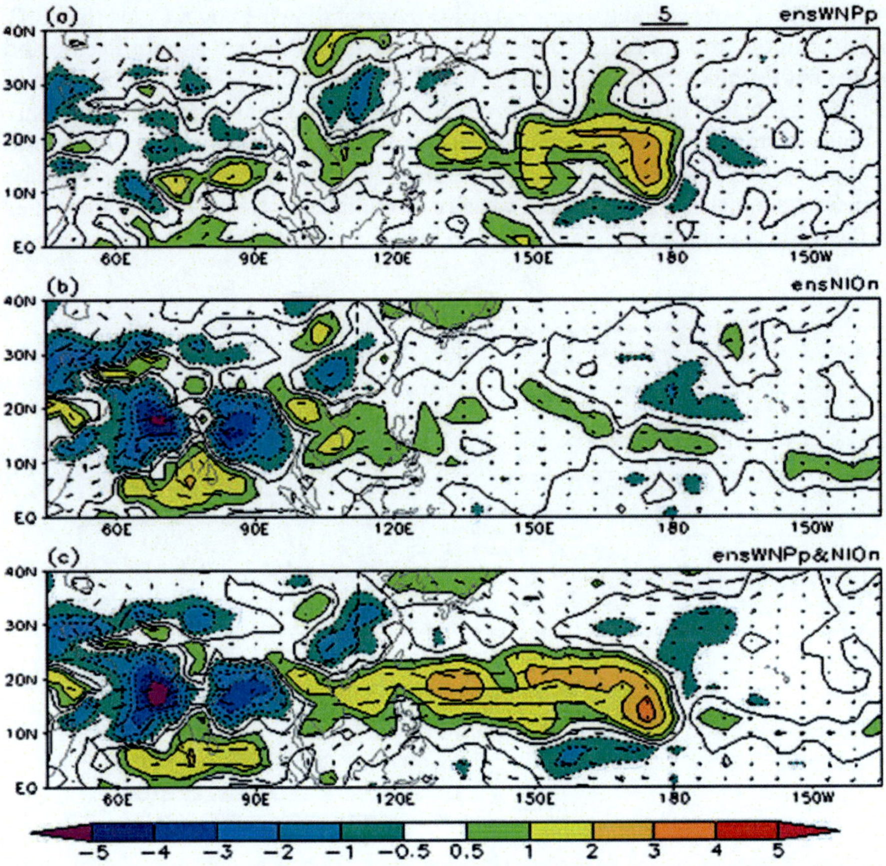

Fig. 3.11 Response of summer rainfall (shading, mm/day) and 850 hPa wind (vector, m/s, scale at the top) to positive SST anomalies specified in the region of 9°–21° N and 147.5°–180° E, negative SST anomalies in the North Indian Ocean between 9°and 21° N, and SST anomalies specified in both regions. The magnitude of SST anomalies is 0.5 °C at the center. (From Wu et al. 2014)

3.2.2 Interannual Variations Of the South China Sea Rainfall During the Spring-To-Summer Transition

This subsection introduces the impacts of SST anomalies in different regions of the tropical Pacific and Indian Oceans on the South China Sea rainfall during the spring-to-summer transition season. Particular attention is on the combined effects of SST anomalies in different regions.

1. Importance of combined effects of SST anomalies in different regions

The South China Sea rainfall during April–June is subject to influences of SST anomalies in the equatorial Pacific, tropical Indian Ocean, and the tropical western North Pacific. Anomalous rainfall in the South China Sea appears usually when SST anomalies coexist in two or three of the above regions and they act coherently (Hu et al. 2014). Corresponding to more April–June rainfall in the South China Sea, SST is lower in the tropical Indian Ocean and the equatorial central–eastern Pacific and higher in the tropical western North Pacific (Fig. 3.12a). Based on the combination of the SST anomalies in the above three regions, the South China Sea rainfall anomaly is divided into three types. In the first type, same sign SST anomalies in the equatorial central–eastern Pacific and the tropical Indian Ocean are accompanied by opposite sign SST anomalies in the tropical western North Pacific. Negative SST anomalies in the tropical southwestern Indian Ocean induce cross-equatorial flows that extend

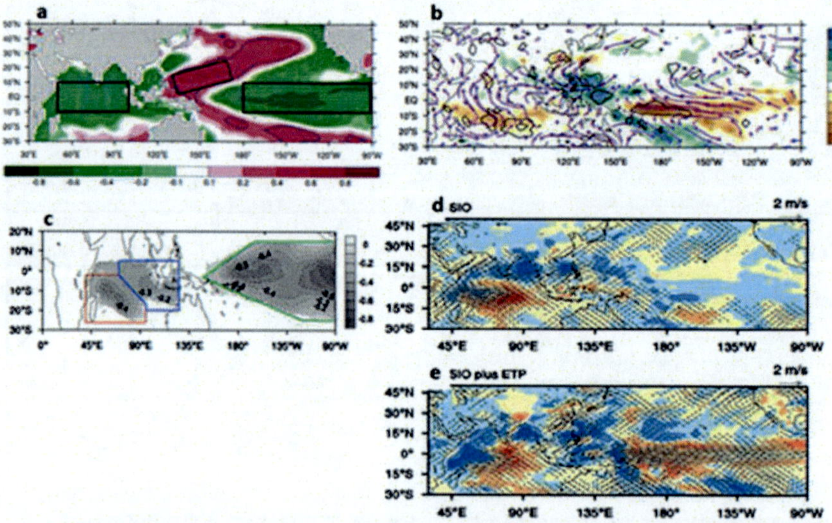

Fig. 3.12 Anomalies of SST (**a**) and precipitation and 850-hPa wind during April–June corresponding to above-normal precipitation over the South China Sea in observations. SST anomalies imposed in the model experiment (**c**), response of precipitation and 850-hPa wind to negative SST anomalies in the South Indian Ocean (**d**) and to negative SST anomalies in the South Indian Ocean and in the equatorial central–eastern Pacific (**e**) in the model. (From Hu et al. 2014)

eastward to the South China Sea. Negative SST anomalies in the equatorial central–eastern Pacific cause an anomalous Walker circulation with the ascending branch over the South China Sea. Their combined effect intensifies the convection over the South China Sea and the nearby regions. Meanwhile, positive SST anomalies in the tropical western North Pacific favor the intensification of convection over the South China Sea through a Rossby wave-type response. In the second type, opposite SST anomalies coexist in the equatorial central–eastern Pacific and the tropical western North Pacific. The latter is induced by the equatorial Pacific SST forcing. Positive SST anomalies in the tropical western North Pacific produce above-normal rainfall over the South China Sea through the Rossby wave-type response. In the third type, same sign SST anomalies coexist in the equatorial central–eastern Pacific and the tropical Indian Ocean. The SST anomalies off the coastal Sumatra produce a cross-equatorial overturning circulation that affects the South China Sea rainfall. When there are negative SST anomalies off the coastal Sumatra, anomalous descending motion is induced over the tropical southeastern Indian Ocean, whereas anomalous ascending motion develops over the South China Sea and western North Pacific.

Numerical experiments of an atmospheric general circulation model validated the impacts of SST anomalies in the tropical South Indian Ocean and the equatorial Pacific. When SST anomalies are imposed in the South Indian Ocean and the equatorial central–eastern Pacific simultaneously (Fig. 3.12c), the model produced circulation and precipitation response over the South China Sea (Fig. 3.12e) is closer to observations (Fig. 3.12b) than when SST anomalies are added only in the South Indian Ocean (Fig. 3.12d). The results demonstrate the importance of the combined effects of SST anomalies in different regions on the South China Sea rainfall variability during the spring-to-summer transition.

2. The relationship of the South China Sea rainfall during the spring-to-summer transition to tropical Indo-Pacific SST in climate models

Analysis of 23 IPCC CMIP5 coupled model simulations show that the ability of simulating the relationship of the South China Sea rainfall to tropical Indo-Pacific SST during the spring-to-summer transition varies largely among the models (Hu and Wu 2015). The spatial pattern of the South China Sea rainfall anomalies during the spring-to-summer transition displays obvious model dependence. The distribution of SST anomalies in the tropical Indo-Pacific region appears very important for models to capture the South China Sea rainfall variability. In some models, a realistic simulation of impacts of the equatorial western Pacific SST anomalies and local SST anomalies is essential for the models to obtain reasonable rainfall variance in the South China Sea. The impacts of the equatorial western Pacific SST anomalies are overestimated in some models, destructing the relationship of the South China Sea rainfall to the tropical Indo-Pacific SST.

Hu and Wu (2015) compared in detail the spatial distribution of tropical Indo-Pacific SST and precipitation anomalies and the temporal variations of SST anomalies in some key regions in models and observations. Based on the similarity of the spatial pattern and the magnitude of SST anomalies in key regions, they classified the 22 models into 5 categories. In the first category (10 models), the South China Sea rainfall during the spring-to-summer transition is subject to a combined influence of tropical Indian and Pacific SST anomalies, similar to observations. In the second category (4 models), the South China Sea rainfall is mainly affected by the tropical Pacific SST anomalies. In the third category (2 models), the South China Sea rainfall is mainly associated with the tropical Indian Ocean SST anomalies. In the fourth category (2 models), the equatorial western Pacific SST anomalies play a major role in the South China Sea rainfall variability. In the fifth category (5 models), the South China Sea rainfall variability is related to local SST variations.

Jiang et al. (2017) diagnosed the relationship between ENSO and the tropical western North Pacific climate in 37 IPCC CMIP5 climate models. They found a large discrepancy among the models. Through a comparative analysis, they showed that the ability of simulating the relationship between ENSO and tropical western North Pacific climate depends to a large extent upon whether the models simulate reasonably the pace of ENSO decay. When the ENSO decay is too slow in the models, warm SST bias in the equatorial western Pacific maintains to the succeeding summer, which is unfavorable to the formation of an anomalous anticyclone over the western North Pacific. This bias is related to the overly westward extension of the cold tongue, leading to an overly large zonal SST gradient. Under the large SST gradient, anomalous westerly winds induce an excessive warm zonal advection, which is in favor of the maintenance of warm SST bias. In turn, the warm SST bias causes anomalous convection and westerly winds. This positive feedback further maintains the warm SST anomalies in the equatorial western Pacific.

3.2.3 Influence Of Local Air–Sea Interaction on the South China Sea Climate During the Spring-to-Summer Transition

During the spring-to-summer transition over the South China Sea, the precipitation anomaly shows prominent month-to-month variations. The analysis shows that the contribution of precipitation anomaly departure is comparable or even larger than that of seasonal mean precipitation anomaly (Hu and Wu 2016). Figure 3.13 compares the standard deviation of April–June mean precipitation anomaly and April, May, and June precipitation anomaly departure (monthly anomaly minus April–June mean anomaly) for the period 1979–2014. Apparently, the standard deviation of monthly precipitation anomaly departure is larger than that of April–June mean precipitation anomaly over the North India Ocean, the South China Sea, and subtropical western North Pacific. This indicates that the monthly precipitation anomaly departure is an important component. Thus, it is necessary to consider both seasonal mean precipitation anomaly and the monthly precipitation anomaly departure over the South China Sea during the spring-to-summer transition seasons. Correlation analysis shows that the April-to-June precipitation anomaly change is a good indicator of summer mean precipitation anomaly over the South China Sea (Wu and Hu 2015).

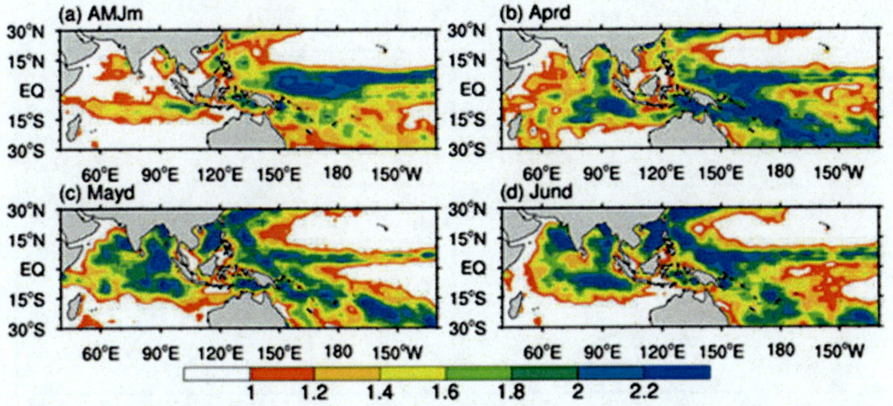

Fig. 3.13 Standard deviation (mm/day) of April–June mean precipitation anomaly (**a**), and April (**b**), May (**c**), and June (**d**) precipitation anomaly departure. (From Hu and Wu 2016)

The analysis reveals that the precipitation anomaly change from April to June over the South China Sea is closely associated with local air–sea interactions (Wu and Hu 2015). Corresponding to an increase in precipitation from April to June, the temporal evolution of regional anomalies is characterized by a sequence of less precipitation-higher SST-more precipitation-SST decrease (Fig. 3.14). Such a temporal relation is illustrated by comparing the area-mean precipitation and SST anomaly evolution over the South China Sea (Fig. 3.15). This differs remarkably from the seasonal mean precipitation anomaly that is mainly a response to remote SST forcing in the equatorial Pacific and tropical Indian Ocean (Wu and Hu 2015; Hu and Wu 2016).

To illustrate the role of air–sea interaction in the precipitation anomaly change over the South China Sea, Wu and Hu (2015) conducted a heat budget analysis of the ocean mixed layer. It shows that the precipitation decrease over the South China Sea in April is accompanied by a reduction in cloudiness and a weakening in surface wind speed, leading to an increase in net surface shortwave radiation and a decrease in latent heat flux. Consequently, the SST increases and positive SST anomalies appear in May. This is followed by an increase in precipitation in June due to the SST forcing of the atmosphere. In turn, this is accompanied by more cloudiness and enhanced surface wind speed, leading to a decrease in net surface shortwave radiation and an increase in latent heat flux. Accordingly, the SST turns to decrease and negative SST anomalies appear in June.

Spatial-temporal evolution of monthly precipitation anomaly departure (obtained as monthly anomaly minus seasonal mean anomaly) during the spring-to-summer transition reveals strong local air–sea coupled processes over the South China Sea. Hu and Wu (2016) compared the difference of the major spatial mode and the associated air–sea interaction processes corresponding to seasonal mean precipitation anomaly and monthly precipitation anomaly departure over the South China Sea and the tropical Indian Ocean during the spring-to-summer transition season. It is found that the remote forcing of the equatorial central Pacific is a major reason of April–June mean

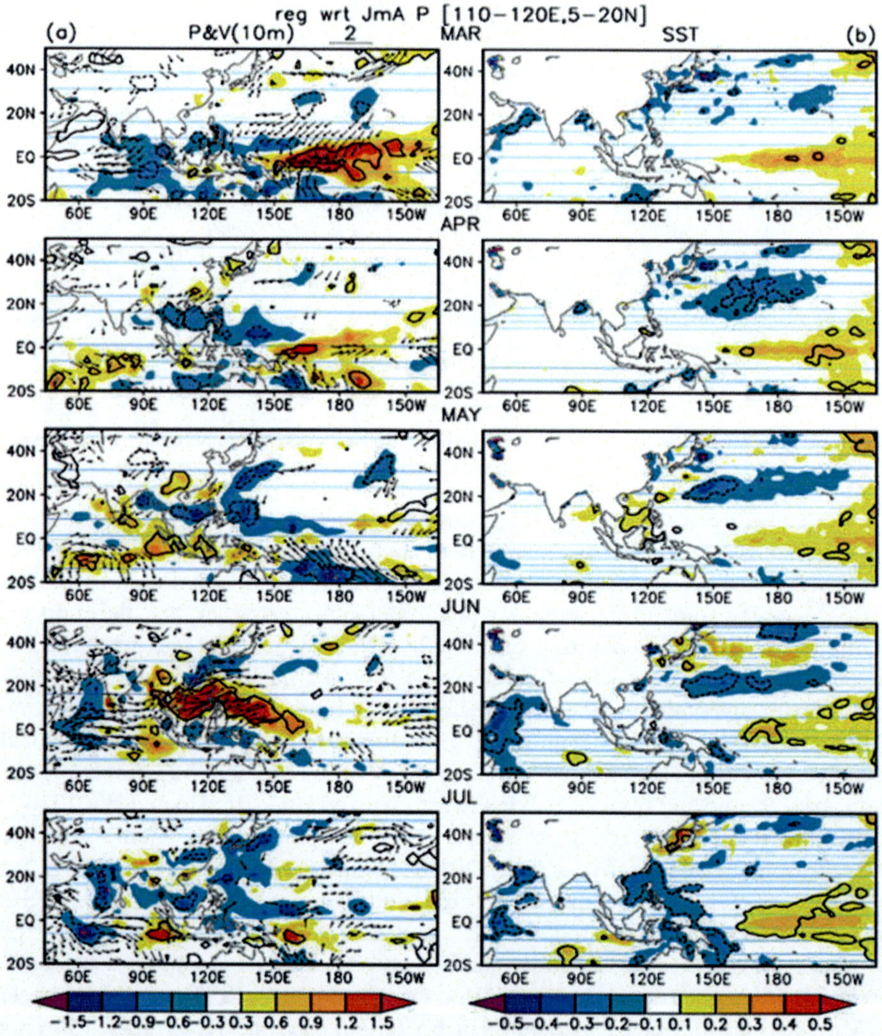

Fig. 3.14 Anomalies of precipitation (shading, mm/day) and surface (10 m) wind (vector, m/s) (**a**) and SST (shading, °C) (**b**) corresponding to precipitation change from April to June over the South China Sea. (From Wu and Hu 2015)

precipitation anomaly over the South China Sea and the tropical Indian Ocean and locally the atmosphere has a negative effect on SST change. A strong local air–sea coupled process is involved in the monthly precipitation anomaly departure, featuring a temporal sequence of less precipitation-SST increase-precipitation increase-SST decrease. Figure 3.16 shows the spatial-temporal evolution of monthly precipitation and surface wind anomaly obtained by regression against the leading mode of monthly precipitation anomaly departure over the South China Sea. The precipitation

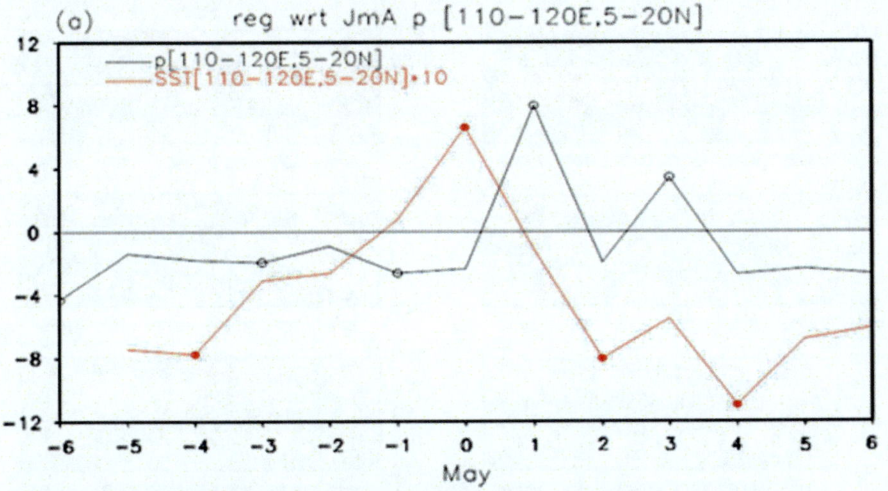

Fig. 3.15 Monthly anomalies of precipitation (black, mm/day) and SST (red, °C) averaged over the South China Sea corresponding to June-minus-April precipitation anomaly change over the region of 5°–20° N and 110°–120° E. (From Wu and Hu 2015)

data used is from GPCP version 2 (Adler et al. 2003), the SST data is from TropFlux (Kumar et al. 2012), and the wind data is from NCEP-DOE reanalysis 2 (Kanamitsu et al. 2002). According to the figure, the temporal evolution in the South China Sea region is characterized by less precipitation in April–May, warmer SST in May, more precipitation in June, and lower SST in June–July (Hu and Wu 2016). Thus, the SST anomaly departure in May serves as an indicator of precipitation anomaly in June.

Wu and He (2017) analyzed the impacts of both April–June mean precipitation anomaly and June-minus-April precipitation anomaly change during the spring-to-summer transition over the South China Sea. The temporal evolution of precipitation, wind, and SST anomalies corresponding to the two types of anomalies reveals two processes leading to anomalous spring-to-summer transition over the South China Sea. One process is associated with the large-scale SST anomalies in the tropical Indo-Pacific region. During the La Niña decaying spring, equatorial central Pacific negative SST anomalies coexist with tropical southwestern Indian Ocean negative SST anomalies and tropical western North Pacific positive SST anomalies. Negative SST anomalies in the equatorial central Pacific force an anomalous Walker circulation, negative SST anomalies in the tropical southwestern Indian Ocean induce cross-equatorial flows that turn to go eastward north of the equator to the South China Sea, and positive SST anomalies in the tropical western North Pacific produce a Rossby wave-type response to the west. Their combined effects lead to enhanced convection and the development of an anomalous lower-level cyclone over the South China Sea. The persistent anomalies induce an earlier transition from spring to summer over the South China Sea. The other process is associated with regional air–sea interaction processes around the Maritime Continent. Preceding positive SST anomalies in

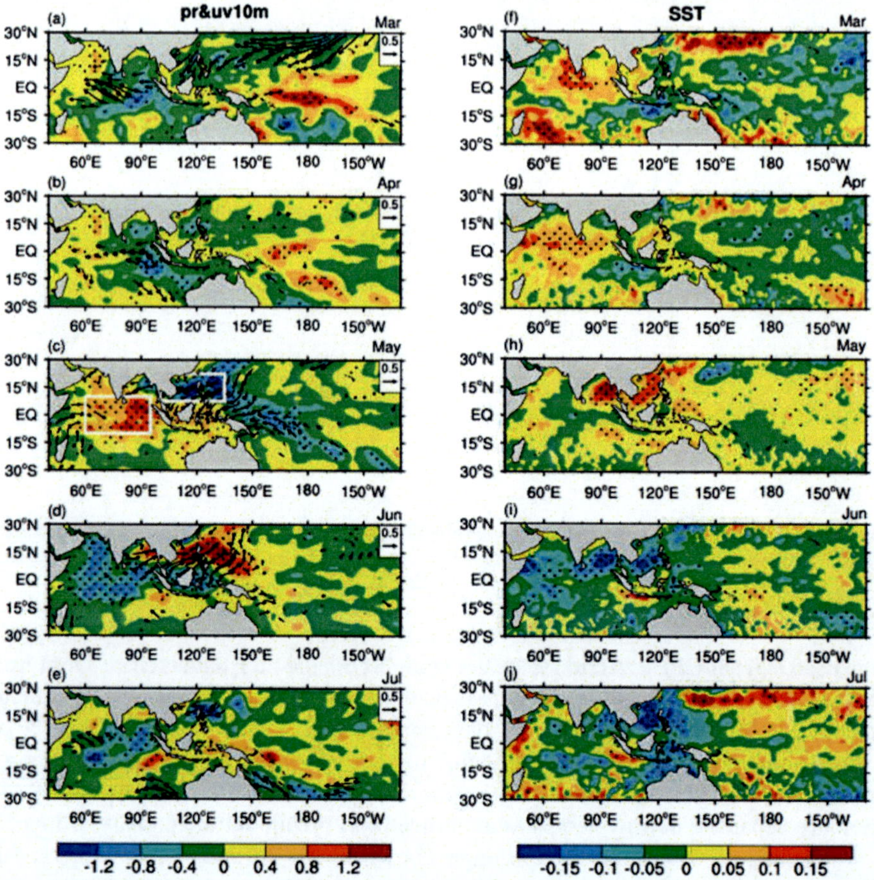

Fig. 3.16 Anomalies of precipitation (shading, mm/day) and surface (10 m) wind (vector, m/s) (**a**) and SST (°C) corresponding to the first EOF mode of monthly precipitation anomaly departure during April through June for the period 1979–2014. Dotted region denotes anomalies significant at the 90% confidence level. Only wind vectors significant at the 90% confidence level are drawn. (From Hu and Wu 2016)

the equatorial central Pacific in March induce anomalous descending over the region surrounding the Maritime Continent, leading to SST increase in both the South China Sea and the tropical southeastern Indian Ocean. Meanwhile, an anomalous convection region moves eastward along the tropical South Indian Ocean and reaches northwest of Australia in May. This induces an anomalous meridional overturning circulation across the Maritime Continent with an ascent over the tropical southeastern Indian Ocean and descent over the South China Sea, leading to further increase of SST in the South China Sea. Following the SST warming, anomalous convection develops over the South China Sea in June, inducing an accelerated transition from spring to summer. The analysis shows that the above two processes are equally important in anomalous spring-to-summer transition over the South China Sea.

3.3 Interdecadal Variation

Obvious interdecadal changes have been detected in climate over East Asia and surrounding the South China Sea (Chan and Zhou 2005; Ding et al. 2008; Wu and Mao 2017a). The interannual relationship between ENSO and climate over East Asia and the region surrounding the South China Sea is nonstationary with a prominent interdecadal difference (Wu and Wang 2002; Gao et al. 2006; Wang et al. 2008; Wu et al. 2012a, b; Chen et al. 2014). An important reason for the unsteady relationship is the modulation of the Pacific Decadal Oscillation (PDO) (Chen et al. 2013; Feng et al. 2014; Wu and Mao 2016, 2017b). This section introduces the interdecadal climate change in the region surrounding the South China Sea, analyzes the interdecadal difference in the interannual relationship, explores the impacts of the PDO, and investigates the physical processes of the modulation of PDO on the ENSO-related atmospheric circulation and eastern China spring rainfall through numerical experiments.

3.3.1 Interdecadal Variability Of Early Summer Monsoon Rainfall Over South China In Association With the Pacific Decadal Oscillation

Climatologically, early summer (May–June) is the major rainy season over South China (Chan and Zhou 2005). Thus, the bimonthly (May–June) rainfall is usually used as an index to represent the South China monsoon rainfall (SCMR) during early summer (Mao et al. 2011). SCMR exhibits strong interannual variations and is subject to modulation by the Pacific Decadal Oscillation (PDO) (Zhou et al. 2006). Mao et al. (2011) identified that the dominant atmospheric teleconnection patterns associated with extremely wet and dry SCMR conditions are remarkably different during the epochs 1958–1976 and 1980–1998, corresponding to negative and positive PDO phases. As suggested by Duan et al. (2013), the rainfall amount in the single month of June is better than the bimonthly (May–June) rainfall amount to represent the SCMR. Wu and Mao (2017a) thus used the June rainfall to measure SCMR. Their analysis revealed both interannual and interdecadal variations of SCMR (Fig. 3.17). The interdecadal fluctuations of SCMR exhibit two negative epochs (1926–1945, 1977–2000) and two positive epochs (1946–1976, 2001–2013). Such interdecadal variations of SCMR show an out-of-phase relationship with interdecadal fluctuations of the wintertime PDO index, as identified by its 11-year running mean. The correlation between the interdecadal SCMR and wintertime PDO indices is up to

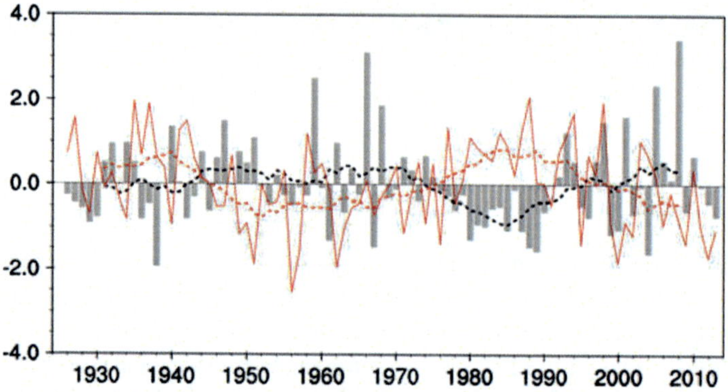

Fig. 3.17 Standardized time series of SCMR (bars) and wintertime (November–March)-averaged PDO index (solid line). The 11-year running mean values of SCMR (wintertime PDO index) are indicated by the dashed (long-dashed) line. (From Wu and Mao 2017a)

−0.80, which is statistically significant at the 99% confidence level according to the calculated effective degree of freedom. Differences of composite June rainfall anomalies between positive PDO epochs (1926–1945 and 1977–2000) and negative PDO epochs (1946–1976 and 2001–2013) confirm that SCMR tends to be below-normal in positive PDO epochs, while SCMR is often above-normal in negative PDO epochs.

The link between the interdecadal variability of SCMR and PDO is through circulation teleconnections in response to PDO-related SST anomalies (Wu and Mao 2017a). As shown in Fig. 3.18 that displays the differences of composite lower tropospheric circulations and SSTs in June, during positive PDO epochs, significant negative SST anomalies are present in the mid-latitude North Pacific, and positive SST anomalies exist in the eastern tropical Pacific and off the west coast of North America and Gulf of Alaska (Fig. 3.18a). Significant positive SST anomalies are also present in the South China Sea and the western North Pacific (Fig. 3.18a). Although these positive anomalies are weak compared to those in the eastern Pacific, they can nevertheless induce an anomalous low pressure at the surface over the South China Sea–western North Pacific (Fig. 3.18b) and an anomalous cyclone in the lower and middle troposphere (Fig. 3.18c and 3.18d). This is accompanied by a positive sea level pressure (SLP) anomaly to the north over eastern China and another negative SLP anomaly over the Lake Baikal (Fig. 3.18b), roughly forming a meridionally arranged vortex (MAV) pattern. Such an MAV pattern is more evident in the lower and middle troposphere (Fig. 3.18c and 3.18d), which is characterized by an anomalous cyclone, an anomalous anticyclone, and an anomalous cyclone alternately over the South China Sea–western North Pacific, eastern China, and the Lake Baikal. As the southern segment of the MAV pattern, the anomalous cyclone over the South China Sea–western North Pacific is forced by both the *in situ* positive SST anomalies and those remotely in the central and eastern equatorial Pacific (Fig. 3.18a).

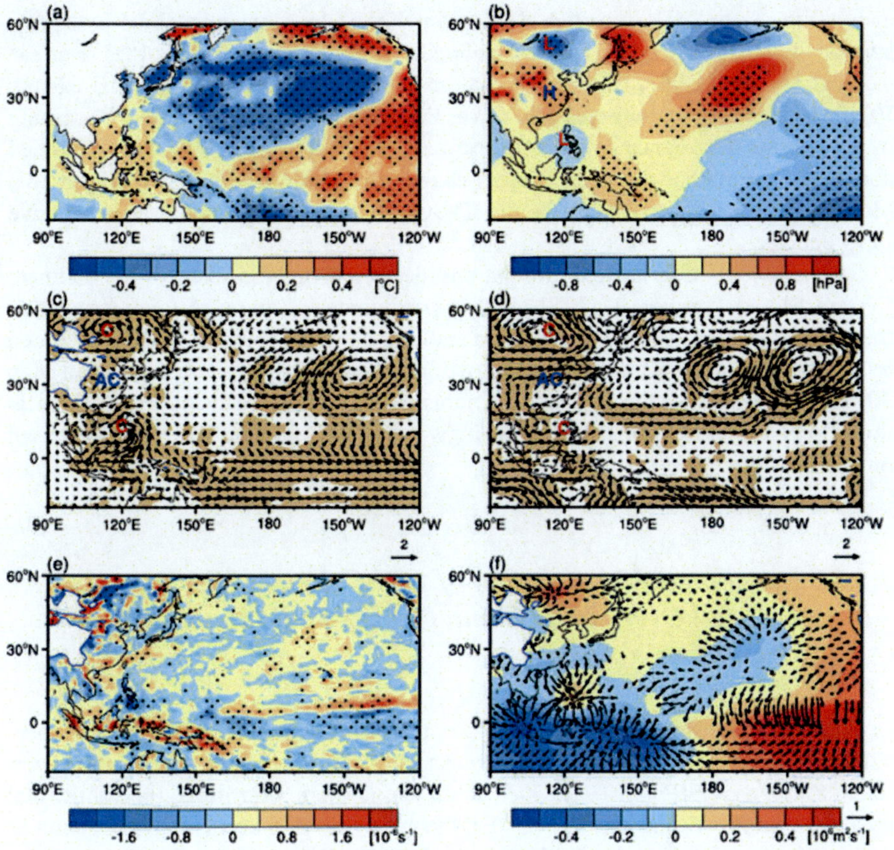

Fig. 3.18 Composite differences in June for positive PDO (1977–2000) minus negative PDO (1958–1976 and 2001–2013) of **a** SST (shading, °C), (b) SLP (shading, hPa), **c** 850-hPa winds (vectors, m s^{-1}), **d** 500-hPa winds (vectors, m s^{-1}), **e** 850-hPa divergence (shading, 10^{-6} s^{-1}), and **f** 850-hPa velocity potential (shading, 10^6 m^2 s^{-1}) along with divergent winds (vectors, m s^{-1}). The stippling in **a**, **b**, and **e** denotes anomalies statistically significant at the 95% confidence level, and the shading in **c** and **d** indicates the areas where at least one of the zonal and meridional wind components is statistically significant at the 95% confidence level. Only the statistically significant anomalous divergent winds are shown in (**f**). The letters L and H in **b** denote the centers of anomalous low and high pressures, respectively. The letters C and AC in **c** and **d** denote the centers of anomalous cyclones and anticyclones, respectively. (From Wu and Mao 2017a)

As shown in Fig. 3.18f, notable divergent winds blow from the equatorial eastern Indian Ocean and western Pacific toward the Philippines. This implies that the anomalous cyclone over the South China Sea–western North Pacific occurs not only in response to local positive SST anomalies, but also is related to the remote SST anomalies in the tropical Pacific. Indeed, significant positive SST anomalies are present in the tropical eastern Pacific, exhibiting a pattern similar to an El Niño event (Fig. 3.18a). The positive SST anomalies induce an anomalous low pressure at the surface (Fig. 3.18b) and anomalous convergence (Fig. 3.18e) over the tropical eastern Pacific. In turn, anomalous convergence over the tropical eastern Pacific induces anomalous westerlies over the tropical Pacific (Fig. 3.18c), resulting in anomalous divergence over the Maritime Continent (Fig. 3.18e). As mentioned above, the anomalous divergent westerlies over the Maritime Continent (Fig. 3.18f) blow toward the tropical eastern Pacific. The opposite situation is observed in negative PDO epochs.

The analyses of historical simulations derived from the Community Climate System Model (version 4, CCSM4) that participates in the CMIP5 indicate that the CCSM4 model reproduces well the anomalous atmospheric circulations forced by the PDO-related SST anomaly pattern during positive PDO epochs (Wu and Mao 2017a). This validates the physical mechanism responsible for the out-of-phase relationship of the interdecadal variations between SCMR and the PDO derived based on observations.

3.3.2 *Interdecadal Modulation of ENSO-Related Spring Rainfall Over South China by the Pacific Decadal Oscillation*

In eastern China, spring is a transitional season from winter to summer monsoons. Climatological spring rainfall rate exhibits a southward-increasing distribution from 0.2 to 8 mm day^{-1} (Fig. 3.19a). Over South China (here referring to the continental area south of 28° N and east of 105° E), the spring rainfall accounts for as much as 30%–40% of annual total rainfall (Wu et al. 2003; Wu and Mao 2016) with large

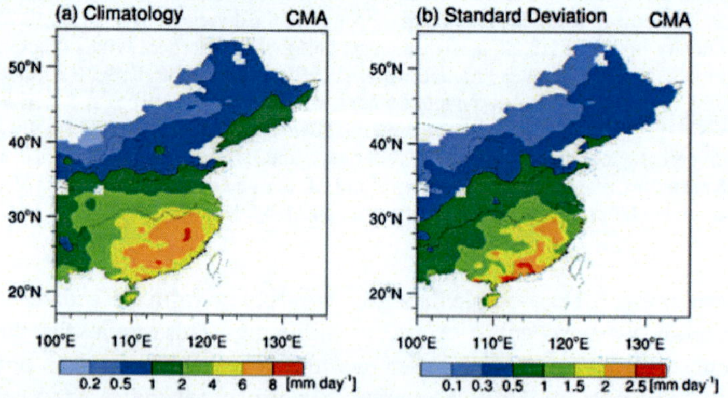

Fig. 3.19 Spring (March–May) mean rainfall over eastern China based on the China Meteorological Administration (CMA) rain-gage station observations for the period 1958–2013. **a** Climatology (shading, mm day^{-1}) and **b** interannual standard deviation (shading, mm day^{-1}). (From Wu and Mao 2017b)

interannual standard deviation exceeding 2 mm day^{-1} (Fig. 3.19b), indicative of frequent severe floods and droughts over South China. North of the Yangtze River Basin, climatological spring rainfall is lower than that over South China (Fig. 3.19a). However, the interannual standard deviation over most of northern China (Fig. 3.19b) is not small compared to climatological spring rainfall, with the magnitude about half of the climate-mean amount (Fig. 3.19b). The relatively large standard deviations over South China imply that anomalous rainfall may have a huge impact on local socioeconomic development.

Two rain-gage (gridded) rainfall data sets with different lengths of temporal coverage from China Meteorological Administration (CMA) and Climate Research Unit (CRU) for the period 1958–2013 (1920–2013) are applied to conduct an empirical orthogonal function (EOF) analysis to identify the spatial patterns of interannual variations of spring rainfall over eastern China. The dominant spatial patterns of the first two leading EOFs of the spring rainfall anomaly percentages over eastern China derived from CMA rain-gage data are shown in Fig. 3.20a and 3.20b. The first mode (EOF1) and the second mode (EOF2) account for about 19.46% and 10.01% of the total variance, respectively. Each of these two modes is separated from the subsequent mode based on the method proposed by North et al. (1982). Similar spatial patterns are derived from the CRU rainfall data with a longer time period (Fig. 3.20c and 3.20d). The EOF1 mode is characterized by a monopole structure (Fig. 3.20a and 3.20c), with coherently positive rainfall anomalies over northern China extending from the Yangtze to northern China (YNC). The EOF2 mode shows a north–south dipole structure with predominant positive rainfall anomalies over South China and negative anomalies over the lower reaches of the Yellow River (Fig. 3.20b and 3.20d). The high similarities of the dominant EOF structures demonstrate that the EOF1 and EOF2 modes indeed represent two fundamental spatial structures of spring rainfall variability over eastern China.

The principal component (PC) time series are examined to investigate the temporal variations of the two leading EOF modes. The longer normalized PC1 and PC2 time series from the CRU data sets are used for this purpose (not shown). Moreover, the longer time series is more suitable for detecting interannual variations of spring rainfall anomalies and their interdecadal modulations associated with the PDO. The temporal evolutions of both PC1 and PC2 exhibit distinct year-to-year fluctuations with large amplitudes greater than one standard deviation in many years. Such strong interannual variability of PC1 is associated with ENSO events. The correlation coefficient between PC1 and the Niño3.4 index is 0.23, which is statistically significant at the 95% confidence level. However, the correlation between PC2 and Niño3.4 index is only 0.08. This, however, does not mean that the interannual variability of the EOF2 mode has nothing to do with ENSO events; rather, the interannual relationship between PC2 and ENSO undergoes interdecadal changes. The spatial pattern of EOF2 (Fig. 3.20b and 3.20d) bears a strong resemblance to the distribution of composite spring rainfall anomalies over South China under the in-phase PDO–ENSO combination, as reported by Wu and Mao (2016), with significant positive rainfall anomalies over South China.

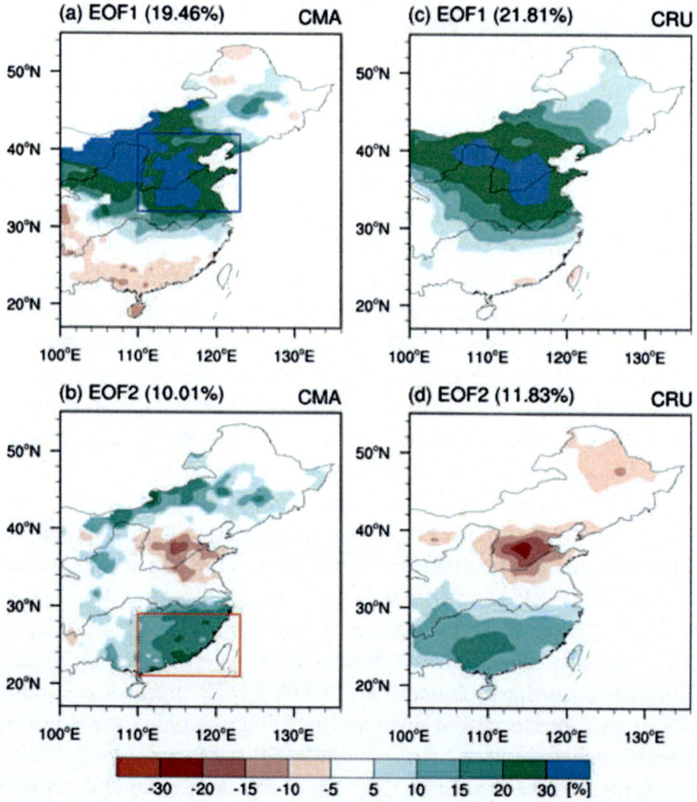

Fig. 3.20 Spatial patterns of the first two leading EOFs of the spring rainfall anomaly percentages over eastern China. **a** EOF1 and **b** EOF2 based on the CMA rain-gage station observations for the period 1958–2013. The loadings (unit %) are scaled by one standard deviation of the corresponding PC. The percentage in parenthesis at the top of each panel refers to the variance explained by the mode. **c** and **d** are the same as **a** and **b** but based on the CRU data sets for the period 1920–2013. (From Wu and Mao 2017b)

The relationship between the interannual variations of the dominant EOF modes with ENSO events under different PDO backgrounds is examined based on the contingency tables. The numbers of positive or negative PC1 or PC2 years are counted up separately for the four PDO–ENSO combinations, as in Wu and Mao (2016). The spring rainfall anomalies of the EOF1 mode is more likely associated with out-of-phase PDO–ENSO events [i.e., El Niño during cold PDO (EN_CPDO) and La Niña during warm PDO (LN_WPDO)], while rainfall anomalies of the EOF2 mode are more likely connected to in-phase PDO–ENSO events [i.e., El Niño during warm PDO (EN_WPDO) and La Niña during cold PDO (LN_CPDO)].

The PDO modulation of the ENSO-related spring rainfall anomalies over eastern China is through anomalous circulations induced by the SST anomaly distributions of the combined PDO–ENSO events (Wu and Mao 2016, 2017b). The typical SST anomaly distribution of an El Niño (La Niña) event is characterized by strong positive (negative) anomalies over the central and eastern equatorial Pacific and weak negative (positive) anomalies over the extratropical North Pacific. A similar SST anomaly pattern occurs in a warm (cold) PDO year except that the magnitude of PDO-based SST anomaly in the mid-latitude North Pacific is larger than that in the equatorial Pacific (Mantua et al. 1997). Therefore, if El Niño events occur in warm PDO phases, the resultant EN_WPDO events would be characterized by both enhanced positive SST anomalies in the equatorial Pacific and enhanced negative SST anomalies in the mid-latitude North Pacific due to in-phase superposition of SST anomalies (Wu and Mao 2017b). Similarly, the LN_CPDO events exhibit a similarly enhanced SST anomaly pattern to the EN_WPDO, but with reversed signs of the SST anomalies. As mentioned above, the EN_WPDO and LN_CPDO in-phase PDO–ENSO events exhibit distinctly opposite polarities in the distribution of SST anomalies so that the EN_WPDO minus LN_CPDO differences simply represent anomalous condition of EN_WPDO events, with the opposite situation for LN_CPDO events. In addition, such differences in SST anomalies between EN_WPDO and LN_CPDO categories highlight the in-phase superposed effects that force anomalous circulations (as discussed below). In contrast, for the out-of-phase PDO–ENSO categories, the ENSO-related SST anomalies and PDO-related SST anomalies tend to cancel each other. Therefore, if El Niño events occur during a cold PDO phase, the resultant EN_CPDO events have weak positive SST anomalies in the equatorial Pacific and weak positive SST anomalies in the North Pacific because the amplitude of SST anomalies in the equatorial Pacific for El Niño events is usually larger than that for cold PDO episodes, but it is opposite in the mid-latitude North Pacific. The LN_WPDO events exhibit a weakened SST anomaly distribution similar to the EN_CPDO events, but with reversed signs of SST anomalies (Wu and Mao 2017b). Thus, the out-of-phase superimposed effects in SST anomalies along with the resultant circulation anomalies (as discussed below) are also highlighted by the composite differences between EN_CPDO and LN_WPDO events.

The circulation anomalies forced by the in-phase and out-of-phase composite SST anomalies are shown in Fig. 3.21 in terms of the SLP and lower tropospheric winds together with moisture conditions (Wu and Mao 2017b). In the in-phase composites, a strong anomalous anticyclone associated with positive SLP anomalies dominates the western North Pacific, accompanied by a simultaneously enhanced Aleutian Low over the North Pacific (Fig. 3.21a and 3.21d). The strong anomalous anticyclone is forced by the enhanced SST anomalies in the central and eastern equatorial Pacific (Fig. 3.21c). As suggested by Wang et al. (2000), the anomalous western North Pacific anticyclone results from a Rossby wave response to suppressed convective heating, which is induced by both the in situ ocean surface cooling and the subsidence forced remotely by the central Pacific warming. The in-phase superposition of PDO and ENSO enhances positive SST anomalies in the central and eastern equatorial Pacific (Fig. 3.21c) that excite a Rossby wave response over the western North Pacific, and then strengthen the anomalous anticyclone over the western North Pacific (Fig. 3.21a and 3.21d). Moreover, associated with the intensified negative SST anomalies in the central North Pacific (Fig. 3.21c), a strong and westward-extended Aleutian Low is

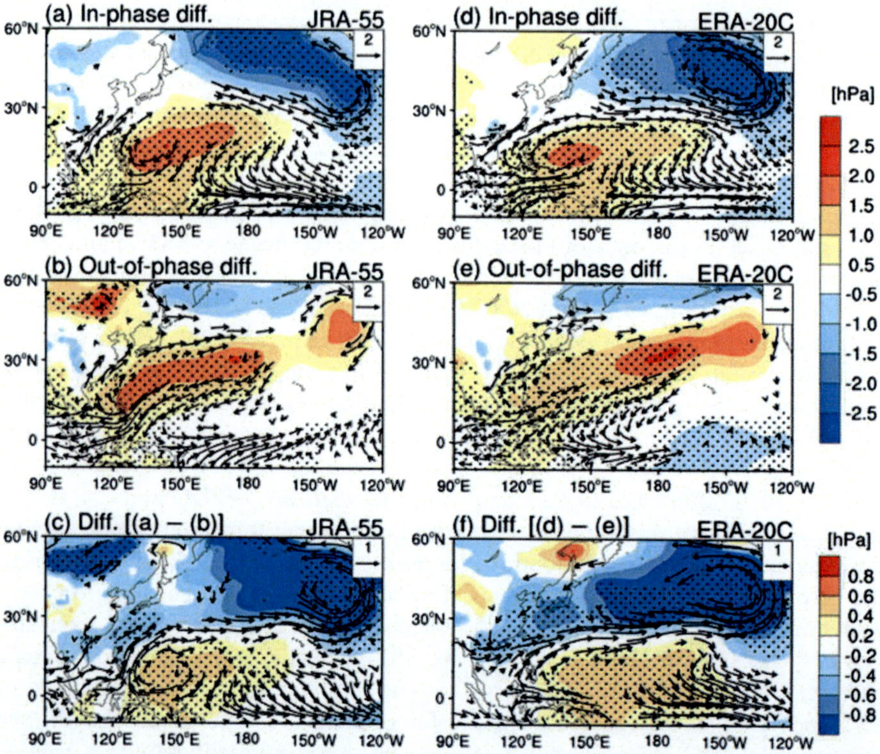

Fig. 3.21 Composite differences of spring (March–May) SLP (shading, hPa) and 850-hPa wind (vectors, m s^{-1}) **a** between the in-phase PDO–ENSO events of EN_WPDO minus LN_CPDO, **b** between the out-of-phase PDO–ENSO events of EN_CPDO minus LN_WPDO, **c** between in-phase and out-of-phase conditions [i.e., (**a**) minus (**b**)], based on the JRA-55 reanalysis for the period 1958–2012. The stippling denotes SLP anomalies statistically significant at the 95% confidence level. Only vectors of which at least one of the zonal and meridional wind components is statistically significant at the 95% confidence level are shown. **d**–**f** are the same as **a**–**c** but based on the ERA-20C reanalysis data for the period 1920–2010. (From Wu and Mao 2017b)

present over the North Pacific (Fig. 3.21a and 3.21d). The enhanced Aleutian Low extends westwards to East Asia and the anomalous anticyclone contracts away from the coastal region of East Asia. As a result, the anomalous western North Pacific anticyclone is confined to the southern part of the South China Sea and east of the Philippines, and the strong anomalous southwesterlies are confined over South China (Fig. 3.21a and 3.21d). Consequently, southwesterly anomalies bring large amounts of water vapor to South China, as indicated by the southwesterly anomaly of the vertically integrated water vapor flux over South China (not shown). The anomalous water vapor flux converges over South China, leading to above-normal rainfall there, in agreement with the EOF2 pattern of spring rainfall anomalies over eastern China (Fig. 3.21b and 3.21d).

The circulation anomalies in the out-of-phase composite are significantly different from those in the in-phase composite. In response to the weakened SST anomalies in the equatorial central Pacific (Fig. 3.21f), there is a weak anomalous anticyclone over the western North Pacific (Fig. 3.21b and 3.21e). Although the strength of the western North Pacific anticyclone in the out-of-phase composite is somewhat weaker than that in the in-phase composite (Fig. 3.21a and 3.21d), it extends northeastward to connect with the positive SLP anomalies over the eastern North Pacific (Fig. 3.21b and 3.21e), which are associated with the cold PDO-related positive SST anomalies over the North Pacific (Fig. 3.21f). More importantly, in response to positive SST anomalies from the coast of East Asia to the central North Pacific, the anomalous western North Pacific anticyclone extends northward to eastern and northern China (Fig. 3.21b and 3.21e). As a result, large amounts of water vapor are transported to the YNC regions as indicated by the anomalous vertically integrated vertical water vapor. Abundant water vapor converges over YNC, resulting in above-normal rainfall over those regions, corresponding to the EOF1 pattern of spring rainfall anomalies over eastern China (Fig. 3.20a and 3.20c).

The differences of SLP and 850-hPa winds between in-phase and out-of-phase composites are shown in Fig. 3.21c and 3.21f. A dipole-like anomalous circulation is present over the Pacific and East Asia, characterized by an anomalous anticyclone over the equatorial western Pacific and an anomalous cyclone over the North Pacific. The strong anomalous cyclone extending southwestwards from the North Pacific to East China and the anomalous anticyclone east of the Philippines are the typical characteristics of warm PDO-related circulation anomalies (Mantua et al. 1997). As a result, anomalous northeasterlies prevail over YNC and converge with anomalous southwesterlies over South China and north of the South China Sea (Fig. 3.21c and 3.21f). This results in abundant water vapor and anomalous convergence over South China, but reduced water vapor and anomalous divergence over YNC, leading to above-normal rainfall over South China and below-normal rainfall over YNC. Such differences in the circulation and rainfall anomalies between in-phase and out-of-phase composites closely resemble the warm PDO-related circulation and rainfall anomalies shown by Wu and Mao (2016). The above analyses demonstrate that the circulation anomalies, as responses to ENSO-related SST anomalies modified by PDO, could actually induce the first two EOF modes of spring rainfall anomalies over eastern China.

The above physical processes about how the resultant SST anomaly patterns of the in-phase and out-of-phase PDO–ENSO events induce spring rainfall anomalies over eastern China are verified through numerical experiments using the NCAR-CAM5 atmospheric general circulation model (Wu and Mao 2017b). Three experiments are conducted with one control experiment and two sensitivity experiments using different SST patterns as varying lower boundary conditions. The control experiment (referred to as CAM5-CTRL) is forced by the climatological mean annual cycle of SST. The first sensitivity experiment is forced by the climatological mean annual cycle of SSTs with superimposed composite monthly varying SST anomalies of EN_WPDO events from December to June to simulate the in-phase PDO–ENSO impact (referred to as CAM5-In-phase). The other sensitivity experiment is forced in the same way except that the SST anomalies of EN_CPDO events are superimposed to simulate the out-of-phase PDO–ENSO situations (referred to as CAM5-Out-of-phase).

The CAM5 model is capable of simulating reasonably well the major features of the spring persistent rainfall belt from southern China to southern Japan though the maximum rainfall belt extends farther northward over eastern China. It also captures well the southwesterlies over East Asia over the western North Pacific and south of the Tibetan Plateau, and the northwesterlies over the mid-latitudes. As such, the differences between the CAM5-In-Phase and CAM5-CTRL simulations are used to represent the in-phase composite anomalies, and the differences between CAM5-Out-of-phase and CAM5-CTRL for the out-of-phase composite anomalies. The differences of simulated rainfall anomaly percentage between the CAM5-In-phase and CAM5-CTRL experiments exhibit positive anomalies over South China, in conjunction with weak negative anomalies north of the Yangtze River, very similar to the in-phase composite rainfall anomalies as indicated by the EOF2 pattern (Fig. 3.20b). In response to the forcing by the SST anomaly pattern of EN_WPDO events, a southwest–northeast-oriented dipole-like anomalous SLP pattern dominates the tropical and North Pacific, with positive SLP anomalies over the South China Sea and subtropical western North Pacific, and strong negative SLP anomalies over the entire North Pacific. The simulated negative SLP anomalies clearly extend southwestward to Japan, the Korean Peninsula, and eastern China, leading to southeastward contraction of the anomalous western North Pacific anticyclone from eastern China. Anomalous southwesterlies prevail over South China, leading to above-normal rainfall there. Thus, the CAM5 simulations forced by EN_WPDO composite SST anomalies reproduce well the EOF2 pattern of spring rainfall over eastern China as well as the circulation anomalies. This indicates that the EOF2 is actually induced by the in-phase PDO–ENSO combinations, with EN_WPDO (LN_CPDO) events favoring positive (negative) rainfall anomalies over South China.

The rainfall anomaly percentages of CAM5-Out-of-phase simulations exhibit positive anomalies over the middle and lower reaches of the Yellow River, capturing the major features of the EOF1 pattern (Fig. 3.20a) though the simulated rainfall anomalies are somewhat weaker than the observations over the regions between the Yangtze River and Yellow River. As a response to the EN_CPDO SST anomalies, the simulated positive SLP anomalies over the western North Pacific extend northward over eastern China and northeastward to the eastern North Pacific though there is some overestimation of negative SLP anomalies near the Aleutian region. Correspondingly, the model simulates well anomalous southwesterlies over the major part of eastern China along with above-normal rainfall over the middle and lower reaches of the Yellow River. With forcing by the EN_CPDO SST anomaly pattern, the simulation reproduces the major features of the EOF1 pattern of spring rainfall over eastern China and the associated circulation anomalies. This confirms that the EOF1 mode is indeed related to the out-of-phase PDO–ENSO combinations, that is, the rainfall anomaly pattern of positive EOF1 usually follows EN_CPDO events while that of negative EOF1 follows LN_WPDO events.

In view of the CAM5 simulation results, the physical mechanisms relating the SST forcing of the in-phase and out-of-phase PDO–ENSO events to the spring rainfall anomalies over eastern China can be summarized as in Fig. 3.22. The simulated southward shrinking of the western North Pacific anomalous anticyclone in in-phase composites and its northward extension in out-of-phase composites are in good agreement with the observations (Fig. 3.21a and 3.21b). The simulated circulation differences between in-phase and out-of-phase composites show typical features of a warm PDO, characterized by a dipole-like pattern with a zonally stretched anomalous anticyclone over the subtropical and equatorial western Pacific and an anomalous cyclone over the entire mid-latitude North Pacific and East Asia. This confirms that the anomalous anticyclone over the western North Pacific is modified by the anomalous cyclone over the mid-latitude North Pacific and East Asia in association with PDO-related SST anomalies in the extratropical Pacific, and then it modulates the spring rainfall anomalies over eastern China.

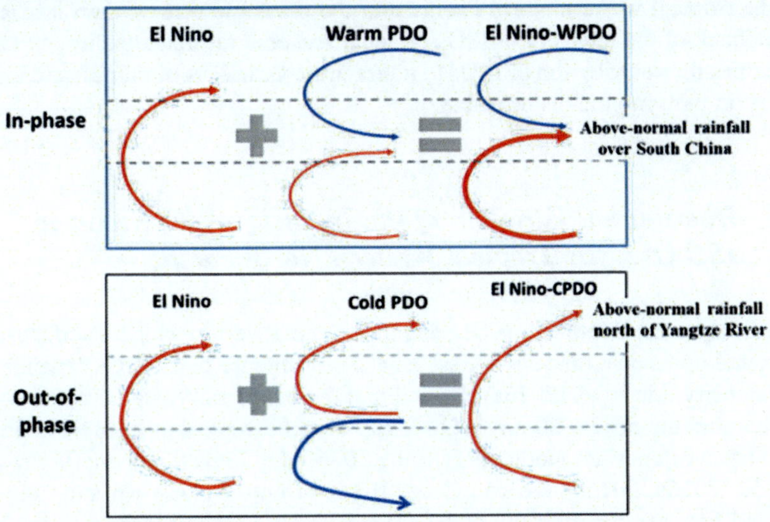

Fig. 3.22 Schematic diagram showing the changes in location and intensity of the El Niño-induced springtime western North Pacific subtropical high under warm and cold PDO epochs for the in-phase PDO–ENSO condition that favors above-normal rainfall over South China (upper panel) and the out-of-phase PDO–ENSO condition that favors the above-normal rainfall north of Yangtze River (lower panel). Red curves in the left column represent the characteristic contour and streamline of the western North Pacific subtropical anticyclone induced by the SST anomaly pattern of an El Niño event. Red (blue) curves in the middle column refer to the characteristic contours and streamlines of the western North Pacific subtropical anticyclone (the anomalous cyclone) induced by the SST anomaly pattern of warm (cold) PDO episode. Red (blue) curves in the right column denote the characteristic contours and streamlines of the resultant western North Pacific subtropical anticyclone (the anomalous cyclone) induced by the SST anomaly pattern of the in-phase PDO–ENSO condition as El Niño-WPDO and the out-of-phase PDO–ENSO condition as El Niño-CPDO

3.4 Cross-Scale Relation

Multiple time-scale variations are observed over the South China Sea and the tropical western North Pacific. There are indications for connections among the different time-scale variations. The intensity of summer ISOs over the tropical western North Pacific is subject to modulation of ENSO (Kajikawa and Yasunari 2005; Yang et al. 2008; Liu et al. 2016a; Wu and Cao 2017). Both interannual anomalies and ISOs affect the formation and movement of tropical cyclones over the western North Pacific (Wang and Chan 2002; Huang et al. 2011). Tropical cyclones form more likely in the wet phase of the ISOs, and the wet phase of the ISOs guide the movement of the tropical cyclones (Zhu et al. 2004; Kim et al. 2008; Chen and Huang 2009; Sun et al. 2009; Mao and Wu 2010; Pan et al. 2010; Huang et al. 2011; Feng et al. 2013; Li and Zhou. 2013a, b). On the other hand, the ISOs may have feedback on interannual anomalies.

This section first introduces the interannual variations of the intensity of ISOs over the tropical western North Pacific and the plausible factors, then it discusses the feedback of the intensity of ISOs on local seasonal mean anomalies, and at last it describes the contribution of ISOs to interannual variations of rainfall and impacts of ISOs on the synoptic disturbances.

3.4.1 Interannual Variation of the Intensity of the Summer ISO Over the Tropical Western North Pacific

The ISOs over the South China Sea and the tropical western North Pacific exert an important impact on climate over East Asia. The influence of the ISOs depends upon their intensity. Liu et al. (2016a) showed that the equatorial eastern Pacific cooling enhances the summer 30–60-day ISO activity over the tropical western North Pacific. ENSO has an important modulation of the 30–60-day ISOs (Teng and Wang 2003; Liu et al. 2016a, 2016b; Wu and Cao 2017). An out-of-phase variation has been identified between the intensity of 10–20-day and 30–60-day ISOs over the South China Sea during June–July (Kajikawa and Yasunari 2005; Yang et al. 2008). This is attributed to the difference in the region of the background field that influences the two ISOs (Yang et al. 2008).

The change in the intensity of the 10–20-day and 30–60-day ISOs over the South China Sea–tropical western North Pacific during summer displays an obvious different relationship to ENSO. Figures 3.23 shows the lead–lag correlation of monthly mean SST anomalies in the NINO3.4 region (5° S–5° N, 170°–120° W) with respect to summer 10–20-day and 30–60-day ISO intensities averaged in the region of 5°–15° N and 110°–150° E. The intensities of 10–20-day and 30–60-day ISOs are measured by the kinetic energy of filtered 850 hPa wind variations (Wu and Cao 2017). The wind data are from the NCEP DOE reanalysis 2 (Kanamitsu et al.

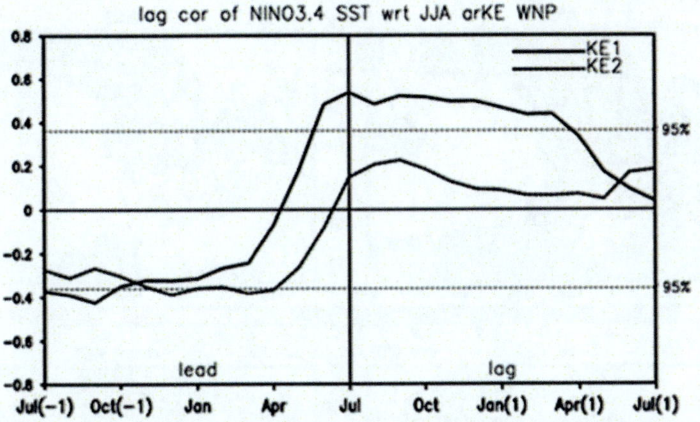

Fig. 3.23 Lead–lag correlation of monthly mean NINO3.4 SST with respect to summer 10–20-day (solid line) and 30–60-day (dashed line) ISO intensities in the region of 5°–15° N and 110°–150° E

2002). From the figure, the 10–20-day ISOs are enhanced during the El Niño developing summer, whereas the 30–60-day ISOs intensify during the La Niña decaying summer (Wu and Cao 2017). Figure 3.24c–d presents the distribution of concurrent correlation of tropical Indo-Pacific SST anomalies with summer ISO intensity in the region of 5°–15° N and 110°–150° E. The region is where the summer mean ISO intensity displays the largest standard deviation (Fig. 3.24a–b) (Wu and Cao 2017). The SST anomalies corresponding to the 10–20-day ISO intensity change feature a La Niña-type distribution (Fig. 3.24c), whereas those corresponding to the 30–60-day ISO intensity change display a west–east contrast between the North Indian Ocean and the central North Pacific (Fig. 3.24d). Such a west–east pattern is a feature during the La Niña decaying summer.

The different relationship of the two ISO intensity changes with ENSO is related to the difference in the source of the two ISOs that affect the tropical western North Pacific. The 10–20-day ISOs are mainly from the equatorial western Pacific and they reach the tropical western North Pacific during the west-northwestward propagation (Yang et al. 2008). During the El Niño developing summer, the source region and the propagation path of the 10–20-day ISOs feature a southeast–northwest band where the vertical wind shear, upward motion, and moisture are subjected to modulation of El Niño SST anomalies (Fig. 3.24e). Thus, ENSO influences the 10–20-day ISO intensity by modulating the background field in the above band. The 30–60-day ISOs over the western North Pacific come mainly from the eastern Indian Ocean and the Maritime Continent (Yang et al. 2008). During the La Niña decaying year, due to the influence of preceding SST anomalies in the equatorial central–eastern Pacific, a west–east contrasting SST anomaly pattern forms in the North Indian Ocean–central North Pacific through air–sea coupled processes. The combined effects of the SST anomalies in the two regions modulate the vertical wind shear and vertical motion over the tropical western North Pacific (Fig. 3.24f). When the 30–60-day ISOs

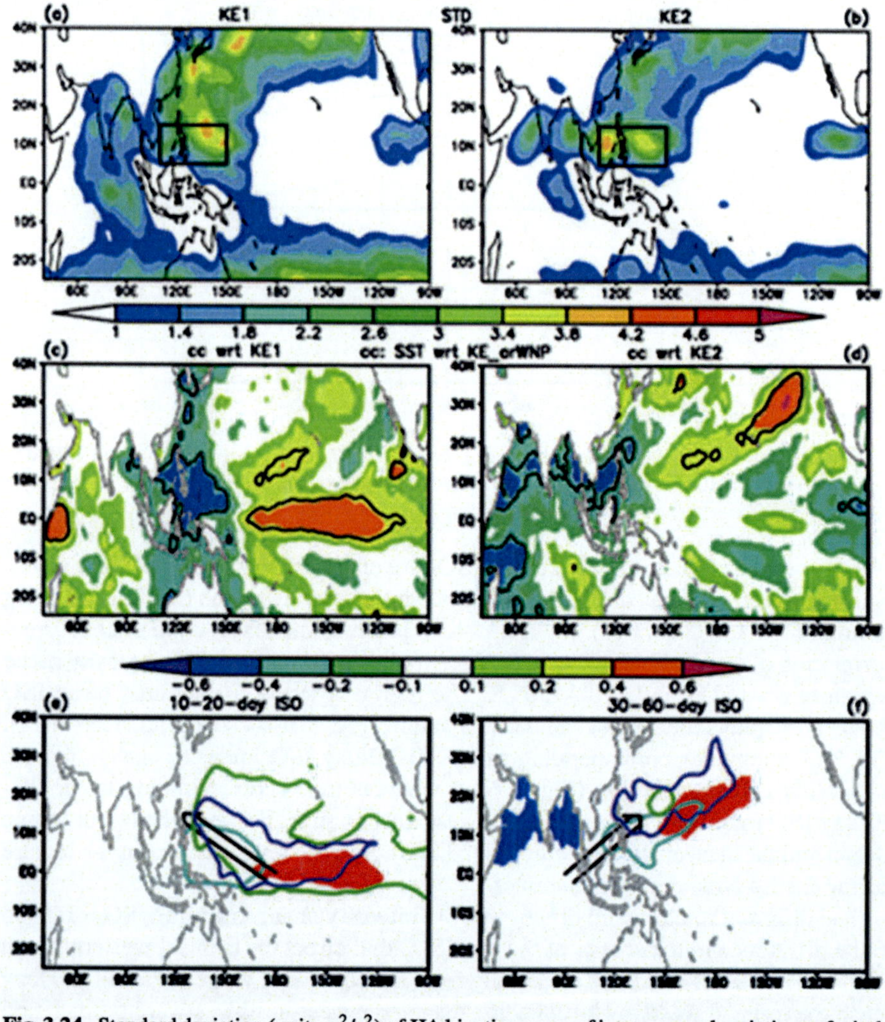

Fig. 3.24 Standard deviation (unit: m^2/s^2) of JJA kinetic energy of intraseasonal variations of winds at 850 hPa on the **a** 10–20-day and **b** 30–60-day time scales for the period 1979–2014. Correlation coefficient of JJA SST with respect to area-mean JJA kinetic energy of intraseasonal variations of winds at 850 hPa over the region of 5°–15° N, 110°–140° E (boxes in a and **b**) on the **c** 10–20-day and **d** 30–60-day time scales for the period 1982–2014. Thick lines denote correlation significant at the 95% confidence level. Schematic diagrams for **e** 10–20-day and **f** 30–60-day **b** ISOs. Black arrows denote the direction of propagation of ISOs, cyan lines denote the region of anomalous easterly vertical shear of zonal wind, green lines denote the region of positive specific humidity, and purple lines denote the region of anomalous upward motion. Red and blue shadings denote positive and negative SST anomalies, respectively

reach the tropical western North Pacific, the modified background field changes their intensity.

3.4.2 The Change of the Intensity of the ISOs Over the Tropical Indo-Pacific Region With the Phase of ENSO

The atmospheric circulation in the tropical Indo-Pacific region is modulated by ENSO and displays temporal evolution with the phase transition of ENSO (Wang et al. 2003; Wu et al. 2003). Due to the effect of the background field, the ISO intensity over the tropical Indo-Pacific displays a systematic spatial-temporal evolution with the phase of ENSO (Wu and Song 2018). Wu and Song (2018) analyzed in detail the 10–20-day and 30–60-day ISO intensity changes between El Niño and La Niña developing and mature phases, between fast and slow decaying El Niño years, and between El Niño and La Niña decaying years.

First, the ISO intensity shows opposite anomalies during the El Niño and La Niña developing and mature phases. During the developing phase, the ISO intensity anomaly displays a southwest–northeast contrast between the tropical southeast Indian Ocean and the tropical western North Pacific (Fig. 3.25a). The magnitude of the ISO intensity anomaly decreases from summer to fall over the tropical western North Pacific, but increases from summer to fall over the tropical southeastern Indian Ocean. During the mature phase, the ISO intensity anomaly features a west–east contrast between the Maritime Continent and the tropical central Pacific. During the El Niño developing summer and fall, the ISO is enhanced over the tropical western North Pacific, but weakened over the tropical southeastern Indian Ocean. During El Niño mature winter, the ISO intensity is enhanced over the tropical central Pacific, but weakened over the Maritime Continent.

Second, the ISO intensity anomaly displays prominent difference between El Niño and La Niña decaying years and between fast and slow El Niño decaying years. During the El Niño decaying year, the ISO is weakened over the Maritime Continent in spring and over the tropical western North Pacific in summer (Fig. 3.25b). During the La Niña decaying years, the 10–20-day ISO intensity anomaly displays a northwest–southeast contrast between the Philippine and the tropical western-central Pacific in spring and a south–north contrast over the western North Pacific in summer (Fig. 3.25c). During the El Niño fast decaying years, the ISO intensity is weakened over the equatorial western Pacific in spring and in the off-equatorial region in summer (Fig. 3.25d).

Third, the ISO intensity anomaly displays an obvious different distribution during La Niña developing and decaying summer. It features a southwest–northeast contrast between the tropical southeastern Indian Ocean and the tropical western North Pacific during the developing summer, whereas it is characterized by a southeast–northwest contrast between the equatorial western Pacific and the South China Sea-Bay of

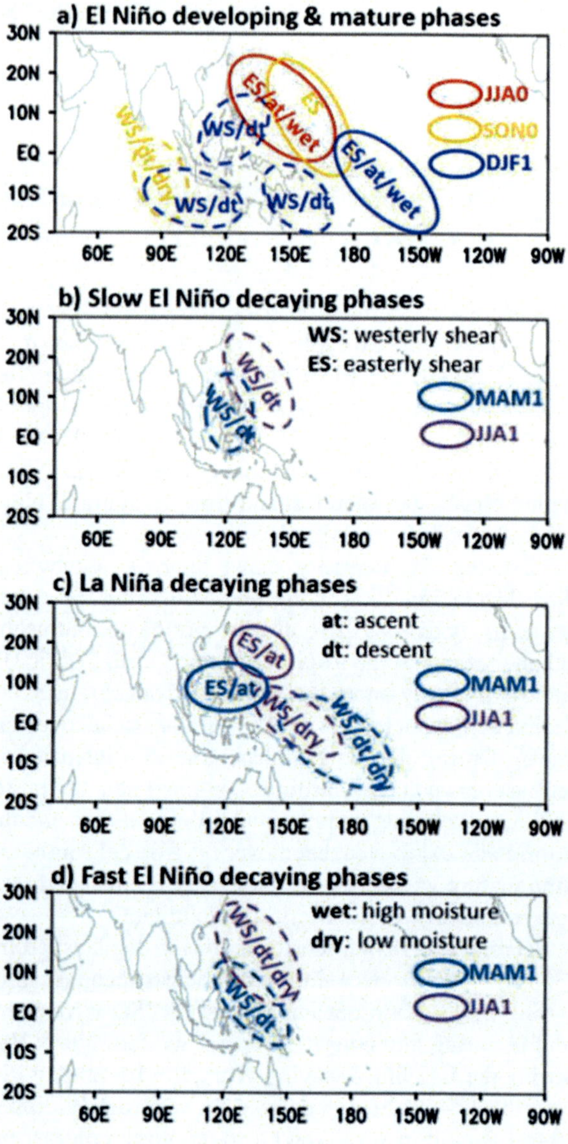

Fig. 3.25 Schematic diagrams illustrating the 10–20-day ISO intensity change and the main factors during El Niño developing and mature phases (**a**), slow El Niño decaying phases (**b**), La Niña decaying phases (**c**), and fast El Niño decaying phases (**d**). Solid and dashed shapes denote intensity increase and decrease, respectively. (From Wu and Song 2018)

Bengal. In addition, the magnitude of the ISO intensity anomaly during El Niño decaying spring and summer depends upon the pace of decay. The magnitude of the anomaly over the tropical western North Pacific is larger during El Niño fast decaying summer than during El Niño slow decaying summer.

The 10–20-day and 30–60-day ISO intensity anomaly shows prominent different features (Wu and Song 2018). First, the magnitude of 10–20-day ISO intensity anomaly is larger than the 30–60-day ISO. Second, the distribution of the ISO intensity anomaly is different between El Niño developing and decaying summers. The 10–20-day ISO intensity anomaly displays a northwest–southeast-oriented distribution, whereas the 30–60-day ISO intensity anomaly is mainly confined to the tropical western North Pacific. Third, during the El Niño developing summer, the 30–60-day ISO intensity anomaly tends to be opposite between the Philippine Sea and the central South China Sea, where such a feature is not seen in the 10–20-day ISO intensity anomaly. Fourth, during the La Niña decaying summer, the 30–60-day ISO intensity anomaly shows a prominent contrast between the South China Sea and the equatorial western Pacific, whereas the 10–20-day ISO intensity displays a south–north contrast over the western North Pacific.

The magnitude of the ISO intensity anomaly depends upon the phase of ENSO. The largest anomaly appears in the regions where several background fields have a consistent influence on the ISO intensity. These regions include the tropical western North Pacific during the ENSO developing summer, the tropical southeast Indian Ocean during the ENSO developing fall, the equatorial central Pacific during the ENSO mature winter, and the tropical western North Pacific during El Niño decaying summer. In these regions, the vertical wind shear, vertical motion, and the lower-level moisture fields have a combined effect on the ISO intensity change (Wu and Song 2018).

3.4.3 Feedback of ISOs on Seasonal Mean SST Anomalies in the Tropical Western North Pacific

The 10–20-day ISO intensity has a prominent negative correlation with local seasonal mean SST tendency over the tropical western North Pacific during spring, summer, and fall. This indicates plausible feedback of the ISO intensity on seasonal mean SST (Wu and Cao 2017). Figure 3.26 presents the point-wise correlation between 10–20-day and 30–60-day ISO intensities and concurrent SST tendency in summer. The data are from the NCEP-DOE reanalysis 2 (Kanamitsu et al. 2002) and the NOAA Optimum Interpolation version 2 (Reynolds et al. 2002). An obvious negative correlation is observed in the tropical western North Pacific (Fig. 3.26a–b). In comparison, the 10–20-day ISO has a larger correlation than the 30–60-day ISO. The lead–lag correlation of local monthly mean SST anomalies with respect to area-mean JJA ISO intensity in the tropical western North Pacific shows a pronounced decrease in SST following the ISO (Fig. 3.26c–d). This indicates an influence of the ISO on seasonal mean SST variation. In comparison, the impact of the 10–20-day ISO appears larger than the 30–60-day ISO.

The feedback of ISO on seasonal mean SST results from a net effect of surface heat fluxes. Surface shortwave radiation and evaporation induced by atmospheric ISOs depend nonlinearly upon the mean state. Their anomalies are asymmetric between active and inactive phases of the ISOs. Thus, the ISO induced surface cooling during

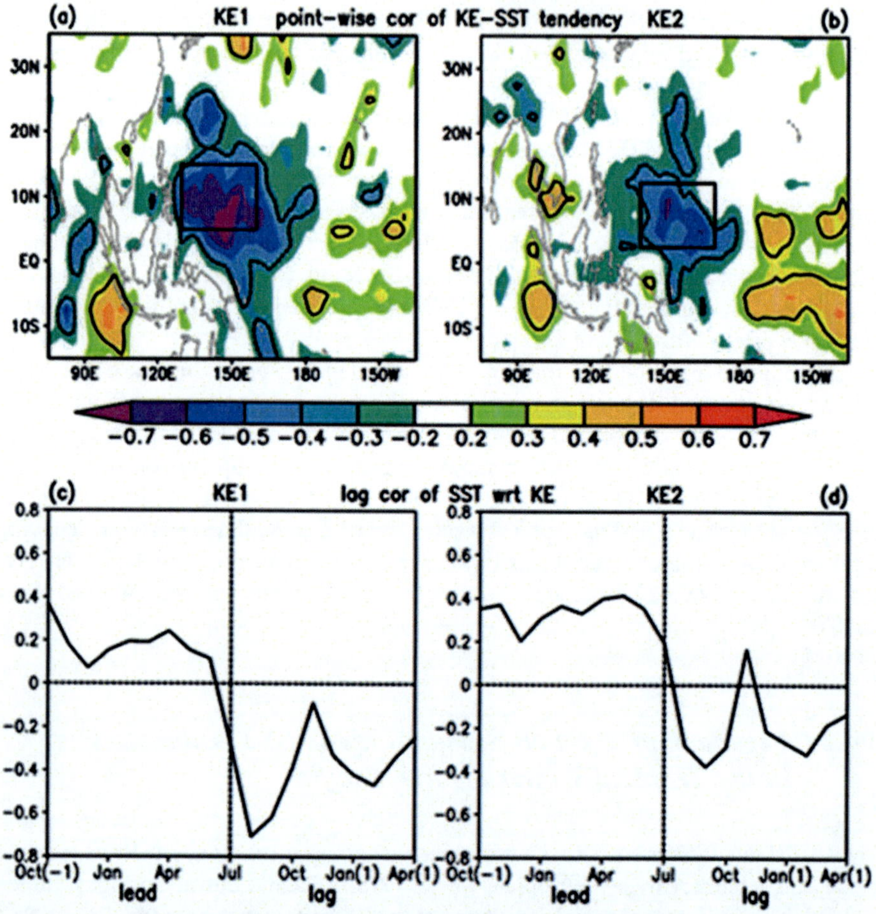

Fig. 3.26 The point-wise simultaneous correlation of the June–July–August (JJA) mean (**a**) 10–20-day and (**b**) 30–60-day ISO intensity (KE1) (measured using 850 hPa kinetic energy) and SST tendency (the difference of SST in the succeeding minus preceding month) for the period 1982–2014. The thick lines denote regions where the correlation is significant at the 95% confidence according to the Student *t*-test. The box denotes the region for area averaging of the 10–20-day ISO intensity. Lead–lag correlation of area-mean monthly mean SST with respect to area-mean (**c**) 10–20-day and (**d**) 30–60-day ISO intensity (KE1) averaged in the region of 5°–15° N and 130°–160° E (the box in (**a**)) and 2.5°–12.5° N and 140°–170° E (the box in (**b**)) in JJA. (From Wu and Cao 2017)

the active phases and warming during the inactive phase is not canceled, leading to seasonal mean SST anomalies. This residual effect is more prominent when the ISOs are strong.

Wu (2018) decomposed the variables, including wind speed, SST, and moisture, into different time-scale components with periods longer than 90 days (denoted as low-frequency component), shorter than 90 days (denoted as high-frequency component), and 10–20-day high-frequency component. Then, the contributions of different components to the seasonal mean surface heat flux are estimated. It is shown that over the tropical western North Pacific during summer, the accumulated surface latent heat flux anomalies have an important contribution to seasonal mean heat flux anomalies. This suggests that the ISO is an important factor in the seasonal mean SST tendency. Figure 3.27 displays the distribution of concurrent anomalies obtained by regression with respect to summer 10–20-day ISO intensity averaged in the region of 5–15° N, 130–160° E. Over the tropical western North Pacific, seasonal mean latent heat flux

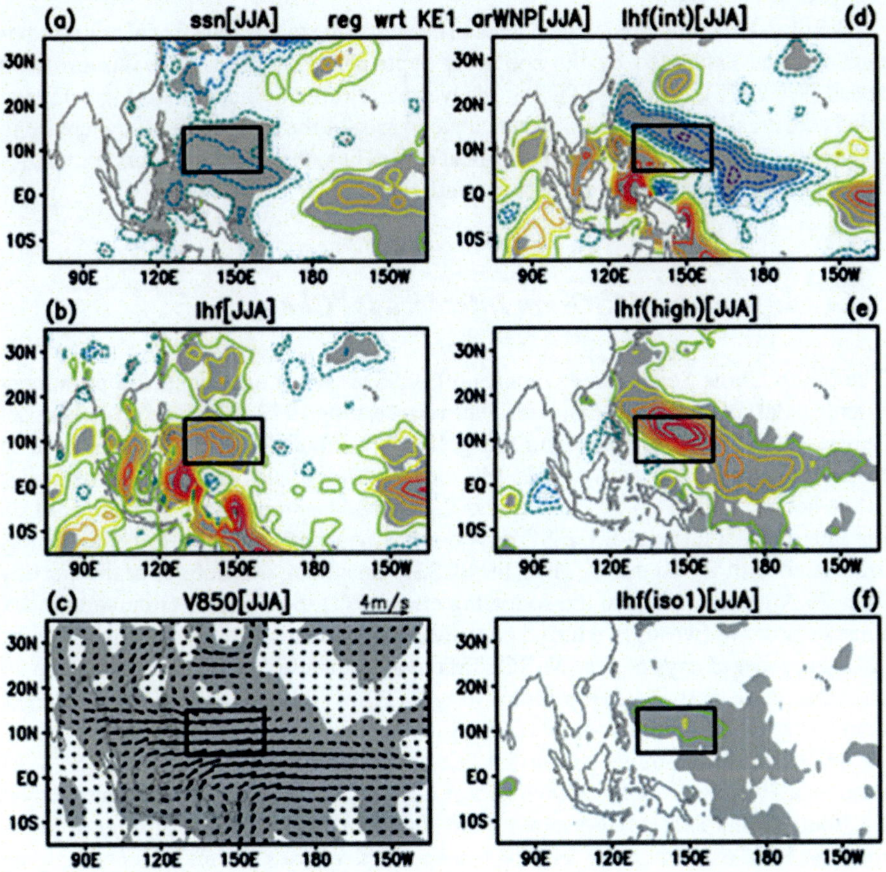

Fig. 3.27 Simultaneous regression with respect to normalized JJA 10–20-day ISO intensity (KE1) in the region of 5°–15° N and 130°–160° E (the box) for the period 1982–2014. **a** SST tendency converted to the unit of heat flux using a mixed-layer depth of 45 m, **b** latent heat flux, **c** 850 hPa wind (m/s, scale at the top-right), **d** interannual component of latent heat flux, **e** high-frequency (<90-day) component of latent heat flux, and **f** 10–20-day component of latent heat flux. The contour interval is 3 W/m^2. The zero contours have been suppressed. Shading denotes that the anomalies are significant at the 95% confidence level according to the Student *t*-test. (From Wu and Cao 2017)

increases with the intensity of the ISO (Fig. 3.27b), which is mainly attributed to the contribution of the high-frequency component (Fig. 3.27e). The low-frequency component (including both seasonal mean and interannual variation) of latent heat flux anomalies displays a southwest–northeast contrast (Fig. 3.27d). This contrast may be explained by the relation between interannual wind anomalies and climatological mean winds that are of the same and opposite sign to the west and east of the intertropical convergence zone (ITCZ), respectively (Fig. 3.27c). This results in opposite wind speed anomalies on the two sides of the ITCZ. Corresponding to more latent heat flux, the SST tendency is negative (Fig. 3.27a). It is estimated that about 1/3 of the high-frequency latent heat flux anomalies is due to the 10–20-day component (Wu 2018).

The above accumulated seasonal mean heat flux due to the high-frequency component may be explained by the nonlinear dependence of latent heat flux on wind speed (Wu 2018). Near the ITCZ of the western North Pacific, climatological mean winds are small. Both the active or inactive phases of the ISOs induce an increase in wind speed, enhancing surface latent heat flux. Thus, the latent heat flux anomalies averaged over a season have a net accumulation effect.

3.4.4 Influence of ISOs on Interdecadal Change

The South China Sea summer monsoon (SCSSM) onset experienced a prominent interdecadal shift around 1993, with a late onset during 1979–1993 and an early onset during 1994–2008. Kajikawa and Wang (2012) attributed this change to enhanced tropical cyclone (TC) activity and ISO on 30–80-day and 10–25-day time scales in the second epoch. In order to assess the individual impacts of TCs and ISO on the interdecadal change of the SCSSM onset, Chen (2015) removed the anomalies associated with TCs and ISO. Then, the SCSSM onset date was defined as the first day after 25 April according to the following criteria: (1) the SCSSM circulation index must be positive (westerly wind) on the onset day and during the subsequent 5 days; (2) the number of days of positive SCSSM circulation index during the subsequent 20 days reaches at least 15; (3) the 20-day mean SCSSM circulation index must be larger than 1 m s^{-1}. The comparison of the SCSSM onset dates with and without ISO- and TC-related wind anomalies shows that TCs have no significant impact on the SCSSM onset in all years except for 2006 in which a strong and long-lived TC occurred over the South China Sea. After removing the 30–80-day anomaly, the difference in the mean SCSSM onset date between the two epochs decreases to some extent, implying that the 30–80-day anomaly may partly play a role in the interdecadal shift of the SCSSM onset. In contrast, the 10–25-day anomaly has an insignificant contribution to the interdecadal shift of the SCSSM onset. The discrepancy of ISO contribution results from the SCSSM background state, the magnitude and spatiotemporal scale of ISO, and the phase relationship between ISO and SCSSM transition from easterly to westerly (Chen 2015).

A pronounced increase occurred in summer rainfall over southern China around 1992/1993. Diagnostic analysis indicates that the boreal summer ISO (BSISO) may contribute to this decadal increase. The BSISO-induced rainfall increase accounts for approximately 17.4% of the observed decadal rainfall increase, with a primary component resulting from the change in rainfall pattern associated with phases 3–5 of the BSISO (Wang et al. 2017). Moreover, the change in rainfall pattern over southern China is mainly ascribed to the change in the spatial structure of anomalous convection associated with the interdecadal change in BSISO tracks. In addition, the change in the frequency of active BSISO phases also contributes to the interdecadal change in summer rainfall over southern China. The interdecadal change in the BSISO tracks and active BSISO phases is likely related to the coherent change in the atmospheric circulation and SST over the Indian Ocean and the western Pacific.

The summer rainfall increase over southern China around 1992/93 is shown to be related to the long (over 14 days) consecutive rainy day events (Liu et al. 2017). Such long consecutive rainy day events resulted from lower-level cyclonic circulation over the South China Sea and southern China. The anomalous cyclone originates from the tropical region north of New Guinea about 30 days ago and moves northwestward. The moving feature is obvious in the 30–60-day filtered wind fields, indicating an important contribution of the northward propagation 30–60-day ISOs from the tropical western North Pacific to the occurrence of long consecutive rainy day events over southern China.

3.4.5 *Influence of Interannual Variation on Synoptic Disturbances*

The activities of tropical waves over the western North Pacific and the South China Sea display obvious interannual variability according to main signals isolated from the westward-propagating tropical waves (tropical depressions, mixed Rossby–Gravity waves, and equatorial Rossby waves) by a wavenumber-frequency spectral analysis. Such interannual variability is linked tightly with the monsoon trough (Wu et al. 2015a, b). When the monsoon trough is enhanced and extends eastward, the activities of tropical waves are strengthened and shift eastward. During El Niño years, eastward-extending and enhanced monsoon trough provides a favorable condition for active tropical waves. Observational and theoretical analyses and numerical experiments indicate that tropical waves in the western Pacific gain the energy from the mean flow of the monsoon trough, leading to the enhancement of intensity, the shrinkage of spatial scale, and the changes of horizontal and vertical structures of waves. Meanwhile, the monsoon trough is beneficial to the transformation from the mixed Rossby–Gravity waves to off-equatorial tropical depression disturbances, which provides the background condition for the formation and development of synoptic-scale disturbances and tropical cyclones. These results reveal that the interannual variability of tropical wave activity over the western North Pacific and the South China Sea is influenced by the change in mean flow.

Wu et al. (2018) identified that differences and similarities between the impacts of different types of tropical Pacific warming (El Niño, El Niño Modoki) on atmospheric circulation over the western North Pacific are associated with the differences in the location and intensity of SST anomalies. Eastern Pacific warming induces pronounced anomalous westerlies and strong ascending motion of anomalous Walker circulation over the equatorial central Pacific. Central Pacific warming leads to anomalous convection over the equatorial central Pacific, and a Gill-type response leads to anomalous westerlies in the off-equatorial region. A strong El Niño event results in simultaneous central and eastern Pacific warming, leading to a poleward extension of anomalous westerlies over the central Pacific and eastward extension of the monsoon trough, which favors the eastward shift in the location of the tropical cyclone genesis and more strong and active tropical cyclones. During the El Niño Modoki events, central Pacific warming induces anomalous westerlies away from the equator, leading to an anomalously broad zone of the eastward-extending monsoon trough and an eastward shift in the location of tropical cyclone genesis. During moderate and weak El Niño (eastern Pacific warming) events, the anomalous equatorial westerlies are weak and the tropical cyclone genesis does not show a dramatic southeastward shift with weak positive anomalies east of the Philippines. During moderate El Niño years, anomalous westerlies along and/or near the equator lead to a narrow zone of the eastward-extending monsoon trough with a slightly southward shift. During weak El Niño events (the same level as the central Pacific warming), the western North Pacific tropical cyclones and atmospheric circulations in response to SST forcing are more dependent on local SST compared to the extreme and moderate El Niño events.

The impact of the Modoki events on the monsoon trough over the western North Pacific has been intensified since 1984, which is accompanied by an enhanced influence of Modoki events on the East Arian climate and weather (Wu et al. 2018). This change is attributed to the tropical Pacific warming caused by global warming. Before 1985, the impacts of the Modoki events were mainly due to western Pacific cooling with a weak central Pacific warming and thus had a weak impact on the monsoon trough. After 1985, both the western and central Pacific becomes warmer. Especially, the pronounced central Pacific warming leads to the strengthening and eastward extension of the monsoon trough. These have a remarkable impact on monsoon and tropical cyclone genesis over the western North Pacific.

Zhang et al. (2018) investigated the change of the monsoon trough and the tropical cyclones during different types of El Niño events during 1948–2015 and the possible influence of these two types of El Niño events on the monsoon trough and tropical cyclones. The results indicate that compared to the years of eastern Pacific warming, the monsoon trough is weaker and shifts northward in the years of central Pacific warming. During the central Pacific warming years, central (western and eastern) Pacific SST warming (cooling) induces anomalous westerlies over the equatorial western-central Pacific and anomalous ascending motion and enhanced convective activities over the equatorial central Pacific. This leads to an enhanced and eastward extended monsoon trough and a weaker subtropical high over the western North Pacific, which in turn result in a northward shift of the monsoon trough. During the eastern Pacific warming years, equatorial eastern (western) Pacific warming (cooling) induces enhanced and eastward-extending anomalous equatorial westerlies, an eastward shift of the ascending branch of anomalous Walker circulation toward eastern Pacific, and an enhanced southwestward-extending subtropical high. This results in enhancement and eastward extension of the monsoon trough. Further analysis indicates that these changes in the strength and location of monsoon trough are accompanied by those in the large-scale environmental factors of the tropical cyclone genesis over the western North Pacific. During central Pacific warming years, intensified low-level anomalous cyclonic vorticity and upper-level divergence increased relative humidity in the lower and mid-troposphere, and reduced vertical wind shear movement northward along with the monsoon trough. During the eastern Pacific warming years, these anomalous regions move southward. The variations in these large-scale environmental factors induce a northward and westward movement in the location of tropical cyclone genesis over the western North Pacific during central Pacific warming years compared to that during eastern Pacific warming years.

3.5 Summary

This chapter reviewed the air–sea relationship on intraseasonal and interannual time scales over the South China Sea and tropical western North Pacific, the influences of tropical Indo-Pacific SST anomalies on rainfall anomalies over the South China Sea during summer and spring-to-summer transition, and the interconnections among different time-scale variations. The main results are as follows.

(1) The intraseasonal air–sea relationship displays obvious scale dependence. During northern summer, the air–sea relationship shows a different horizontal structure over the South China Sea–tropical western North Pacific between the 10–20-day and 30–60-day time scales. The correlation between SST and precipitation, surface shortwave radiation, and surface latent heat flux is characterized by a southwest–northeast distribution on the 10–20-day time scale, but a zonal distribution on the 30–60-day time scale.

(2) The East Asian winter monsoon has an important impact on the intraseasonal SST variations in the South China Sea. During the weak (strong) stage of the East Asian winter monsoon, surface wind speed decrease (increase) reduces (enhances) surface latent heat flux, leading to warming (cooling) of the ocean surface water. The intraseasonal signals in SST, surface wind, and latent heat flux display a coherent southward propagation over the South China Sea.

(3) The interannual air–sea relationship displays obvious seasonality and regional feature in the South China Sea. Atmospheric forcing is dominant in the central and southern South China Sea during the warm season. Oceanic forcing is

present in the northern South China Sea during the cold season. In the lower-latitude tropical western North Pacific, there is obvious oceanic forcing on the atmosphere during winter and spring, whereas atmospheric change plays an important role in the SST change during summer and fall. During the spring-to-summer transition, the air–sea relationship shows a unique interactive feature over the North Indian Ocean and the South China Sea.

(4) ENSO influences the South China Sea summer rainfall both directly and indirectly. The North Indian Ocean SST anomalies act as a medium for the indirect influence of ENSO. The rainfall anomalies are more prominent over the South China Sea when the direct and indirect influences of ENSO coexist. In addition, the Indian Ocean affects the South China Sea summer rainfall variability independent of ENSO.

(5) The South China Sea rainfall variability during April–June is subject to combined influences of SST anomalies in the equatorial Pacific, tropical Indian Ocean, and tropical western North Pacific. Rainfall anomalies in the South China Sea often appear when SST anomalies coexist and work cooperatively in two or three of the above regions.

(6) Local air–sea interaction leads to a month-to-month change of rainfall anomalies in the South China Sea during the spring-to-summer transition, which contributes largely to the pace of transition from spring to summer.

(7) The interdecadal variation of early-summer rainfall in South China has a close association with the PDO. Less and more early summer rainfall periods correspond to positive and negative PDO phases, respectively. The physical connection between PDO and South China early-summer rainfall is achieved by oceanic and atmospheric teleconnections.

(8) PDO has an important modulation of the influence of ENSO on South China spring rainfall variability. Circulation anomalies during in-phase PDO and ENSO years lead to more prominent spring rainfall anomalies over South China. Under the modulation of warm PDO SST anomalies, equatorial central–eastern Pacific warm SST anomalies and North Pacific cold SST anomalies are both enhanced during the El Niño decaying spring. This leads to the development of a very strong anomalous anticyclone over the western North Pacific, accompanied by a strong and large anomalous cyclone over the North Pacific and East Asia, forming a north–south dipole pattern. The moisture convergence within the southwesterly flows is located over South China, leading to more rainfall there. In contrast, when La Niña coexists with the negative PDO phase, less rainfall is induced over South China.

(9) The 10–20-day and 30–60-day ISO intensity change has a different relationship to ENSO. During northern summer, the 10–20-day ISOs intensify during the El Niño developing summer, whereas the 30–60-day ISOs are enhanced during the La Niña decaying summer. This difference is attributed to the relationship between the ENSO-induce background field change region and the source and propagation path of the ISOs. Equatorial central–eastern Pacific SST anomalies modulate the vertical wind shear, lower-level moisture, and vertical motion in a southeast–northwest band extending from the equatorial central Pacific to

the tropical western North Pacific. This is the region where the 10–20-day ISOs originate and propagate. Thus, ENSO can modulate the 10–20-day ISO intensity. Preceding equatorial central–eastern Pacific SST anomalies induce the North Indian Ocean SST anomalies in the succeeding summer. The latter modulates the vertical wind shear and vertical motion over the tropical western North Pacific. The modulated background field affects the intensity of the 30–60-day ISOs when they propagate to the tropical western North Pacific from the equatorial region.

(10) The ISOs near the ITCZ over the tropical western North Pacific have prominent feedback on seasonal mean SST anomalies in boreal summer. The feedback is due to an accumulation of ISO-induced surface latent heat flux anomalies. As climatological mean winds are weak in this region, surface wind anomalies enhance surface latent heat flux during both dry and wet phases of ISOs. Thus, the atmospheric ISOs make an important contribution to the seasonal mean SST variations.

References

Adler RF, Huffman GJ, Chang A et al (2003) The Version 2 Global Precipitation Climatology Project (GPCP) monthly precipitation analysis (1979-present). J Hydrometeor 4:1147–1167

Alexander MA, Bladé I, Newman M, Lanzante JR, Lau NC, Scott JD (2002) The atmospheric bridge: The influence of ENSO teleconnections on air–sea interaction over the global ocean. J Climate 15:2205–2231

Annamalai H, Slingo JM (2001) Active/break cycles: diagnosis of the intraseasonal variability of the Asian summer monsoon. Clim Dyn 18:85–102

Camargo SJ, Emanuel KA, Sobel AH (2007) Use of a genesis potential index to diagnose ENSO effects on tropical cyclone genesis. J Climate 20:4819–4834

Cao X, Li T, Peng M, Chen W, Chen GH (2014) Effects of monsoon trough interannual variation on tropical cyclogenesis over the western North Pacific. Geophys Res Lett 41:4332–4339

Cao X, Wu R, Chen SF (2017) Contrast of 10-20-day and 30-60-day intraseasonal SST propagation during summer and winter over the South China Sea and western North Pacific. Clim Dyn 48:1233–1248. https://doi.org/10.1007/s00382-016-3138-z

Cao X, Wu R (2017) Origins of intraseasonal rainfall variations over the southern South China Sea in boreal winter. Atmos Ocean Sci Lett 10:44–50. https://doi.org/10.1080/16742834.2017.1232584

Cao X, Wu R (2018) Origins and interrelationship of intraseasonal rainfall variations around the Maritime Continent during boreal winter. Theor Appl Climatol 132:543–554. https://doi.org/10.1007/s00704-017-2106-9

Chan JCL, Ai WX, Xu JJ (2002) Mechanisms responsible for the maintenance of the 1998 South China Sea summer monsoon. J Meteor Soc Japan 80:1103–1113

Chan JCL, Zhou W (2005) PDO, ENSO and the early summer monsoon rainfall over South China. Geophys Res Lett 32:L08810. https://doi.org/10.1029/2004gl022015

Chang C, Chen JM (1992) A statistical study of winter monsoon cold surges over the South China Sea and the large-scale equatorial divergence. J Meteor Soc Japan 70:287–302

Chen GH (2015) Comments on "Interdecadal change of the South China Sea summer monsoon onset". J Climate 28:9035–9209

Chen J, Wen Z, Wu R, Chen ZS, Zhao P (2014) Interdecadal changes in the relationship between southern China winter–spring precipitation and ENSO. Clim Dyn 43:1327–1338
Chen TC, Chen JM (1995) An observational study of the South China Sea monsoon during the 1979 summer: Onset and life cycle. Mon Wea Rev 123:2295–2318
Chen TC, Weng SP, Yamazaki N, Kiehne S (1998) Interannual variation in the tropical cyclone formation over the western North Pacific. Mon Wea Rev 126:1080–1090
Chen W, Feng J, Wu R (2013) Roles of ENSO in the link of the East Asian winter monsoon to the following summer monsoon. J Climate 26:622–635
Chen W, Graf HF, Huang RH (2000) The interannual variability of East Asian winter monsoon and its relation to the summer monsoon. Adv Atmos Sci 17:48–60
Chen W (2002) Impacts of El Niño and La Niña on the cycle of the East Asian winter and summer monsoon. Scientia Atmospheric Sinica (in Chinese) 26(5):595–610
Chen GH, Huang RH (2009) Dynamical effects of low frequency oscillation on tropical cyclogenesis over the western North Pacific and the physical mechanisms. Scientia Atmospheric Sinica (in Chinese) 33(2):205–214
Ding YH, Wang ZY, Sun Y (2008) Inter-decadal variation of the summer precipitation in East China and its association with decreasing Asian summer monsoon. Part I: observed evidences. Int J Climatol 28:1139–1161
Duan W, Song L, Li Y, Mao J (2013) Modulation of PDO on the predictability of the interannual variability of early summer rainfall over south China. J Geophys Res Atmos 118:13008–13021
Feng J, Wang L, Chen W (2014) How does the East Asian summer monsoon behave in the decaying phase of El Niño during different PDO phases? J Climate 27:2682–2698
Feng X, Wu R, Chen J, Wen ZP (2013) Factors for interannual variations of September–October rainfall in Hainan, China. J Climate 26:8962–8978
Fu X, Wang B, Tao L (2006) Satellite data reveal the 3-D moisture structure of tropical intraseasonal oscillation and its coupling with underlying ocean. Geophys Res Lett 33:L03705
Fukutomi Y, Yasunari T (1999) 10–25-day intraseasonal variations of convection and circulation over East Asia and western North Pacific during early summer. J Meteor Soc Japan 77:753–769
Gao H, Wang YG, He JH (2006) Weakening significance of ENSO as a predictor of summer precipitation in China. Geophys Res Lett 33:L09807. https://doi.org/10.1029/2005gl025511
Gao RZ, Zhou FX (2002) Monsoonal characteristics revealed by intraseasonal variability of sea surface temperature (SST) in the South China Sea (SCS). Geophys Res Lett 29:1222. https://doi.org/10.1029/2001GL014225
He Z, Wu R (2013) Seasonality of interannual atmosphere-ocean interaction in the South China Sea. J Oceanography 69(6):699–712
He Z, Wu R (2014) Indo-Pacific remote forcing in summer rainfall variability over the South China Sea. Clim Dyn 42(9):2323–2337
He Z, Wu R, Wang W (2016) Signals of the South China Sea summer precipitation variability in the Indian Ocean. Clim Dyn 46(9–10):3181–3195
Hu W, Wu R, Liu Y (2014) Relation of the South China Sea precipitation variability to tropical Indo-Pacific SST anomalies during spring-to-summer transition. J Climate 27(14):5451–5467
Hu W, Wu R (2015) Relationship between South China Sea precipitation variability and Tropical Indo-Pacific SST anomalies in the IPCC CMIP5 models during spring-to-summer transition. Adv Atmos Sci 32(9):1308–1318
Hu W, Wu R (2016) Air-Sea Interaction in association with monthly anomaly departure over the South China Sea and Tropical Indian Ocean during the spring-to-summer transition. J Climate 29(6):2095–2108
Hu W, Li T (2017) Local intraseasonal air–sea relationship over the North Indian Ocean and western North Pacific during the spring-to-summer transition. Atmos Oceanic Sci Letters 10:65–72
Huang P, Chou C, Huang R (2011) Seasonal modulation of tropical intraseasonal oscillations on tropical cyclone genesis in the western North Pacific. J Climate 24:6339–6352
Huang R, Sun F (1992) Impact of the tropical western Pacific on the East Asian summer monsoon. J Meteor Soc Japan 70:243–256

Jiang W, Huang G, Hu K, Wu R, Gong HN, Chen XL, Tao WC (2017) Diverse relationship between ENSO and Northwest Pacific summer climate among CMIP5 models: Dependence on the ENSO decay pace. J Climate 30(1):109–127
Kajikawa Y, Wang B (2012) Interdecadal change of the South China Sea summer monsoon onset. J Climate 25:3207–3218
Kajikawa Y, Yasunari T (2005) Interannual variability of the 10–25- and 30–60-day variation over the South China Sea during boreal summer. Geophys Res Lett 32:L04710. https://doi.org/10.1029/2004GL021836
Kanamitsu M, Ebisuzaki W, Woollen J et al (2002) NCEP–DOE AMIP-II reanalysis (R-2). Bull Amer Meteor Soc 83:1631–1643
Kemball-Cook S, Wang B (2001) Equatorial waves and air-sea interaction in the boreal summer intraseasonal oscillation. J Climate 14:2923–2942
Kikuchi K, Wang B (2009) Global perspective of the quasi-biweekly oscillation. J Climate 22:1340–1359
Kim JH, Ho C, Kim HS, Sui CH, Park SK (2008) Systematic variation of summertime tropical cyclone activity in the western North Pacific in relation to the Madden–Julian oscillation. J Climate 21:1171–1191
Klein SA, Sode BJ, Lau NC (1999) Remote sea surface temperature variations during ENSO: Evidence for a tropical atmospheric bridge. J Climate 12:917–932
Kumar BP, Vialard J, Lengaigne M, Murty VSN, McPhadon M (2012) TropFlux air-sea fluxes for the global tropical oceans-Description and evaluation. Clim Dyn 38:1521–1543
Lau KM, Sui CH (1997) Mechanisms of short-term sea surface temperature regulation: observations during TOGA COARE. J Climate 47:465–472
Li CY, Li GL (1997) Evolution of intraseasonal oscillation over the tropical western Pacific/South China Sea and its effect to the summer precipitation in Southern China. Adv Atmos Sci 14:246–254
Li R, Zhou W (2013a) Modulation of Western North Pacific Tropical Cyclone Activity by the ISO. Part I: Genesis and Intensity. J Climate 26(9):2904–2918
Li R, Zhou W (2013b) Modulation of western North Pacific tropical cyclone activity by the ISO. Part II: Tracks and landfalls. J Climate 26(9):2919–2930
Li CY, Pan J, Song J (2013) Progress on the MJO research in recent years. Scientia Atmospheric Sinica (in Chinese) 37(2):229–252
Liang JY, Wu SS (2000) The multi-time scale variations of summer monsoon over South China Sea and its interaction with SST anomaly. Q J Appl Meteorol (in Chinese) 11(1):95–104
Liu F, Li T, Wang H, Deng L, Zhang YW (2016a) Modulation of boreal summer intraseasonal oscillations over the western North Pacific by ENSO. J Climate 29:7189–7201
Liu F, Zhou L, Ling J, Fu X, Huang G (2016b) Relationship between SST anomalies and the intensity of intraseasonal variability. Theor Appl Climatol 124:847–854
Liu G, Wu R, Wang H (2017) Contribution of intraseasonal oscillation to long-duration summer precipitation events over southern China. Atmos Ocean Sci Lett 10(1):82–88
Mantua NJ, Hare SR, Zhang Y, Wallace JM, Francis RC (1997) A Pacific interdecadal climate oscillation with impacts on salmon production. Bull Am Meteorol Soc 78(6):1069–1079. https://doi.org/10.1175/1520-0477
Mao JY, Chan JCL, Wu GX (2011) Interannual variations of early summer monsoon rainfall over South China under different PDO backgrounds. Int J Climatol 31:847–862
Mao JY, Chan JCL (2005) Intraseasonal variability of the South China Sea summer monsoon. J Climate 18:2388–2402
Mao JY, Sun Z, Wu GX (2010) 20-50-day oscillation of summer Yangtze rainfall in response to intraseasonal variations in the subtropical high over the western North Pacific and South China Sea. Clim Dyn 34(5):747–761
Mao JY, Wu GX (2010) Intraseasonal modulation of tropical cyclogenesis in the western North Pacific: a case study. Theor Appl Climatol 100(3–4):397–411

Mu MQ, Li CY (2000) On the outbreak of South China Sea summer monsoon in 1998 and activity of atmospheric intraseasonal oscillation. Climatic and Environmental Research (in Chinese) 5(4):375–387

Nitta T (1987) Convective activities in the tropical western Pacific and their impacts on the Northern Hemisphere summer circulation. J Meteor Soc Japan 65:165–171

North GR, Moeng FJ, Bell TL, Cahalan RF (1982) The latitude dependence of the variance of zonally averaged quantities. Mon Wea Rev 110:319–326

Pan J, Li CY, Song J (2010) The modulation of Madden-Julian Oscillation on typhoons in the Northwestern Pacific Ocean. Scientia Atmospheric Sinica (in Chinese) 34(6):1059–1070

Ren X, Yang X, Sun X (2013) Zonal oscillation of western Pacific subtropical high and subseasonal SST variations during Yangtze persistent heavy rainfall events. J Climate 26:929–8946

Reynolds RW, Rayner NA, Smith TM, Stokes DC, Wang W (2002) An improved in situ and satellite SST analysis for climate. J Climate 15:1609–1625

Roxy M, Tanimoto Y, Preethi B, Terray P, Krishnan R (2013) Intraseasonal SST-precipitation relationship and its spatial variability over the tropical summer monsoon region. Clim Dyn 41:45–61

Roxy M, Tanimoto Y (2012) Influence of sea surface temperature on the intraseasonal variability of the South China Sea summer monsoon. Clim Dyn 39:1209–1218

Sengupta D, Goswami B, Senan R (2001) Coherent intraseasonal oscillations of ocean and atmosphere during the Asian summer monsoon. Geophys Res Lett 28:4127–4130

Sengupta D, Ravichandran M (2001) Oscillations of Bay of Bengal Sea surface temperature during the 1998 summer monsoon. Geophys Res Lett 28(10):2033–2036

Stephens GL, Webster PJ, Johnson RH, Engelen R, L'Ecuyer T (2004) Observational evidence for the mutual regulation of the tropical hydrological cycle and tropical sea surface temperatures. J Climate 17:1124–2213

Sun Z, Mao JY, Wu GX (2009) Influences of intraseasonal oscillations on the clustering of tropical cyclone activities over the western North Pacific during boreal summer. Scientia Atmospheric Sinica (in Chinese) 33(5):950–958

Tao L, Fu XH, Lu WS (2009a) Moisture structure of the quasi-biweekly mode revealed by AIRS in Western Pacific. Adv Atmos Sci 26(3):513–522

Tao L, Fu XH, Wang B (2009b) The Moisture Structure of ISO in Western North Pacific Revealed by AIRS. J Meteor Res 23(2):191–205

Tao S, Chen LX (1987) A review of recent research on the East Asian summer monsoon in China. In: Chang C-P, Krishnamurti TN (eds) Monsoon Meteorology. Oxford University Press, New York, pp 60–92

Teng H, Wang B (2003) Interannual variations of the boreal summer intraseasonal oscillation in the Asian-Pacific Region. J Climate 16:3572–3584

Wang DX, Zhou FX, Qin ZH (1997) Study on the air-sea interaction on the interannual time scale in the South China Sea. Acta Meoteorological Sinica (in Chinese) 11:45–56

Wang B, Wu R, Fu X (2000) Pacific-East Asian teleconnection: How does ENSO affect East Asian climate? J Climate 13:1517–1536

Wang B, Wu R, Lau KM (2001) Interannual variability of the Asian summer monsoon: Contrasts between the Indian and western North Pacific-East Asian monsoons. J Climate 14:4073–4090

Wang B, Chan JCL (2002) How strong ENSO events affect tropical storm activity over the western North Pacific. J Climate 15:1643–1658

Wang B, Wu R, Li T (2003) Atmosphere-warm ocean interaction and its impacts on the Asian-Australian monsoon variation. J Climate 16:1195–1211

Wang B, Yang J, Zhou TJ (2008) Interdecadal changes in the major modes of Asian-Australian monsoon variability: strengthening relationship with ENSO since the late 1970s. J Climate 21:1771–1789

Wang B, Huang F, Wu Z, Yang J, Fu X, Kikuchi K (2009) Multi-scale climate variability of the South China Sea monsoon: A review. Dyn Atmos Ocean 47:15–37

Wang JB, Wen ZP, Wu R, Lin AL (2017) The impact of tropical intraseasonal oscillation on the summer rainfall increase over eastern China around 1992/1993. Clim Dyn 49:1847–1863. https://doi.org/10.1007/s00382-016-3425-8

Wen ZP, Huang RH, He HY, Lan GD (2006) The influences of anomalous atmospheric circulation over mid-high latitudes and the activities of 30-60-day low frequency convection over low latitudes on the onset of the South China Sea summer monsoon. Scientia Atmospheric Sinica (in Chinese) 30(5):952–964

Wentz FJ, Gentemann C, Smith D, Chelton D (2000) Satellite measurements of sea surface temperature through clouds. Science 288:847–850

Woolnough SJ, Slingo JM, Hoskins BJ (2000) The relationship between convection and sea surface temperature on intraseasonal timescales. J Climate 13:2086–2104

Wu L, Wen Z, Huang R, Wu R (2012a) Possible linkage between the monsoon trough variability and the tropical cyclone activity over the western North Pacific. Mon Wea Rev 140:140–150

Wu L, Wen Z, Wu R (2015a) Influence of the monsoon trough on westward-propagating synoptic-scale disturbances over the western North Pacific. Part I: Observations. J Climate 28(18):7108–7127

Wu L, Wen Z, Wu R (2015b) Influence of the monsoon trough on westward-propagating synoptic-scale disturbances over the western North Pacific. Part II: Energetics and numerical experiments. J Climate, 28(23):9332–9349

Wu L, Zhang H, Chen JM, Feng T (2018) Impact of two types of El Niño on tropical cyclones over the western North Pacific: Sensitivity to location and intensity of Pacific warming. J Climate 31:1725–1742

Wu R, Wang B (2000) Interannual variability of summer monsoon onset over the western North Pacific and the underlying processes. J. Climate 13(14):2483–2501

Wu R, Wang B (2001) Multi-stage onset of the summer monsoon over the western North Pacific. Clim Dyn 17:277–289

Wu R (2002) Processes for the northeastward advance of the summer monsoon over the western North Pacific. J Meteor Soc Japan 80:67–83

Wu R, Wang B (2002) A contrast of the East Asian summer monsoon and ENSO relationship between 1962-77 and 1978-93. J. Climate 15(15):3266–3279

Wu R, Hu Z, Kirtman BP (2003) Evolution of ENSO-related rainfall anomalies in east Asia. J Climate 16:3741–3757

Wu R, Kirtman BP, Pegion K (2006) Local air-sea relationship in observations and model simulations. J Climate 19(19):4914–4932

Wu R, Kirtman BP (2007) Regimes of local air-sea interactions and implications for performance of forced simulations. Clim Dyn 29(4):393–410

Wu R (2010) Subseasonal variability during the South China Sea summer monsoon onset. Clim Dyn 34:629–642

Wu R, Yang S, Wen ZP, Huang G, Hu KM (2012b) Interdecadal change in the relationship of southern China summer rainfall with tropical Indo-Pacific SST. Theor Appl Climatol 108(1):119–133

Wu R, Huang G, Du Z, Hu K (2014) Cross-season relation of the South China Sea precipitation variability between winter and summer. Clim Dyn 43(1–2):193–207. https://doi.org/10.1007/s00382-013-1820-y

Wu R, Chen Z (2015) Intraseasonal SST variations in the South China Sea during boreal winter and impacts of the East Asian winter monsoon. J Geophys Res 120(12):5863–5878

Wu R, Hu W (2015) Air-sea relationship associated with precipitation anomaly change and mean precipitation anomaly over the South China Sea and the Arabian Sea during the spring-to-summer transition. J Climate 28(18):7161–7181

Wu R, Cao X, Chen SF (2015b) Co-variations of SST and surface heat flux on 10-20-day and 30-60-day time scales over the South China Sea and western North Pacific. J Geophys Res 120(24):12486–12499

Wu R (2016) Coupled intraseasonal variations in the East Asian winter monsoon and the South China Sea-western North Pacific SST in boreal winter. Clim Dyn 47(7):2039–2057

Wu R, He Z (2017) Two distinctive processes for abnormal spring to summer transition over the South China Sea. J. Climate 30:9665–9678. https://doi.org/10.1175/JCLI-D-17-0215.1

Wu R, Cao X (2017) Relationship of boreal summer 10-20-day and 30-60-day intraseasonal oscillation intensity over the tropical western North Pacific to tropical Indo-Pacific SST. Clim Dyn 48(11–12):3529–3546

Wu R (2018) Feedback of 10-20-day intraseasonal oscillations on seasonal mean SST in the tropical western North Pacific during boreal spring through fall. Clim Dyn, online, https://doi.org/10.1007/s00382-016-3362-6

Wu R, Song L (2018) Spatiotemporal change of intraseasonal oscillation intensity over the tropical Indo-Pacific Ocean associated with El Niño and La Niña events. Clim Dyn 50:1221–1242. https://doi.org/10.1007/s00382-017-3675-0

Wu X, Mao J (2016) Interdecadal modulation of ENSO-related spring rainfall over South China by the Pacific Decadal Oscillation. Clim Dyn 47:3203–3220. https://doi.org/10.1007/s00382-016-3021-y

Wu X, Mao J (2017a) Interdecadal variability of early summer monsoon rainfall over South China in association with the Pacific Decadal Oscillation. Inter J Climatol 37:706–721. https://doi.org/10.1002/joc.4734

Wu X, Mao J (2017b) Spatial and interannual variations of spring rainfall over eastern China in association with PDO-ENSO events. Theor Appl Climatol. https://doi.org/10.1007/s00704-017-2323-2

Xie Q, Wu XY, Yuang WY et al (2007a) Life cycle of intraseasonal oscillation of summer SST in the western South China Sea. Acta Meteor Sinica (in Chinese) 3:1–8

Xie SP, Chang CH, Xie Q et al (2007b) Intraseasonal variability in the summer South China Sea: Wind jet, cold filament, and recirculations. J Geophys Res 112:C10008. https://doi.org/10.1029/2007JC004238

Xie SP, Hu K, Hafner J et al (2009) Indian Ocean capacitor effect on Indo-Western Pacific climate during the summer following El Niño. J Climate 22:730–747

Yang H, Li CY (2003) The Relation between Atmospheric Intraseasonal Oscillation and Summer Severe Flood and Drought in the Changjiang-Huaihe River Basin. Adv Atmos Sci 20(4):540–553

Yang J, Wang B, Wang B (2008) Anticorrelated intensity change of the quasi-biweekly and 30-50-day oscillations over the South China Sea. Geophys Res Lett 35:L16702. https://doi.org/10.1029/2008GL034449

Ye KH, Wu R (2015) Contrast of local air-sea relationship between 10–20-day and 30–60-day intraseasonal oscillations during May-September over the South China Sea and the western North Pacific. Clim Dyn 45(11–12):3441–3459

Zhang HJ, Wu L, Huang RH (2018) Possible impacts of two types of El Niño events on the western North Pacific monsoon trough and tropical cyclogenesis. Climatic and Environmental Research (in Chinese) 23(2):150–160

Zhou W, Chan JCL (2005) Intraseasonal oscillations and the South China Sea summer monsoon onset. Int J Climatol 25:1585–1609

Zhou W, Li C, Chan JCL (2006) The interdecadal variations of the summer monsoon rainfall over South China. Meteorol Atmos Phys 93:165–175

Zhu C W, Nakazawa T, Li JP (2003) The 30–60 day intraseasonal oscillation over the western North Pacific Ocean and its impacts on summer flooding in China during 1998. Geophys Res Lett 30:1952, https://doi.org/10.1029/2003gl017817

Zhu CW, Nakazawa T, Li JP (2004) Modulation of tropical depression/cyclone over the Indian-western Pacific Oceans by Madden-Julian Oscillation. Acta Meteorological Sinica (in Chinese) 62(1):42–50

Chapter 4
Land–Atmosphere Interaction and Climate Variability in Southeast Asia and Its Surrounding Area

Land covers nearly 30% of the Earth's surface. It not only provides a place for humans to live, but also bonds all layers of the Earth system organically. Human activities change land-surface processes, leading to variation of land-atmosphere processes, which affects global climate and environment (Lee et al. 2011). Therefore, it is of great significance to study land-surface processes and land-atmosphere interaction characteristics, to understand the causes of climate and environmental changes, and to improve climate and environment for humans.

Land-surface processes are divided into aboveground processes and underground processes. The former includes thermodynamic process, hydrological process, and biological process; and the latter includes soil thermal conduction process as well as water and heat transport processes. Land–atmosphere interaction refers to the constant momentum, energy, and material exchanges between geo-gases occurring inside the atmospheric boundary layer (ABL) due to the forcing associated with atmospheric conditions and solar radiation (Wang 1999). As turbulence is the main characteristic of air motion inside the ABL, learning about land-surface processes and land-atmosphere interaction plays a pivotal role in studying the characteristics of turbulent motion inside the ABL.

Numerical simulation is an essential way for predicting future climate and environment, one of the objectives and cores of the research plans included the World Climate Research Programme (WCRP) and the International Geosphere-Biosphere Program (IGBP). In recent 30 years, studies have been focused on the parameterizations of ABL mode and land-surface process mode based on large-scale field observations, and these studies showed that the predictability of climate model can be enhanced dramatically by improving land-surface process model (Avissar and Pielke 1991). Although great achievements have been made in numerical simulations along with the improvement of computer performance and optimizations of numerical methods, the simulation of land-surface process in the model is far from meeting the needs of many practical problems, due to the heterogeneity and complexity of the underlying surface. Land-surface process model needs further improvement.

S. Yang et al., *Climate Change in Southeast Asia and Surrounding Areas*,
Springer Climate, https://doi.org/10.1007/978-981-15-8225-7_4

Local observations, which serve as the basis for current land process parameterization, provide the basis for the improvement of the model's performance. Compared with the long-term observations in Heihe of Inner Mongolia, the Qinghai–Tibet Plateau, Huaihe River Basin, and other places in North China, there are only a few long-term observations on different underlying surfaces in South China where the East Asian summer monsoon prevails.

4.1 Field Observations of Typical Underlying Land-Surface Process and Relevant Theoretical Analysis

4.1.1 Field Observations of Typical Underlying Surface in the South China

1. Introduction of stations and observations

a. Subtropical Broad-leaved Forest Site in the Phoenix Mountain of Zhuhai

The Phoenix Mountain of Zhuhai, located south of the Tropic of Cancer and north of the urban area of Zhuhai (Fig. 4.1), is a coastal hilly area covered 90% by forest vegetation, which is a south subtropical evergreen broad-leaved forest community. Characterized by various species of trees, the Phoenix Mountain area has been classified as a national-level ecological forest. There are diversified landforms and topographies, many mountains and peaks, rich water resources and plenty of rivers in the Phoenix Mountain, where water and soil are well conserved.

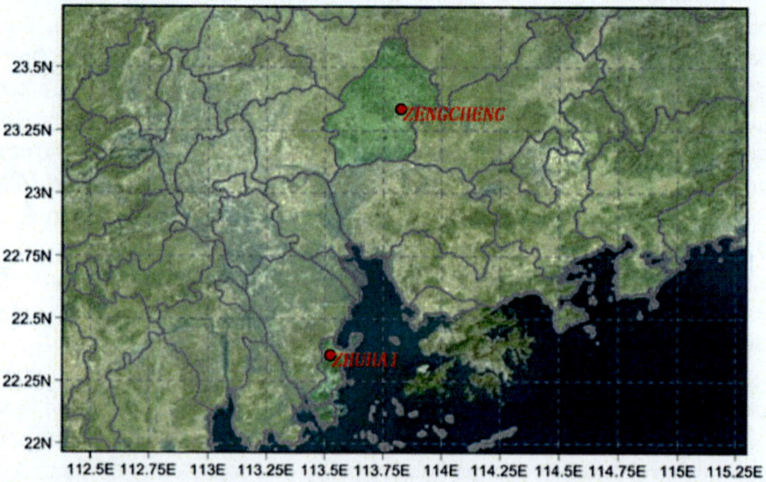

Fig. 4.1 Geographical locations of subtropical broad-leaved forest site at the Phoenix Mountain of Zhuhai and hilly shrub forest site at Zengcheng of Guangzhou

Located in the university park of the Tangjiawan Hi-tech Zone north of Zhuhai, Beijing Normal University's Zhuhai Campus is surrounded by the Phoenix Mountain, which is mainly hills and slopes covered by forest as the typical secondary forest surrounding cities or villages in the Lingnan region. After making observations and argumentations, we built a 60-m iron observation tower to observe land-atmosphere interaction and carbon flux in the low sloping land where canopies are relatively flat in the foothill forest of the Phoenix Forest inside Beijing Normal University's Zhuhai Campus (Fig. 4.2). At (113°31′34.2″ E, 22°21′15.5″ N), the altitude of this tower's base is 38.5 m above the sea level. The underlying surface of the tower is a typical subtropical forest consisting of acacia forests, eucalyptus forests, and shrub-grass slopes; the average height of forest canopies is 18 m. Three sets of observation systems are mounted on the tower to observe land-atmosphere interaction and CO_2 flux over a long term.

b. Hilly Shrub Forest Site in Zengcheng of Guangzhou

The Hilly Shrub Forest Site in Zengcheng of Guangzhou is located inside the China Meteorological Administration (CMA) Zengcheng National Climate Datum Station (113°49'39" E, 23°20'05" N) on the hilly area and 38.5 m above the sea level. It was built in May 2014 and improved thereafter; new equipment was installed and debugged in October 2014, and was put into operational use in the first 10 days of November 2014.

Fig. 4.2 The observation tower at Zhuhai

Fig. 4.3 The observation tower at Zengcheng

Located in the mid-east of Guangdong province and northeast of Guangzhou, Zengcheng is next to Dongguan in its south across the river. High in the north and low in the south, it mainly consists of hills that distribute in the mid-south and account for 35.1% of its urban area. The observation site is 100 m south of the CMA Zengcheng National Climate Datum Station; its underlying surface is subtropical hilly shrub forest and the average canopy height is 8 m. The Zengcheng Observation Tower is a 70-m tall iron tower provided with a triangle Sect. (540 * 540 * 540 mm), as shown in Fig. 4.3. Observation instruments are mounted on the tower to observe wind, temperature and humidity gradient, radiation and momentum, sensible, latent heat, and CO_2 flux. Four levels of soil temperature and humidity and 1 level of soil heat flux also observed in the soil.

4.1.2 Data Processing Method

First, we make some improvements over existing data quality control methods. The period of the data is from January 2015 to December 2015. We use a simple threshold calculation method based on robust statistical theories, to control the quality of meteorological data from the carbon flux observation tower of the Phoenix Mountain in Zhuhai, and to analyze land-atmosphere interaction. While performing the coherency test, we utilize spatial correlation and time dependence among all levels of data and provide a method combining spatial correlation and temporal coherency to judge coherency. Note that the stiff value test, temporal coherency test, and spatial coherency test of wind speeds at 8 and 15 m are not carried out as instruments are installed inside canopies and measurements are greatly affected by canopies at these two levels. The stiff value test and spatial coherency of the internal temperature test of soil areas are also not performed due to small temperature variation.

Here, data quality is controlled following the procedures below according to Wang (2004), Wang et al. (2007, 2010) and Wang and Liu (2012): extreme test → stiff value test → coherency test → comprehensive test. Only the data from the levels above canopy (18 m) were tested for their coherency in this study. The data control codes used in this study are as follows: 0–normal, 1–doubtful, and 2–abnormal. All procedures are described below.

1. Extreme and threshold calculation methods

On the basis of robust statistics (David et al. 1983), we use a simple threshold calculation method in this study, which is not easily affected by extremes. The prior judgment of variable distribution is not needed for this method, which is referred to as "two-point method."

We sort the original data or the data calculated based on raw data in ascending order, with median M, upper quartile Q1, lower quartile Q2, and the upper and lower extremes. We calculate fourth quartiles with the formula $H = Q2 - Q1$. After summarizing lots of cases, the statists show that the data outside (Q1–1.5H, Q2 + 1.5H) can be deemed outliers in general.

2. Stiff value and threshold calculation methods

Stiff values are generated by observation records that remain unchanged for a long time. Stiff value test is carried out to see if observation values are not changed for a long time for the purpose of quality control. If the changes of factors are within the set threshold among 24 records (12 h), these factors should be deemed stiff values. To determine stiff values, such factors as the local meteorological conditions and instrument precision should be taken into consideration. Considering the extremely small change of soil temperature inside, no stiff value test is made for soil temperature.

3. Temporal coherency and spatial coherency test methods

According to the temporal coherency principle, data measured by the same instrument at the same location and time will not fluctuate significantly in general. Spatial coherency principle assumes that data measured by the same instrument at adjacent locations at the same time will not fluctuate significantly generally. Data not conforming to coherency requirement cannot be eliminated completely during the stiff value test, and need to be judged whether they are abnormal according to temporal coherency and spatial coherency principles.

Temporal coherency test method: In this quality control, the coherency of data is obtained using a time-varying test method. The detail of the method is as follows: calculating the standard deviation σ of a meteorological element within the whole time period, and expressing the meteorological data at adjacent times of x_{i-1}, x_i and x_{i+1}, respectively. If $|x_{i+1} - x_i| > 2\sigma$ or $|x_i - x_{i-1}| > 2\sigma$ at a certain time, xi is considered as suspicious data, which should be marked with 1. The data corresponding to the starting time could not be tested using this method, and are thus marked as 1. Then, these data marked with 1 are re-inspected using the spatial coherency test method mentioned below. For the data passing the spatial coherency test, their quality code is marked as 0; and for those failing the test, they are marked as 2.

Spatial coherency test method: In this study, we compare data of adjacent levels. First, the threshold of each kind of observation data is determined. If the absolute value of the difference between two adjacent observation values $|xi + 1 - xi| > N$, after determining threshold N, x_{i+1} should be marked as 1. When determining threshold N, we calculate the absolute value of the difference between meteorological data of adjacent levels at the same time and sort them in an ascending order. For the meteorological data at 8 and 60 m, only data of one adjacent level are used to calculate the difference. However, for the data at other levels, the data at the levels above and below are needed to calculate the differences. Then, these data are marked with 1 and re-inspected with the time coherency method mentioned above. The data passing the temporal coherency test are marked with 0, and those failing the test are marked with 2. In this paper, we test temperature, humidity and wind speed at 40, 47.5, 55, and 60 m using this method.

4. Aerodynamic roughness and zero plane displacement calculation methods

Aerodynamic roughness and zero plane displacement of underlying surfaces are calculated with the fitting method and roughness element method, respectively. In the northwest and northeast of the observation tower in the Phoenix Mountain of Zhuhai, there are dense contour lines. In its southwest and southeast, there are gentle and open slopes. Besides, there is one hill in the northwest of the tower, and there is another in the northeast of the tower. The altitude of the hill in the northeast is higher than 120 m, while the one in the northwest is only higher than 90 m.

After quality control, there are a total of 521 sets of data under neutral stratification, including 414 sets of data in autumn and winter (September–February of the following year) and 107 sets of data in spring and summer (March-August). According to the characteristics of contours around the tower, the underlying surface is divided into four sectors clockwise centered at the tower station, with the due north as 0°. Table 4.1 shows the range of angle of each sector, the times that neutral

Table 4.1 Statistics of occurrence frequency of neutral stratification at Zhuhai

Section division	Autumn and winter	Spring and summer	Total
315°–5°	329	38	367
45°–135°	78	31	109
135°–225°	5	33	38
225°–315°	3	5	8

stratification data of different seasons are contained in each sector, and occurrence frequency of neutral stratification in each sector within the whole time period.

Table 4.1 shows that neutral stratification is mainly in the north of the northwest area, the north of the northeast area, and is also distributed densely between 45° and 135° in autumn and winter. In spring and summer, neutral stratification is distributed uniformly in the east and southwest areas. The distribution of neutral stratification is basically consistent with the seasonal wind direction distribution, i.e., prevailing northeasterly wind and easterly wind, and occasional southeasterly wind in autumn and winter (September–February of the following year), and prevailing easterly wind and southwesterly wind in spring and summer (March–August), as observed over the years in Zhuhai.

5. Calculation of flux source area

The relatively mature K-M analytical footprint model by Kormann and Meixner (2001) is selected to analyze the distribution characteristics of the single-level flux source area first. Then, the flux source areas at 55 and 32.5 m are compared and analyzed, respectively.

The K-M model is applied to the areas where underlying surface vegetation is basically uniform and topographic relief is small. As the flux tower is 1,200 m, 240 m, 1200 m, and 1,000 m from the forest borders in the east, west, south, and north, respectively, and the vegetation there is mainly subtropical evergreen broad-leaved forest.

To determine the topographic relief in this area, within the circular scope of 1–km radius centered at the flux tower, the surrounding area of the flux tower is divided into four sectors: A1, B1, C1, and D1. Within each sub-area, sector areas of 500-m radius, including A2, B2, C2, and D2, and sub-areas of 300–m radius, including A3, B3, C3, and D3, are divided (Fig. 4.4). The projected density of sub-area is defined by

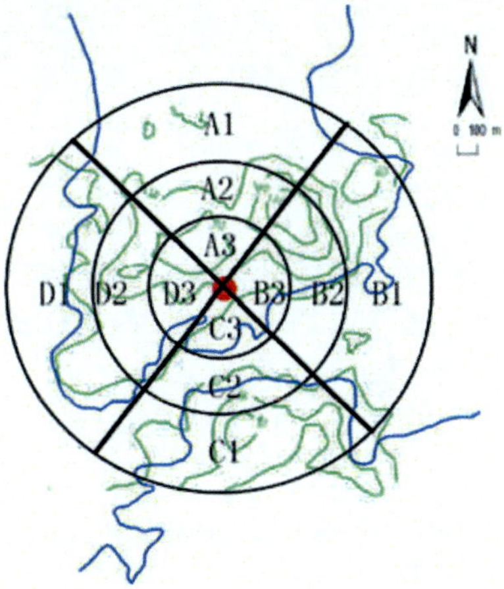

Fig. 4.4 Sub-area division surrounding the tower station at Zhuhai

Table 4.2 λ_f' of All Sub-areas

	λ_f'		λ_f'		λ_f
A1	0.97	A2	0.92	A3	0.99
B1	0.97	B2	0.97	B3	0.95
C1	0.99	C2	0.99	C3	0.92
D1	0.99	D2	0.97	D3	0.91

$$\lambda_f' = \frac{A_f}{A_d} \tag{4.1}$$

where A_f represents the projection area of each sub-area on the surface and A_d represents the superficial area of each sub-area (excluding vegetation). With the area calculation tools provided with Local Space Viewer, λ_f' of sub-area is calculated according to DEM data, as listed in Table 4.2.

Table 4.2 shows that within the divided sub-areas of different sizes, λ_f' is larger than 0.9. It indicates that surface undulation in each direction of this area is not large according to the analysis on all sizes, meeting the applicable conditions of the K-M model.

To calculate the scope of flux source area at 55 m, we set up a retrieving table to establish relations between wind direction and corresponding sub-area, to establish the correspondence between aerodynamic roughness and zero plane displacement of different wind direction scopes and different underlying surfaces. The sub-area of 1,000-m radius is divided in the middle by two means, as shown in Fig. 4.5, forming sub-areas A1 (315°–5°), B1 (45°–135°), C1 (135°–225°), and D1 (225°–315°), and sub-areas E1 (0°–90°), F1 (90°–180°), G1 (180°–270°), and H1 (270°–360°). With the sub-areas of A1, B1, C1, and D1, and of E1, F1, H1, and G1, corresponding aerodynamic roughness and zero plane displacement are calculated, and the retrieving table of sub-area and wind direction corresponding to each other is set up (Table 4.3).

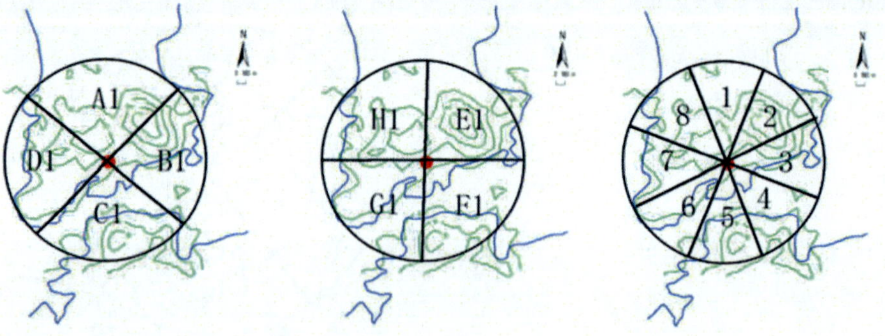

Fig. 4.5 Sub-area division around the tower at Zhuhai, and division of wind direction range (1,000-m radius)

Table 4.3 Retrieving Table of the Wind Direction of Observation Level at 55 m, Corresponding Sub-areas and Parameters of Relevant Underlying Surfaces

	Wind direction	Corresponding sub-area	H (m)	λ_f	L	λ_p	d (m)	z_0 (m)
1	337.5°–2.5°	A1	45.26	0.18	2.2	0.40	30.29	1.80
3	67.5°–112.5°	B1	13.17	0.18	2.0	0.36	8.24	0.67
5	157.5°–202.5°	C1	26.31	0.18	2.0	0.36	16.46	1.35
7	247.5°–292.5°	D1	30.11	0.18	2.2	0.40	20.15	1.20
2	22.5°–67.5°	E1	41.31	0.18	2.2	0.40	27.64	1.65
4	112.5°–157.5°	F1	33.81	0.18	2.0	0.36	21.15	1.73
6	202.5°–247.5°	G1	18.12	0.18	2.0	0.36	11.33	0.93
8	292.5°–337.5°	H1	41.58	0.18	2.2	0.40	27.82	1.66

If the wind direction at a time is known, corresponding sub-areas can be searched from the retrieving table for calculation.

In Table 4.3, the projection density of roughness elements λ_f and the projection density of vegetation λ_p are defined as follows:

$$\lambda_f = \frac{A_f}{A_d} \tag{4.2}$$

$$\lambda_p = \frac{A_p}{A_d} \tag{4.3}$$

where A_d is the surface area of the zone, A_p is the vegetation coverage, and A_f is the projected area of the surface.

Fig. 4.6 Sub-areas Corresponding to 337.5°–2.5° wind direction and the coverage of 70% flux source area in due-north Direction

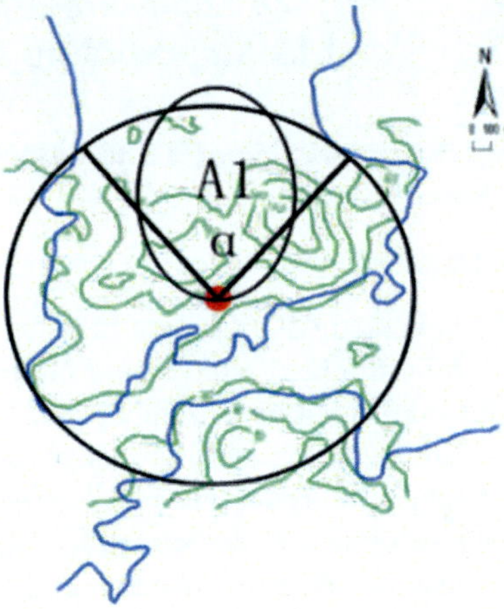

Here, the data quality is controlled, and the data are evaluated by the EddyPro flux processing software. Basic quality control steps include eliminating anomalous values of the original data, coordinate rotation, humidity correction, frequency response correction, and WPL correction (correction for density fluctuations). Flux test steps include turbulent stability test, turbulent development sufficiency test, and flux representativeness test.

The data under neutral condition are also eliminated as the K-M model is derived under non-neutral conditions. The specific quality control steps are as follows: ① eliminating the times when CO_2 flux >50 or <−100 μmol/(m^2 s) in the daytime (6:30–18:30), and the times when CO_2 flux <0 or >50 μmol/(m^2·s) at night (19:00–6:00 of the next day); ② eliminating the data close to neutral stratification (using criterion $-0.1 < z\text{-}d/L < 0.1$); ③ eliminating the data of sensible heat flux >300 W/m^2 and the data of latent heat flux <−50 W/m^2 or >650 W/m^2.

To study single-level (55-m) flux source area related problems, the data outside 1,000 m of flux source area are eliminated in this study. The elimination of such data is for two reasons. First, as the flux tower is 1,200 m, 240 m, 1,200 m, and 1,000 m from the forest borders in the east, west, south, and north, respectively, sources areas that are >1,000 m may be beyond forest borders. If that is the case, the situation of underlying surfaces may not satisfy the conditions necessary for the model, and thus the model may no longer be applicable. Second, as the main purpose of single-level research is to discuss the spatial representativeness of the flux of subtropical evergreen broad-leaved forest in the Lingnan region, the data recorded by the instruments cannot reflect the flux variation of forest underlying surfaces accurately if the flux calculated is beyond the forest borders.

4.1.3 Analysis on Characteristics of Typical Underlying Surface Land–Atmosphere Interaction in South China and Its Surrounding Area

1. Characteristics of Land–atmosphere Interaction in Phoenix Mountain of Zhuhai

a. Data Quality

We collected 650,528 data measured at 22,432 times from December 2014 to May 2016 by the land-atmosphere interaction and carbon flux observation tower in the Phoenix Mountain of Zhuhai.

Results of the quality control showed that abnormal data account for 0.62% of the total data, reflecting measurement instruments being reliable. For the abnormal data, extreme data accounted for 74.57% and stiff value data accounted for 25.43%. To be more specific, most extreme data occurred at the levels of low height, and stiff value data were mostly humidity data. For example, there were 2,980 extreme data among humidity and temperature data at 8 m, accounting for 98.48% of the extreme

Table 4.4 Results Obtained with Fitting Method

Area	d (autumn and winter, m)	z_0 (autumn and winter, m)	d (spring and summer, m)	z_0 (spring and summer, m)	d (full time, m)	z_0 (full time, m)
315°–5°	26.78	3.59	28.03	4.69	27.00	3.64
45°–135°	19.03	1.73	20.11	1.70	19.40	1.72
135°–225°					28.99	2.80
225°–315°					22.39	2.50

Table 4.5 Results Obtained with Roughness Element Method

	H (m)	λ_f	L	λ_p	d (m)	z_0 (m)
A1	45.26	0.18	2.2	0.40	30.29	1.80
B1	13.17	0.18	2.0	0.36	8.24	0.67
C1	26.31	0.18	2.0	0.36	16.46	1.35
D1	30.11	0.18	2.2	0.40	20.15	1.20

data; there were 984 stiff value data among humidity data, accounting for 95.35% of the stiff value data. Overall, the data quality was high, and abnormal values were stiff values and extremes that occurred most frequently in the initial instrument stage and low mounting height, implying good data quality at this site. Data measured at certain times failed to pass regular temporal coherency test of spatial coherency test. Further manual re-check also showed that such data are not outliers.

b. Aerodynamic Roughness and Zero Plane Displacement Calculation Results

The calculation results obtained with the fitting method and roughness element method are shown in Tables 4.4 and 4.5.

Calculation errors may be generated from data, method, and assumption made for the fitting method used. Errors of the roughness element method are mainly from the simplification and parameters used in the calculation process. Compared with the calculation results obtained with the roughness element method, the calculation results obtained with the fitting method are higher in 315°–45° directions, and lower in the other directions. We analyzed the calculation results obtained with the two methods and found they are similar when the radius of the tower station surrounding scope is set to 500 m. It can thus be inferred that the spatial representativeness of the fitting method is 500-m radius or so. This method shows big defects when dealing with complicated landforms. The roughness element method can give the specific scope of underlying surfaces calculated, compared with the fitting method. As the vegetation of the region around the tower station is subtropical evergreen broad-leaved forest, the vegetation coverage changes with the season. The representativeness scope of the calculated results obtained with the roughness element method is much better than that with the fitting method; therefore, the roughness element method can better reflect the overall situation of the aerodynamic roughness and zero plane displacement of underlying surfaces in this area. Besides, the roughness element method can be used to calculate the aerodynamic roughness and zero plane displacement of areas for which direct calculation using wind profiles is not available due to insufficient wind data.

For the underlying surfaces that are mainly evergreen broad-leaved forest, it is apparently advantageous to use the roughness element method, which can give specific calculation scope and calculate areas that cannot be calculated with the fitting method due to insufficient wind data. According to the calculation results, seasonal factors change vegetation growth conditions, affecting aerodynamic roughness and zero plane displacement, and topographical factors affect the calculation of zero plane displacement more significantly than seasonal factors.

c. Characteristics of Flux Source Area

To understand the temporal and spatial variation of flux in this area, the scope of flux source area should be paid attention to first. Figure 4.7 gives the rose map of wind direction frequency statistics from January to December 2015 centering at the meteorological tower with 22.5° step. It also shows the frequency of wind speed in all directions based on wind speed classification.

Figure 4.7 shows that wind direction shifted clockwise from winter to early summer. In December, January and February, there were mainly northerly and easterly winds. Since March, the wind started to shift toward the southeasterly direction. In this process, wind direction and large wind speed distributed uniformly on the east of the flux tower. Since April, wind direction started to centralize in southwesterly and southeasterly directions due to summer monsoon. Almost all the largest wind speeds in May and June were in southwesterly direction, related to the summer monsoon onset. Wind direction was uniformly distributed in all directions when mid-summer came (July and August), and started to centralize in the easterly and northeasterly directions gradually starting in September.

We calculated the source areas with 70% contribution under unstable and stable atmospheric conditions, respectively. In Fig. 4.8, blue dots represent the farthest sites in the scope represented by flux source area in 30 min. Moreover, stability is judged by $(z\text{-}d)/L$; and the data are classified as stable and unstable using $(z\text{-}d)/L > 0.1$ and $(z\text{-}d)/L < -0.1$, respectively.

The results show that the scope of the flux source area was consistent with main wind directions without showing significant difference only when the atmosphere was unstable. When the atmosphere was stable, flux source areas were distributed in both main wind directions and other wind directions, not consistent with main wind directions. Moreover, the scope of flux source area under stable conditions was larger than that under unstable conditions. This is because under unstable atmospheric environment, air moves vertically, materials are transported quickly upward, concentration changes quickly and can be detected by sensor immediately. Therefore, the flux measured by sensor is from places close to main wind directions. If the atmosphere is stable, turbulence is weak, less material is transported by atmosphere and flux disperses slowly; and sensor can only detect the concentration change of flux transported slowly from far away. In such a case, there were large flux source areas

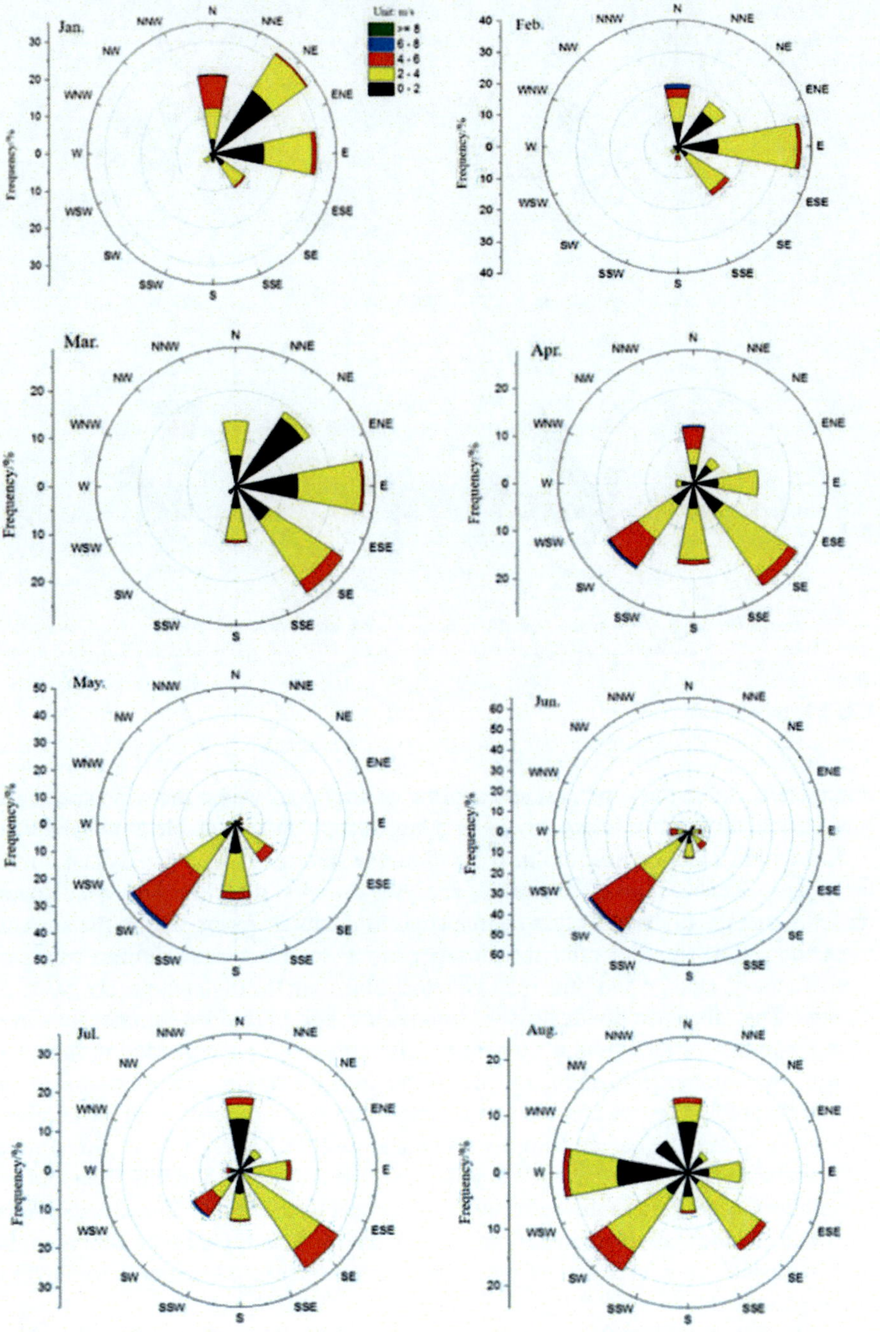

Fig. 4.7 Rose Map of Wind Direction, Wind Speed and Their Distribution Frequencies from January to December 2015

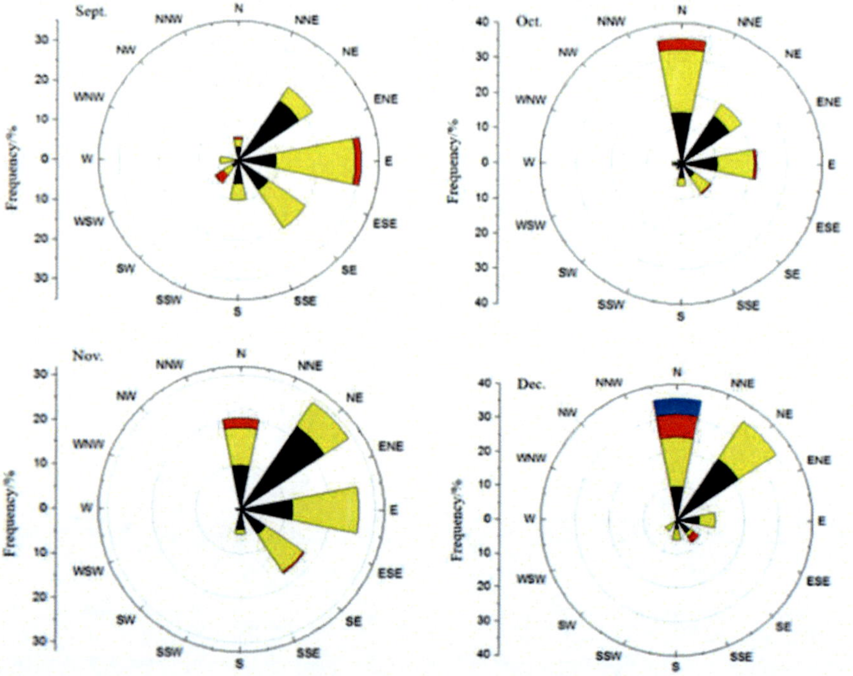

Fig. 4.7 (continued)

distributed even in the southeast and southwest directions where the wind speed was low. Hence, flux source areas can cover a large scope under a stable atmosphere.

Located in a unique geographical location, the summer monsoon plays an important role in Zhuhai climate. From the analyses on wind direction and wind speed, we learn that the division of flux source areas into winter, spring before the summer monsoon onset, summer after the summer monsoon onset, and autumn based on wind direction change and time is more reasonable than the division purely based on months. Therefore, we divide seasons into winter, spring (before summer monsoon onset), summer (after summer monsoon onset), and autumn, considering the onset time of the summer monsoon. In this subsection, we discuss the variation of the distribution of flux source areas in winter before summer monsoon onset (December–February), in spring (before summer monsoon onset, 1 March–1 April, in summer (after summer monsoon onset, 2 April–31 August), and in autumn (September–November), according to the definition of the onset of the South China Sea summer monsoon. As the stability is lower than zero at daytime but higher than zero at nighttime in most cases, source areas for each season are discussed separately for daytime and nighttime.

Figure 4.9 shows that flux source areas are mainly in the north and east directions. In spring before the summer monsoon onset, the distribution of flux source areas is similar to that in winter. In summer after the summer monsoon onset, flux source

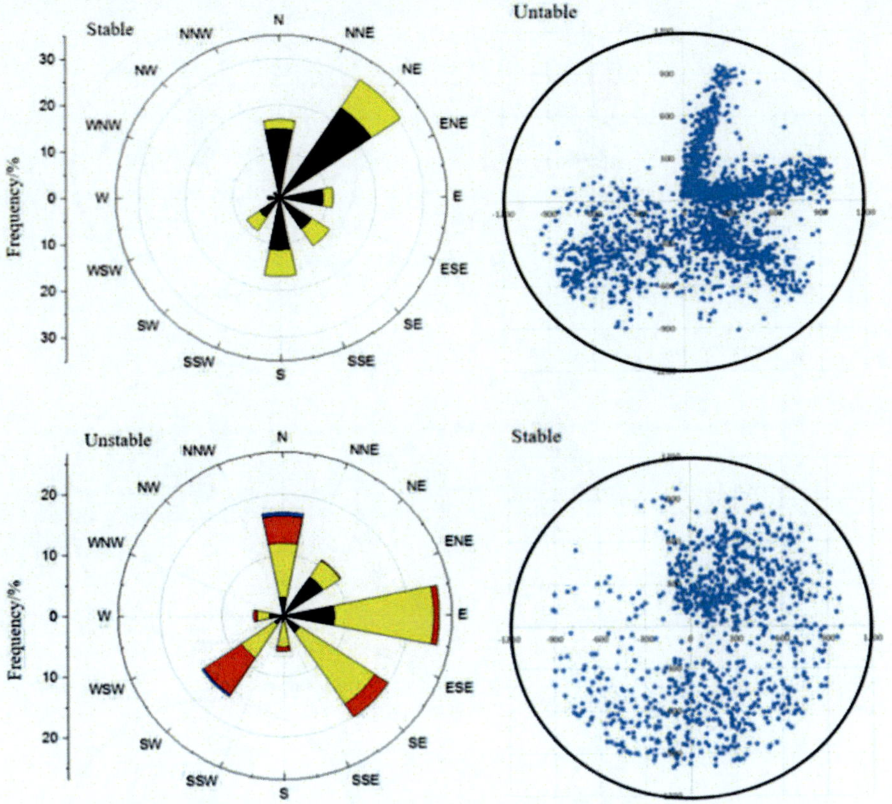

Fig. 4.8 Wind Rose Map under Different Stability and Variation Chart of 70% Flux Source Area along with Wind Direction

areas are in the southwest and southeast directions. In autumn, flux source areas are in the east direction. This is consistent with the wind direction distribution in Zhuhai, i.e., northerly wind and easterly wind prevail before the summer monsoon onset, and southwesterly wind and southeasterly wind prevail for a long time after the summer monsoon onset. There are barely any flux source areas within a large scope in the northwest direction of Zhuhai, as northwesterly wind occurs rarely.

We calculated the average daily change of flux source areas in winter, spring (before summer monsoon onset), summer (after summer monsoon onset), and autumn. The calculation method is as follows. We calculated the size of average flux source areas (not shown) at the same times (time interval of one hour) in each season. Generally speaking, the daily variation of the size of flux source areas is similar in all four seasons, and the scope of flux source areas is larger at nighttime than in daytime. The largest flux source areas appear at the following times: around 8:00 in winter, 2:00 in spring, 4:00 in summer, and 18:00 in autumn. Along with the summer monsoon onset, the largest flux source areas in summer appear at a time (of the day) later than that in spring, probably caused by the different time that the atmosphere reaches its highest stability. The smallest flux source areas occur between

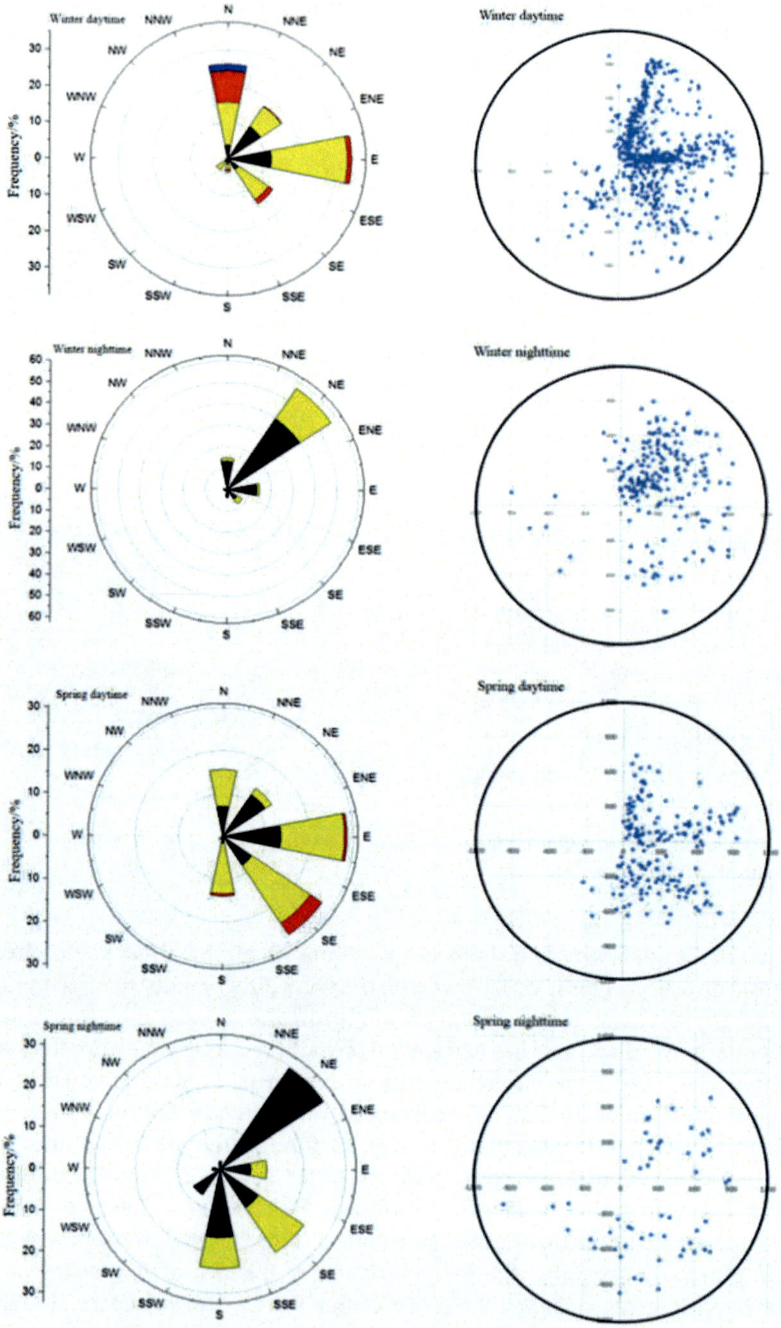

Fig. 4.9 Wind rose map in different seasons and variation chart of 70% flux source area along with wind direction

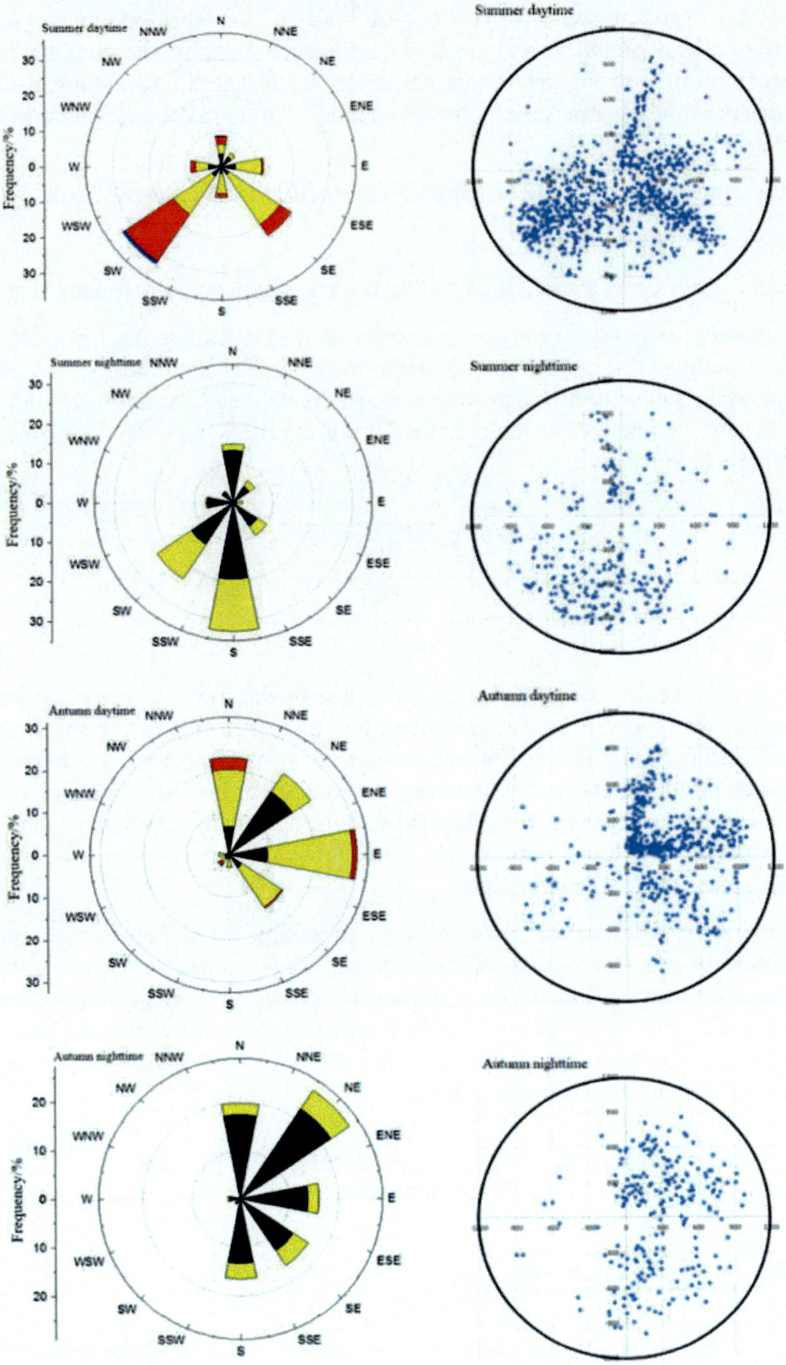

Fig. 4.9 (continued)

6:00–10:00 when turbulence is the strongest. We also found that the scope of source areas is smaller at nighttime and larger at daytime occasionally, showing that turbulence may become strong due to thermal instability at nighttime and thus make the atmosphere unstable; atmospheric inversion may occur to make stratification stable at nighttime.

2. Characteristics of Land–atmosphere Interaction in Zengcheng Guangzhou

a. Quality control/evaluation

(1) Inspect original data and eliminate the data with anomaly ratio greater than 1%.

(2) Turbulence stability inspection is mainly used to eliminate the intensive non-turbulent motions that cannot be eliminated by detrending and filtering. In general, this test is completed through the comparison of the average covariance within whole stage and sub-stages, and the main method is the algorithm in Foken (2003), such as the formula below:

$$\mathrm{IST} = \left| \frac{\frac{1}{6} * \sum_{i=1}^{6} \overline{(\omega' x')_5} - \overline{(\omega' x')_{30}}}{(\omega' x')_{30}} \right| * 100\% \tag{4.4}$$

where ω' is vertical speed perturbation and x' is the pulsation of a vector component or a scaler, with the subscript representing the number of minutes. Equation (4.4) is the deviation of the 30-min covariance of vertical speed and x from the average covariance of all sub-stages, representing the stability of main statistics within one time. If it is not larger than 30%, the quality of turbulent stability data is good; if it is large, the data are less representative. Table 4.6 is the turbulence stability quality classification given by Foken (2003).

(3) Test of overall turbulence characteristics (inspection of turbulence development) is to make sure full development of turbulence conforms to the similarity theory and the normalized standard deviation of turbulence components is a stability function or constant. In general, turbulence standard deviation measured and the "standard" value obtained through simulation are compared with Foken and Wichura (1996) method, as shown in the formula below:

$$\mathrm{ITC} = \left| \frac{(\delta_x / X_*)_{\bmod el} - (\delta_x / X_*)_{measured}}{(\omega' x')_{30}} \right| * 100\% \tag{4.5}$$

Table 4.6 Data quality classification based on turbulence stability test

CLASS	1	2	3	4	5	6	7	8	9
IST(%)	0–15	16–30	31–50	51–75	76–100	101–250	251–500	501–1000	>1000

Table 4.7 Data quality classification based on turbulence stability test

CLASS	1	2	3	4	5	6	7	8	9
ITC(%)	0–15	16–30	31–50	51–75	76–100	101–250	251–500	501–1000	>1000

Table 4.8 Overall quality classification of turbulence data

CLASS	1	2	3	4	5	6	7	8	9
IST class	1	2	1–2	3–4	1–4	5	≤6	≤8	9
ITC class	1–2	1–2	3–4	1–2	3–5	≤5	≤6	≤8	9
Simplified-CLASS	0		1				2		

Note Class 0 represents high-quality data that can be used for the basic study. Class 1 represents middle-class data that can be used for long-term data processing and analysis. Class 2 represents data of poor quality.

where δ_x is the standard deviation of turbulence parameter, X^* is the non-dimensional characteristic factor of turbulence parameter x and ITC equation is the deviation from the non-dimensional standard deviation obtained from the actual measurement and "standard" nondimensional standard deviation. Standard nondimensional standard deviation is the value calculated with the similar function fitted according to flux variance similarity, and parameters of fitting functions for different underlying surfaces in different regions are different, which will be described in detail in neat section. Tables 4.7 and 4.8 are the turbulence stability quality classification given by Foken and Wichura (1996), Foken 2003. In general, when ITC $\leq$ 30%, turbulence is considered to be fully developed.

(4) Flux quality evaluation: After turbulence stability and development inspection of sensible heat flux, latent heat flux, momentum flux, and all calculation components of CO_2 flux, according to Foken and Wichura (1996) and the simplified classification method proposed for Carbo-Europe in 2003, flux quality evaluation was conducted (Table 4.8).

b. A preliminary study on flux observation data about the humid and rainy environment

(1) Issues of concern

Due to the high frequency and high accuracy of Eddy Covariance (EC) observation system, it is sensitive to the environment or weather process. Most researchers removed the data during precipitation and analyzed the time periods before and precipitation, and then interpolated adjacent non-precipitating time period data using the mean diurnal variation (MDV) method, or discard data. Such data processing is feasible for the observation stations in North China where it is dry and rains a little, but not for the South China where it is wet and rains a lot, as there will be lots of outliers that cannot be eliminated.

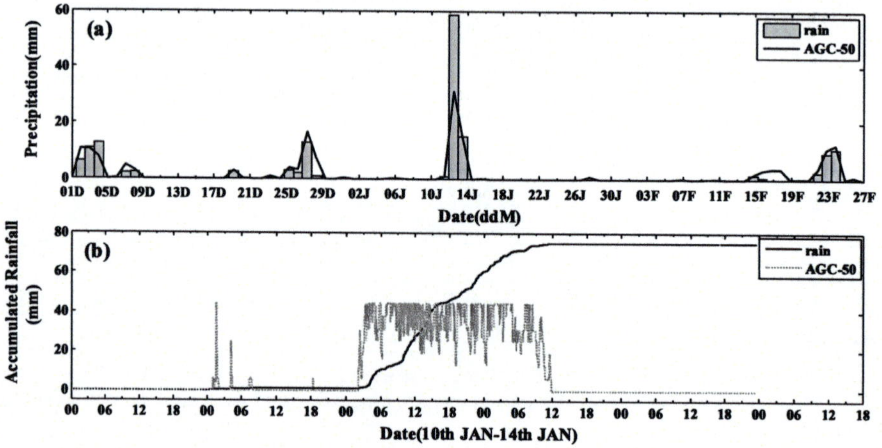

Fig. 4.10 Precipitation and AGC comparison diagram of Zengcheng in winter. **a** daily average precipitation and AGC-50, and **b** accumulative precipitation per min and average AGC-50

(2) Approaches

Open-circuit EC observation system, consisting mainly of CAST3 ultrasonic wind speed and temperature tester and LI-7500 H_2O–CO_2 gas analyzer, is widely used at many flux stations for its good frequency response and low power consumption. However, the open-circuit system has a common problem, that is, its probes are very sensitive and can easily be affected by precipitation, dew, snow, etc. when they have contact with the environment directly and measure intermediate environment (as shown in Fig. 4.10). In addition to certain tilt that is set during installation, which is needed to allow attached water drops to fall in, automatic gain control (AGC) value is added to measure whether there are water drops that may affect mirror cleanness and through-optical length, to determine whether these factors may have affected the measurements.

AGC, as a kind of output control, samples the amplitude of input signals with negative feedback to obtain a control voltage and adjusts the amplification times of amplifier reversely so as to realize automatic control. A typical value of AGC clean window of LI-7500 is between 50 and 60% (LI-COR 2001). AGC of new instruments should be recorded when they are used for the first time as the base value of the clean window. AGC will increase along with dust accumulation and droplet impact, and LI-7500 allows certain window droplets with accuracy ensured; the accuracy will be affected as AGC continues increasing. It has reference meaning to control data quality with AGC in South China where it is humid and rains a lot.

Taking Zengcheng as an example, its AGC base value is 50 and the relation between its daily precipitation in winter and AGC change along with time is as shown in Fig. 4.10a. It clearly indicates that AGC fluctuates in rainy days and its fluctuation degree is positively correlated with rainfall intensity. The correlation coefficient is

0.92. Figure 4.10b shows 1-min precipitation accumulation and average AGC fluctuation per minute from January 10 to 14. We can see that AGC is very sensitive to precipitation, and it fluctuates whenever accumulative precipitation increases. At the meantime, AGC is recorded on someday when it does not rain or during time periods before and after precipitation (5–7 min before precipitation and 12–47 min after precipitation). Zhang et al. (2016) suggested that the precipitation particles that appear before and after precipitation cannot be recorded by precipitation measurement instruments due to its accuracy, but these particles do affect the measurement accuracy of eddy covariance system. As sensitive photoelectric switching, AGC can acquire these abnormal signals in the air. Moreover, water particles accumulated around the probe will affect the measurement accuracy of ultrasonic air thermometer for a period of time after the instrument finishes recording data. Thus, it is used to control quality with the AGC that outputs synchronously with the ultrasonic signal in the data processing process.

(3) Materials and methods

The EC observation system at Zengcheng consists of CSAT3 and LI-7500. Take the observational data of Zengcheng for February 1–28, 2015, and do calculations step by step according to the normal data processing flow. The calculation results reject not only the time of precipitation and the two time periods before and after precipitation (30-min average is used) and over 1% data miss for the original 30 min. As described above, we reject the wild points based on a criterion four times as great as the standard deviation during data processing.

c. Characteristics of turbulence transport in the typical surface layer of underlying surface of the hilly shrub forest in winter in Zengcheng, Guangzhou

(1) Characteristics of the mean field of basic meteorological elements in the surface layer

Figure 4.11a shows a time series of solar radiation at Zengcheng in winter. Clearly, the downward shortwave radiation (S↓) and reflected shortwave radiation (S↑) have clear diurnal changes and are the same in phase. The downward shortwave radiation in winter is 129.8 W/m^2 on average and peaks at 811.76 W/m^2, but in rainy days it is less than 300 W/m^2. The reflected shortwave radiation ranges in 0–112 W/m^2. The downward longwave radiation (L↓) and upward longwave radiation (L↑) have small diurnal changes. The average downward longwave radiation of 366.25 W/m^2 is less than the average upward longwave radiation of 393.57 W/m^2.

Figure 4.11b shows diurnal changes in the four components of winter radiation. The downward shortwave radiation and reflected shortwave radiation are enhanced gradually from 0 since sunrise and reach their peaks around 13:00. Then, they drop to zero around 18:00. These times coincide with sunrise and sunset (07:05 ± 06 min and 17:56 ± 16 min) at Zengcheng in winter. The outward longwave radiation of canopy, high in daytime and low at night, has small diurnal change and is basically stable at night. Meanwhile, the average downward longwave radiation of the atmosphere almost has no diurnal change and is less than the outward longwave radiation of

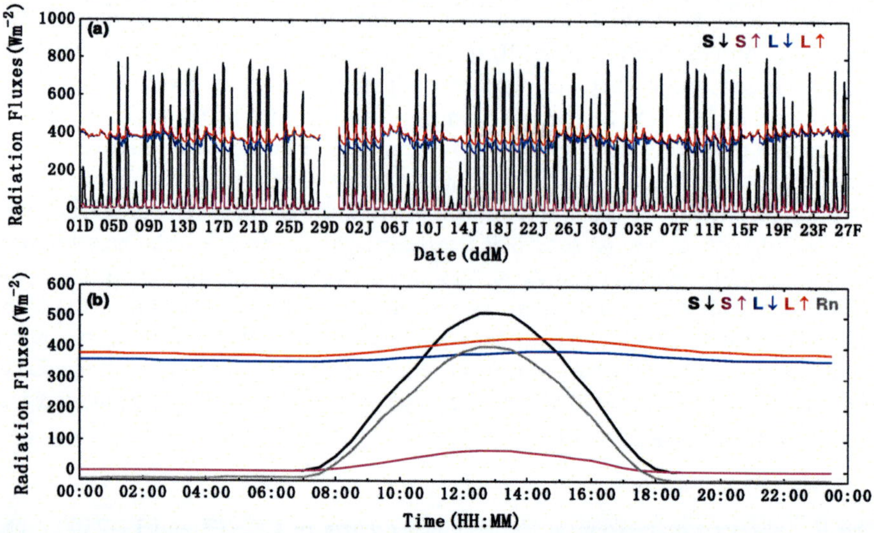

Fig. 4.11 **a** Diurnal variation of the four components of winter radiation at Zengchen. **b** diurnal variation of the four components and net value of winter radiation at Zengcheng. S: shortwave radiation; L: longwave radiation. Arrows represent directions. Rn represents net radiation

canopy. The net radiation (Rn) has a trend in common with the solar shortwave radiation. They are both negative at night and positive in the daytime. Negative-to-positive transition is later than global radiation, and positive-to-negative transition earlier than global radiation, under the influence of longwave radiation. Here, Rn > 0 is considered as the boundary between daytime and nighttime.

The winter distributions of wind speed and direction are shown in Fig. 4.12. In winter, northerly wind prevails. In the daytime, the northeasterly wind is high; but at night, northerly wind is dominant. The wind speed is 1.90 m/s in the daytime and 1.78 m/s at night, with a daily mean of 1.83 m/s. The wind speed is between 1 and 4 m/s. The background wind in winter is breeze, because the strong wind is rare.

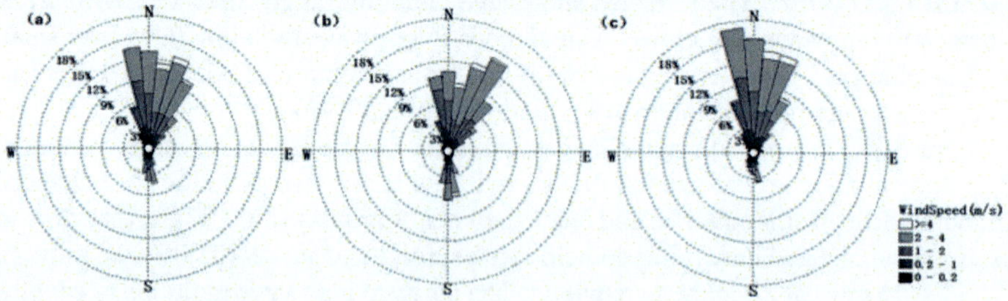

Fig. 4.12 Diagram of the Rose statistics of wind speed and direction at Zengcheng in winter **a**: winter, **b** daytime, and **c** nighttime

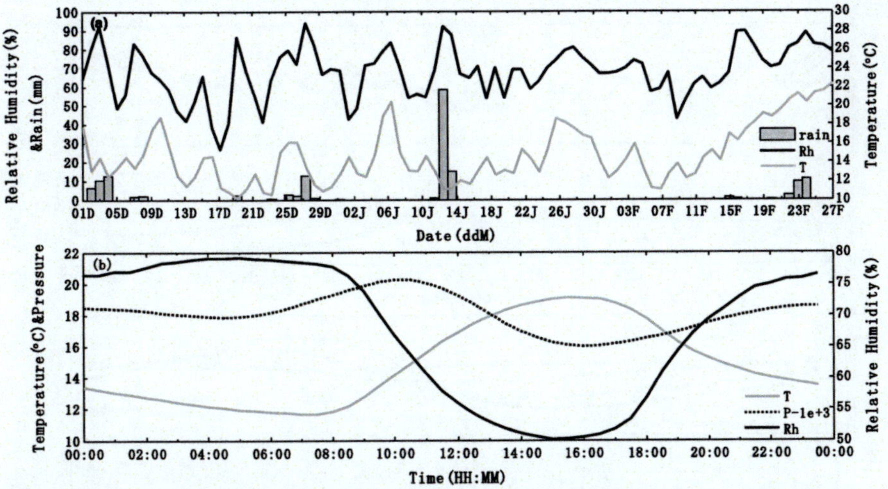

Fig. 4.13 Basic meteorological conditions of Zengcheng in winter. **a** seasonal variation of daily average values of temperature (gray), relative humidity (black) and precipitation (columnar). **b** diurnal changes in temperature (gray), relative humidity (black solid line) and air pressure (black dotted line)

Figure 4.13a shows the diurnal variations in daily average air temperature, relative humidity and precipitation at Zengcheng in winter. In the surface layer, the average temperature is 14.8 °C and the standard deviation is 3.09 °C. Affected by cold air, temperature fluctuations are big. Even in winter, precipitation is frequent (occurred in 21 days) in this region and the total amount is up to 154.3 mm. On January 12, the daily precipitation peaks at 58.6 mm. The relative humidity is 67.94% on average and ranges from 26.67% (on December 17) to 93.81% (on January 12). Figure 4.13b shows diurnal changes in air temperature, relative humidity and air pressure in winter. The air temperature peaks around 16:00 and 08:00. Rn reaches a minimum just before negative-to-positive transition. The diurnal change in relative humidity is contrary to that in air temperature. The relative humidity peaks at dawn and reaches a minimum around 15:00. The air pressure varies slightly, peaks in the morning and reaches a minimum around 16:00.

Figure 4.14a shows the seasonal variation of daily mean soil temperature. In the first half of the winter, soil temperature rises with the increase in depth. in February, with the sustained rise in temperature, the daily mean topsoil temperature exceeds the daily mean deep soil temperature. According to Fig. 4.14b, topsoil humidity is affected by precipitation significantly. Each precipitation process raises soil humidity suddenly, and such rise becomes less and less significant as depth increases. Light precipitation almost has no effect on the soil humidity at a depth of 20 cm or below. After precipitation, topsoil humidity drops to a stable value quickly. On the whole, the soil humidity at Zengcheng in winter rises with the increase in depth. Figure 4.14c shows the mean diurnal variation of soil temperature. Evidently, the shallow soil temperature has a great diurnal change. The temperature is low at night and in the

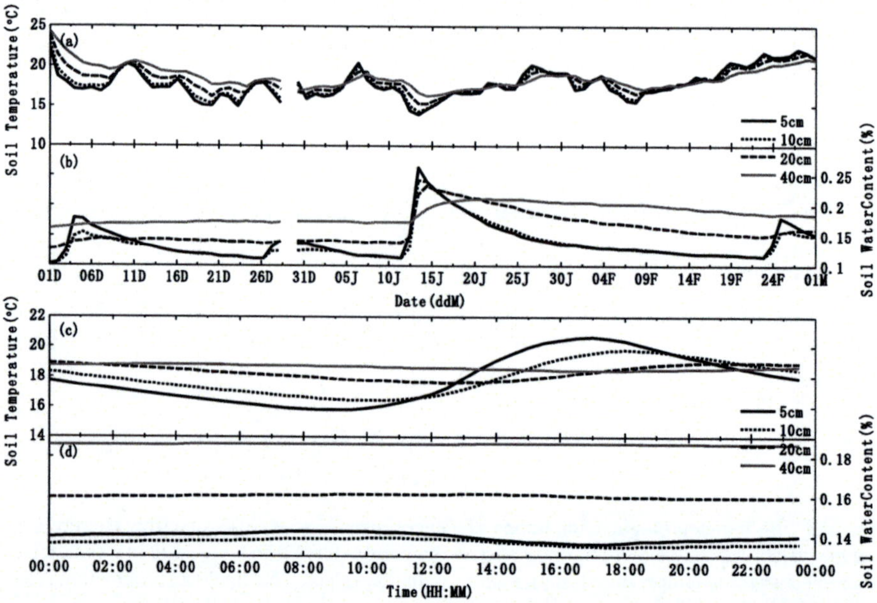

Fig. 4.14 Seasonal variation of the daily average values of soil temperature and soil humidity (volumetric water content) at Zengcheng in winter (**a**), (**b**) and mean diurnal variations (**c**), (**d**)

morning, but is high at noon and in the afternoon. It reaches a minimum around 9:00 and peaks around 17:00, lagging air temperature. With the increase in depth, the peak temperature occurs later and its diurnal fluctuation gets smaller. At a depth of 40 cm, soil temperature shows no diurnal change. The average temperature of all layers rises with the increase in depth, which is associated with dense vegetation, limited direct solar radiation absorbed by soil, low air temperature in winter, and global heat loss of soil. The mean diurnal variation of soil humidity (Fig. 4.14d) is significant at 5 cm only. The humidity peaks around 10:00 and then drops to the minimum around 17:00, which is contrary to changes in soil temperature.

Figure 4.15a shows the probability density distribution of atmospheric stability for different time periods at Zengcheng in winter. On the whole (black solid line), the atmospheric stability presents normal distribution (−0.5 to 0.5) in winter. Its peak is near the value of 0 and slightly stable. This indicates that during observation, settlement is frequent in the near-neutral layers at Zengcheng in winter but strong unstable or strong stable conditions are rare. Considering the daytime (black dotted line) only, the unstable condition is dominant, and weak stable is possible. The peak is 0 and the stability frequency range is wider than that at night. At night (gray dotted line), however, stable atmosphere is dominant and instability is rare.

Figure 4.15b shows the daily variation of atmospheric stability. Atmosphere is unstable in daytime but stable at night. In the morning, as sunrise brings more solar radiation, the atmosphere in the surface layer gains more energy and gets unstable

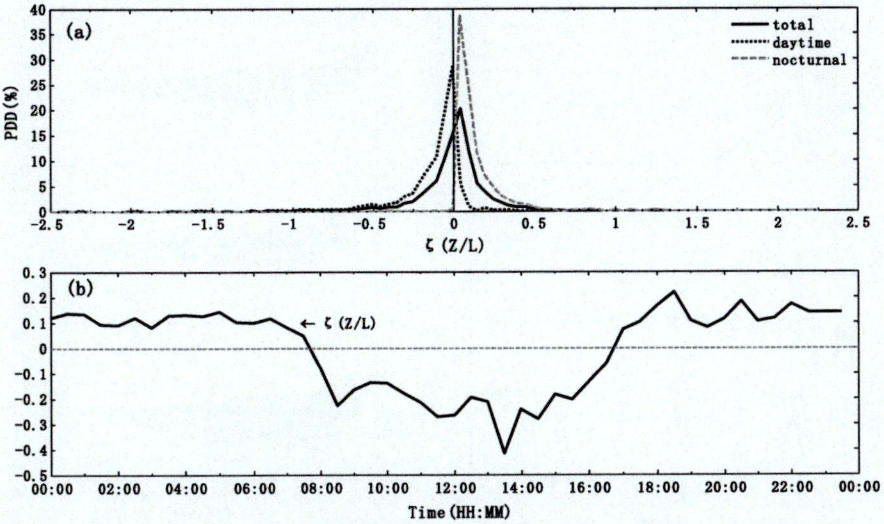

Fig. 4.15 Characteristics of atmospheric stability at Zengcheng in winter. **a** probability density distribution atmospheric stability frequency; **b** mean diurnal variation of atmospheric stability

in the development of heat convection. In the afternoon, as solar radiation decreases, air temperature drops and atmosphere turns stable gradually.

Turbulence intensity is a quantity expressing turbulence being strong or weak in airflow. It is equal to the ratio of the root mean square (or standard deviation) of eddy velocity to the mean wind speed for the same period, i.e., $I_x = \delta_x/\bar{U}$, where x represents any of the three components of wind speed u, v, and w. Figure 4.16 shows the probability density distribution of turbulence intensity of the three components and the variation of turbulence intensity with mean wind speed. Table 4.9 lists the average turbulence intensity and peak frequency related turbulence intensity for different periods (full-day, daytime and nighttime). Clearly, on the whole, the turbulence intensity distribution of vertical wind speed is narrower than that of horizontal wind speed. I_u presents the widest distribution. The turbulence intensity of the three components is mostly less than 0.5, which is deemed to satisfy the 'frozen' turbulence hypothesis proposed by Stull (1988). In the daytime, both average turbulence intensity and frequency peak of each component are higher than those throughout the winter, meaning that strong turbulence is more often seen in the daytime. At night, the peak-related turbulence intensity of each component coincides with strong turbulence throughout the winter, except that when the intensity range concentrates, the probability is generally larger, and the turbulence intensity is obviously reduced.

The wind speed makes a difference in turbulence development. Seen from Fig. 4.16 (a1-c1), when wind speed is <1 m/s, turbulence development is booming, during which the wind speed is low and turbulence is mainly caused by buoyancy convection. The boundary layer stays in the state of free convection. I_u and I_v are

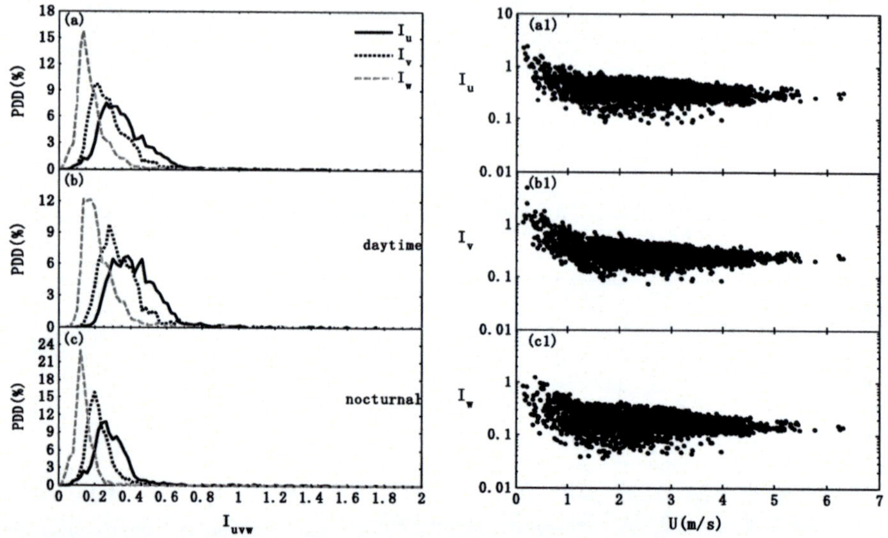

Fig. 4.16 Probability density distribution of turbulence intensity: **a** winter, **b** daytime, and **c** nighttime. variation of turbulence intensity with wind Speed (a1, b1, and c1 correspond to the three components of u, v, and w, respectively)

Table 4.9 Average turbulence intensity and peak frequency related turbulence intensity for different periods

	Average turbulence intensity			Peak frequency related turbulence intensity		
	Winter	Daytime	Nighttime	Winter	Daytime	Nighttime
I_u	0.37	0.43	0.31	0.26	0.38	0.26
I_v	0.31	0.36	0.25	0.18	0.28	0.20
I_w	0.19	0.23	0.15	0.14	0.14	0.12

>1, and I_w is >0.5. With the increase in wind speed, turbulence intensity becomes weaker and the stratification becomes more stable.

In the surface layer, the turbulence velocity is related to atmospheric stability. In the case of full turbulence development, based on the Monin–Obukhov similarity theory, the normalized dimensionless standard deviation in wind components in the surface layer is nothing but a function of atmospheric stability (Z/L). The normalized standard deviation in wind components is reduced with the decrease in |Z/L| (Fig. 4.17). In the uniform underlying surface of neutral stratification ($Z/L \rightarrow 0$), the normalized standard deviation in horizontal turbulence is constant (2.286, 1.757 and 1.174 in this site), complying with the general rule of $\delta_u/u_* > \delta_v/u_4 > \delta_w/u_*$.

As indicated in Table 4.10, δ_w/u_* varies slightly between different observation stations, suggesting the large-scale terrain and underlying surface type make a small difference to this quantity. The values on the Qinghai–Tibet Plateau are significantly

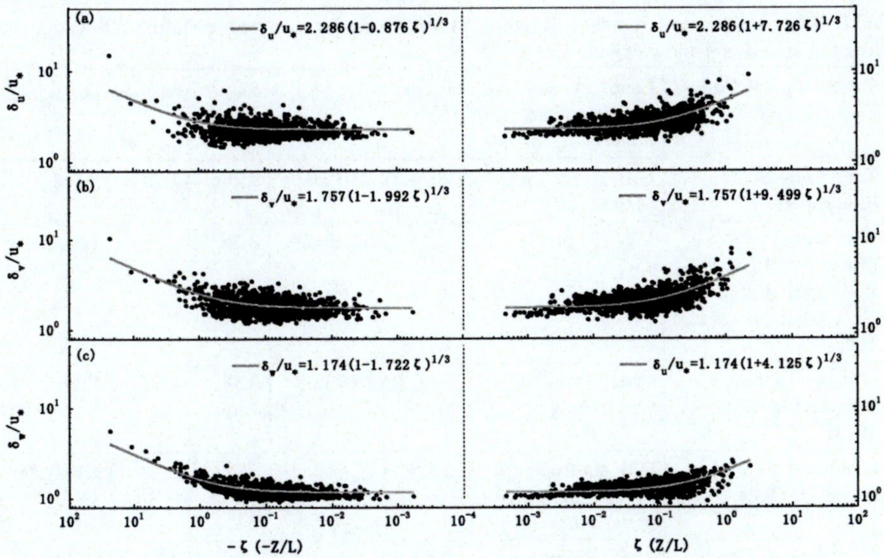

Fig. 4.17 Variation of normalized standard deviation in wind speed with stability

larger than those on plains and hills, which is related to the perennial high wind speed on highlands. For Zengcheng with topographic relief in a hilly area, the difference between U and V in normalized standard deviation is bigger than that over flat terrains, which is possibly caused by the underlying surface in different wind directions.

Figure 4.18 shows the daily mean variation of energy and CO_2 flux at Zengcheng in winter. As shown in Fig. 4.18a–c, the latent and sensible heat fluxes in winter show significant diurnal variation, like the net solar radiation. Same as Rn, the sensible heat flux is positive in the daytime and negative at night; meanwhile, the latent heat flux is generally positive. This is embodied both on sunny days and on cloudy days. The net solar radiation received at Zengcheng in winter mainly compensate for the losses of sensible heat and latent heat. The sensible heat loss is more significant, and the overall average Bowen ratio (H/LE) is 1.337 in the daytime. When it is sunny, the difference between the consumption of sensible heat flux and that of latent heat flux is even bigger, and the Bowen ratio is 1.612. When it is cloudy, however, the consumption of sensible heat flux is almost equal to that of latent heat flux, and the Bowen ratio is 1.059.

Figure 4.18d shows the mean diurnal variation of CO_2 concentration, ranging in 711.0–769.8 mg/m^3. The concentration peaks around 08:00; then, it drops gradually as Rn turns positive and reaches a minimum around 16:00. After that, respiration prevails over photosynthesis while CO_2 concentration is elevated again. By comparison, CO_2 concentration peaks earlier on sunny days than on cloudy days, and the concentration changes more quickly on sunny days than on cloudy days. As shown in Fig. 4.18e, the CO_2 flux vegetation released into the atmosphere in the daytime is negative; at noon when the solar radiation intensity peaks, the flux reaches a

Table 4.10 Normalized standard deviation in wind speed under near-neutral stratification on different underlying surfaces

Observation station	Landform and type (height) of the underlying surface	Period	δ_u/u_*	δ_v/u_*	δ_w/u_*
Zengcheng Station (this paper)	Hill; shrub forest (8 m)	2014.12–2015.02	2.286	1.757	1.174
Environment Monitoring Center Roof Station in Guangzhou (Xu et al. 1993)	City (15 m)	1988.07.01–07.09	2.35		1.4
Observation base at Bohe (Jiang et al. 2013)	Coastal area of Guangdong	2007.05–2008.08	2.38	1.97	1.27
Qianyanzhou Station in Jiangxi (Wen 2005)	Hill; subtropical conifer plantation (11–12 m)	2003			1.17/1.28
Changbai Mountain (Xu et al. 2014)	Plain; Pinus koraiensis (26 m)	2003.08	1.89	1.99	1.21
		2003.09	2.47	2.47	1.47
Party School Station in Nanjing (Peng et al. 2008)	City (<20 m)	2006.02.17–03.02	1.95	1.95	1.2
Pukou Station at Nanjing University (Peng et al. 2008)	Plain; suburb (2 m)		2.15	2.15	1.18
Gerze Station on Qinghai-Tibet Plateau (Liu et al. 2000)	Plain; sparse alpine grassland (<10 cm)	1998.06–07	3.21	2.69	1.46
Station at the upper reaches of Shule River Basin (Wu et al. 2013)	Plain; alpine meadow	2011.05–09	3.5	3.3	0.96
Nuoergai Station (Chen et al. 2014)	Plain; alpine meadow	2010	3.11	2.74	0.96

minimum. The flux on sunny days is significantly less than that on cloudy days in the daytime, whereas at night the two values are both positive and not that different. Comparing Fig. 4.18d with Fig. 4.18e, the variations in CO_2 concentration and CO_2 flux are not the same. The minimum CO_2 concentration occurs earlier than the flux reversal, and is very close to the maximum air temperature and minimum relative humidity. Maybe it is because what the CO_2 flux reflects primarily is the difference between photosynthesis and respiration of vegetation, and the CO_2 concentration is affected by vegetation respiration, photosynthesis, air temperature, and humidity (Guo et al. 2007).

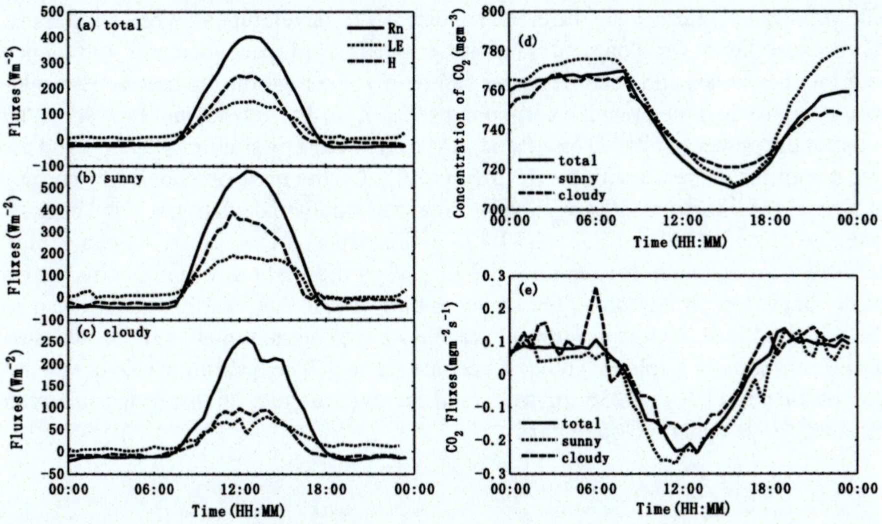

Fig. 4.18 Characteristics of daily mean variation of energy and CO_2 flux at Zengcheng in winter. **a** winter; **b** on sunny days; **c** on cloudy days; **d** diurnal change of CO_2 concentration; **e** diurnal variation of CO_2 flux

Based on the observational data of subtropical hilly shrub forest at Zengcheng during December 2014–February 2015, basic characteristics of winter meteorological elements, including wind, temperature, pressure, humidity, and atmospheric stability in the surface layer of the underlying surface of the typical hilly shrub forest in South China, as well as turbulence and flux characteristics, are analyzed. The main conclusions are as follows. (1) At Zengcheng in winter, the total incident solar radiation is 1009.3 MJ/m^2; northerly wind is dominant, with a wind speed of 1–4 m/s and a mean of 1.83 m/s. The background wind is a breeze. The air temperature is 14.8 °C on average, and is affected by cold air, showing large fluctuation. The average relative humidity is 67.94%. The precipitation amount is 154.3 mm. The soil humidity is significantly affected by precipitation, showing small diurnal variation; the average humidity rises with the increase in depth. The topsoil temperature shows significant diurnal variation, which is reduced with the increase in depth and is near 0 at 40 cm. Compared with air temperature, the peak of soil temperature appears later; and this lagging is more obvious with the increase in depth. (2) At Zengcheng in winter, the average turbulence intensities Iu, Iv and Iw are, respectively, 0.37, 0.31, and 0.19, conforming to the general rule of atmospheric turbulence intensity Iu > Iv > Iw; and the turbulence intensity is much higher in the daytime than at night. The average value of turbulence intensity is not as big as in the urban area of Guangzhou (Xu et al. 1993) and not as small as in the coastal area of Guangzhou (Jiang et al. 2013).

The wind speed makes a big difference to turbulence development: when wind speed <1 m/s, turbulence development is booming; and as wind speed increases, turbulence intensity decreases and becomes more stable. (3) The relationship between normalized standard deviation in wind components ($\delta_u/u_*, \delta_v/u_*, \delta_w/u_*$) and stability (Z/L) satisfies the power law of 1/3 based on the Monin–Obukhov similarity theory; and the two quantities decrease with the decline in |Z/L|. On the uniform underlying surface of neutral stratification (Z/L → 0), the normalized standard deviation in wind components is constant (2.286, 1.757, and 1.174 in this paper). (4) At Zengcheng in winter, sensible heat accounts for the majority of energy dissipation, which is obvious on sunny days. On cloudy days, the consumption of sensible heat is almost equal to that of latent heat. CO_2 concentration reaches a maximum around sunrise and then declines gradually as photosynthesis becomes strong. The maximum rate of vegetation absorbing CO_2 is 0.236 mg/m^2/s, and the daytime rate of absorption is higher on sunny days than on cloudy days.

4.2 Influence of Land–Atmosphere Interaction on Climate Change of the South China Sea and Its Surrounding Areas as a Response to Global Climate Change

With the constant improvement of computing technology and broad recognition of the importance of land-surface processes, land-surface models have been developed rapidly, which contributes to more frequent applications of land-atmosphere coupled model in the studies on climate changes than ever. We can learn from the above two sections that the parameters of land-surface processes of surrounding areas around the South China Sea are distinctive, and the vegetation and soil humidity affects the climate of the surrounding areas obviously. Therefore, we will discuss the influence of underlying surface vegetation and soil humidity surrounding the South China Sea on climate from the perspective of land-surface models and coupled land-atmosphere models, and explore their possible mechanisms.

In this section, we use land-surface models (SSiB, CLM, etc.) and coupled land-atmosphere models, including global coupled land-atmosphere models and regional climate models (RegCM, WRF, etc.) to discuss the importance of land-atmosphere interaction surrounding the South China Sea.

4.2.1 Correction, Optimization, and Improvement of Important Land-Surface Process Parameters of the Typical Underlying Surfaces in the South China Sea and the Surrounding Area

In meteorology, observations are essential. These observation data are the basis for analyses. For the purpose of studying the influence of land-atmosphere interaction on the climate change of the South China Sea and its surrounding areas, we set up the Phoenix Mountain Observation Station to collect precious land-atmosphere flux data, laying a solid foundation for numerical simulation studies. In consideration of the unique underlying surfaces in the South China Sea and its surrounding areas and of limited observation data, we also select the commonly used intensive data from AsiaFlux for further analysis and the optimization of land-atmosphere model.

AsiaFlux (http://www.asiaflux.net/) is a regional flux observation and research network established in the 1990s. Composed of ChinaFlux, JapanFlux, and KoFlux, it includes over 100 sites (as of the end of 2017) observing the underlying surfaces of tropical rainforest, grassland, farmland, as well as land-atmosphere fluxes such as vapor flux, heat flux, carbon flux, etc. In this study, we select the Pasoh Forest Reserve (PSO) (102°18′ E, 2°58′ N, 75-m elevation), whose underlying surface is a tropical rainforest in Malaysia.

We select the Simplified Simple Biosphere (SSiB; Xue et al. 1991), which consists of three layers of soil. The first layer is vegetation canopy, which can forecast surface temperature, canopy temperature, and moisture and carbon dioxide fluxes. SSiB has been used for studying desertification and other subjects in the Amazon Rainforest, Inner Mongolian Grassland, East Asia, Africa, and the semi-arid regions; it can simulate surface features of underlying surfaces of different lands reasonably (Xue et al. 1991, 2004a, b, 2010). In this section, we present the simulation analysis on the underlying surfaces of the South China Sea and its surrounding areas using SSiB model. We also correct and optimize the parameters, including surface albedo and vegetation characteristics.

1. Correction of land-surface process parameters of typical underlying surfaces surrounding the South China Sea in SSiB

a. Albedo

In SSiB, the surface albedo for rainforest model is 0.15, but the albedo obtained according to the analysis of the observation data of PSO is about 0.12. Therefore, we reduced the model albedo from 0.15 to 0.12 (by 3%), and the root mean square error (RMSE) simulated based on upward shortwave radiation was decreased from 9.41 to 3.51 W/m^2, and the RMSE of the net radiation was decreased from 8.39 to 2.51 W/m^2.

b. Leaf Area Index (LAI) and the Maximum carboxylation efficiency of photosynthesis (Vmax)

In SSiB, LAI is set as 5.01 and Vmax is set to 0.0001 for rainforest model. Considering observation data, the model LAI is optimized as 6.05 and Vmax, as 0.00005; accordingly, the simulated RMSE of carbon dioxide flux is decreased from 3.2 to 2.1 micro mol/(m^2)/s.

2. Improvement and optimization of the simulation of moisture transport parameterization of soil in land-surface process model

The simulation of surface water and heat flux plays a pivotal role in a land-surface model. But for most of the current land-surface process models, including SSiB, many deficiencies exist in the simulation of water and heat flux.

Soil moisture simulation and the parameterization of relevant soil-water characteristic curve (SWCC) play vital roles in a land-atmosphere interaction model. Soil shows a strong vertical heterogeneity, but the currently used land-surface process model provides only simple simulations of the moisture transport of vertically inhomogeneous soil.

The equation established based on soil moisture conservation and used to describe the vertical motion of the unsaturated flow of soil is as follows:

$$\frac{\partial \theta_l}{\partial t} = -\frac{\rho_i}{\rho_l}\frac{\partial \theta_i}{\partial t} - \frac{\partial (q_l)}{\partial Z} \tag{4.6}$$

In the case of unfrozen soil, the equation can also be written as

$$\frac{\partial \theta_l}{\partial t} = -\frac{\partial (q_l)}{\partial Z} \tag{4.6a}$$

where θ_l is the soil volumetric content of water (m^3 m^{-3}; hereinafter referred to as "volumetric water content" or "moisture content"); θ_i is soil volumetric ice content (m^3 m^{-3}); ρ_i and ρ_l are the densities of water and liquid water (kg $^{-3}$), respectively; t is time (s); and z is soil depth (m) and also the vertical coordinate (positive downward). q_l is the liquid water flux (positive downward). Following Darcy's law promoted for unsaturated media (i.e., Richard equation), it can be written as

$$q_l = K_l\left[-\frac{\partial \psi}{\partial z} + 1\right] \tag{4.7}$$

where ψ is soil matrix potential, and K_l is the hydraulic conductivity of soil (m s^{-1}). Substituting the above formula into Eq. (1), the equation of variable volumetric water content becomes:

$$\frac{\partial \theta_l}{\partial t} = -\frac{\rho_i}{\rho_l}\frac{\partial \theta_i}{\partial t} - \frac{\partial}{\partial Z}\left(-K_l\frac{\partial \psi}{\partial z} + K_l\right) \tag{4.8}$$

a. Soil-Water Characteristic Curve (SWCC)

If neglecting the influence of the osmotic potential on unsaturated soil, SWCC characterizes the constitutive relation between soil moisture potential and soil moisture content/conductivity. This constitutive relationship is correlated with the size and distribution of soil particles and the structure of interstitial pore space, and varies with the change of soil texture. As shown by lots of field and laboratory sampling tests, SWCC for different soil textures can be concluded. At present, there are three rule-of-thumb relations used to characterize SWCCs, i.e., Brooks–Corey relation (Brooks and Corey 1966), Van Genuchten relation (Van Genuchten 1980), and Clapp–Hornberger relation (Clapp and Hornberger 1978), all of which are concise and convenient mathematical relations capable of fitting with many soil characteristic curves perfectly.

The Brooks–Corey relation:

$$\psi(S_e) = \psi_s S_e^{-B}$$

$$K_l(S_e) = K_{sat} S_e^{3+2B}$$

The Van Genuchten relation:

$$\psi(S_e) = \frac{1}{\alpha}\left(S_e^{-\frac{1}{m}} - 1\right)^{1-m}$$

$$K_l(S_e) = K_{sat} S_e^{\frac{1}{2}}\left(1 - \left(1 - S_e^{-\frac{1}{m}}\right)^m\right)^2$$

The Clapp–Hornberger relation is a simplification and approximation of the Brooks–Corey relation:

$$\psi(\theta_l) = \psi_s\left(\frac{\theta_l}{\theta_s}\right)^{-B}$$

$$K_l(\theta_l) = K_{sat}\left(\frac{\theta_l}{\theta_s}\right)^{3+2B}$$

Among these relations, K_{sat} is the saturated hydraulic conductivity of soil (m s^{-1}), and ψ_s and θ_s are, respectively, saturated moisture potential of soil (m; a negative value) and soil porosity (m^3 m^{-3}). S_e is the effective porosity of soil. Parameter $B, \alpha, and\, m$ are the constants that vary with soil texture (Burdine 1953). Parameter B is soil pore size distribution index, as pore size and its distribution decide the relationship between soil adsorption and moisture. The experiments of Clapp and Hornberger indicated that B Parameter for sand is the smallest while B parameter for clay is the largest. The Clapp–Hornberger relation and Brooks–Corey relation come to existence only when $\psi < \psi_s$; when $\psi > \psi_s, \theta_l = \theta_s$. The Van Genuchten relation can better describe the relationship between soil water potential and soil moisture content, especially when soil is approximate to saturation. However, this relation has not been adopted by many land-surface process models due to its complicated nonlinear relationship. By contrast, the more concise Clapp–Hornberger relation is used more widely.

When soil is frozen, ice will affect soil moisture potential and conductivity. Therefore, in the case of frozen soil, the Clapp–Hornberger relation can be used (Jame and Norum 1980; Koren et al. 1999):

$$\psi = \psi_s \left(\frac{\theta_l}{\theta_s}\right)^{-B} (1 + C_k \theta_i)^2 \tag{4.9}$$

$$K_l(\theta_l) = 10^{-E\theta_i} K_{sat} \left(\frac{\theta_l}{\theta_s}\right)^{3+2B} \tag{4.10}$$

where C_k is an empirical constant (with certain value range) and taken as 8 here. E is determined according to the saturated hydraulic conductivity relevant rule-of-thumb relation brought up by Shoop and Bigl (1997).

It can be thus seen that when soil texture varies in the vertical direction and shows obvious inhomogeneity, the soil texture relevant parameters (such as K_{sat}, ψ_s, θ_s, B, Ck, E, etc.) in SWCC will change correspondingly. Note that Eq. (4.7) in many land-surface process models (Engelmark and Svensson 1993; Koren et al. 1999; Mölders and Walsh 2004) is written as

$$q_l = \left[-D(\theta_l)\frac{\partial \theta_l}{\partial z} + K_l\right] \tag{4.7a}$$

where $D(\theta_l)$ is moisture diffusivity of unsaturated soil (m s^{-2}) and $D = K_l \frac{\partial \psi}{\partial \theta_l}$; and liquid water flow of unsaturated soil is characterized by moisture gradient. It can be said that moisture potential and moisture content of homogenous and unfrozen soil vary continuously in the same direction. Equation (4.7a) can describe the variation of the flow speed of liquid water reasonably. For inhomogeneous soil or frozen oil, however, the moisture potential is continuous in spatial distribution and water content is not continuous at the interface of different soil textures and will jump; moisture potential increases (decreases) might be in a direction different from moisture content increasing (decreasing) direction. Therefore, the water flow direction determined by Eq. (4.7) may be different from soil moisture decreasing direction, and the water flow decided by moisture content gradient (i.e., decided by Eq. (4.7a) cannot reflect the flow of the liquid water of soil correctly. Equation (4.9) also shows that moisture potential of inhomogeneous soil is not only related to the liquid water content of soil, but also to ice content when soil is frozen, and will also be significantly affected by saturated moisture potential and parameter B. Hence, when soil moisture diffusivity is used to characterize the flow of liquid water in soil, Eq. (4.7) should be written as

$$q_l = K_l\left[-\left(\frac{\partial\psi}{\partial\theta_l}\frac{\partial\theta_l}{\partial z}+\frac{\partial\psi}{\partial\theta_i}\frac{\partial\theta_i}{\partial z}+\frac{\partial\psi}{\partial\psi_s}\frac{\partial\psi_s}{\partial z}+\frac{\partial\psi}{\partial B}\frac{\partial B}{\partial z}+\frac{\partial\psi}{\partial C_k}\frac{\partial C_k}{\partial z}\right)+1\right] \tag{4.7b}$$

For simplification, (4.7b) can be written as

$$q_l = K_l\left[-\left(\frac{\partial\psi}{\partial\theta_l}\frac{\partial\theta_l}{\partial z}+\frac{\partial\psi}{\partial\theta_i}\frac{\partial\theta_i}{\partial z}+\frac{\partial\psi(B, C_k, \psi_s)}{\partial z}\right)+1\right] \tag{4.7c}$$

The first term of (4.7c) represents the moisture flow caused by the gradient of liquid water content in soil. The second term represents the liquid water flow caused by the influence of vertical inhomogeneity of ice content on moisture potential. The third terms characterize the influence of inhomogeneity on liquid water flow caused by different soil textures (B, C_k, ψ_s) when liquid water content and ice content are given. Therefore, the vertical motion of soil moisture described by Darcy's law is not purely given by $\frac{\partial\psi}{\partial z} = \frac{\partial\psi}{\partial\theta_l}\frac{\partial\theta_l}{\partial z}$. In consideration of inhomogenous soil and frozen soil in a land-surface process model, it is only reasonable to describe liquid water flow using soil moisture potential gradient $\left(\frac{\partial}{\partial z}\left(K_l\frac{\partial\psi}{\partial z}\right)\right)$ or the expansion Eq. (4.7b). From the perspective of model simplicity and application, however, the expression based on soil moisture potential gradient will be more reasonable.

2. Influence of various parameters on the hydraulic characteristics of soil in Clapp-Hornberger relation

In the SWCC relations (4.9) and (4.10) characterizing unsaturated soil, parameters includ ing K_{sat}, ψ_s, θ_s, B, Ck, and E are crucial for SWCC. Since the influences of Ck and E on the water and heat transfer in freezing and thawing soils were studied by Zhang et al. (2007), we will not repeat them there. As B is found affecting the hydraulic characteristics of soil in an exponential form through observations, B is taken as the very example to describe the influence on soil moisture potential and conductivity quantitatively. We discuss how soil moisture potential and soil hydraulic conductivity change with B under different soil humidity.

Figure 4.19a shows that the logarithm ((lg($-\psi$)) of the absolute value of soil moisture potential varies with B changes under different liquid water contents when the total soil moisture content (θ_τ) is 0.4 and the extent of porosity (θ_s) is 0.45. We can see that the absolute value of soil moisture potential increases as B increases. However, it increases more quickly when soil is dry or contains much ice. At the same time, the absolute value of soil moisture potential of clay (B = 11.4) is larger than that of sand (B = 4.05) when their water contents are the same. Hydraulic conductivity decreases as B increases (Fig. 4.19b). Moreover, it decreases more obviously when soil is dry or deeply frozen. A larger B leads to more significant hydraulic conductivity changes with moisture content, which shows the hydraulic conductivity of clay is obviously lower than that of sand, and the change of soil moisture content can cause a greater change of hydraulic conductivity when soil texture is clay. In addition to B, saturated hydraulic conductivity can affect soil hydraulic conductivity to some extent. Obviously, soil hydraulic conductivity increases as saturated hydraulic conductivity

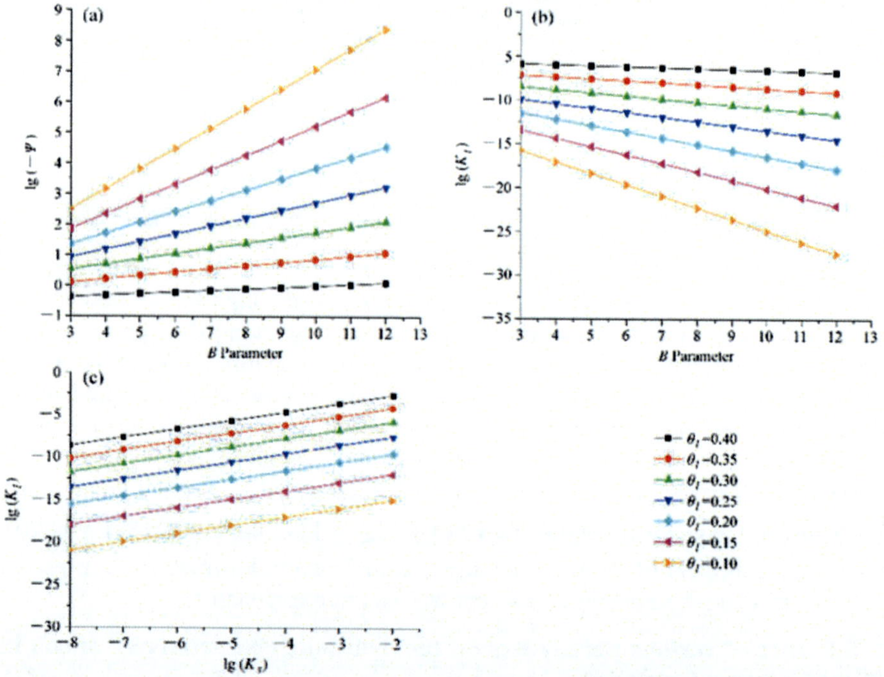

Fig. 4.19 **a** Relationship between the logarithm $((\lg(-\psi))$ of the absolute value of soil moisture potential and B ($\theta_\tau = 0.40$, $\theta_s = 0.45$, $\psi_s = -0.3$). **b** Relationship between the logarithm $(\lg(K_l))$ of soil hydraulic conductivity and B ($\theta_\tau = 0.40$, $\theta_s = 0.45$, $K_{sat} = 4e^{-6}$). **c** Relationship between the logarithm $(\lg(K_l))$ of soil hydraulic conductivity and the logarithm $(\lg(K_s))$ of saturated hydraulic conductivity ($\theta_\tau = 0.40$, $\theta_s = 0.45$, $B = 5$)

increases (Fig. 4.19c); that is to say, the hydraulic conductivity of sand is higher than that of loam or clay when they have the same moisture content.

By observing the relation (4.9), we also find that the saturated hydraulic conductivity of different soil texture also affects soil moisture potential obviously. It can be clearly seen that the larger the absolute value of soil saturated hydraulic conductivity is, the larger the absolute value of soil hydraulic potential is, which means under the same temperature, soil particle is more absorptive to soil moisture, and soil moisture will start to freeze when the temperature becomes lower. Therefore, we can judge the cause of the deviation more easily when simulating vertical inhomogeneous soil with land-surface process model so long as we know about the influences of basic soil property parameters on soil moisture simulation.

The above discussion gives the theoretical analysis on the promoted Darcy's law that is commonly used in land-surface process model to describe the vertical flow of inhomogeneous soil moisture. It points out the reasonability and superiority to characterize moisture flow with soil moisture potential gradient, and the irrationality to characterize moisture flow with soil moisture content gradient in some currently used land-surface process models. It also points out that the simulation of soil humidity is affected significantly by soil moisture potential and hydraulic conductivity depending on soil texture. To be specific, saturated moisture potential, saturated hydraulic conductivity, and soil pore size and distribution index B play a critical role in the simulation of soil humidity. B is especially important as its increase causes a significant drop of hydraulic conductivity, thus affecting the vertical distribution of moisture in soil greatly. The above results reveal that it is of great importance to consider the vertical inhomogeneity of soil in land-surface process model and adopt effective soil characteristic parameters.

4.2.2 Influences of Land–Atmosphere Interaction on Spring and Summer Climate Variability in Southeast Asia and Its Surrounding Areas, and Mechanism

1. Climatic effects of biophysical processes of underlying surface vegetation on the South China Sea and its surrounding areas

The vegetation of tropical regions is regarded as the main source of global surface evaporation, and also one of the factors that drive global atmospheric circulation and water cycle (Numaguti 1993; Larson et al. 1999; Werth and Avissar 2004). Besides, the response of the latent heat of vaporization in tropical regions to climate changes is quite important (Hulme and Viner 1998). Moreover, the tropical ecological system generates half of the Gross Primary Productivity GPP and contains 40% carbon of the terrestrial biosphere. There are many types of land underlying surfaces in South China, the South China Sea, and their surrounding areas, mainly dominated by tropical and subtropical rainforests. As the biophysical process of vegetation (including radiation transfer inside vegetation canopy, vegetation roots pumping, vegetation evapotranspiration, and moisture interception by canopy) plays an important role in land–atmosphere interaction, we conducted sensitivity test using land–atmosphere coupled models to further discuss the influence of Vegetation Biophysical Process (VBP) on the South China Sea and its surrounding area in spring and summer.

a. Model and experimental design

The atmospheric circulation model used in the study is the AGCM (Mechoso et al. 2000) of the University of California, Los Angeles (UCLA), which has a horizontal resolution of 2.5° and 29 layers in the vertical direction from the ground up to the top layer at 1 hpa. Its predictor variables include horizontal wind, potential temperature, water-vapor mixing ratio, cloud liquid water, cloud ice water, height of planetary boundary layer, surface pressure, and surface temperature.

We coupled the UCLA AGCM to two different land-surface process models, including SSiB (Xue et al. 1991) and a simple land-surface parameterization scheme. The difference between the two land-surface process models is given in Table 4.11. The model experiment involving SSiB is called LA-VBP, which takes VBP into

Table 4.11 Important Differences between Two Land-surface Process Schemes

	Simple land-surface process scheme	*SSiB*
Surface albedo	Specified monthly average	Two-stream approximation with diurnal changes
Surface roughness	Specified monthly average	The specified monthly average based on vegetation characteristics
Soil moisture	Specified monthly average	Soil humidity of three layers, interactive
Aerodynamic surface impedance	Refer to Deardorff (1972)	3, depending on vegetation status
Stomatal resistance	No	Yes
Snow process	Snow albedo is the specified monthly average	Calculate according to surface energy balance
Vegetation parameter (LAI, VCR, etc.)	No	Yes

consideration. The model experiment involving a simple land-surface parameterization scheme is called LA-NOVBP, which does not take VBP into consideration. Each model is integrated for one year using five different initial values from the reanalysis data.

b. Analysis of experimental results

In terms of surface energy balance, the net shortwave in summer of LA-VBP decreases as albedo increases in South China, but it increases when the cloud decreases. The net radiation tends to increase although net longwave radiation decreases. Sensible heat is dominant. In the Indochinese Peninsula, albedo increases but cloud cover does not change obviously. Therefore, the net shortwave decreases obviously, and the net longwave radiation decreases, leading to reduced net radiation. Latent heat increases and sensible heat decreases, so sensible heat is dominant. Around the Malay Archipelago, the net radiation decreases. Latent heat decreases and sensible heat increases, and latent heat is dominant. In spring, the net radiation at surface tends to decrease and latent heat decreases obviously although surface temperature and sensible heat increase due to VBP.

In terms of the water cycle, the summer precipitation decreases in South China and the Malay Archipelago but increases in the Indochinese Peninsula according to LA-VBP simulation results (Fig. 4.20a). The main reason for decreased precipitation is the reduced moisture flux convergence and divergence in the areas of South China and Southeast Asia. Although surface evaporation decreases as well, precipitation is controlled by moisture flux convergence and divergence in South China and Southeast Asia. In spring, precipitation also decreases in South China considering VBP. However, it increases in the Malay Archipelago but decreases in the Indochinese

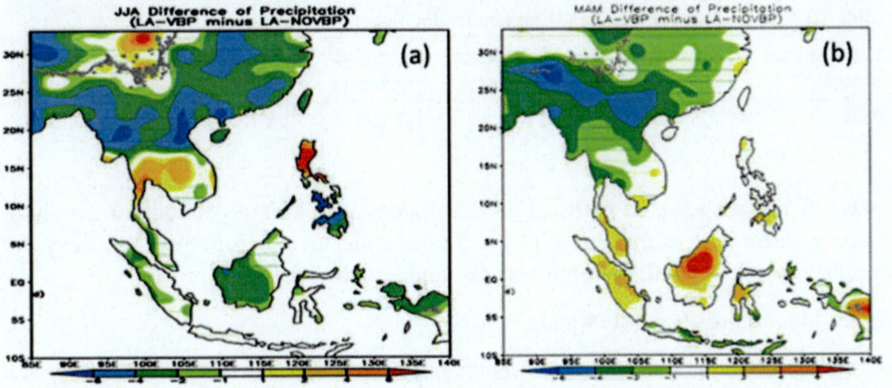

Fig. 4.20 Precipitation difference between the South China Sea and its surrounding areas in Summer **a** and Spring **b** simulated by LA-VBP and LA-NOVBP, respectively. Units: mm/day

Peninsula (Fig. 4.20b). Considering VBP, the precipitation in the above regions is also caused by the moisture flux convergence and divergence.

For atmospheric circulation, LA-VBP shows the positive vorticity at 850 hpa weakened. Both the positive vorticity of the Indochinese Peninsula and the negative vorticity of the Malay Archipelago at 850 hpa strengthen, compared with LA-NOVBP results. Changes of these circulations also result in precipitation changes in different regions (Fig. 4.21).

2. Influences of land–atmosphere interaction on Spring and Summer climate variability over South China and Its surrounding areas, and mechanism analysis

To analyze the influences of land–atmosphere interaction on climate in South China, the South China Sea, and their surrounding areas and understand corresponding

Fig. 4.21 Wind (vector) and vorticity (shade) fields of South China, the South China Sea, and their surrounding areas at 850 hpa

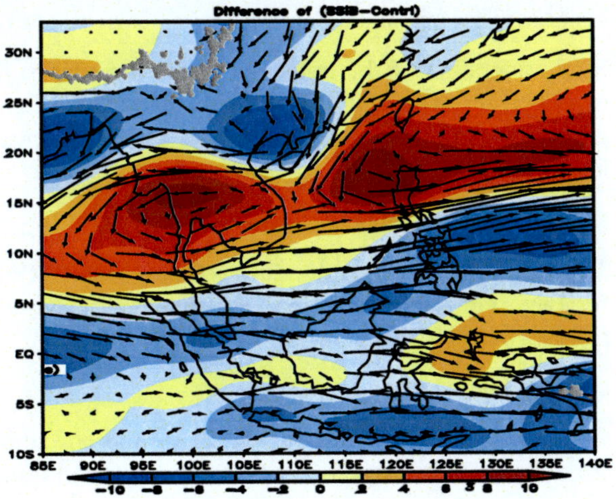

Table 4.12 Main Physical Process Parameterization Schemes

Land-surface process	Microphysical process	Shortwave radiation	Longwave radiation	Boundary layer	Cumulus convection
SSiB	WSM6	CAM	CAM	YSU	Kain-Fritsch

mechanisms, we adopted regional climate models (RCMs) to conduct longer simulation experiments in the South China Sea and its surrounding area. We carry out analysis based on results of better spatial and temporal resolution.

a. Introduction to regional climate model

As an important tool for studying atmospheric science, RCMs have been widely used in dynamical downscaling, including the downscaling of global reanalysis data and GCM. In addition to numerical weather prediction, RCMs have been widely used in climatic modeling in recent years. However, many factors may affect dynamical downscaling performance, such as the selection of an RCM, lateral boundary condition, the scope and resolution of the simulation area, and the selection of physical process parameterization scheme.

Land-surface processes are an important member of the climatic system. Especially in the regions with strong land-atmosphere coupling, the simulation of land-atmosphere interaction process is a key part of regional climate simulation. The RCM that is coupled with SSiB land-surface parameterization scheme, WRF/SSiB model (Advanced Weather Research & Forecasting Model, WRF-ARW, Version 3.6.1) can better simulate the basic characteristics of winter and summer monsoons in East Asia and the Qinghai–Tibet Plateau snow cover, according to Li et al. (2016). Therefore, we adopted the WRF/SSiB model to conduct simulation experiment on South China and Its surrounding areas. The parameterization schemes used in the WRF/SSiB model are shown in Table 4.12.

b. Experimental basics

(1) Model domain and resolution

In consideration of the influence of the boundaries of the RCM, we selected the model center at (115° E, 35° N), and adopted the Lambert projection that covers a domain of 9,250 km × 6,350 km with a horizontal resolution of 50 km. The model has 28 layers.

(2) Forcing data and boundary conditions

The initial and boundary conditions to drive the WRF are from the National Centers for Environmental Prediction/Department of Energy (NCEP/DOE) Reanalysis 2, referred to as NCEP2.

c. Experimental design

Control experiment: We simulated 21 springs and summers from 1995 to 2015, and from April 23 to August 31 of each year, covering the whole spring and summer seasons. The first 8-point accumulation days each year were regarded as the "stationary phase" of the model, and June–August were regarded as summer. The vegetation types in the control experiment are shown in Fig. 4.22 (top). One can conclude that the underlying surface of the South China region is mainly dominated by shrub forest, broad-leaved forest, and coniferous forest.

Sensitivity test: The importance of land-surface processed in the climatic system is mainly shown by their influences on surface energy balance and atmospheric circulation above the land surface. The surface energy balance is mainly shown by the sensible and latent heat fluxes at the surface. The sensible and latent heat fluxes transported by land surface upward can affect atmospheric circulation significantly. To analyze the influence of land-atmosphere interaction on climate characteristics in South China region, we replaced the vegetation type of the underlying surface in this region (107–125° E, 18–25° N) with bare soil, and carried out the simulation experiment for 21 springs and summers as we did in the control experiment. The comparison

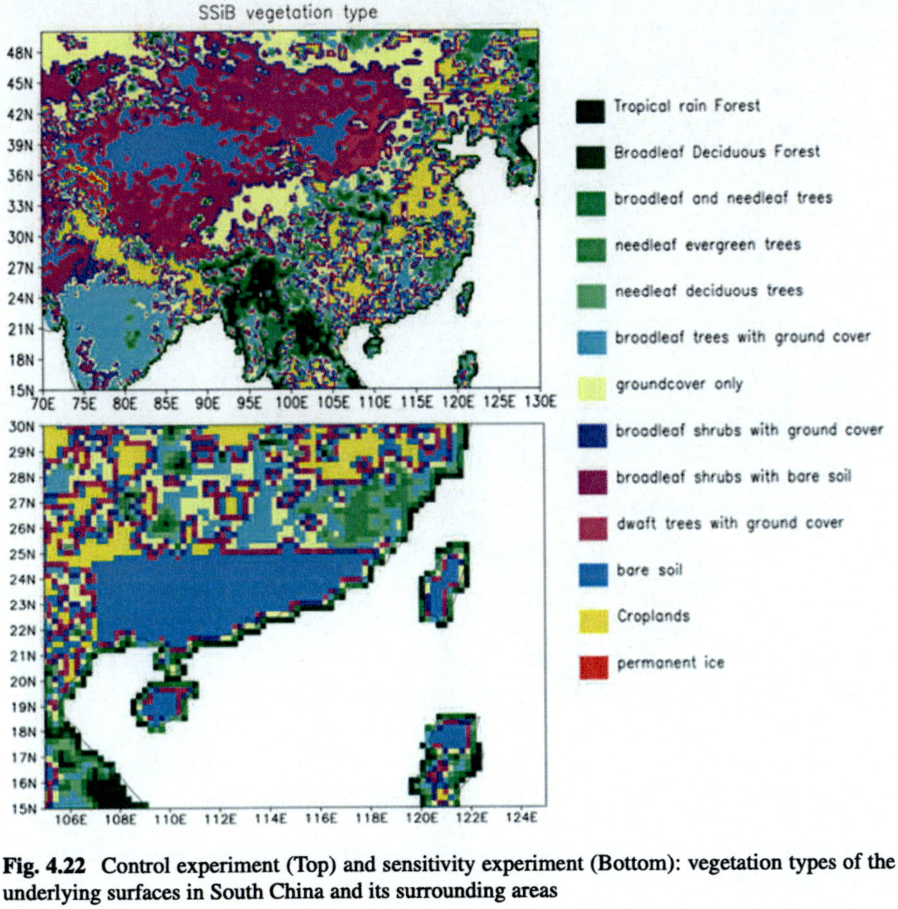

Fig. 4.22 Control experiment (Top) and sensitivity experiment (Bottom): vegetation types of the underlying surfaces in South China and its surrounding areas

clearly shows the influence of land-surface processed in spring and summer climate variability and corresponding mechanism. Figure 4.22 (bottom) shows the vegetation of South China region is replaced with bare soil in the sensitivity test.

d. Analysis of experiment results

(1) Analysis of control experiment results

To inspect the basic climate characteristics in spring and summer in East Asia by the WRF/SSiB, we evaluated several main factors preliminarily. During the East Asian summer monsoon, continuous warm temperature weathers and hot waves are typical meteorological disasters endangering most areas in the East Asia continent. So, 2-m temperature is an important diagnosis variable.

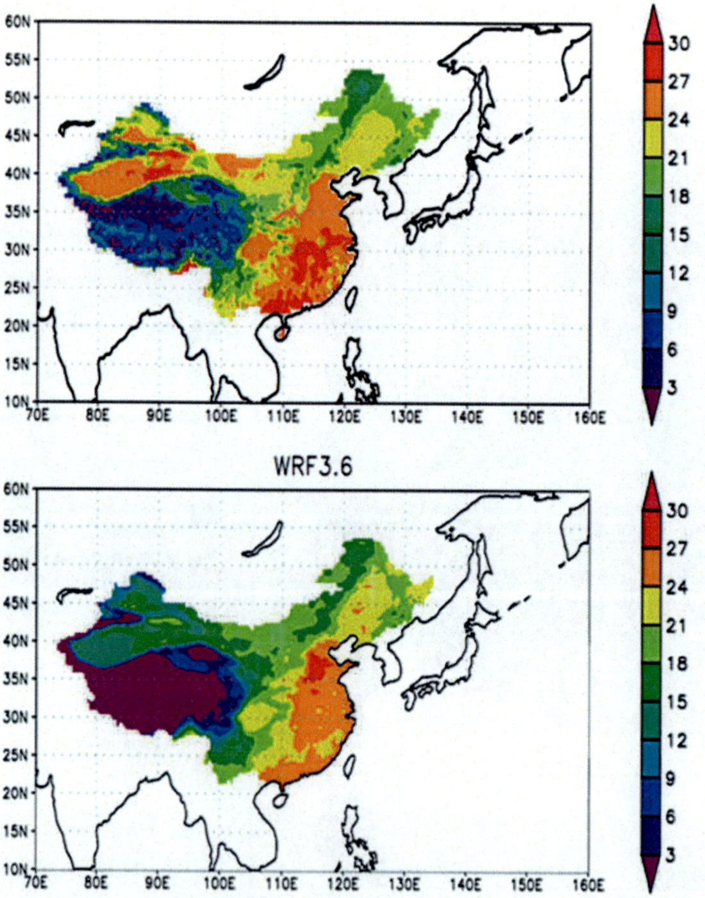

Fig. 4.23 Observation (yop) and simulation (bottom) of the average 2-m temperature (°C) in summer (June–August) from 1995 to 2015

Figure 4.23 shows the average 2-m temperature observed and simulated in the summers of 1995–2015. The WRF/SSiB simulated fairly well the spatial distribution characteristics of 2-m temperature, especially in Southeast and Northeast China.

Precipitation is also one of the most fundamental and important climate diagnosis factors. Figure 4.24 shows the average precipitation in the summers of 1995–2015. The WRF/SSiB simulated fairly well the spatial distribution characteristics of precipitation for summer monsoon rainfall, especially in Southeast China and Southwest China. The WRF/SSiB is capable of simulating the north-south movement of the rain band during the monsoon onset period according to Li et al. (2016), laying a solid foundation for our studies on the climate characteristics of South China region using the WRF/SSiB.

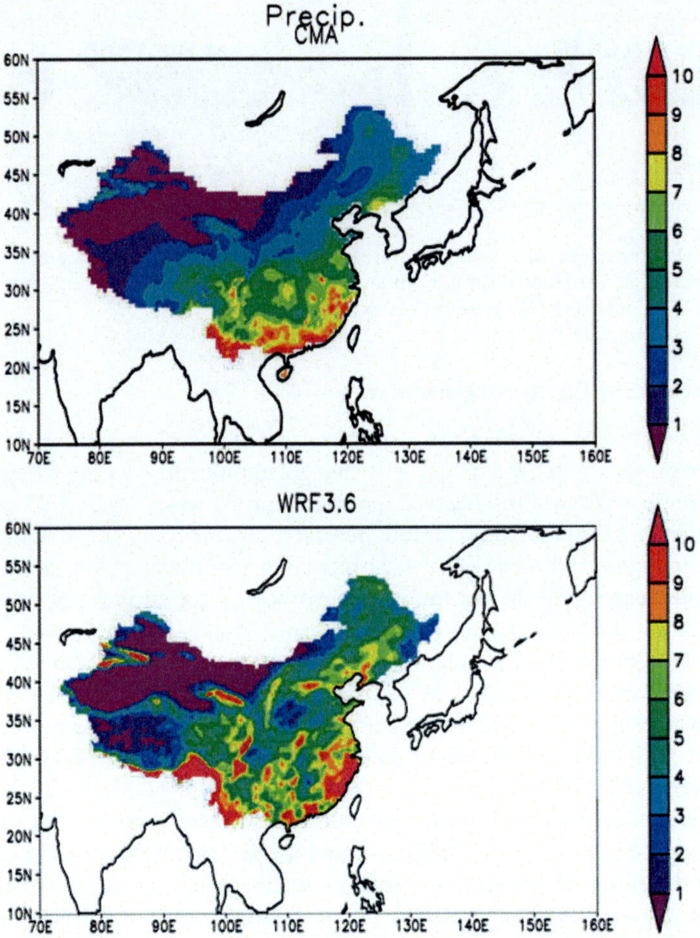

Fig. 4.24 Observation (top) and simulation (bottom) of precipitation (mm/day) in summer (June–August) from 1995 to 2015

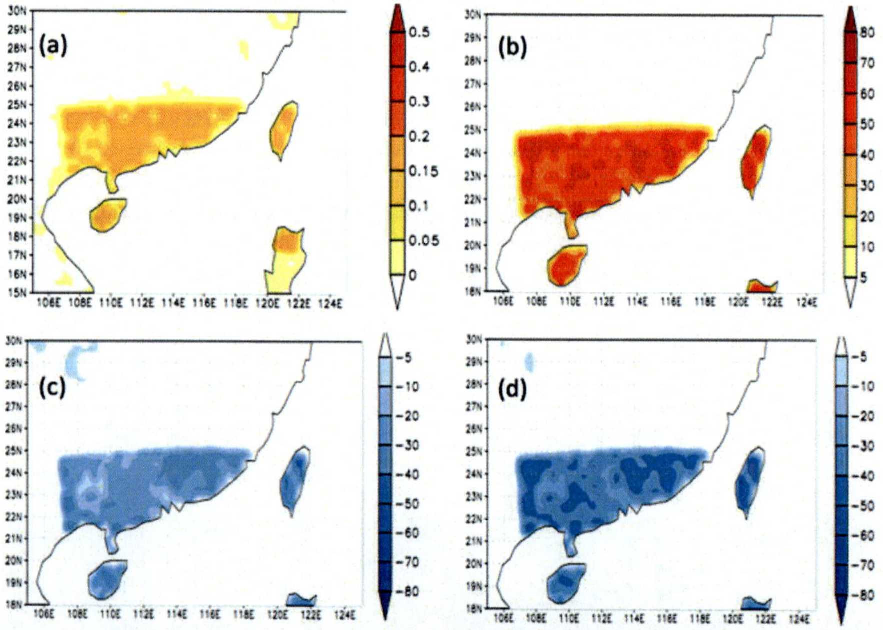

Fig. 4.25 Differences between summer averages in 1995–2015 by sensitivity experiment and control experiment: (1) albedo, (2) upward shortwave radiation flux (W/m^2), (3) net shortwave radiation flux (W/m^2), and (4) net radiation flux (W/m^2)

(2) Analysis of sensitivity experiment results

Figure 4.25 indicates that when the vegetation type of underlying surface is changed, the albedo simulated by the model will change significantly, as the albedo of bare soil is larger than that of broad-leaved forest and shrub forest. Figure 4.25 also shows that the albedo of South China region increased obviously by 0.2–0.4, making the upward shortwave radiation of surface increased significantly, the net shortwave radiation decreased and the net radiation absorbed by the surface decreased correspondingly. As the surface energy balance characteristics are significantly affected, the net radiation of surface decreases, and the upward sensible and latent heat fluxes also change. Figure 4.26 indicates that the sensible heat flux of surface increases slightly as a result of the temperature increase caused by the replacement of surface vegetation with bare soil. However, latent heat flux changes most significantly; it decreases by 40–50 W/m^2.

There is no doubt that a significant decrease of surface latent heat flux will affect the atmospheric circulation above. So, we analyze the wind fields of 850 and 500 hpa above South China and its surrounding areas. Obviously, the decrease of surface latent heat flux in the sensitivity experiment leads to abnormal anticyclonic circulation over South China and Southeast China (Fig. 4.27), which causes an obvious decrease (3–4 mm/day) of precipitation and increase in temperature (2–3 °C) (Fig. 4.28).

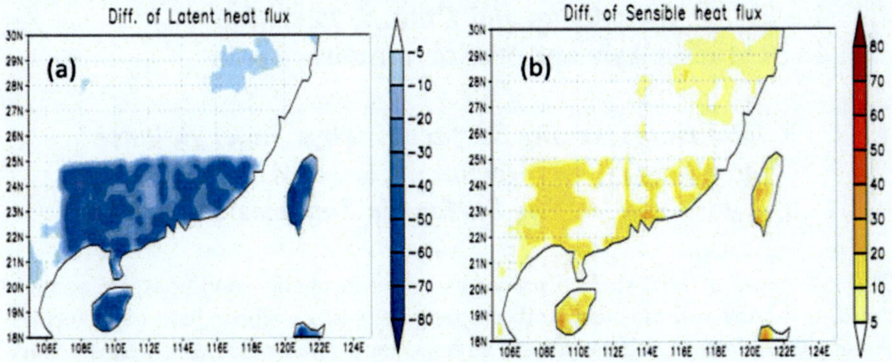

Fig. 4.26 Differences between summer averages from 1995 to 2015 by sensitivity experiment and control experiment: **a** latent heat flux and **b** sensible heat flux. Units: W/m^2

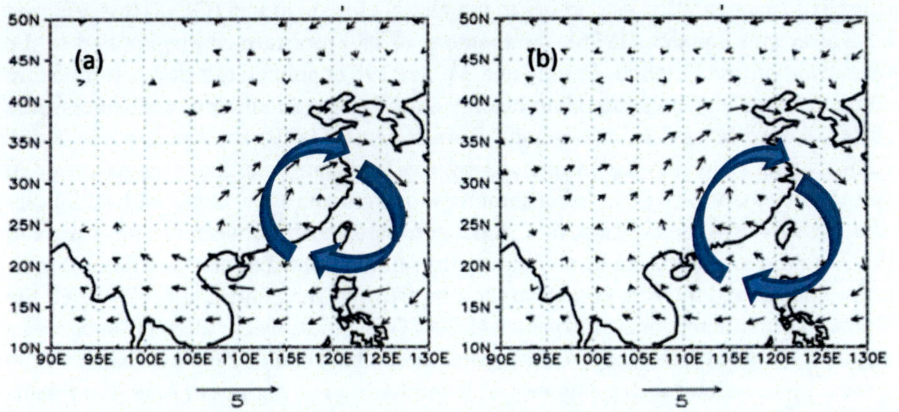

Fig. 4.27 Differences between summer averages from 1995 to 2015 by sensitivity experiment and control experiment: (Left) 850 hpa wind and (right) 500 hpa wind. Units: m/s

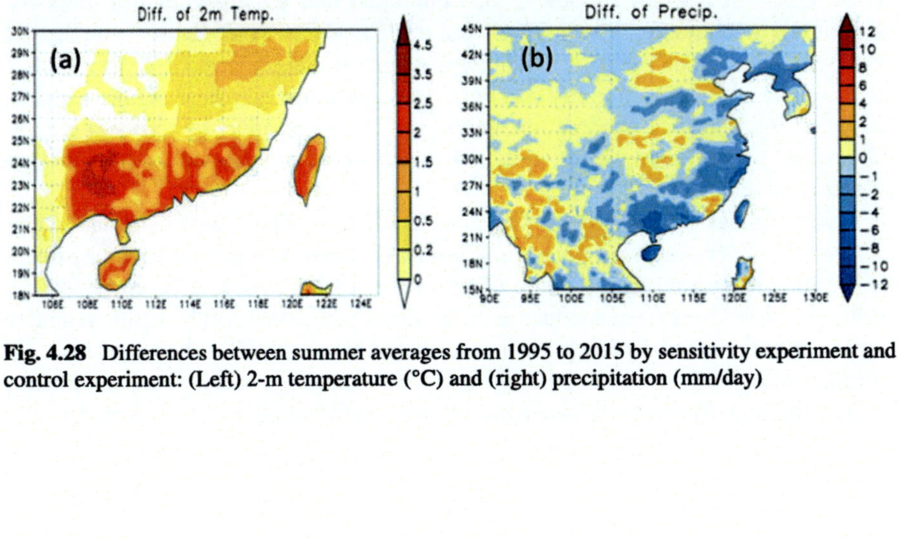

Fig. 4.28 Differences between summer averages from 1995 to 2015 by sensitivity experiment and control experiment: (Left) 2-m temperature (°C) and (right) precipitation (mm/day)

4.3 Land–Air Interaction and Climate Variation in Southeast Asia and Its Surrounding Areas

4.3.1 Relationships Among Surface Energy, Water Balance Characteristics, and Regional Climate Anomalies of Typical Underlying Surface in Southeast Asia

The influences of land-surface processes on atmospheric conditions are complicated, involving soil moisture, soil temperature, surface albedo, and other factors. Among these factors, soil moisture is an important variable that affects land-surface processes. Over the past several decades, a substantial body of research has provided valuable insight into soil moisture. Delworth and Manabe (1988) proposed the concept of Soil Moisture Memory, which suggested that soil moisture has a certain degree of sustainability. According to the global climate model (GCM) test reported by Koster and Squarez (1996), the memory of soil moisture can be applied to the time scale of precipitation. Shukla and Mintz (1982) pointed out that soil moisture anomalies can last long enough to change atmospheric circulation on different time scales, from seasonal to interannual. Barnett et al. (1989) showed that the Asian summer monsoon may be influenced by the Eurasian continent's snow cover in spring, thus associating the Asian summer monsoon and snow cover with soil moisture. Meehl (1994) derived a positive feedback relationship between soil moisture and precipitation in South Asia by comparing the relative contributions of external condition (albedo) and internal condition (soil moisture). Yang and Lau (1998) used the Goddard Laboratory for Atmospheres (GLA) GCM to study the interaction between tropical SST and certain land-surface processes, which could affect the change of the westerly jet in winter and spring seasons in the subtropical regions of Asia. The results of 12 coupled atmospheric circulation models in the global land–air coupling experiment were used to analyze the intraseasonal interrelationship between soil moisture and atmosphere. Orlowsky and Seneviratne (2010) statistically analyzed the results of 15 IPCC AR4 coupled models, and concluded that the effect of soil moisture on summer precipitation is as much as that of SST. Koster et al. (2004) investigated areas with optimal soil moisture and precipitation coupling, referred to as "hot spots," most of which are located in dry and humid transition zones. These studies emphasized the importance of evaporation, and found that the evaporation process greatly influences changes in precipitation, which is an important land-surface process that affects atmospheric conditions.

Several studies in China have shown that the warming of the Qinghai–Tibet Plateau can lead to an increase in temperature difference between sea and land, which affects the summer monsoons in the South China Sea (Li 1996; Wu and Zhang 1998). In addition, studies have been conducted on the Indochinese Peninsula, which indicates that its latitudinal temperature gradient and that of the South China Sea can affect the summer wind induced by temperature gradients on a large scale (Liu et al. 2003a, b). The thermal difference between the Indochinese Peninsula and the South China

Sea is an additional factor that influences large-scale sea and continental thermal differences, and it plays a critical role in the onset of the South China Sea monsoon (Wang and Qian 2001; Ren and Qian 2002). Meng and Zheng (2006) pointed out that there is a relationship between the thermal difference in the South China Sea and the Indochinese Peninsula's land-surface processes. More specifically, the dry anomalies of soil moisture and precipitation lead to a decrease in the difference of temperature between the central and southern regions of the peninsula and the South China Sea, and the monsoon intensity is weakened. The difference between SST in the South China Sea and near-ground temperature in the Indochinese Peninsula is a necessary condition for the summer monsoon in South Asia (Liu et al. 2009).

The above research indicates the importance of soil moisture in land-surface processes, which has an obvious influence on the South China Sea summer monsoon. In this subsection, we first analyze the relationship between summer precipitation in the South China Sea and its surrounding areas and the spring soil moisture in the Indochinese Peninsula. Then, by analyzing the difference among atmospheric circulation, land-surface processes, and land–air energy in spring, we discuss soil moisture anomalies and the mechanisms that drive summer precipitation.

1. Data and Methods

The soil moisture data used here were obtained from the ERA-Interim reanalysis data of the European Centre for Medium-Range Weather Forecasts (ECMWF), with a resolution of 1° × 1°. Precipitation data are the Global Precipitation Climatology Project version 2.2 (GPCP-v2.2), which consist of satellite data, exploration data, and precipitation data from ground observatory stations with a resolution of 2.5° × 2.5°. The NCEP/NCAR Reanalysis Project provides monthly precipitation data from 1979 to 2014.

In a related analysis, the vector field of correlation coefficient was used to analyze the correlation between wind and near-surface temperature. The definition of the correlation coefficient vector field is

$$\mathrm{R} = \mathrm{Ru\&t}\,\mathrm{i} + \mathrm{Rv\&t}\,\mathrm{j} \tag{4.11}$$

where Ru&t = Correlation (u, t) is the correlation coefficient field of u wind and time series t; Rv&t = Correlation (v, t) is the correlation coefficient field of v wind and time series t; and i and j are unit vectors of the x and y directions, respectively.

2. Correlation between Soil Moisture and Summer Precipitation

Spring soil moisture data from 1979 to 2014 (March to May, or MAM) in the 0.1-m depth layer of the Indochinese Peninsula and summer (June–August, or JJA) precipitation in the areas surrounding the South China Sea were selected for SVD analysis (Fig. 4.29). The first mode accounts for 56% of all the variance contribution. In spring, soil moisture in most areas of the Indochinese Peninsula is significantly correlated with summer precipitation over the South China Sea and its surrounding areas. When the spring soil moisture in the Indochinese Peninsula has a positive (negative) anomaly, the summer precipitation in the eastern tropical Indian Ocean

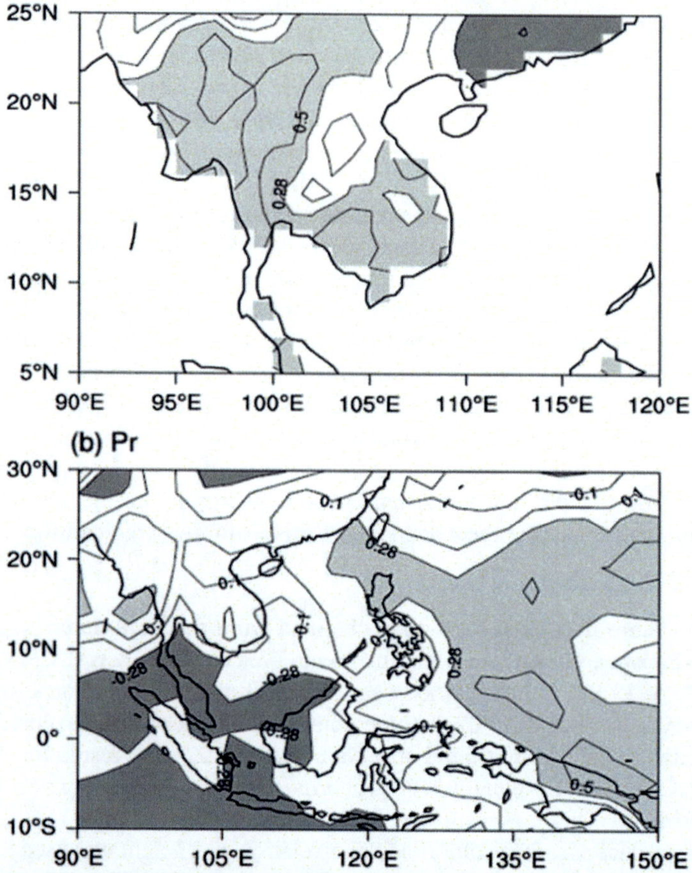

Fig. 4.29 Heterogeneous correlation coefficient of 0.1-m depth layer soil moisture and precipitation using SVD analysis of the first mode. The shading indicates the significance level of the heterogeneous correlation over 0.1: the dark shading indicates a negative correlation, and the light shading indicates a positive correlation

(EIO) is negative (positive), with less (more) precipitation, and the summer precipitation in the western tropical Pacific (WP) has a positive (negative) anomaly, with more (less) precipitation.

The Indochinese Peninsula region(10° N–20° N,95° E–110° E,ICP)was selected as the key area affecting the summer precipitation in the South China Sea and its surrounding areas. The distribution pattern of correlation coefficient (Fig. 4.30) is similar to that in Fig. 4.29. There are negative correlation in the EIO and positive correlation in the WP. The soil moisture at a 0.1-m depth in the ICP has a significant influence on summer precipitation in the EIO and WP.

The soil moisture data from the depths of 0.1, 1, and 2 m represent the land surface; they indicate that soil moisture in the middle and root regions is roughly the same in terms of temporal variation of soil moisture in the three layers (not shown). The soil

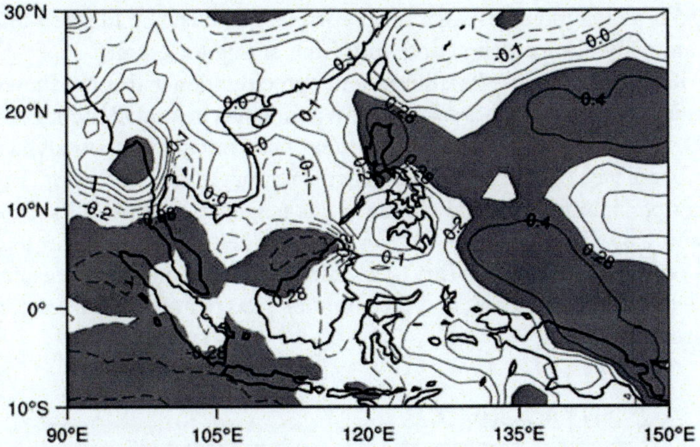

Fig. 4.30 Correlation coefficient distribution of precipitation and the average soil moisture in the 0.1-m depth of the Indochinese Peninsula. The shading indicates a significant level of over 0.1

moisture correlation coefficient is above 0.94 between the middle and root regions, thus, soil moisture data from the 0.1-m depth layer are used from now on.

3. Precipitation and Circulation Characteristics of Soil Moisture Anomaly Years

The composite analysis was carried out for the years with the absolute value of the standardization of soil moisture in spring greater than 1 (Fig. 4.31a). A total of seven years showed high soil moisture (HSM), namely, soil moisture greater than 1: 1985, 1989, 1999, 2000, 2001, 2008, and 2011. A total of six years with low soil moisture (LSM), namely, soil moisture less than 1: 1983, 1992, 1995, 1998, 2005, and 2010. Figure 4.31b is the normalized time series of soil moisture in summer. From spring and summer, the HSM years demonstrated a positive anomaly in summer soil moisture, whereas the LSM years had a negative anomaly in summer soil moisture. The positive correlation of the soil moisture time series between spring and summer

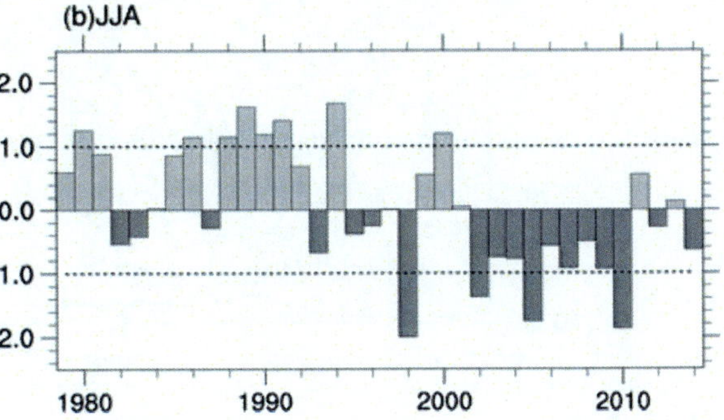

Fig. 4.31 Averaged, normalized spring (**a**) and summer (**b**) soil moisture time series in the ICP region

reached 0.52, which indicates significant correlation. It can be inferred that the soil moisture anomaly has a certain continuity from spring to summer.

In the HSM years (Fig. 4.32a), the summer precipitation in the WP shows positive anomalies in most areas, with weak negative anomalies in a small number of regions. In the EIO, summer precipitation in most areas shows negative anomaly, with only a small number of weak positive anomalies in the region. In the LSM years (Fig. 4.32b), the summer precipitation anomaly in the WP is negative, but the EIO shows obvious positive anomaly. This suggests that in spring when the soil in the Indochinese Peninsula region is more humid (dry), the summer precipitation in the eastern Indian Ocean is less (more) abundant, and the summer precipitation in the western Pacific is more (less) abundant.

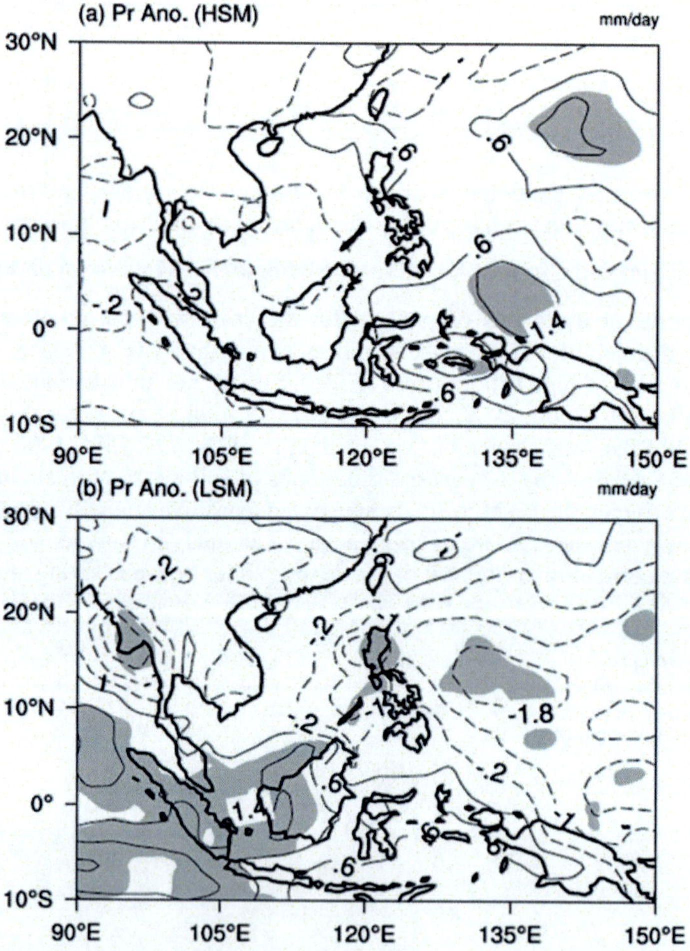

Fig. 4.32 Summer precipitation anomaly in HSM years (**a**) and LSM years (**b**). The shading indicates the 0.05 significance level. Units: mm

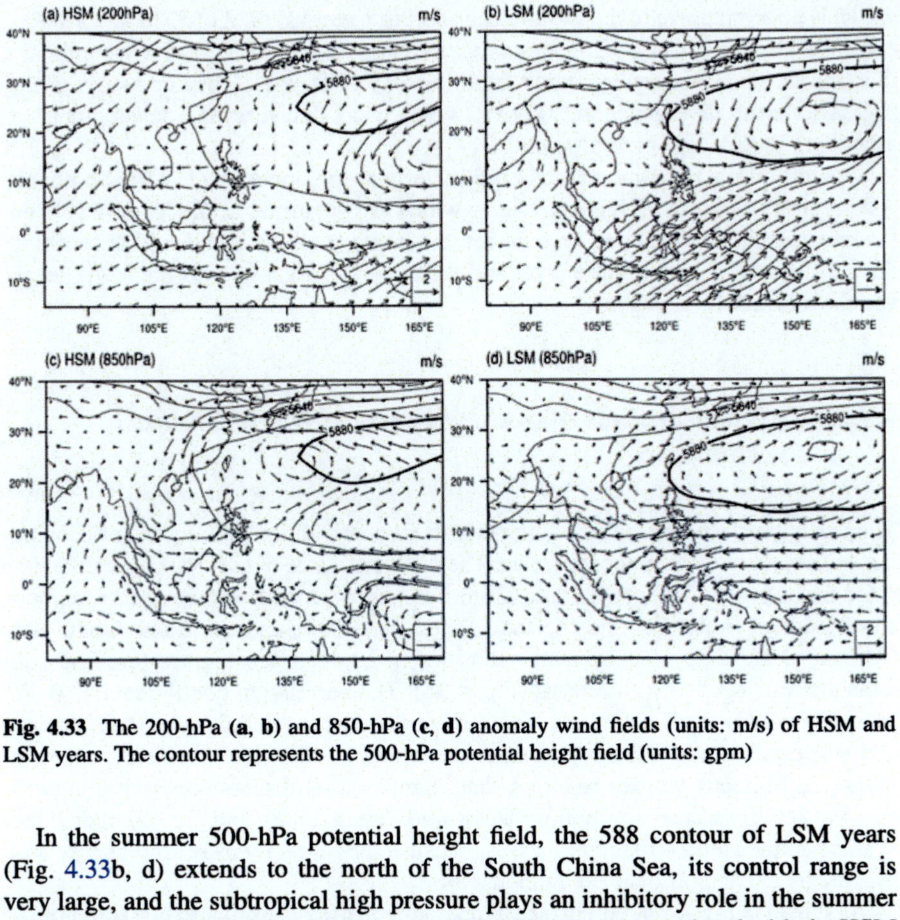

Fig. 4.33 The 200-hPa (**a**, **b**) and 850-hPa (**c**, **d**) anomaly wind fields (units: m/s) of HSM and LSM years. The contour represents the 500-hPa potential height field (units: gpm)

In the summer 500-hPa potential height field, the 588 contour of LSM years (Fig. 4.33b, d) extends to the north of the South China Sea, its control range is very large, and the subtropical high pressure plays an inhibitory role in the summer precipitation of the WP in the LSM years. The 588 contour associated with the HSM years (Fig. 4.33a, c) withdraws by 15° longitudes from the east of the LSM years and shrinks by 10° latitudes to the north. The control area is reduced. This subtropical high has little impact on the monsoon rainfall in the Western Pacific. In the HSM years, the anomalous cyclonic circulation at 850 hPa occurs along the eastern coast of China (Fig. 4.33c), which is conducive to the occurrence of precipitation in the region. In the WP, the convergence of anomalous westerly and easterly winds is a potential cause of the high level of precipitation associated with the summer monsoon. In the EIO, there is a southwesterly wind anomaly in the equatorial region and a positive horizontal shear in the EIO, indicating that the wind field is divergent, which is not conducive to precipitation. At 200 hPa (Fig. 4.33a), an anticyclonic anomalous wind field appears on the east coast of China.

In the WP, there is an anomaly of divergence, which is conducive to the occurrence of precipitation. For the LSM years, there is a large-scale anticyclonic anomaly at the lower level of East China (Fig. 4.33d), which overlaps with the 588 contour. The whole area is controlled by an anticyclone anomaly (Fig. 4.33b), prone to subsidence, which is not conducive to the upward transport of water vapor. At 850 hPa, the WP is controlled by the easterly wind anomaly, stretching into the EIO; the wind anomaly then meets the northwesterly wind from the Bay of Bengal. They converge in the EIO, with wind divergence at 200 hPa, which is conducive to the development of ascending motion and precipitation.

In addition, there is anomalous northeasterly at the upper level in the Southeast Asia region in the HSM years. westerly winds are abnormal at the lower level and the monsoon is strengthened. In contrast, the larger southwesterly wind anomaly in the LSM years is in the upper level and the easterly anomaly is in the lower level, which weakens the monsoon.

4.3.2 *Physical Process of Soil Moisture Affecting Summer Precipitation*

The Earth's surface thermal condition can affect precipitation amount by influencing the development of summer monsoon. The different states of the Indochinese Peninsula surface process affect the South China Sea summer monsoon through the temperature gradient in and around the Indochinese Peninsula and the South China Sea (Amenu et al. 2005). The annual variations in land-surface temperature and soil moisture are negatively correlated (Fig. 4.34). The correlation coefficient is −0.65, passing the 99% significance test. Surface temperature is related to ground radiation and evaporation, and surface processes can affect atmospheric conditions through radiation. Previous studies reported that changes in soil moisture have a significant impact on surface energy flux (Meng and Zheng 2006), and are considered the primary influence in surface processes, because they directly affect the energy and hydrologic exchanges between land and air.

As shown in Fig. 4.34, the surface temperature in May is similar to that throughout the spring. The correlation coefficient of land-surface temperature in May and soil

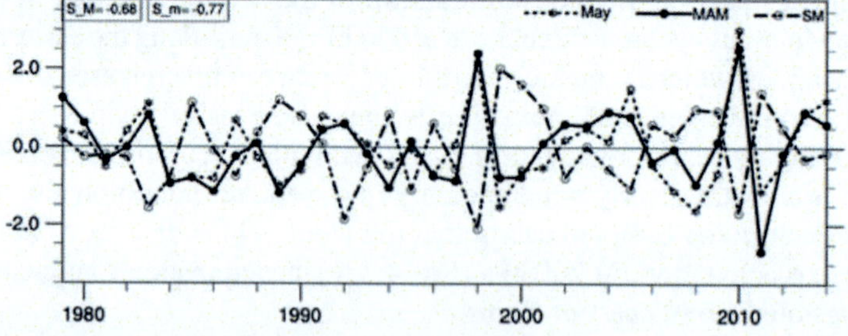

Fig. 4.34 Averaged, normalized time series of spring soil moisture, surface temperature in May, and spring surface temperature in the ICP region. Correlation coefficients of surface temperature in May, spring surface temperature and spring soil moisture are given in the upper left corner

moisture is −0.77 (Fig. 4.34). May is also the key period for the South China Sea summer monsoon. Therefore, the conditions in May can affect the spring soil moisture and summer precipitation in the later period, while also influencing the physical process of precipitation during the summer months.

a. Terrestrial Radiation

The longwave and shortwave radiation are the main processes of land–air energy exchange. Figure 4.35 presents the difference between longwave and shortwave radiation in May of the HSM and LSM years, respectively. The longwave radiation difference is negative in the northern region of the Indochinese Peninsula (Fig. 4.35a), and

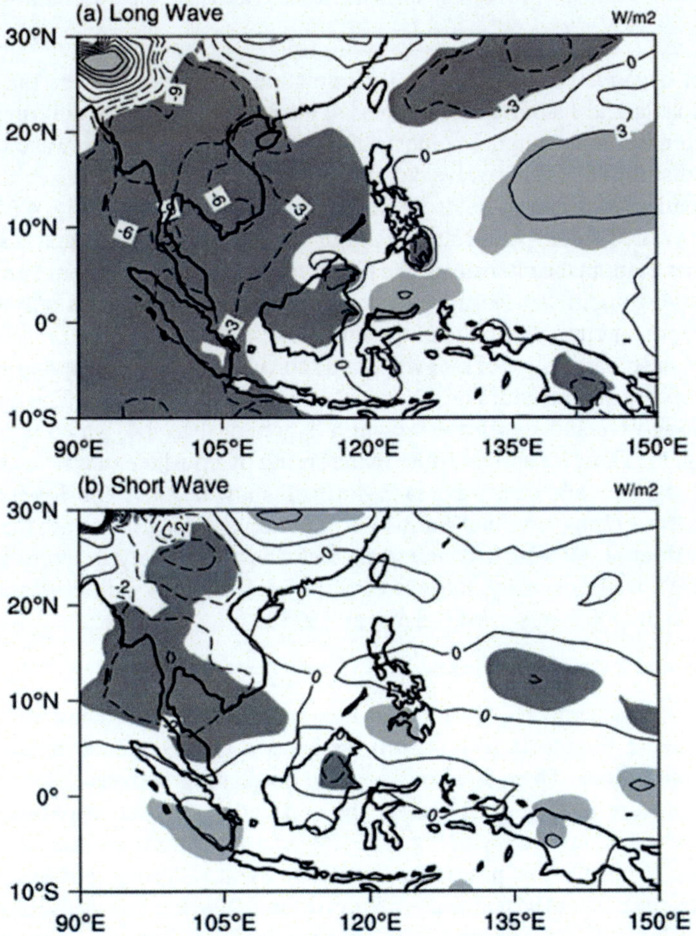

Fig. 4.35 Diagram of longwave radiation (**a**) and shortwave radiation (**b**) during May of HSM and LSM years (contour; units: W/m^2). The shading indicates that the correlation at 0.1 significance level (the dark shading is a negative correlation, and the light shading is a positive correlation)

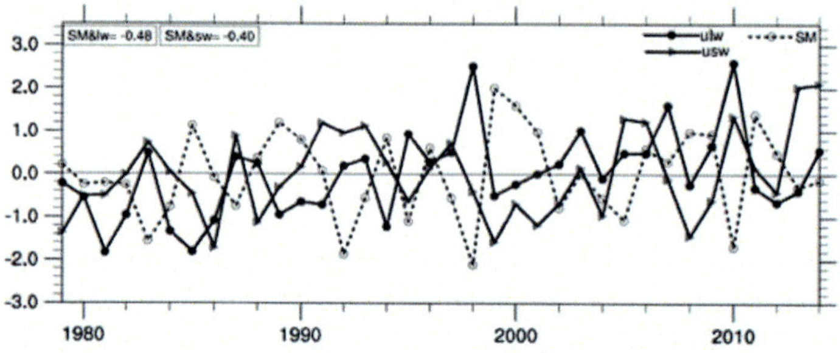

Fig. 4.36 Normalized time series of upward longwave radiation, upward shortwave radiation, and spring soil moisture in the ICP region in May. Correlation coefficients of upward longwave radiation, upward shortwave radiation, and spring soil moisture are given at the upper left corner

significant negative correlation with soil moisture completely covers the Indochinese Peninsula and its surrounding areas. This indicates that the soil moisture on the Indochinese Peninsula has a substantial influence on local longwave radiation. Different distribution of shortwave radiation on the Indochinese Peninsula also shows the distribution of different fields of similar longwave radiation (Fig. 4.35b), with the center located on the peninsula. Although the Indochinese Peninsula is an area of negative correlation, the intensity of the negative correlation center is not particularly strong, and the significant negative correlation with soil moisture is only observed in the western and northern parts of the Indochinese Peninsula.

In May, the relationships of longwave and shortwave radiation in spring soil moisture with spring soil moisture are negatively correlated (Fig. 4.36). The correlation between longwave and shortwave radiation is greater than 0.4, which is consistent with Fig. 4.35. Thus, for soil moisture in the spring of a positive (negative) abnormal year, the longwave and shortwave radiation will show negative (positive) anomaly. The responses of longwave and shortwave radiation are consistent with the findings by Delworth et al. (1993). Because surface albedo is inversely proportional to the surface soil's water capacity, the wetter (drier) soil reduces soil albedo and more shortwave radiation is absorbed (Idso et al. 1975).

b. Sensible and Latent Heat Fluxes

In May, the zero contour of latent heat flux surrounds the southeastern region of the Indochinese Peninsula (Fig. 4.37a), whereas a positive center appears in the northwestern region. There is also a positive correlation between latent heat flux and soil moisture in spring. Moreover, the correlation coefficient passes the 90% significance test. Compared with Fig. 4.29, the area of the northwestern region of the Indochinese Peninsula that passes the significance test also corresponds to the most obvious change in latent heat. The distribution of sensible heat flux on the Indochinese Peninsula (Fig. 4.37b) is similar to that of latent heat flux, but was a negative value. A more intense negative center appears in the northwest. The significant areas of negative correlation basically coincide with negative centers. When the soil moisture on the Indochinese Peninsula is positive (negative) during spring, the latent heat flux

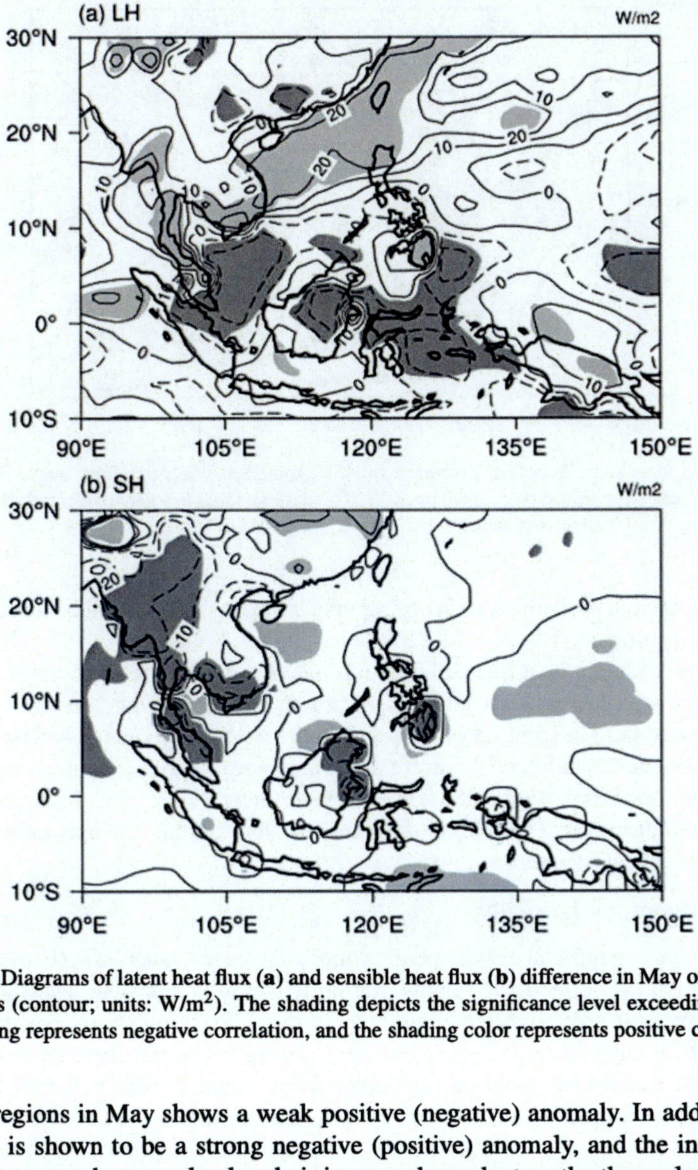

Fig. 4.37 Diagrams of latent heat flux (**a**) and sensible heat flux (**b**) difference in May of HSM and LSM years (contour; units: W/m^2). The shading depicts the significance level exceeding 0.1 (the dark shading represents negative correlation, and the shading color represents positive correlation)

in most regions in May shows a weak positive (negative) anomaly. In addition, the heat flux is shown to be a strong negative (positive) anomaly, and the interannual change in energy between land and air is more dependent on the thermal effect.

Studies have shown that convection in the northern South China Sea is enhanced by convective activity during the Bay of Bengal summer monsoon. This convection promotes the release of condensation latent heat in the northern South China Sea during the Bay of Bengal summer monsoon, which affects the South China Sea summer monsoon (Liu et al. 2003b). The enhancement of convective activity translates into increases in precipitation and soil moisture. Therefore, although soil moisture change is not the primary factor affecting the change in latent heat flux at

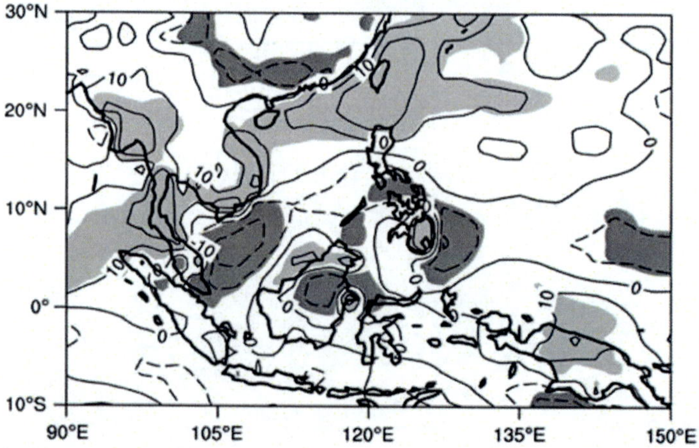

Fig. 4.38 Spring latent heat flux difference in HSM and LSM years (contour; units: W/m^2). The shading indicates the significance level over 0.1. The dark shading is a negative correlation, and the light shading is a positive correlation

the end of spring, soil moisture in spring may correlate well with the latent heat flux throughout spring.

A center of latent heat flux difference is also located in the northwest of the ICP region (Fig. 4.38). In contrast to the latent heat flux in late spring, the zero contour of the difference field in spring does not exist in the ICP region, which is dominated by positive values. In addition, the correlation between the averaged normalized time series of soil moisture field and the area of latent heat flux reaches 0.54, passing the 99% significance test. The release of latent heat flux on the ICP may coincide with the change in soil moisture.

c. Temperature and Humidity

Soil moisture can influence atmospheric condition due to evaporation (Dirmeyer et al. 2009), which also affects the relative humidity of the atmosphere near the ground. In May, the near-ground relative humidity in the ICP region shows positive values (Fig. 4.39), and the center is in the northwest. Owing to the correlation distribution of soil moisture in spring, local near-ground relative humidity has a positive response to changes in soil moisture, and more surface evaporation is caused by more soil moisture.

The transformation of energy between land and air causes a change in the near-surface temperature, which is the most direct reflection of soil moisture in spring. For the near-surface temperature (Fig. 4.39b) in the HSM and LSM years, the entire Indochinese Peninsula is covered by negative values. The center of difference is also in the northwestern region of the Indochinese Peninsula, which is also negatively correlated with what, and the correlation coefficient of local near-surface temperatures and soil moisture is −0.73.

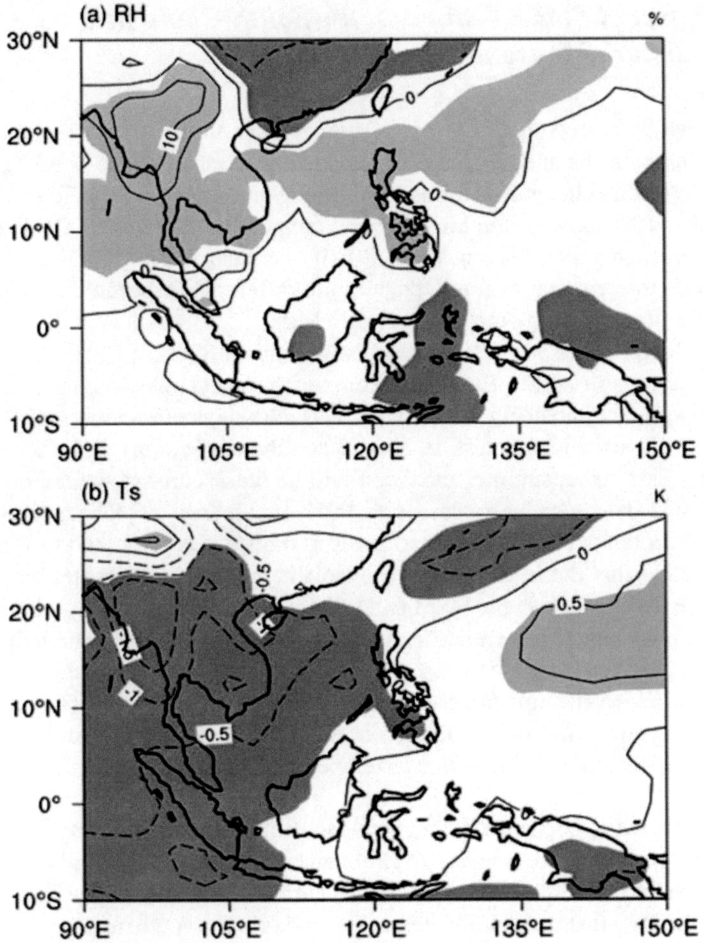

Fig. 4.39 Differences (contour) of relative humidity (**a**; units: %) and surface temperature (**b**; units: K) in May. The shading indicates the significance level exceeding 0.1. The dark shading is a negative correlation, and the light shading is a positive correlation

To sum up, the spring soil humidity anomaly in the ICP region has two effects. On the one hand, the positive (negative) anomaly of soil moisture in spring leads to a decrease (increase) in albedo in late spring, more (less) shortwave radiation absorption, corresponding to upward shortwave radiation decrease (increase), and energy transfer from the land to the atmosphere decrease (increase). On the other hand, the increase (decrease) of surface moisture content leads to more (less) evaporation. Therefore, at the end of spring, there will be a corresponding decrease (increase) in the temperature of the near-surface atmosphere, and both upward shortwave radiation and heat flux will decrease (increase).

4.3.3 Effect of Soil Moisture on the South China Sea Summer Monsoon

Studies have indicated that the lower-level temperature difference between the Indochinese Peninsula and the South China Sea has an effect on the South China Sea summer monsoon (Zheng and Meng 2006; Meng and Zheng 2006). To evaluate the relationship between monsoon intensity and temperature difference, the East Asian summer monsoon index (Zhang et al. 2003) was selected to conduct correlation analysis with the regional average temperature difference between the Indochinese Peninsula region and the South China Sea in May.

For the South China Sea, the selected range was (110°–120° E, 10°–20° N). The normalized time series of the East Asian summer monsoon index and the temperature difference were obtained (Fig. 4.40), with a correlation coefficient of −0.47, which passed the 99% significance test. In May, when the temperature difference is large (small), the East Asian summer monsoon will be weak (strong), with a correlation coefficient of −0.44, which also passes the 99% significance test. The negative correlation between temperature difference and soil moisture is presented in Fig. 4.39b, which indicates that the Indochinese Peninsula is the difference center between the near-surface temperature in the HSM and LSM years. Although the South China Sea area has the same negative correlation, its absolute value is relatively small, indicating that the negative feedback effect on soil moisture is stronger for local near-ground temperature. When the soil moisture is increased (decreased), the decrease in the near-ground temperature on the Indochinese Peninsula is larger than that over the South China Sea, and the difference in temperature between the two places will be small (large).

Because there is a good correlation between surface temperature and the monsoon index, surface temperature also has an effect on circulation. The potential height over the large area of the South China Sea and its surrounding areas in the HSM years is much smaller than that in the LSM years (Fig. 4.41a). The southeastern part of China is the low-value center of the difference field, indicating that cyclonic activity in this

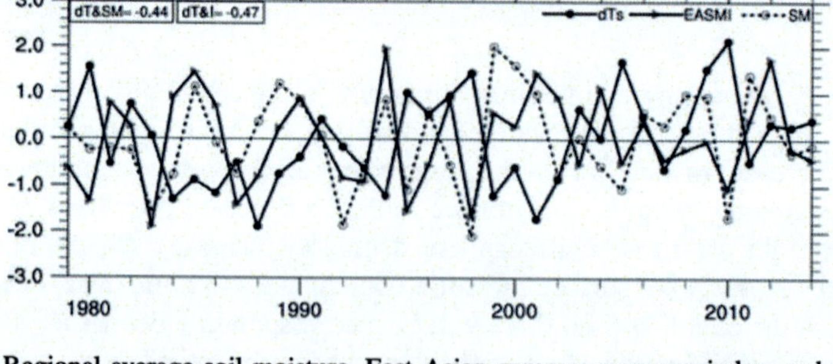

Fig. 4.40 Regional average soil moisture, East Asian summer monsoon index, and normalized time series of the average temperature difference between the Indochinese Peninsula and the South China Sea in May. The correlation coefficients are provided at the upper left corner

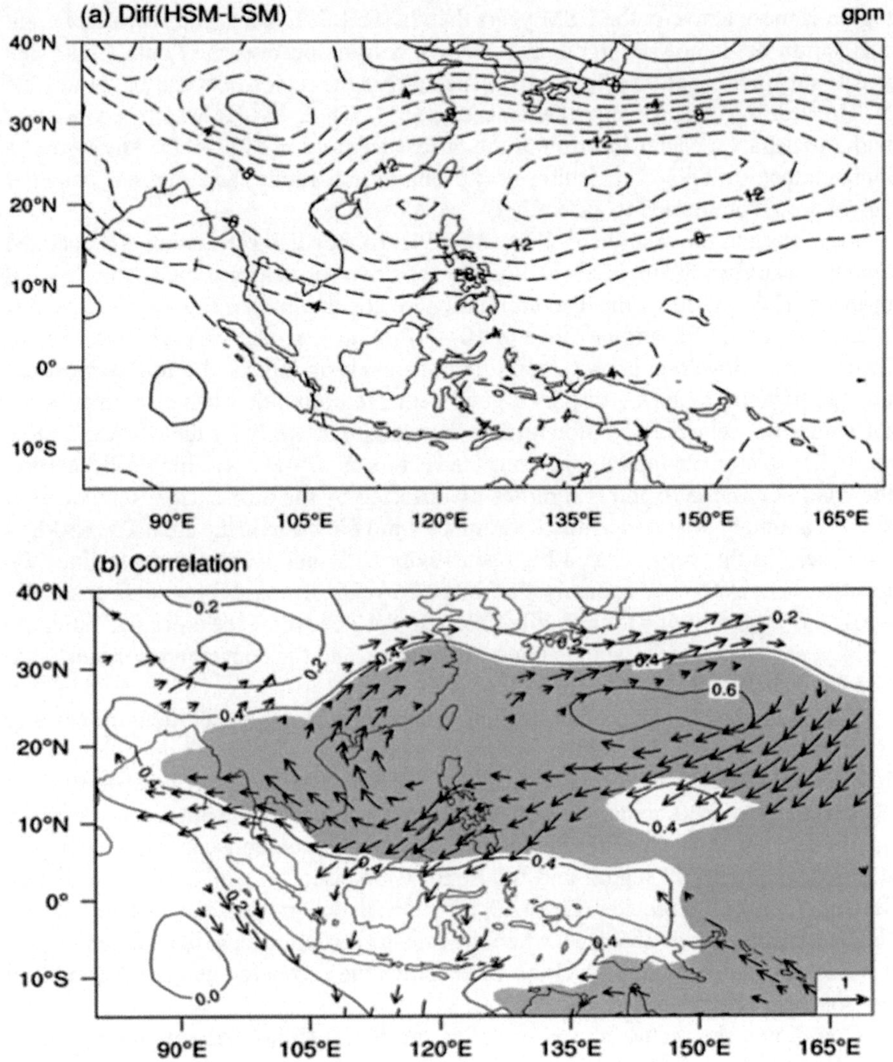

Fig. 4.41 Summer 850-hPa potential height difference in HSM and LSM years (units: gpm) (**a**) and correlation of near-surface temperature and summer 850-hPa potential height (contour; the shading indicates where the significance level is more than 0.01), correlation of near-surface temperature and 850-hPa wind (the vector field shows only the vector where the significance level is more than 0.1) (**b**)

region is more active in the HSM years than in the LSM years. The potential height correlation field shows a significant positive correlation over the South China Sea and its surrounding areas (Fig. 4.41b), especially in the region between 10° N and 30° N. The location of the positive center is close to (150° E, 25° N), which is consistent with the negative center location of the difference field in Fig. 4.41a. These results indicate that spring soil moisture, near-ground temperature, and summer potential height are all connected to each other.

According to the relation (4.11) and Fig. 4.41b, positive correlation is associated with the easterly wind in the zonal direction, positive correlation is the southerly wind in the meridional direction. The correlation vector that passes the significance test is also located in the latitude range of 10°–30° N; as a whole, the correlation vector field of an anticyclone is presented. Previous analysis indicated that soil moisture and temperature in spring display a good negative correlation, which results in the difference between precipitation level in the vicinity of 30° N for the HSM and LSM years. There is an obvious unity among the vectors of 10°–20° N, which indicates that the wind belt in the region is significantly affected by the near-surface temperature, and the anomaly of soil moisture determines wind direction in the area. The 850-hPa wind field in this region has a high sensitivity to humidity, which determines the location of convergence anomaly. For the HSM years, the easterly wind is unusually weak, the westerly wind is unusually strong, and the convergence is eastward. For the LSM years, the easterly wind is unusually strong, and the convergence is westward and reaches the EIO.

This creates a difference in precipitation near the equator. In addition, the easterly wind belt covers the ICP region, which indicates that when the temperature of the ICP is increased because of the decrease in soil moisture, the local lower-level wind becomes an easterly anomaly that hinders the monsoon development and weakens the southwest monsoon. Considering the negative correlation between the temperature difference in the ICP region and the South China Sea, as well as the East Asian summer monsoon depicted in Fig. 4.40, the temperature gradient between the ICP and the South China Sea is abnormal when soil moisture is reduced, so that the sea surface wind is unusually strong, which also intensifies the formation and development of the southwest monsoon.

Combining the methods used to define the South China Sea summer monsoon onset index (He et al. 2000; Liang et al. 1999), the onset superiority pentad is derived. A sequence of high-frequency onsets in the same year is selected as the dominant onset interval sequence. Figure 4.42 indicates that the onset superiority pentads of the South China Sea summer monsoon in the HSM years are the 30th, 28th, 24th, 26th, 26th, 25th, and 29th pentad, and the average onset time is the 26.8th pentad, earlier than the climate average of the 28th pentad. The corresponding onset times in the LSM years are the 29th, 30th, 27th, 28th, 29th, and 29th pentad, with an average onset time of the 28.7th pentad, and later than the average onset time of climatology. In addition, the correlation coefficient between the onset time and spring soil moisture is obtained, and the correlation coefficient reaches −0.34, passing the 95% significance test. The thermal difference between the ICP and SCS is a necessary condition for the onset of the South China Sea summer monsoon. Further, in the HSM years the ICP's

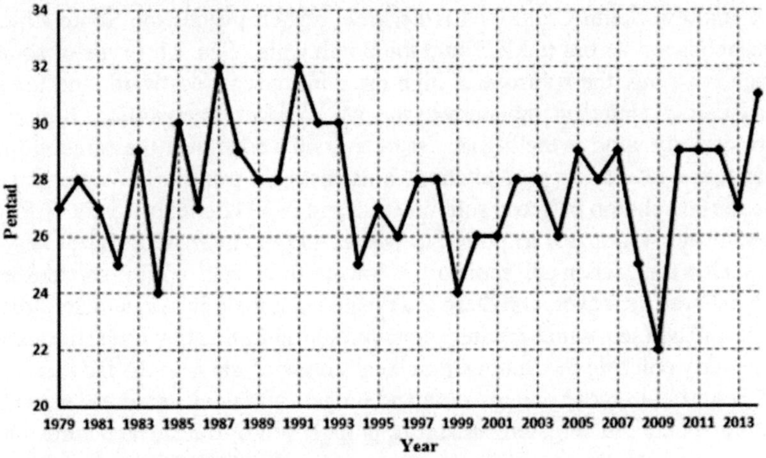

Fig. 4.42 The onset superiority pentad of South China Sea summer monsoon

near-ground temperature decreases, and the thermal differences between the ICP and SCS change from positive to negative more easily. In the LSM years, the ICP's near-ground temperature is high, which is not good for the difference of thermal difference between the two regions. Therefore, owing to the influence of thermal difference, the South China Sea summer monsoon onset occurs early in the HSM years, whereas it is late in the LSM years.

By means of SVD analysis and correlation analysis, we obtained the mode of summer precipitation in the South China Sea and the ICP region. When soil moisture in spring has anomaly, precipitation in the Western Pacific near the South China Sea is higher than normal, and that in the eastern Indian Ocean is lower. In contrast, dry soil decreases the summer precipitation in the Western Pacific, whereas it increases the summer precipitation in the eastern Indian Ocean, and extends to the Western Pacific region. According to the positive and negative dipole mode of precipitation, seven years were selected as HSM (dry), and six years, as LSM (wet). The composite analysis shows that the configuration of circulation fields is the direct cause of the summer precipitation dipole mode.

Higher soil moisture in the ICP region during spring decreases the near-ground temperature in late-spring, because evaporation is more active, surface moisture increases, and the latent heat flux between the ground and the atmosphere increases. Higher soil moisture content also reduces albedo, and the absorption of solar shortwave radiation is more intense, resulting in a corresponding decrease in the amount of upward shortwave radiation. In short, the near-ground temperature, heat flux, and longwave radiation decrease, and the opposite occurs when soil moisture is low.

May is a critical period for the onset of the South China Sea summer monsoon. The surface temperature in the ICP region during late spring plays an influential role in atmospheric circulation. When soil moisture is abnormal, the temperature of the lower atmosphere in spring is low. The lower- level temperature difference between the ICP and the South China Sea is negative, which pushes the South China Sea summer monsoon across the ICP into the South China Sea. This type of southwest monsoon is strong, the subtropical high pressure retreats eastward, and the southwesterly carries abundant moisture to the WP to form precipitation. It also meets with the easterly wind at the high-pressure system's edge near the equator, forming an updraft and increasing precipitation. This eastward transport of water vapor can explain the correlation between soil moisture and local latent heat in the ICP region. Increased water vapor is transported to where, and the increase in latent heat is not obvious. However, when soil moisture is low on the ICP in spring, the temperature of the local area increases, and there is a positive anomaly in temperature difference between land and sea, which creates a unfavorable condition that limits the amount of southwesterly entering the South China Sea. The wind stays in the EIO for a longer period, resulting in more precipitation and longer rainfall events in the region.

Finally, when soil moisture in spring is high (low), the surface temperature is low (high), and the thermal difference between ICP and SCS is reduced, which is (is not) conducive to the strengthening of the summer monsoon. The South China Sea summer monsoon starts early (late), which is advantageous (disadvantageous) to the EIO moisture advance to the WP. Thus, the summer precipitation in the EIO is less (more) than normal, and the summer precipitation in the WP is more (less) than normal.

References

Amenu GG, Kumar P, Liang XZ (2005) Interannual variability of deep-layer hydrologic memory and mechanisms of its influence on surface energy fluxes. J Clim 18(23):5024–5045

Avissar R, Pielke RA (1991) The impact of Plant stomtal control on mesoscale atmospheric circulations. Agric For Metror 54:353–372

Barnett TP, Menil LD, Schlese U, Roeckner E, Latif M (1989) The effect of Eurasian snow cover on regional and global climate variations. J Atmos Sci 46:661–685

Brooks RH, Corey AT (1966) Properties of porous media affecting fluid flow. J Irrig Drain Div ASCE 92:61–90

Burdine NT (1953) Relative permeability calculations from pore size distribution data. J Petrol Technol 5(3):71–78

Chen Y, Zhang Y, Wang S et al (2014) Seasonal variation of turbulence characteristics over alpine meadow ecosystem (in Chinese). Plateau Meteorol 33(03):585–595

Clapp RB, Hornberger GM (1978) Empirical equations for some soil hydraulic properties. Water Resour Res 14:601–604

Deardorff JW (1972) Numerical investigation of neutral and unstable planetary boundary layers. J Atmos Sci 39:91–115

Delworth TL, Manabe S (1988) The influence of potential evaporation on the variabilities of simulated soil wetness and climate. J Clim 1(5):523–547

Delworth TL, Manabe S, Stouffer RJ (1993) Interdecadal variations of the thermohaline circulation in a coupled ocean-atmosphere model. J Climate 6:1993–2011

Dirmeyer PA, Schlosser CA, Brubaker KL (2009) Precipitation, recycling, and land memory: an integrated analysis. Journal of Hydrometeorology 10(1):278–288

David CH, Frederick M, John WT (1983) Understanding robust and exploratory data analysis[M]. Wiley, pp 127–139

Engelmark H, Svensson U (1993) Numerical modelling of phase change in freezing and thawing unsaturated soil. Nord Hydrol 24:95–110

Foken T, Wichura B (1996) Tools for quality assessment of surface-based flux measurements. Agric Forest Meteorol 78:83–105

Foken T (2003) Angewandte meteorologie. Mikrometeorologische Methoden, Springer, Berlin, Heidelberg, p 289

Guo J, Bian L, Dai Y (2007) Measured CO_2 concentration and flux at 16 m Height during corn growing period on the north China plain (in Chinese). Chin J Atmospheric Sci (Chin) 31(4):695–707

He HY, Wen ZP, Jian MQ (2000) A Study on the Onset of the Asian Summer Monsoon During 1982~1996, PARTI: Basic features of the Asian Summer Monsoon circulation and its onset index(in Chinese). Acta Scientiarum Naturalium Universities Sunyatseni 39(1):91–96

Hulme M, Viner D (1998) A climate change scenario for the tropics. Climat Change 39(2–3):145–176

Idso SB, Aase JK, Jackson RD (1975) Net radiation-soil heat flux relations as influenced by soil water content variations. Bound-Lay Meteorol 9(1):113–122

Jame YW, Norum DI (1980) Heat and mass transfer in a freezing unsaturated porous medium. Water Resour Res 16:811–819

Jiang D, Huang F, Huang J (2013) Observational study on surface-layer turbulent parameters over the South China coast. Periodical Ocean Univers China (Chin) 43(12):007–015

Koren V, Schaake J, Mitchell K et al (1999) A parameterization of snowpack and frozen ground intended for NCEP weather and climate models. J Geophys Res 104:19569–19585

Kormann R, Meixner F (2001) An analytical model for non-neutral stratification. Bound Lay Meteorol 103:205–224

Koster RD, Dirmeyer PA, Zhichang G et al (2004) Regions of strong coupling between soil moisture and precipitation. Science 305(5687):1138–1140

Koster RD, Suarez MJ (1996) The Influence of Land Surface Moisture Retention on Precipitation Statistics. J Clim 9(9):2551–2567

Larson K, Hartmann DL, Klein SA (1999) The role of clouds, water vapor, circulation, and boundary layer structure in the sensitivity of the tropical climate. J. Climate 12:2359–2374

Lee X, Michael LG et al (2011) Observed increase in local cooling effect of deforestation at higher latitudes. Nature 479:384–387

Li W, Guo W, Xue Y et al (2016) Sensitivity of a regional climate model to land surface parameterization schemes for East Asian summer monsoon simulation. Climate Dyn. 47:2293–2308

Li C (1996) The onset and interannual variability of the Asian summer monsoon in relation to land-sea thermal contrast. J Clim 9(2):358–375

Liang JY, Wu SS, You JP (1999) The research on variations of onset time of the SCS summer monsoon and its intensity. Acta Meteorologica Sinica 15(02):97–105 (in Chinese)

Liu Huizhi and Hong Zhongxiang (2000) Turbulent Characteristics in the Surface Layer over Gerze Area in the Tibetan Plateau (in Chinese). Chin J Atmospheric Sci 24(3):289–300

Liu XF, Li Q, He JH, Ping Wang (2009) Effects of the thermal contrast between Indo-China Peninsula and South China Sea on SCS monsoon onset. Acta Meteorologica Sinica 67(1):100–107 (in Chinese)

Liu YM, Chen ZL, Mao JY (2003a) Impacts of the onset of the Bay of Bengal monsoon on the onset of the South China Sea monsoon. Part I: A case study. Acta Meteorologica Sinica, 61(1):1–9 (in Chinese)

Liu YM, Chen ZL, Wu GX (2003b) Impacts of the onset of the Bay of Bengal monsoon on the onset of the South China Sea monsoon. Part II: Numerical experiments. Acta Meteorologica Sinica 61(1):10–19 (in Chinese)

Mechoso CR, Yu J-Y, Arakawa A (2000) A coupled GCM pilgrimage: from climate catastrophe to enso simulations. book chapter in general circulation model development: past, present, and future., D. A. Randall, Eds., Academic Press, pp 539–575

Meehl GA (1994) Influence of the land surface in the asian summer monsoon: external conditions versus internal feedbacks. J Clim 7(7):1033–1049
Meng WG, Zheng B (2006) The features of south china sea summer monsoon onset of 2004 and the possible effects of land-surface process in Indo-China Peninsula II: numerical simulation. Acta Meteorologica Sinica 64(1):81–89 (in Chinese)
Mölders N, Walsh JE (2004) Atmospheric response to soil-frost and snow in Alaska in March. Theor Appl Climatol 77:77–105
Numaguti A (1993) Dynamics and energy balance of the Hadley circulation and the tropical precipitation zones: significance of the distribution of evaporation. J Atmos Sci 50(13):1874–1887
Orlowsky B, Seneviratne SI (2010) Statistical analyses of land-atmosphere feedbacks and their possible pitfalls. J Clim 23(14):3918–3932
Peng J, Wu X, Jiang Z, Liu H (2008) Characteristics analysis of energy budget over urban and suburban underlying surfaces in Nanjing (in Chinese). *SCIENTIA METEOROLOGICA SINICA* 28(1):21–29
Ren XJ, Qian YF (2002) Numerical simulation experiments of the impacts of local sea-land thermodynamic contrasts on the SCS summer monsoon onset. J Tropical Meteorol 18(4):327–334 (in Chinese)
Shukla J, Mintz Y (1982) Influence of land-surface evapotranspiration on the earth's climate. Science 215(4539):1498–1501
Shoop SA, Bigl SR (1997) Moisture migration during freeze and thaw of unsaturated soils: modeling and large scale experiments. Cold Reg Sci Technol 25:33–45
Stull RB (1988) An Introduction to Boundary Layer Meteorology. Kluwer Academic Publishers, Dordrecht, Boston and London, p 666
Van Genuchten MT (1980) A closed-form equation for predicting the hydraulic conductivity of unsaturated soils. Soil Sci Soc 44:892–898
Wang J (1999) Land surface process experiments and interaction study in China—From HEIFE to IMGRASS and GAME-TIBET/TIPEX(in Chinese). Plateau Meteorol 18(3):80–294
Wang BM (2004) A study on synthetic differentiation method for basia meteorological data quality control (in Chinese). J Appl Meteorological Sci 15(Suppl):50–59
Wang HJ, Yang ZB, Yang DC et al (2007) The method and application of automatic quality control for real time data from automatic weather stations (in Chinese). Meteorological Monthly 33(10):102–109
Wang C, Wei ZG, Li ZC et al (2010) A quality control routine for dunhuang gobi meteorology tower data (in Chinese). J Arid Meteorol 28(2):121–127
Wang HJ, Liu Y (2012) Comprehensive consistency method of data quality controlling with its application to daily temperature(in Chinese). J Arid Meteorol 23(1):69–76
Wang SY, Qian YF (2001) Basic characteristic of surface heat field in 1998 and the possible connections with the SCS summer monsoon onset (in Chinese). Acta Meteorologica Sinica 59(1):31–40
Wu G, Zhang Y (1998) Tibetan plateau forcing and the timing of the monsoon onset over South Asia and the South China Sea. Mon Weather Rev 126(126):913–927
Wen Xuefa, Yu Guirui, Sun Xiaomin, Liu Yunfen, 2005: Turbulence flux measurement above the overstory of a subtropical Pinus plantation over the hilly region in southeastern China., *Science in China Ser. D*, 48 Supp.I: 63–73
Werth D, Avissar R (2004) The regional evapotranspiration of the Amazon. J Hydrometeorol 5:100–109
Wu H, Ye B, Wu J et al (2013) Analysis on turbulent feature of apline meadow in the upper reach of shule River (in Chinese). Plateau Meteorol 32(2):2368–2376
Xu Y, Zhou C, Li Z et al (1993) Microstructure and spectral characteristics of turbulence in the surface-layer atmosphere over guangzhou. Chin J Atmospheric Sci (in Chin),17(3):338–348
Xu J, Sui B, Chen X (2014) Observational analysis on turbulent characteristics of the atmospheric surface layer above Forest (in Chinese). J Arid Meteorol 32(1):1–9

Xue Y, Fernando DS, Ratko VC et al (2010) Global and seasonal assessment of interaction between climate and vegetation biophysical process: a GCM study with different land-vegetation representations. J. Climate 6(23):1411–1433

Xue Y, Sellers PJ, Kinter JL et al (1991) A simplified biosphere model for global climate studies. J Climate 4:345–364

Xue Y, Juang HMH, Li WP et al (2004a) Role of land surface processes in monsoon development: East Asia and West Africa. J Geophys Res 109(D3):215–229

Xue Y, Juang HMH, Li WP et al (2004b) Role of land surface processes in monsoon development: East Asia and West Africa. J Geophys Res Atmospheres 109(D3):215–229

Yang S, Lau KM (1998) Influences of sea surface temperature and ground wetness on Asian summer monsoon. J Clim 11(12):3230–3246

Zhang QY, Tao SY, Chen LT (2003) The inter-annual variability of East Asian summer monsoon indices and its association with the pattern of general circulation over East Asia. Acta Meteorologica Sinica 61(5):559–568 (in Chinese)

Zhang X, Sun SF, Xue YK (2007) Development and testing of a frozen soil parameterization for cold region studies. J Hydrometeor 8(4):690–701

Zhang R, Huang J, Wang X et al (2016) Effects of precipitation on sonic anemometer measurements of turbulent fluxes in the atmospheric surface layer. J Ocean Univ China 15(3):389–398

Zheng B, Meng WG (2006) The features of south china sea summer monsoon onset of 2004 and the possible effects of land-surface process in Indo-China Peninsula *I*: Diagnostic study (in Chinese). Acta Meteorologica Sinica 64(1):72–80

Chapter 5
Effects of the Tibetan Plateau on Climate

5.1 Introduction

The Tibetan Plateau (TP) is located in the northwest of Southeast Asia, which is the highest plateau in the world. It covers about 25% of the land territory of China, with an average elevation of >4 km. There is a significant contrast between the western and eastern TP with regard to land surface features, vegetation, and meteorological characteristics (Ye and Gao 1979). The air mass above the TP (in a unit column) is only about 60% of that above the land at sea level. Because of its distinct geographical features, the TP exerts a significant impact on the weather and climate over many places of the world through both mechanical and thermal-dynamic effects (Yanai and Wu 2006). In winter, the mechanical effect of the TP is the dominant driving force influencing atmospheric circulation. From winter to summer, however, the importance of surface sensible heating increases rapidly due to the strong solar radiation over the TP (Ye and Gao 1979), and effect of elevated plateau heating plays an important role in the evolution of the Asian summer monsoon (ASM).

The thermal forcing over the TP regulates the onset time and location of the ASM, and influences the precipitation and atmospheric circulation during both South Asian monsoon and East Asian monsoon. Many studies have investigated the characteristics of the heating source over the TP through the analyses of surface sensible heat (SH) or the total diabatic heating (Q_1) (Luo and Yanai 1984; Yanai and Wu 2006; Yanai et al. 1992). According to the theory of the TP "SH-driven air pump" (TP-SHAP) (Wu et al. 1997), SH on the surface of the TP can cause low-level moisture in surrounding regions to converge rapidly over the TP, resulting in considerable condensation and precipitation over the southern TP and northern parts of South Asia. However, the spatial distribution of precipitation over the southern slope of the TP is not uniform. Moreover, the water vapor transport along the Somali jet from the Southern Hemisphere does not ascend uniformly over the southern slope of the TP. One branch of the moisture transport deviates northward over the northern Bay of Bengal (BoB). It then ascends over the southern slope of the TP, forming a center of heavy rainfall there in summer. Conversely, less precipitation appears over the

S. Yang et al., *Climate Change in Southeast Asia and Surrounding Areas*,
Springer Climate, https://doi.org/10.1007/978-981-15-8225-7_5

western TP (Wang et al. 2017). Therefore, the dominant summer diabatic heating over the TP should include two different types: surface SH and atmospheric latent heating (LH) that is influenced by the SH. In addition, the precipitation over the TP can in turn change the surface heat budget and the SH. Therefore, it is necessary to reveal the characteristics and interactions of these two different types of heating over the TP.

On the other hand, although the thermal effect of the TP has long been concerned, most of the previous studies focused on the climatic impacts in the downstream regions or South Asia (e.g., Kitoh 2004; Wu et al. 2007, 2012). The potential influences of TP thermal forcing on climate over the upstream regions (like Central Asia, North Africa, southern Europe) have received attention rarely. In Sect. 5.3 of this chapter, the upstream effects of surface heating over the TP on weather and climate will be investigated using an air–sea-coupled global model, and the associated physical mechanism will be clarified. Moreover, the South Asian high (SAH, also called the TP high), which is closely related to the TP thermal forcing, will be discussed in this chapter.

5.2 Effects of the TP Summer Heating on Climate in Regional Models

Compared with the atmospheric general circulation models used previously, the weather research and forecasting (WRF) regional model has a higher resolution that can represent topography and land surface processes more accurately. It is a useful tool for studying weathers and climate over the TP. The WRF model has been employed as a regional model to simulate the variation of the ASM and the climatic effects of TP forcing (e.g., Yang et al. 2012; Ma et al. 2014; Wang et al. 2014a). In this section, the characteristics and interactions of the summer surface SH and LH over the main body of the TP (where the terrain is higher than 2000 m in the area of (70°–105° E, 23°–40° N)) are investigated using the WRF model. The impacts of these two types of TP heating on local vertical motion and monsoonal meridional circulation are compared. To further the understanding of the mechanisms via which these two types of heating over the TP affect the ASM, their influences on the configuration of temperature and circulation in the upper troposphere are also explored. Furthermore, the potential regulation on the climatic effect of TP heating by tropical air–sea coupling is investigated in this section.

5.2.1 Models and Experimental Design

The WRF model version 3.4.1 (Skamarock et al. 2008) is used here. The physical packages incorporated in the WRF model used here include the WRF Single-Moment

6-Class microphysics scheme, the Grell–Devenyi convective scheme, the Noah land surface model, the BouLac planetary boundary layer scheme, the Goddard shortwave scheme, and the Rapid Radiative Transfer Model for longwave radiation. More details of these physical schemes can be found in the WRF user's guide (http://www2.mmm.ucar.edu/wrf/users/docs/user_guide_V3/). The spatial domain of the simulations covers most parts of Asia and its adjacent oceans with the Lambert projection (195 × 133 grids) centered at (95° E, 30° N). The model has 45-km horizontal resolution and 35 vertical layers with a terrain-following sigma coordinate and a prescribed model top at 50 hPa. The initial state of the atmosphere and the lateral boundary conditions (updated every 6 h) are obtained from the Final Analysis of the National Centers for Environmental Prediction (NCEP, http://rda.ucar.edu/datasets/ds083.2/). The sea surface temperature (SST) dataset used in the WRF model is the Optimum Interpolation SST (OISST; Reynolds et al. 2007) from the National Oceanic and Atmospheric Administration (NOAA), which is updated daily during model integration.

To investigate the air–sea interaction involved in the process of TP thermal effect, we embed a one-dimensional (1-D) ocean mixed-layer model (OMLM; Noh and Kim 1999) into the WRF model (Wang and Duan 2012), which is referred to as the WRF-OMLM hereafter. This regional coupled model has shown good performances on simulating typhoons (Wang and Duan 2012), and seasonally modeling the East Asian summer monsoon (EASM; Wang et al. 2014a). More detailed information about the WRF-OMLM can be found in Wang and Duan (2012) and Wang et al. (2018a).

Based on the WRF and WRF-OMLM models, seven ensemble experiments are performed; each covers six summers (2003–2008) with the initial conditions at 0000 UTC on 1 May. The model is integrated for four months in each summer, i.e., every experiment ends at 1800 UTC 31 August, and the ensemble output in the past 3 months (June–August, or JJA) is analyzed. Table 5.1 gives more details of experimental design. The base experiment with the WRF model is named the control experiment (CTL). Studies have demonstrated that the WRF model can reproduce both monsoon circulation and precipitation reasonably well during the ASM (e.g., Wang et al. 2014a, b, 2016), indicating that CTL can serve as the base experiment of the summer monsoon. Figure 5.1 shows the spatial distributions of the JJA mean surface SH and atmospheric LH over the TP and its surrounding areas in CTL. In agreement with the pattern of observed summer precipitation (Wang et al. 2017), the LH center is located over the southern slope and southeastern areas of the TP with a maximum of >300 W m^{-2}. Conversely, SH is the weakest (<40 W m^{-2}) over the southeastern TP but is the strongest (>100 W m^{-2}) over the northwestern TP. We also design several sensitivity experiments. In Exp TP_NL, the LH of the entire vertical atmospheric column over the main body of the TP (where terrain is 2000 m over the region (70°–105° E, 23°–40° N)) is removed (Table 5.1). In Exp TP_NS, the surface SH released over the main TP area is suppressed (Table 5.1). Their differences from CTL can reflect the climatic effects of the two types of TP heating, respectively.

Table 5.1 Experimental design

Exp (model used)	Experimental description
1. CTL (WRF)	Background, climate mean
2. TP_NL (WRF)	No latent heating over the main body of the TP (h ≥ 2000 m)
3. TP_NS (WRF)	No surface sensible heating over the main body of the TP (h ≥ 2000 m)
4. STP_NS (WRF)	No surface sensible heating over the southern slope of the TP
5. ALLTP_NS1 (WRF)	No surface sensible heating over both the main body and southern slope of the TP
6. CTL-O (WRF-OMLM)	Background, climate mean
7. ALLTP_NS2 (WRF-OMLM)	No surface sensible heating over both the main body and southern slope of the TP

The main body of the TP is defined as the area of (70°–105° E, 23°–40° N) with the terrain higher than 2000 m

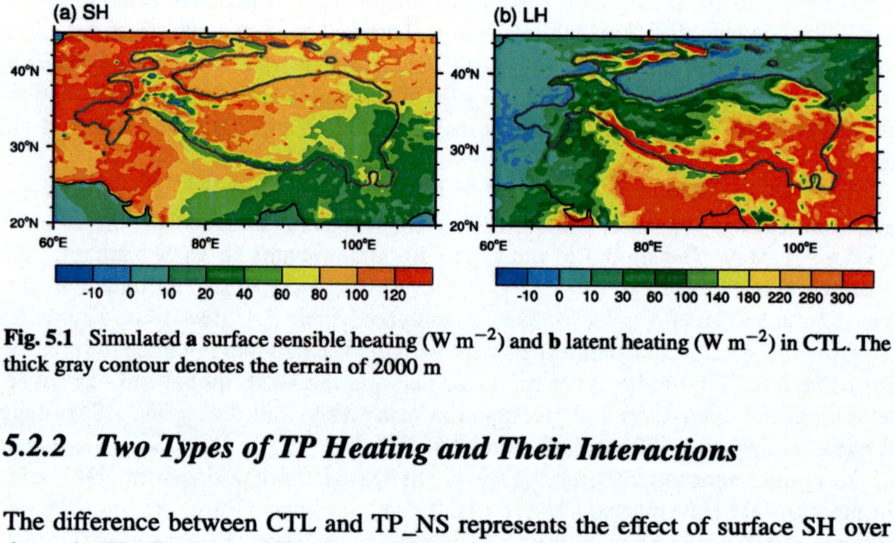

Fig. 5.1 Simulated **a** surface sensible heating (W m^{-2}) and **b** latent heating (W m^{-2}) in CTL. The thick gray contour denotes the terrain of 2000 m

5.2.2 *Two Types of TP Heating and Their Interactions*

The difference between CTL and TP_NS represents the effect of surface SH over the main TP. The thermal forcing of SH by the TP forms an evident surface cyclonic convergent circulation, which is centered over the TP, with air converging from the lower regions surrounding the TP (Fig. 5.2a). The SH also induces increased precipitation over the central, eastern, and southern TP, with a maximum over the southeastern TP that exceeds 6 mm day^{-1}. Precipitation also increases over the BOB and East Asia, but decreases over the South Asia continent. In the areas outside the main body of the TP, the precipitation anomalies lead to different responses of surface SH. The weakened precipitation over Northwestern India results in intensified local SH, while the enhanced precipitation over the eastern and southeastern TP leads to decreased local SH (Fig. 5.2c). On the other hand, the responses of low-level

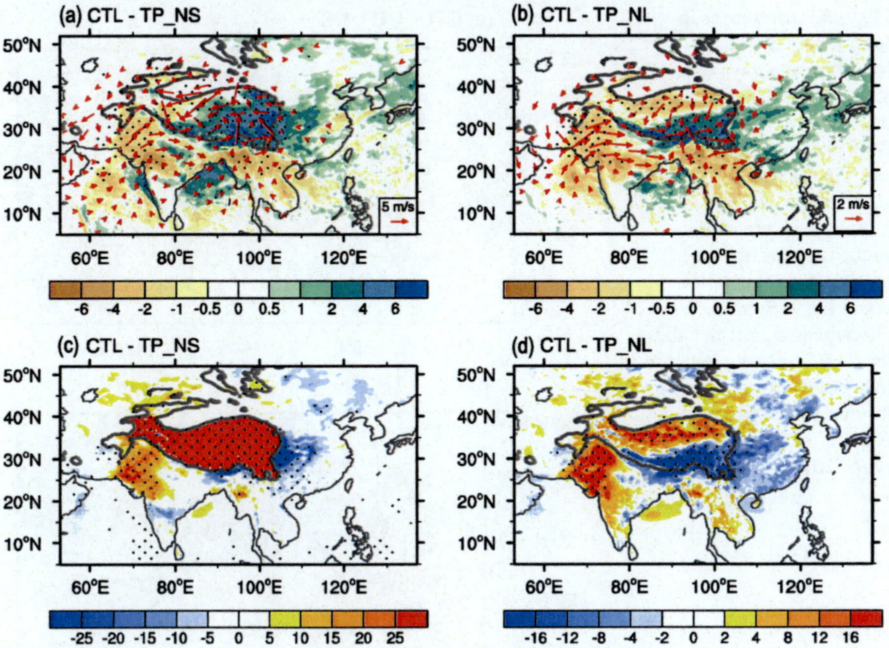

Fig. 5.2 **a** Summer precipitation (shading; mm day^{-1}) and surface wind at $\sigma = 0.98$ level (vector; m s^{-1}) for the difference between CTL and TP_NS. **b** Same as (**a**), except for the difference between CTL and TP_NL. **c** Same as (**a**), except for surface sensible heat (W m^{-2}). **d** Same as (**c**), except for the difference between CTL and TP_NL. Stippling denotes statistical significance above the 95% level. Only the wind difference with statistical significance above the 95% level is plotted in (**a**) and (**b**). The thick gray contour denotes the terrain of 2000 m. Modified from Wu et al. (2016)

circulation and precipitation outside the main body of the TP due to TP LH are similar with the results achieved in the experiment without surface SH (Fig. 5.2a, b). But within the TP area, the effect of TP LH increases precipitation in the south and reduces it in the north, in good agreement with the east–west-oriented convergence zone in the surface wind field located over the southern TP (Fig. 5.2b). The difference in the spatial pattern of precipitation has a strong feedback on the distribution of SH, which decreases in the southern TP where precipitation is enhanced and increases in the north where precipitation is reduced (Fig. 5.2d).

It should be noted that the aforementioned relationship between precipitation (namely, LH) and SH over the TP is not a one-way causality. To understand their complex interaction and feedback, we conduct another sensitivity experiment (STP_NS), in which the surface SH over the southern slope of the TP was removed. The northern boundary of the TP's southern slope is defined by the highest elevation at each latitude. The differences in the results between CTL and STP_NS show the consequence of the effect of surface SH over the southern slope of the TP, and they are used for comparison with the results induced by diabatic heating over the main body of the TP. Figure 5.3 shows that the responses induced by surface SH on the TP's

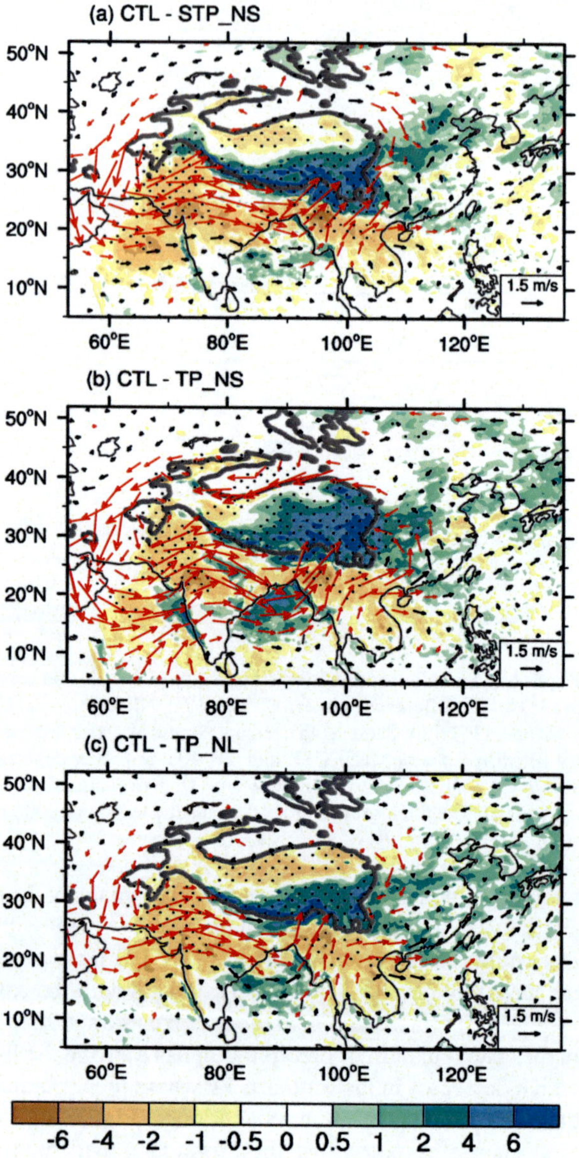

Fig. 5.3 Differences in summer precipitation (shading; mm day^{-1}) and 850-hPa wind (vector; m s^{-1}) between CTL and STP_NS (**a**), between CTL and TP_NS (**b**), and between CTL and TP_NL (**c**). Stippling denotes the statistical significance of precipitation anomalies above the 95% level. Wind differences significant above the 95% level are plotted in red. The thick gray contour denotes the terrain of 2000 m. After Wu et al. (2016)

southern slope (Fig. 5.3a) are similar to those induced by the LH over the main body of the TP (Fig. 5.3c). They both induce cyclonic circulation in the lower troposphere surrounding the TP, enhance summer precipitation over East Asia and the BoB, but reduce precipitation over the South Asia continent. Meanwhile, they lead to positive rainfall over the southern TP and negative rainfall over the northwestern TP. Despite these similarities, the precipitation difference presented in Fig. 5.3a is the consequence of the surface SH over the southern slope of the TP, while that presented in

Fig. 5.3c leads to decreased surface SH in the south and increased surface SH in the north of the TP (Fig. 5.2d). In other words, the surface SH over the southern slope of the TP can result in more precipitation over the southern TP. Conversely, the enhanced precipitation over the southern TP can further weaken surface SH over the southern slope, leading to reduced local precipitation. Thus, we can infer that precipitation (i.e., LH) and surface SH over the TP interact strongly and that their distributions, as observed or seen in a numerical model, represent the quasi-equilibrium state of such a complex interaction and feedback mechanism. Additionally, when surface SH exists over the main body of the TP (Fig. 5.3b), although the distributions of both low-level circulation and precipitation are similar to those in Fig. 5.3a, c, their intensities are remarkably intensified in Fig. 5.3b.

5.2.3 Effects on Meridional Circulation and on Temperature and Circulation in the Upper Troposphere

Figure 5.4 shows the meridional circulation over the BoB (averaged over 85°–95° E). In CTL, two distinct branches of ascending motion are located over the tropical ocean and over the southern slope of the TP (Fig. 5.4a). They correspond to the southern and northern branches of the South Asia summer monsoon, respectively (Wu et al. 2012). From the differences between CTL and TP_NL/TP_NS, we can see that the surface SH over the main body of the TP significantly enhances the ascending motion over the entire TP (including the main body and its southern slope area) (Fig. 5.4b). But for the atmospheric LH over the main body of the TP, this type of heating enhances the ascending motion over the southern TP and its southern slope, corresponding with the descending motion over the northern BOB and northern TP. The descending motion over the northern TP is associated with weakened local precipitation (Fig. 5.2b) and enhanced surface SH (Fig. 5.2d). It can be concluded that both SH and LH over the main body of the TP can remarkably enhance the southerly wind and ascending motion over the southern TP, which increases precipitation. However, the effect of SH is stronger than the effect of LH. That is because SH can trigger the occurrence of precipitation over the main body of the TP (Fig. 5.3b), and the condensational heating associated with precipitation thus overlays on the effect of SH. As a result, it is not surprising that the pumping effect due to SH over the TP is stronger than that triggered by LH.

Figure 5.5a illustrates the summer mean air temperature and circulation at 300 hPa in CTL. We can see the huge upper-level warm anticyclone located over the southern TP and the west of strong precipitation center of the TP. The diabatic heating of the main body of the TP not only intensifies the warm anticyclone but also shifts its center northwestward, resulting in downstream energy dispersion and the formation of a cold cyclonic circulation over the region of Mongolia and Northeast China (Figs. 5.5c, d). These two warm anticyclonic anomalies are located to the west of the corresponding precipitation centers (Fig. 5.3b, c), which agrees with the T-Q_z relation

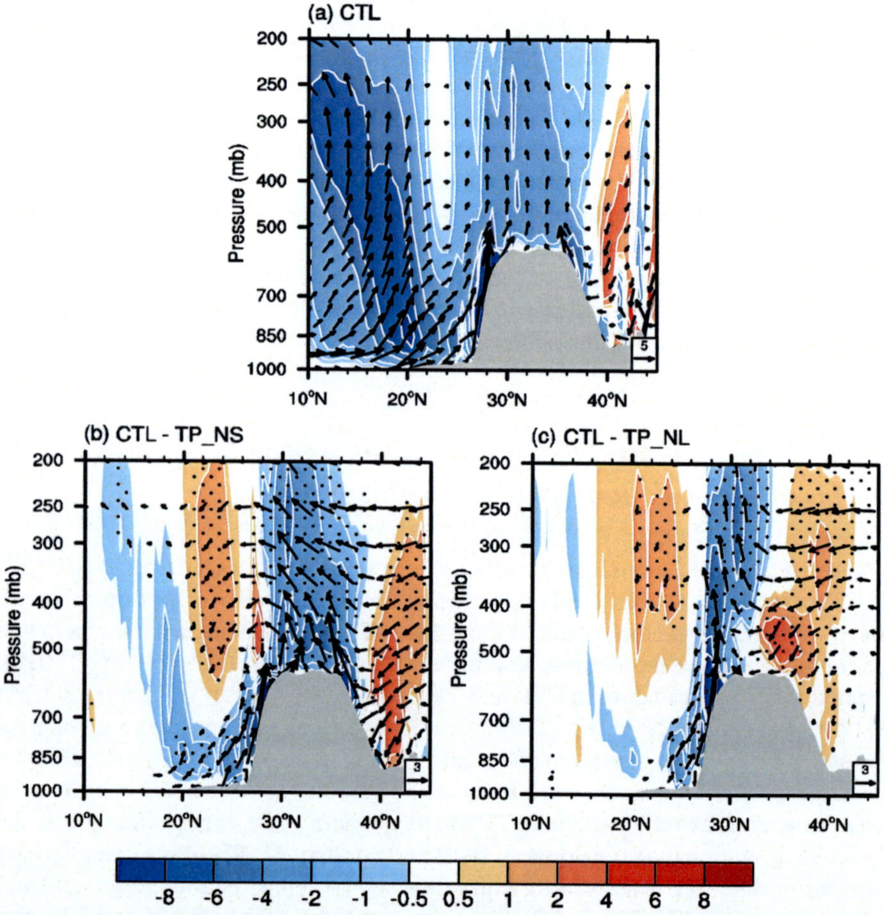

Fig. 5.4 Pressure–latitude cross sections of circulation (vector; v and −50*ω) and vertical velocity (shading; 0.02 Pa s^{-1}) averaged between 85° E and 95° E for the CTL (**a**), and for the differences between CTL and TP_NS (**b**), and between CTL and TP_NL (**c**). Stippling in (**b**) and (**c**) denotes the statistical significance of vertical velocity difference above the 95% level. Only vertical circulation differences in (**b**) and (**c**) with statistical significance above the 95% level are plotted. Modified from Wu et al. (2016)

(temperature and diabatic heating vertical gradient; Wu et al. 2015), implying that the variation of the upper-tropospheric warm center is a consequence of the difference in LH induced by diabatic heating over the main body of the TP. Meanwhile, the effect of surface SH over the main body of the TP on the warm anticyclone is more significant than the effect of LH (Fig. 5.5c, d). The distributions of summer mean air temperature and wind at 100 hPa in CTL are presented in Fig. 5.5b. The South Asian anticyclonic circulation at 100 hPa is similar to that at 300 hPa, but the distributions of temperature at these two levels are completely different. The warm center over southern TP disappears and the temperature increases polewards. The difference in

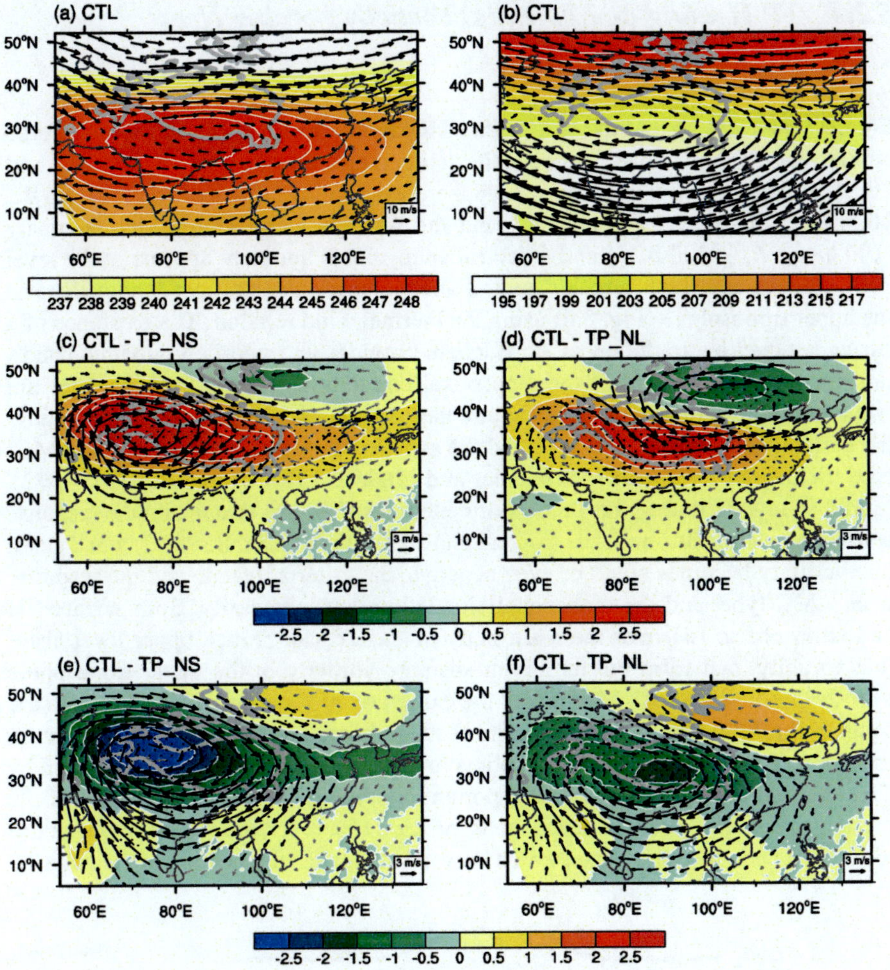

Fig. 5.5 Summer air temperature (shading; K) and wind field (vector; m s^{-1}) at 300 hPa (a, c, d) and 100 hPa (**b**, **e**, **f**). **a** and **b** are from CTL; **c** and **e** are from the difference between CTL and TP_NS; **d** and **f** are from the difference between CTL and TP_NL. Stippling in **c**–**f** denotes the statistical significance of temperature anomalies above the 95% level. Wind differences in **c**–**f** significant above the 95% level are plotted in black. The thick gray contour denotes the terrain of 2000 m. Modified from Wu et al. (2016)

the circulation at 100 hPa (Fig. 5.5e, f) is similar to that at 300 hPa but the intensity is greatly intensified. Most importantly, this 100-hPa anticyclonic center has a cold core, which contrasts with the warm feature of the anticyclone anomaly at 300 hPa. This implies that diabatic heating over the main TP can have a remarkable influence on the spatial configurations of temperature and circulation near the tropopause.

5.2.4 TP Heating and Potential Vorticity Forcing Near the Tropopause

Figure 5.5c–f reveal that the two types of diabatic heating over the main TP force similar anticyclonic flows but opposite temperature distributions at 100 hPa and 300 hPa respectively. Thus, we assume there exists a level (named P_c) between the 100-hPa cold anticyclonic anomaly and the 300-hPa warm anticyclonic anomaly (100 hPa $< P_c <$ 300 hPa), and a neutral anticyclonic anomaly appears at the level P_c. Then, we can show that this neutral anticyclone is the strongest anticyclone in the upper troposphere (Fig. 5.6) using the thermal wind relation. The existence of a strong anticyclone in the upper troposphere provides an important dynamic condition for the development of the monsoonal circulation. That is because there are two types of atmospheric meridional circulation responses to axisymmetric heating: the thermal equilibrium (TE) type and the angular momentum conservation (AMC) type (Schneider 1977, 1987; Schneider and Lindzen 1977; Plumb and Hou 1992). In mid-to-high latitudes, the atmospheric inertial stability is strong and the response is a TE-type circulation. In the tropical and subtropical areas, however, when absolute vorticity becomes small or even negative, the criteria deviate and the response is an AMC type, and the monsoonal circulation develops easily. Both surface SH and atmospheric LH over the main body of the TP can reduce upper-level absolute vorticity, and form the minimum absolute vorticity in the upper troposphere over the TP (not shown). This favors the excitation of monsoonal circulation to the south of the TP, corresponding with the northern branch of the South Asian summer monsoon. Therefore, the strongest anticyclonic response at the P_c level triggered by TP's thermal forcing can provide important conditions for the development of the meridional circulation in the Asian monsoon area. Moreover, the P_c level is very close to 100 hPa near the tropopause, which shows that the diabatic heating of the main TP

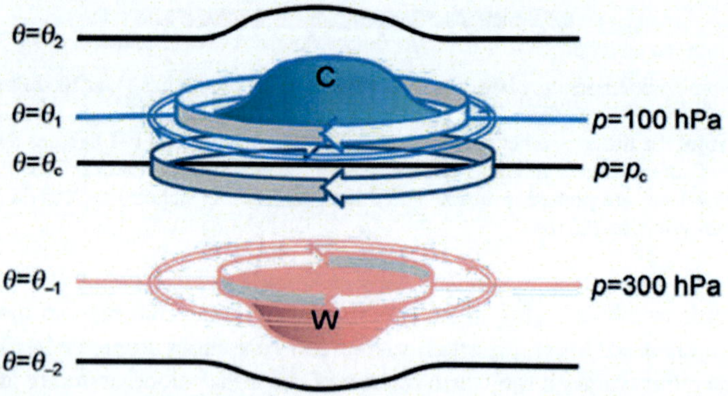

Fig. 5.6 Schematic for the formation of minimum potential vorticity forcing near the tropopause due to thermal forcing over the main TP. After Wu et al. (2016)

in summer can induce the strongest anticyclonic circulation in the tropopause (Wu et al. 2016).

More interesting is that the thermal characteristics of the anticyclone (cold aloft and warm below) lead to a minimum static stability (Fig. 5.6). As a result, the potential vorticity at the anticyclonic center ($P = P_c$) shows a minimum, namely,

$$P = \alpha(f + \zeta)\frac{\partial\theta}{\partial z} = Minimum, \quad p = p_c \tag{5.1}$$

Figure 5.7 shows the potential vorticity responses to the two types of TP heating. It indicates that either surface SH or atmospheric LH over the main body of the TP can trigger a center of minimum potential vorticity at the level near the tropopause (100–150 hPa). Compared with the minimum potential vorticity induced by LH (Fig. 5.7a), the intensity of that induced by SH is stronger, and the center is located further westward (Fig. 5.7b). Moreover, because this forcing is located within the westerly flow, it can trigger a downward-propagating Rossby wave train, as shown in Fig. 5.5c–f. We conclude that diabatic heating over the TP can influence the circulation and climate in the Northern Hemisphere, mainly through changing the configuration of temperature and circulation in the upper troposphere and forming a minimum potential vorticity forcing near the tropopause.

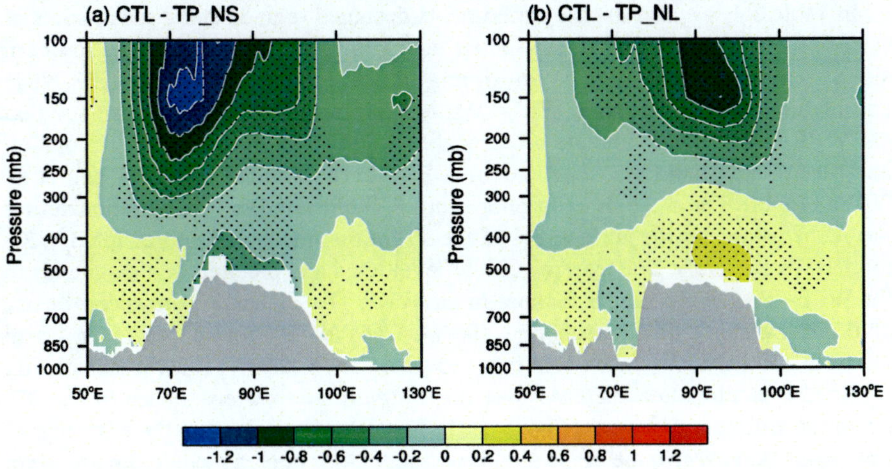

Fig. 5.7 Potential vorticity ($PVU = 10^{-6}\ \mathrm{K\,m^2\,s^{-1}\,kg^{-1}}$) difference induced by TP heating. **a** CTL minus TP_NS along 35° N due to surface sensible heating; **b** CTL minus TP_NL along 32.5° N due to latent heating. Stippling denotes the statistical significance of the difference above the 95% level. After Wu et al. (2016)

5.2.5 *Potential Regulation by Tropical Air-Sea Coupling*

In the above, we show the effects of two types of TP heating on summer monsoon based on the WRF regional model. In most of previous modeling studies (Boos and Kuang 2010; Wu et al. 2012; Duan et al. 2013; Wang et al. 2014b), where atmospheric general circulation models (AGCMs) were used, the SST was prescribed as the lower boundary condition. However, in a study of the effect of TP mountain uplift on the East Asian summer climate, Kitoh (2004) revealed that there exist obviously different responses between coupled general circulation model (CGCM) and AGCM experiments. Overall, the CGCM showed a larger sensitivity of climate response to mountain uplift than the AGCM, suggesting that SST changes in a coupled model be the cause of the difference. If we also apply a regional air–sea-coupled model to revisit the effects of TP heating on climate, will the response of atmospheric circulation to the TP forcing induce changes in SST over the adjacent oceans? The changed SST, especially that in the tropics, may in turn produce a feedback to the atmosphere and regulate the thermal forcing of the TP on the ASM. Consequently, there may be some evident differences between the WRF and WRF-OMLM models when we conduct the same TP thermal forcing sensitivity experiments. Therefore, similar experiments are designed next but use the coupled model of the WRF-OMLM (Wang and Duan 2012), and the potential regulation of air–sea interaction on TP's climatic effects will be discussed.

In Table 5.1, we also list the experiments designed to understand the impacts of surface SH over the entire TP (including the main body of the TP and its southern slope area) on the ASM, which are simulated using the WRF (Exp ALLTP_NS1) and its air–sea-coupled mode WRF-OMLM (Exp ALLTP_NS2). We can see that the WRF-OMLM can simulate a reasonable ASM (both the monsoon circulation and precipitation) and associated summer SST in the tropical oceans (Wang et al. 2018a). Figure 5.8a and b show the anomalous 850-hPa wind and precipitation during the ASM forced by the surface SH of the TP in the uncoupled and coupled WRF models, respectively. The features of low-level wind and precipitation anomalies in the WRF-OMLM are similar to those in the WRF. The TP heating enhances the dry cold northwesterly flow surrounding the west of the TP. Pronounced moist warm air from the tropical oceans converges toward the southern TP, with statistically significant precipitation increase over the southern and eastern slopes of the TP. From the difference between Fig. 5.8a, b, however, it seems that the response of monsoon circulation to the TP heating is weaker in the coupled model. The low-level wind anomalies become weaker after the air-sea coupling is considered (Fig. 5.8c). Specially, the difference of moist warm air flow converged toward the southern TP from the northern BoB is more evident. This weak moist warm airflow directly influences the water vapor transport to the TP, which induces negative precipitation response due to the heating over the TP. In addition, the well-known feature that the thermal forcing over the TP can enhance the warm high-pressure center in the upper troposphere (e.g., Duan and Wu 2005; Wang et al. 2016) is verified by the WRF and WRF-OMLM simulations (Fig. 5.9). Such upper-level anticyclonic anomaly

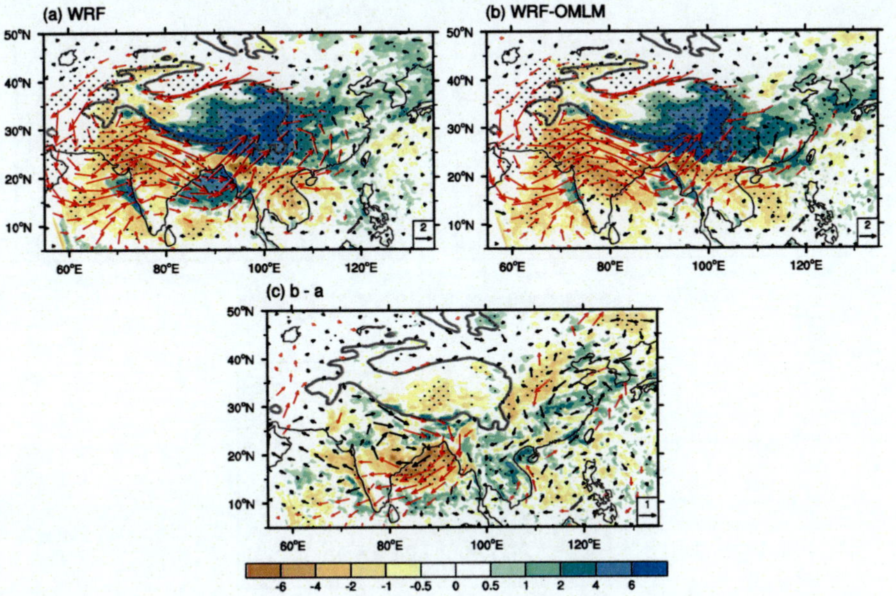

Fig. 5.8 **a** Differences in summer mean precipitation (shading; mm day^{-1}) and 850-hPa wind (vector; m s^{-1}) between CTL and ALLTP_NS1 using the WRF model. **b** Same as (**a**), except between CTL-O and ALLTP_NS2 using the WRF-OMLM model. **c** Same as **a**, except between (**b**) and (**a**). Stippling denotes the statistical significance of precipitation anomalies above the 90% confidence level. The wind differences significant above the 90% confidence level are plotted in red. The thick gray contour denotes the terrain of 2000 m. After Wang et al. (2018a)

induced by the TP heating also becomes weaker in the coupled run of the WRF-OMLM, characterized by an obvious cold cyclonic anomaly (Fig. 5.9c). The above analyses indicate that the air–sea coupling in the adjacent oceans of the TP can weaken the summer mean effects of the TP thermal forcing on climate.

How does the air–sea coupling regulate the TP heating effect? We know that in the uncoupled experiments, observational SST is prescribed in CTL and ALLTP_NS1. While the SST in the WRF-OMLM is simulated by the ocean model, and it can be affected by the variation of atmospheric conditions as well. Figure 5.10 shows the SST difference between the control run of CTL-O and the no TP SH experiment of ALLTP_NS2 using the WRF-OMLM. We can clearly see that obvious cold SST anomalies appear in most of the oceanic areas, especially in the BoB and the Arabian Sea. The primary reason for such cold SST anomalies is the intensified southwesterly monsoon circulation over the tropical regions induced by the TP thermal forcing (Fig. 5.8b). The strong surface wind stress takes more surface turbulent heat fluxes away from the ocean (not shown) and cools the sea surface mainly through the wind-evaporation process under the TP thermal forcing.

Importantly, the cold SST anomalies induced by the TP thermal forcing in the WRF-OMLM further suppress monsoon convective activity over the tropical oceans. The weakened convective upward motion leads to decreased precipitation over and

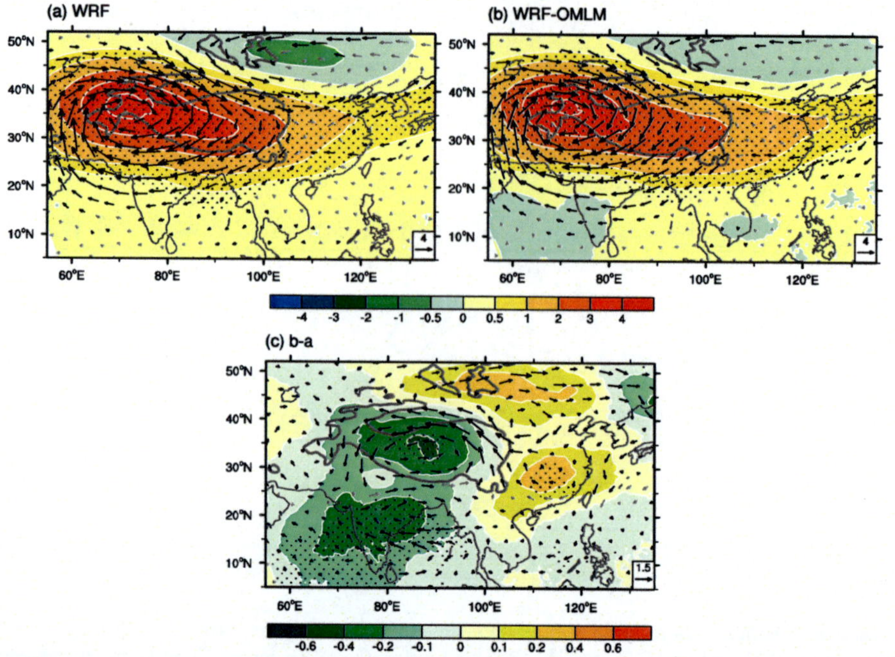

Fig. 5.9 Same as Fig. 5.8, except for the mass-weighted vertical-mean temperature (shading; K) and wind (vector; m s^{-1}) for 400–200 hPa. After Wang et al. (2018a)

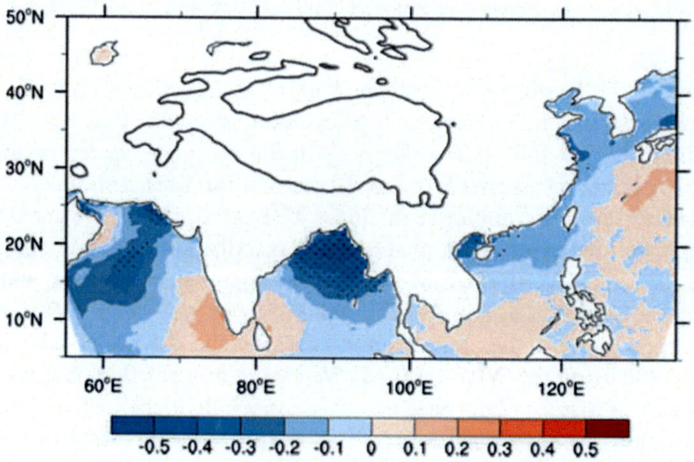

Fig. 5.10 Difference of summer mean SST (units: °C) between CTL and ALLTP_NS2 using the WRF-OMLM model. Stippling denotes the statistical significance of difference above the 90% confidence level. The thick gray contour denotes the terrain of 2000 m. After Wang et al. (2018a)

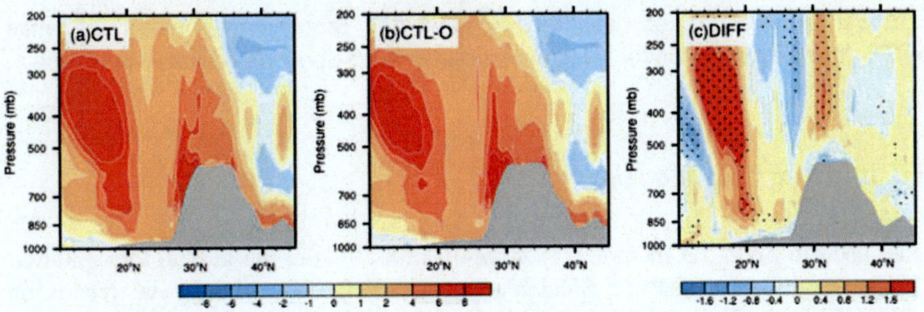

Fig. 5.11 Pressure–latitude cross sections of diabatic heating (K day^{-1}) in CTL using the WRF model (**a**), the CTL-O using the WRF-OMLM (**b**), and the difference between CTL and CTL-O (**c**). Stippling in (**c**) denotes the statistical significance of diabatic heating anomalies above the 90% confidence level. Modified from Wang et al. (2018a)

around the BoB (Fig. 5.8c), along with the weakened southerly wind over the southern slope of the TP and decreased precipitation over the TP's platform. The decreased rainfall over the TP can in turn influence the local heating source of the TP, because the atmospheric latent heating associated with rainfall dominates the total diabatic heating over the TP in summer. Compared with the WRF control experiment (CTL) in which a higher SST is prescribed persistently (Fig. 5.11a), not only the diabatic heating corresponding to the southern branch of the South Asian summer monsoon is weakened, but also the atmospheric heating source over the TP is reduced in the coupled run (CTL-O; Fig. 5.11b). The above analyses show that the influence of air–sea coupling over the adjacent oceans is significant on the TP heating effect, namely, the TP local heating source is significantly weakened (Fig. 5.11c). As a result, the weakened TP heating has a subdued impact on the ASM system under the framework of air–sea interaction. We suggest that air–sea coupling should be fully considered in sensitivity experiments for the effect of the TP heating on climate.

5.3 Key Effect of TP on Eurasian Teleconnection

The North Atlantic Oscillation (NAO) is known as a large-scale seesaw in atmospheric mass between the subpolar low and the subtropical high over and around the North Atlantic Ocean. The winter NAO has the strongest signal, while the summer NAO (SNAO) possesses a relatively smaller spatial extent, with a farther northward location, and its southern node is over northwestern Europe. Most studies have focused on the climate impact of wintertime NAO. Folland et al. (2009) reported the characteristics of SNAO and their effects on climate, and pointed out that the SNAO also has an important impact on the climate of northern Europe, West Africa, and even East Asia. Recently, it was reported that the SNAO has a close connection with the climate variation in China (e.g., Sun et al. 2008, 2012; Linderholm et al. 2011). The TP, which is located between Europe and East Asia, serves as a huge heating source in summer. So, our question here is whether there is any potential influence by TP thermal forcing on the teleconnection between SNAO and East Asian climate.

5.3.1 Data and Model

Data used in Sect. 5.3 include: (1) monthly rainfall from the Global Precipitation Climatology Centre (GPCC; Schneider et al. 2014) with a horizontal resolution of 1.0° × 1.0°; (2) monthly atmospheric data is employed from Japanese 55-year reanalysis (JRA-55; Kobayashi et al. 2015), which has a horizontal resolution of 1.25° × 1.25° and extends from 1000 to 1 hPa with 37 vertical pressure levels; (3) monthly SNAO index calculated following the definition of Folland et al. (2009), which is represented by the first principal component (PC1) time series corresponding to the leading Empirical orthogonal function (EOF) mode of mean sea level pressure over the extratropical North Atlantic–European sector (70° W–50° E, 25–70° N) using the JRA-55 data; (4) monthly SST from the Hadley Centre (HadISST; Rayner et al. 2003) for calculating the El Niño–Southern Oscillation (ENSO) index, which also has a horizontal resolution of 1.0° × 1.0°. In this study, we focus on boreal peak summer (July and August) for the period of 1958–2014. All analyses are performed using 8-year high-pass-filtered time series, to focus on the interannual time scale.

The AGCM used in this study is the Finite-volume atmospheric model (FAMIL) developed by the State Key Laboratory of Numerical Modeling for Atmospheric Sciences and Geophysical Fluid Dynamics, the Institute of Atmospheric Physics of the Chinese Academy of Sciences (CAS) (Zhou et al. 2015). This model has been successfully used in the studies of East Asian climate and ASM, and of the climatic effect of TP (e.g., Hu and Duan 2015). For the model's physical parameterizations, readers are referred to the introduction in Hu and Duan (2015). The horizontal resolution of C48 (1.875° × 1.875°) and 32 vertical levels with the top level at about 2.16 hPa are chosen here.

5.3.2 The Bridge Effect of TP on Eurasian Teleconnection

Figure 5.12 shows the regression patterns of summer precipitation and 850-hPa wind against the SNAO index. To compare the regression results to the dominant mode of the EASM more easily, the SNAO index is reversed (multiplied by −1) here before the calculation. Corresponding to the negative SNAO phase, apparent negative rainfall anomalies occur in Southeastern China near the coast, and positive anomalies appear from the Yangtze River basin to South Korea and Japan. Such a spatial pattern of rainfall anomalies over East Asia is similar to the leading mode of rainfall variation associated with the interannual variability of the EASM (Wang et al. 2008b). For the low-level circulation, an anomalous cyclone is formed north of East Asia, while an anticyclonic response appears south of East Asia, which enhances the southerly

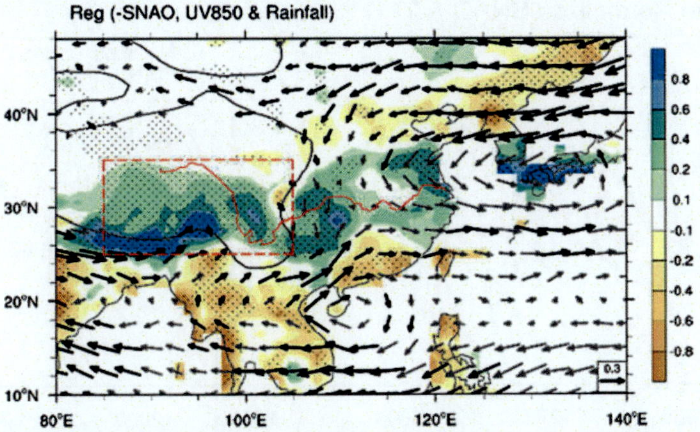

Fig. 5.12 Patterns of regression of summer rainfall (shading; mm day^{-1}) and 850-hPa wind (vector; m s^{-1}) upon the negative SNAO time series. The impacts of Niño 3.4 SST were excluded before these regression analyses were conducted. Stippling denotes the significance exceeding the 90% confidence level for rainfall, and black vector indicates the 90% confidence level for wind. The bold gray contour denotes the topographic height of 1500 m, and the red curve represents the Yangtze River over central China. The red rectangular box (85–105° E, 25–35° N) denotes the averaging area for the TP rainfall index. Modified from Wang et al. (2018b)

wind anomalies in Southeastern China. As a result, the cold dry airflow from the mid-to-high latitudes converges with the anomalous southerlies along the Yangtze River basin, leading to positive rainfall anomaly.

Another important feature is that the rainfall anomaly over the southeastern TP is significantly positive under the negative SNAO phase (Fig. 5.12). Here, we define a TP rainfall index as the area-averaged summer (July–August) rainfall over the southeastern TP (85–105° E, 25–35° N), where the topography is above 1500 m. On the interannual time scale, the correlation coefficient between SNAO and the TP rainfall index is −0.58, which becomes −0.57 after removing the effect of ENSO (Fig. 5.13a). Because the atmospheric latent heating associated with the rainfall over southeastern TP dominates the local total diabatic heating in summer (Ye and Gao 1979; Hu and Duan 2015), it suggests that the SNAO possibly regulates the heating source over the TP. On the other hand, previous studies have shown that TP heating can directly influence the summer rainfall variation in East China (e.g., Wang et al. 2014b). Figure 5.13b shows the regression patterns of 200-hPa stream function and wave-activity flux upon the negative SNAO time series. We can see that an evident wave train extending from the North Atlantic–European region to its downstream regions. Apparent divergence of wave-activity flux appears at the negative anomaly center of stream function. Such center corresponds to the south node of the SNAO, indicating a wave source region. The wave-activity flux diverges southeastward from the source region, and converges around the west of the TP. As a result, a positive stream function anomaly is formed in the upper troposphere over the southern TP (Fig. 5.13b). In other words, the SAH is intensified due to the influence of upstream

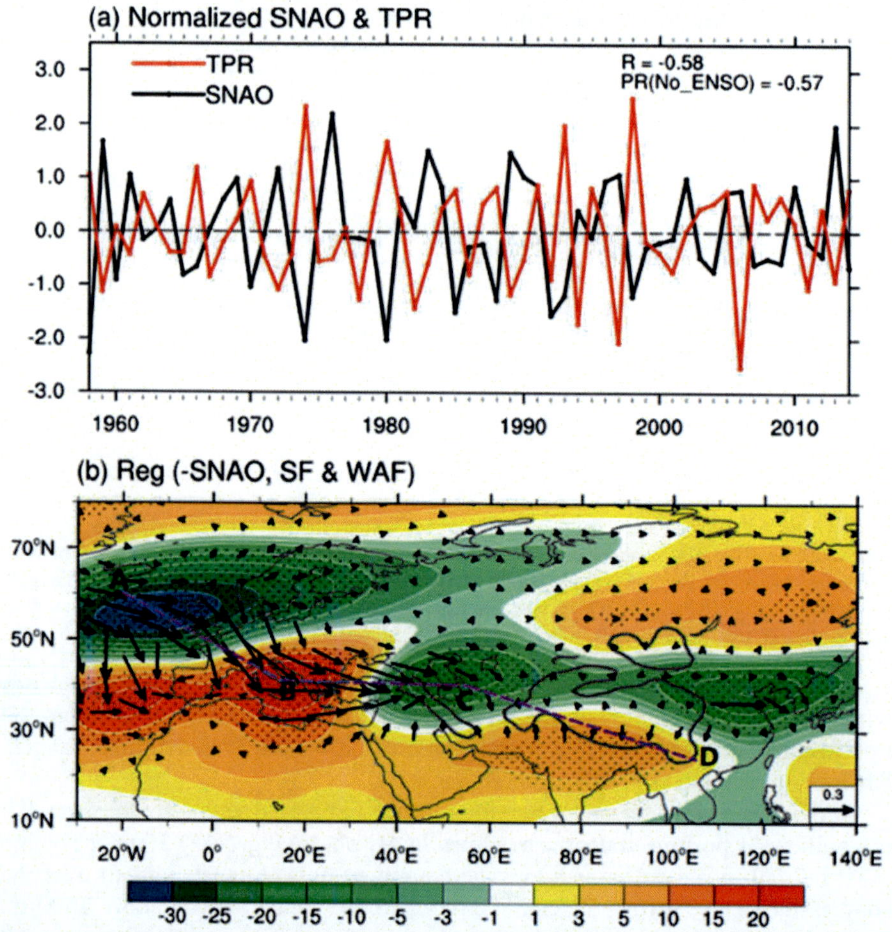

Fig. 5.13 **a** Time series of normalized SNAO (black) and rainfall over the southeastern TP (red) during the period of 1958–2014 using the filtered data to highlight the interannual variability, *R* is the correlation coefficient between the two curves, while *PR* is the partial correlation coefficient with the influence of Niño 3.4 SST excluded. **b** Patterns of regression of 200-hPa stream function (shading; 10^5 m^2 s^{-1}) and wave-activity flux (vectors are omitted for the zonal wind less than 1 m s^{-1}; m^2 s^{-2}) against the negative SNAO index. Stippling denotes the significance exceeding the 90% confidence level for stream function in (**b**). The bold gray contour in (**b**) denotes the topographic height of 1500 m. Modified from Wang et al. (2018b)

SNAO, which enhances the baroclinic vertical structure and the pumping effect of the TP. Such atmospheric circulation structure thus benefits the increase of rainfall over the TP. Overall, the SNAO can directly modulate the interannual variability of the TP summer rainfall through large-scale wave disturbance, and modulate the corresponding TP thermal forcing.

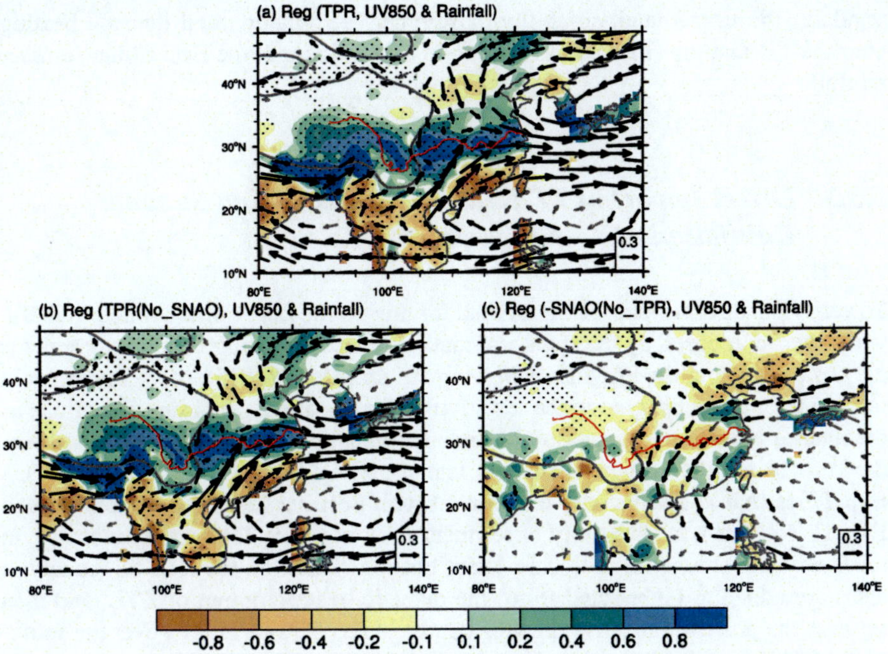

Fig. 5.14 **a** Patterns of regressions of summer rainfall (shading; mm day^{-1}) and 850-hPa wind (vectors; m s^{-1}) against the TP rainfall index. **b** Same as (**a**), except for the partial regression patterns with the impact of SNAO excluded. **c** Same as Fig. 5.1a, except for the partial regression patterns with the impact of TP excluded. The effects of Niño 3.4 SST were excluded before these regression analyses were conducted. Stippling denotes the statistical significance of rainfall exceeding the 90% confidence level, and black vector indicates the wind exceeding the 90% confidence level. The bold gray curve denotes the topographic height of 1500 m. After Wang et al. (2018b)

Furthermore, we show the regression patterns of summer precipitation and 850-hPa wind upon the TP rainfall index (which can represent the interannual variation of TP thermal forcing) in Fig. 5.14a. This regression result is similar to that in Fig. 5.12, strong TP summer heating corresponds to low-level wind convergence and positive rainfall anomaly over central East China. Thus, the question becomes whether or not the teleconnection between SNAO and East China rainfall is influenced by the TP heating. To answer this question, partial regression analysis is used here to better understand the relationships of the SNAO and TP thermal forcing with the variations of East China climate. When the impact of SNAO is removed, the regressions of rainfall and 850-hPa wind over East Asia upon the TP rainfall index (Fig. 5.14b) show little change from the original results (Fig. 5.14a), suggesting that the impact of the TP on East China summer rainfall is not strongly dependent on the effect of SNAO. However, when the influence of TP heating is excluded, the partial regressions of rainfall and low-level wind upon the SNAO index (Fig. 5.14c) are significantly different from the original results (Fig. 5.12). Therefore, the TP does exert an intermediate bridge effect on the Eurasian teleconnection, in which the SNAO primarily regulates the interannual variability of summer precipitation and diabatic heating over the TP, and the changed TP heating in turn influences the East China summer rainfall.

5.3.3 *Direct Impact of TP Heating on East China Summer Rainfall in AGCM Experiments*

To verify the direct impact of TP thermal forcing on the East China summer rainfall, which was diagnosed by the statistical methods discussed above, three experiments are designed using the FAMIL. Following Hu and Duan (2015), the summer latent heating anomaly employed in the sensitivity runs has almost the same magnitude as the natural interannual variability of total TP diabatic heating. Figure 5.15a shows the vertical profiles of condensational latent heating from the control run (CTL), Exp TPHs that is a sensitivity experiment with positive TP heating anomaly, and Exp TPHw that is a sensitivity experiment with negative TP heating anomaly. In this study, the anomalies with 2.5σ latent heating (based on the JRA-55 reanalysis data) are added to (subtracted from) the diabatic heating output of CTL, and then serve as the positive (negative) heating source in Exp TPHs (TPHw) over the southeastern TP. From the differences between Exp TPHs and Exp TPHw, we can see that positive heating over the TP can substantially enhance the SAH (Fig. 5.15b). To the northeast of the TP, a cyclonic anomaly exists from 850 to 200 hPa, suggesting a barotropic response in the troposphere. This cyclonic anomaly can be viewed as part of a Rossby wave train originated from the anticyclonic anomaly over the TP due to its thermal forcing (Wang et al. 2008a, 2014b). Meanwhile, the TP heating also induces an anomaly of southerly flow at 850 hPa over southern China. Subsequently, the northerly and southerly anomalies converge over East China, increasing the summer rainfall along the convergence belt (Fig. 5.15c). These responses of both atmospheric circulation and rainfall to the TP heating in the sensitivity experiments are similar to the regression patterns upon the TP rainfall index in observations (Fig. 5.14a, b), illustrating the direct impact of the TP thermal forcing on the East China summer rainfall.

Using the data diagnose and modeling studies, we have confirmed the intermediate bridge effect of TP thermal forcing on the Eurasian teleconnection. The overall physical processes can be clearly illustrated by a schematic diagram (Fig. 5.16). The SNAO primarily regulates the interannual variability of summer precipitation over the southeastern TP through the southeastward propagation of large-scale wave disturbance. In other words, a negative SNAO always increases TP rainfall, accompanied by enhanced local diabatic heating that subsequently excites a Rossby wave propagating northeastward in the upper troposphere, which then generates low-level northerly wind anomalies over northern China. Meanwhile, the TP heating also generates low-level southerly wind anomalies over southern China. These northerly and

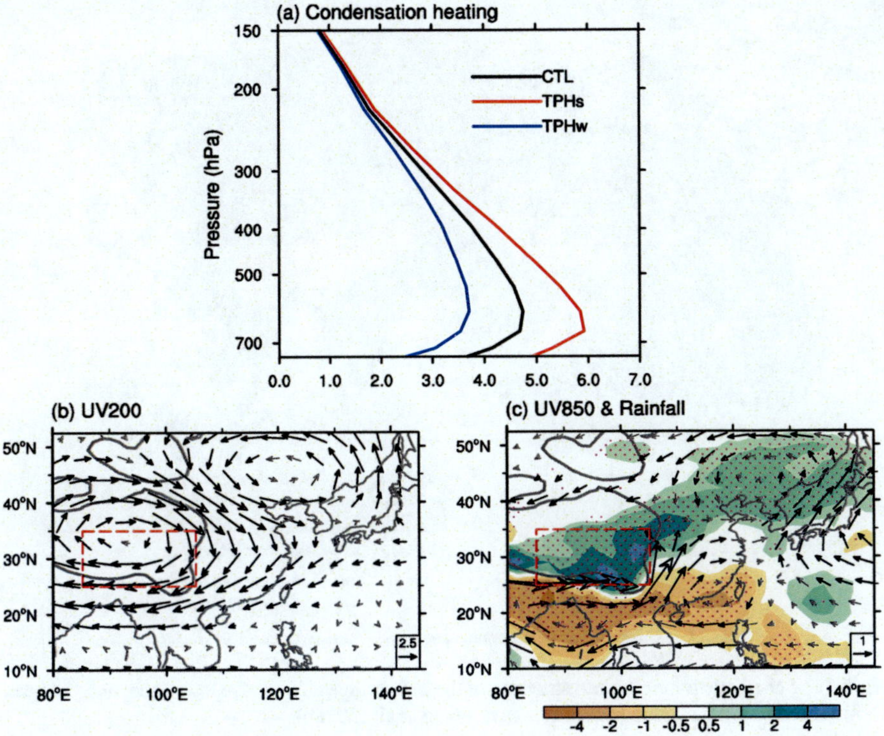

Fig. 5.15 **a** Vertical profiles of condensational latent heating (units: K day^{-1}) in different experiments over the southeastern TP (see the red rectangular box in (**b**) or (**c**)). **b** Difference in 200-hPa wind (units: m s^{-1}) between the positive (TPHs) and negative (TPHw) TP heating experiments. **c** Same as (**b**), except for 850-hPa wind and rainfall (shading; mm day^{-1}). Black vector in (**b**) and (**c**) indicates the wind exceeding the 90% confidence level, and stippling in (**c**) denotes the statistical significance of rainfall above the 90% confidence level. The bold gray curve in (**b**) and (**c**) denotes the topographic height of 1500 m. After Wang et al. (2018b)

southerly wind anomalies converge in the lower troposphere, increasing the summer rainfall across the whole central East China.

5.4 Thermal Effect of TP on Climate Variation Over the Upstream Regions

It is well known that the dynamic and thermal effects of the TP are important for regional and global climate (e.g., Hahn and Manabe 1975; Ye and Wu 1998; Kitoh 2004; Wu et al. 2014). In summer, the TP is viewed as a powerful heating source, and its role in general circulation cannot be ignored (Flohn 1957; Yeh et al. 1957). There are a great number of studies on the effects of the TP on climate in South,

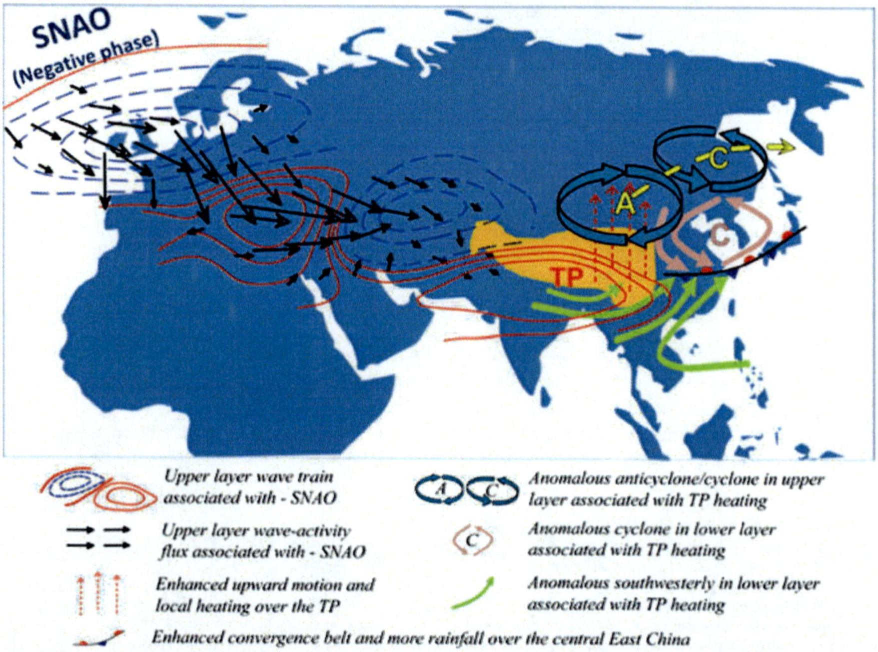

Fig. 5.16 Schematic for the overall structure of the TP's bridge effect in the teleconnection between SNAO and East China summer rainfall. After Wang et al. (2018b)

East, and Southeast Asia (Wu et al. 1997, 2007, 2012; Duan et al. 2008; Abe et al. 2013; Wang et al. 2014b).

Regarding the influence of Asian monsoon heating on the climate in its "upstream" region, studies have shown that the North African desert-Mediterranean climate is closely linked to the Asian monsoon heating via the zonal vertical circulation caused by the difference in thermal conditions between the Asian monsoon and the desert regions over Africa and by Rossby wave induced by monsoon heating (Yang et al. 1992; Rodwell and Hoskins 1996, 2001). Zhao et al. (2012) clarified that the source of atmospheric circulation and SST variation in the North Atlantic can be traced to the Asian continent. Over the past decades, the surface temperature over the TP has shown a significant warming trend, with a value of approximately 1.0 °C/50 years (Fig. 5.17). Wang et al. (2008a) concluded that the future warming of the TP could lead to an increase in summer rainfall over East Asia. Duan and Xiao (2015) pointed out that despite global warming hiatus, the TP has experienced an evident warming trend since the late 1990s.

However, relatively less work has been done on the effects of TP surface heating on the upstream climate over West Asia, South Europe, North Africa, and the North Atlantic (Wu et al. 2009). Therefore, we use a fully coupled climate model to investigate the effect of TP surface heating on the summer climate over the upstream regions and try to understand possible physical processes and mechanisms involved.

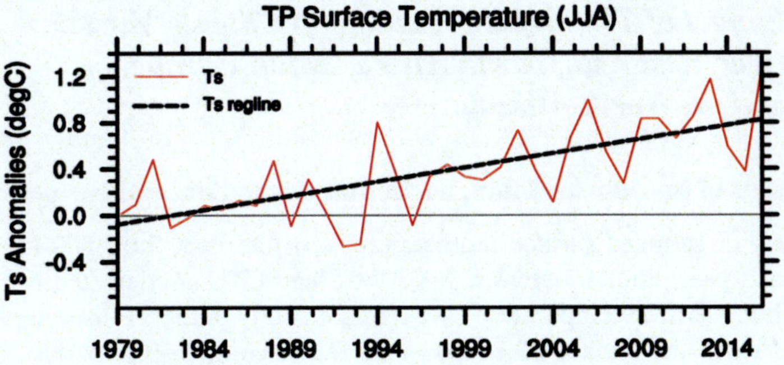

Fig. 5.17 Time series of summer surface temperature anomalies over the TP (red solid line; °C) and corresponding trend (black dashed line). After Lu et al. (2018)

In addition, we discuss the relative importance of TP heating compared to the Asian continent heating in Sect. 5.4.

5.4.1 Data, Model, and Experimental Design

Surface air temperature used here is from the Goddard Institute for Space Studies (GISS/NASA) surface temperature analysis (Hansen et al. 2010).

We conduct several numerical experiments to investigate TP's climate effects on the upstream climate and its relative importance to the Asian continent by using the Community Earth System Model Version 1.2.2 (CESM 1.2.2). This model is a fully coupled climate model with good performance (Neale et al. 2013). Our control run (CTL) is a free run, which is integrated for 300 years. Studies have shown that land processes have important effects on regional and even global climate variation (Yeh et al. 1984; Yang and Lau 1998; Wu Z. et al. 2012; Si and Ding 2013). The surface albedo effect, as an important component of land processes, can affect the surface temperature and modulate the ASM by changing thermal contrast between land and ocean (Barnett et al. 1989). Changing albedo to understand the surface thermal effect on atmospheric circulation has become a method used in numerous studies (Wang et al. 2008a; Boos and Kuang 2010; Liu et al. 2012). For example, when local albedo decreases, the land surface absorbs more shortwave solar radiation and its temperature will rise accordingly. Therefore, in our sensitivity experiment, we reduce the surface albedo over the TP. We restart a TP heating experiment (TP_Alb0.5) from the 251st year of CTL for a 50-year integration, in which external forcing is consistent with that used in CTL, but the surface albedo over the TP domain (62° – 105° E, 23° – 45° N; terrain height greater than 1500 m) is reduced by half. Years 261 – 300 of outputs are selected for analysis.

5.4.2 Impact of TP Surface Heating on Climate Variation Over West Asia, North Africa, South Europe, and the North Atlantic

1. Variations of sensible heat flux, horizontal circulation, and surface climate

Differences in summer surface temperature, sensible heat flux, 500-hPa vertical velocity, and precipitation between TP_Alb0.5 and CTL are depicted in Fig. 5.18. In TP_Alb0.5, surface temperature over the main body of the TP increases by more than 2 °C when the surface albedo over the TP decreases (Fig. 5.18a). Temperatures increase over the regions to the west of the TP, especially over South Europe, the Middle East, and the extratropical North Atlantic, but decrease over Sahara and the subtropical North Atlantic. The apparent change in sensible heat flux mainly occurs over the Eurasian continent (Fig. 5.18b), which is generally consistent with the change in surface temperature.

As the TP surface temperature increases, the local sensible heating increases accordingly, thereby strengthening the local ascending motion. Due to the important role of TP thermal forcing in summer precipitation over East Asia (Wang et al. 2008a; Wu et al. 2012), there exists a significant increase in rising motion and rainfall over East Asia (Fig. 5.18c and 5.18d). However, it is interesting to note that subsidence motion enhances and precipitation decreases in the regions to the west of the plateau, even to the extratropical North Atlantic. According to Rodwell and Hoskins (1996, 2001), the heating in the remote monsoon region can induce Rossby wave response in the regions to the west. Similarly, we find that the variations of vertical motion and precipitation over the regions to the west of the TP could be the result of the atmospheric response to the intensified diabatic heating over the TP. In addition, the

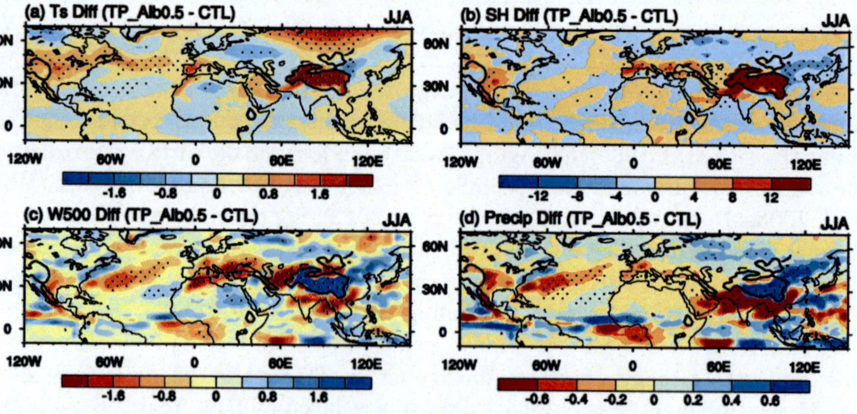

Fig. 5.18 Differences in **a** surface temperature (units: °C), **b** sensible heat flux (units: W m^{-2}), **c** 500 hPa vertical velocity (multiplied by 200; units: Pa s^{-1}) and **d** precipitation (units: mm day^{-1}) between TP_Alb0.5 and CTL. Stippling indicates the value is significantly above the 95% confidence level. The thick black contour indicates the terrain is greater than 1500 m. After Lu et al. (2018)

rising motion and precipitation over the Sahel increase significantly, while those over northern India decrease.

Figure 5.19 shows the climatological distributions of atmospheric circulation and tropospheric temperature during summer and the corresponding differences between TP_Alb0.5 and CTL. In climatology, the anticyclonic circulation and high pressure in the lower atmosphere dominate over the subtropical North Atlantic, centered around (45° W, 30° N) (Fig. 5.19a). Strong low pressure and cyclonic circulation form over the Asian continent, southwesterly and southerly winds prevail in south and east of the cyclonic circulation, respectively. In the mid-troposphere, the giant anticyclonic circulation over the mid-latitude North America–North Atlantic–North Africa can be seen, and a relatively low pressure with cyclonic circulation still exists over the Asian continent (Fig. 5.19c). The most obvious high-level atmospheric circulation system is the SAH, with its center near northern India (Fig. 5.19e). We can see that a tropospheric warm center corresponds well to the SAH.

When the TP surface temperature rises, the increase in sensible heating causes an abnormal rising motion (Fig. 5.18c), which strengthens the local low-level cyclonic

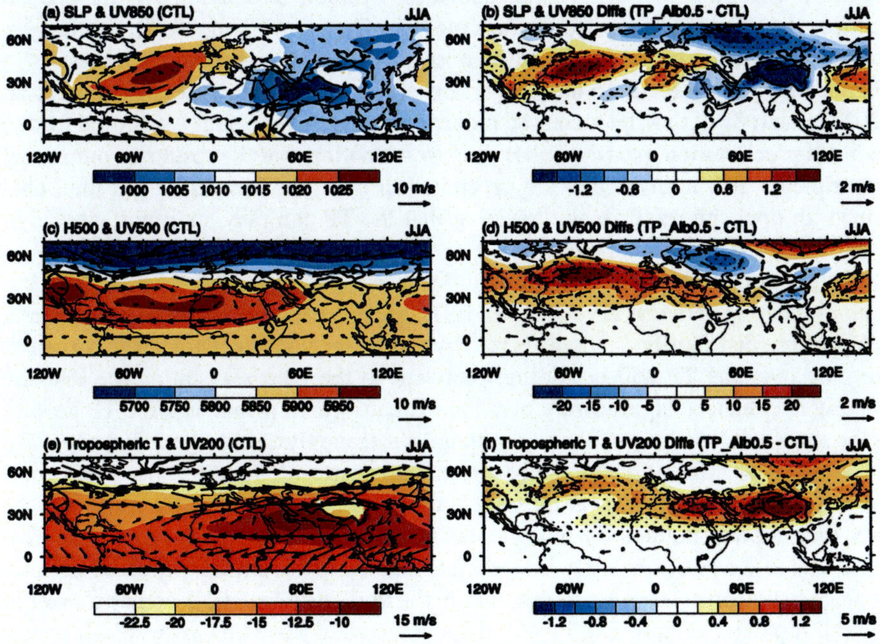

Fig. 5.19 Differences in **a** sea level pressure (shading; units: hPa) and 850-hPa wind (vector; units: m s^{-1}), **c** 500-hPa geopotential height (shading; units: m) and wind (vector; units: m s^{-1}), **e** tropospheric temperature (shading; units: °C) and 200-hPa wind (vector; units: m s^{-1}) in CTL. **b, d,** and **f** are the same as (**a**), (**c**), and (**e**), except for differences between TP_Alb0.5 and CTL. Stippling indicates the value exceeds the 95% confidence level. In **b, d,** and **f**, only the wind differences above the 95% confidence level are plotted. The thick black contour indicates the terrain is greater than 1500 m. After Lu et al. (2018)

circulation and high-level SAH, and the high extends westward (Fig. 5.19b, d, f), resulting in enhancement of both low-level convergence and high-level divergence. These characteristics are consistent with those of air pumping effect of the TP proposed by Wu et al. (1997, 2007). Abnormal rising motion induces an increase in condensation heating, which in turn strengthens large-scale convergent airflow, producing a positive feedback between large-scale convergence and small-scale convective activities. In the regions to the west of the TP, significant anticyclonic anomalies form over the North Atlantic–Mediterranean Sea regions in the entire troposphere, indicating a quasi-barotropic structure. Tropospheric temperature warms up accordingly. These tropospheric anticyclonic anomalies are likely to be the Rossby wave response to the strengthened TP heating. Correspondingly, the Atlantic subtropical high (ASH) strengthens and the center shifts northwestward, located near (50° W, 35° N). The high-level SAH moves westward. However, a weak cyclonic anomaly at 850-hPa forms over the eastern subtropical Atlantic (Fig. 5.19b), indicating that the southeast portion of the ASH weakens. Westerly wind anomalies on the south side of the cyclonic circulation can transport more water vapor from the tropical Atlantic to the Sahel, causing increased precipitation over the Sahel. The westerly wind anomaly on the south side of the cyclonic circulation over the plateau reduces the water vapor transport from the BoB (Jiang and Ting 2017), resulting in a significant reduction in precipitation over northern India (similar to the results in Fig. 4f of Wang et al. 2016 and in Fig. 3b of Wu et al. 2016). Recently, He et al. (2017) pointed out that the increased monsoon heating over South Asia could lead to reduced precipitation over the Sahel. It is worth noting that the results of our model experiments also indicate that the precipitation over northern India and the Sahel shows an opposite relationship, meaning that the TP can play an important role in connecting Asian–African climate.

Height–longitude cross section of differences in tropospheric temperature and diabatic heating between TP_Alb0.5 and CTL are given in Fig. 5.20. The tropospheric temperature distribution is characterized by a uniform warming with the warm center forming over the TP and extending westward to the North Atlantic. The diabatic heating is enhanced significantly over the plateau, which means that the TP heating is the source of Rossby wave response to the west, and the atmospheric Rossby wave response causes the high temperature to extend westward. To verify the presence of the Rossby wave response over the west of the TP, we also calculate the wave-activity flux (not shown) defined by Takaya and Nakamura (2001). In the climatological state, due to the position of the westerly jet stream in summer, the wave-activity flux appears to uniformly propagate eastward in the mid-latitudes and becomes weaker in the subtropical region. In Exp TP_Alb0.5, the eastward wave flux in the mid-latitudes weakens. It can be said that the TP heating results in the weakening of Rossby wave that originally propagates eastward, forming the Rossby wave response in the mid-latitude Eurasia–North Atlantic regions, which is manifested as the quasi-barotropic structure with anticyclonic anomalies in the entire troposphere, leading to the change in the upstream climate.

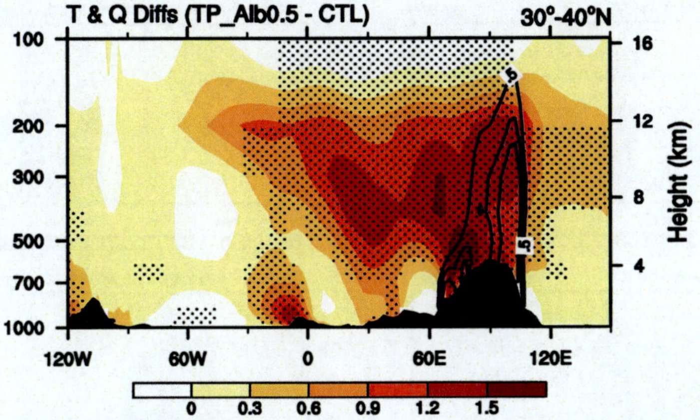

Fig. 5.20 Height–longitude cross section of differences averaged over 30°–40° N of tropospheric temperature (shading; units: °C) and diabatic heating (contour; units: K s^{-1}) between TP_Alb0.5 and CTL. Stippling indicates the value is above the 95% confidence level. The contours are −1.5, −1.0, −0.5, 0.5, 1.0, and 1.5. Black shading indicates terrain. After Lu et al. (2018)

2. Variation of vertical circulation

Since the TP is located in the subtropics, a transition zone between tropics and mid-to-high latitudes. In the tropics, the divergence component of the atmospheric circulation is the most important, while the rotational component dominates in the mid-high latitudes. Therefore, when studying climatic effects of the TP, we must consider not only the Rossby wave characteristics of atmospheric responses to the plateau, but also the vertical circulation features driven by the TP heating, which are discussed next.

Figure 5.21 shows the climatological distributions of zonal and meridional circulations in CTL and their changes in Exp TP_Alb0.5. Climatologically, there is a strong ascending motion in the Asian monsoon region, while subsidence occurs in the regions to the west, as far as the North Atlantic (Fig. 5.21a). As the TP diabatic heating increases, both the ascending motion over the Asian monsoon region and subsidence to the west are strengthened, forming an anomalous zonal vertical cell over the mid-latitude Atlantic–European–African sector (Fig. 5.21b). In other words, the anomalous zonal circulation excited by the TP heating induces the deepening of the subsidence from the west of the TP to the North Atlantic, resulting in the formation and development of an anomalous warm center between the anomalous rising motion over the east and the subsidence over the west. The above characteristics are basically consistent with the temperature-heating vertical gradient (T-Q_z) theory described by Wu et al. (2015). The theory was derived from the thermal wind theory and Sverdrup vorticity balance:

$$\frac{\partial v}{\partial \ln p} = -\frac{R}{f}\left(\frac{\partial T}{\partial x}\right) \tag{5.2}$$

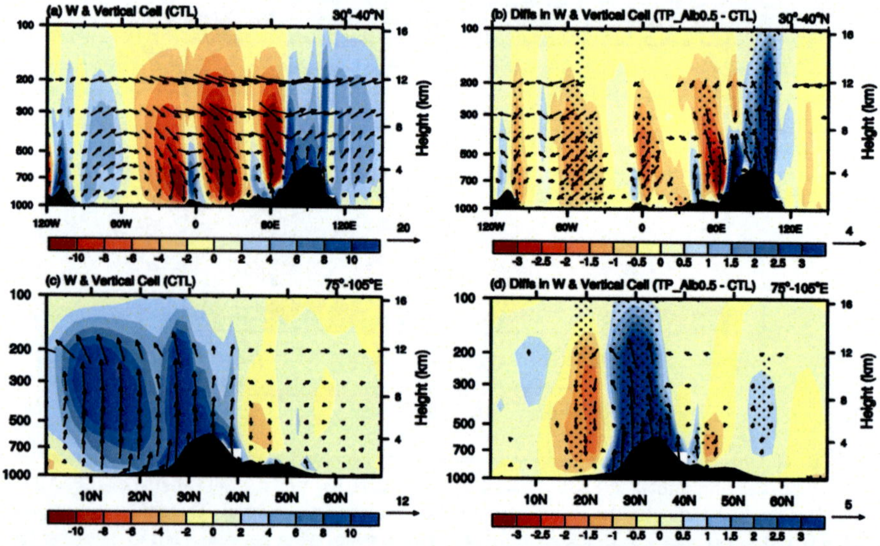

Fig. 5.21 **a** Vertical velocity (shading; units: Pa s^{-1}) and zonal circulation (vector; zonal wind in m s^{-1}; vertical velocity is in Pa s^{-1} and is multiplied by 200) averaged over 30°–40° N in CTL, and **b** differences between TP_Alb0.5 and CTL. **c** Vertical velocity (shading; units: Pa s^{-1}) and meridional circulation (vector; meridional winds in m s^{-1}; vertical velocity is in Pa s^{-1} and is multiplied by 200) averaged over 75°–105° E in CTL, and **d** differences between TP_Alb0.5 and CTL. Stippling indicates it exceeds the 95% confidence level. In **b** and **d**, only the values of the vertical cell above the 95% confidence level are plotted. Black shading indicates terrain. After Lu et al. (2018)

$$\beta v \approx (f + \zeta)\theta_z^{-1}\left(\frac{\partial Q}{\partial z}\right), \ \theta_z \neq 0, \ \vec{V} \cdot \nabla\zeta \to 0. \tag{5.3}$$

A causal relationship has been obtained between temperature and heating vertical gradient, which is applicable to upper troposphere in the subtropics (see Wu et al. 2015 for details): In the upper troposphere, deep convection heating is dominant, i.e., the heating vertical gradient is negative, while the cooling is mainly radiation cooling, i.e., the heating vertical gradient is positive. From Eq. (5.3), there exists northerly vertical shear in the heating region and southerly vertical shear in the cooling region. Then, we can see from Eq. (5.2) that the warm (cold) center must appear over the west (east) side of the heating zone and the east (west) side of the cooling zone. By applying the above theories to our results, anomalous diabatic cooling occurs over the west due to the strengthened TP diabatic heating, leading to an anomalous warm center and an anticyclonic circulation between abnormal heating and cooling regions. In addition, Fig. 5.21b shows that the vertical velocity distribution between 60° W and 120° E is significantly enhanced compared to the climatological one, implying that in the regions to the west the Rossby wave response to the diabatic heating demonstrated by Rodwell and Hoskins (1996) intensifies.

Therefore, Fig. 5.21b reflects the integrated results of anomalous zonal circulation and atmospheric Rossby wave response excited by TP heating, and it also reflects their mutual effects on the upstream climate. Figure 5.21c shows strong ascending motion over the Asian monsoon region and the existence of two rising centers in climatology. In Exp TP_Alb0.5, an anomalous meridional vertical cell occurs with an ascending branch over the plateau and a subsidence branch over northern India, implying an increase in precipitation over the TP and a decrease in precipitation over northern India.

From the precipitation variation shown earlier (Fig. 5.18d), we can see that a precipitation dipole appears over the eastern tropical Atlantic and West Africa, indicating that the Atlantic Intertropical Convergence Zone (ITCZ) shifts northward due to the northward shift of the ASH. The Sahel rainfall increases due to the weakening of trade winds and local evaporation caused by the northward shift of the ASH. So, what happens to the local Hadley circulation if the Atlantic ITCZ moves northward? The climatology of local Hadley circulation and associated differences between TP_Alb0.5 and CTL averaged over 10° W–30° E are given in Fig. 5.22. In the climatological state, the rising center of the Hadley cell is located near 10° N and the sinking center is near 35° N (Fig. 5.22a). Obviously, the anomalous rising and sinking branches of the local Hadley circulation in Exp TP_Alb0.5 are shifted approximately 5° northward (Fig. 5.22b). Therefore, when the ITCZ moves northward, the local Hadley cell also moves northward, weakening the subsidence over the desert regions in North Africa and enhancing the sinking motion in the Mediterranean region. This anomalous meridional cell also contributes to the drier condition in the Mediterranean region.

3. Possible physical processes and mechanisms

In short, the physical processes and mechanisms for the impact of TP surface heating on the upstream climate can be summarized as follows. In the TP region, the surface temperature increases with the decrease of surface albedo, enhancing both surface sensible heating and ascending motion. The low-level convergence intensifies, and the SAH strengthens and extends westward. At the same time, the release of latent heat increases, which further exacerbates the low-level convergence and ascending motion. In the regions to the west of the TP, Rossby wave response appears due to the strengthened TP heating. As a result, the anomalous anticyclonic circulation and the warm tropospheric temperature control the entire troposphere, and the relationship between anomalous warm center and anomalous zonal circulation fits the temperature-heating vertical gradient theory. In the lower atmosphere, the ASH is reinforced with eastward extension, and the center moves northwestward, strengthening the subsidence over the extratropical North Atlantic. Meanwhile, the Atlantic ITCZ and the local Hadley cell also move northward, resulting in increased ascending motion and precipitation over the Sahel, weakening of the subsidence over the desert regions of North Africa, and causing an increase in the subsidence and decrease in rainfall over the Mediterranean Sea region. The ASH moves northward, accompanying the weakening of its southeastern portion. Westerly anomalies appear over the south side of the corresponding cyclonic anomalies, which increases the water vapor

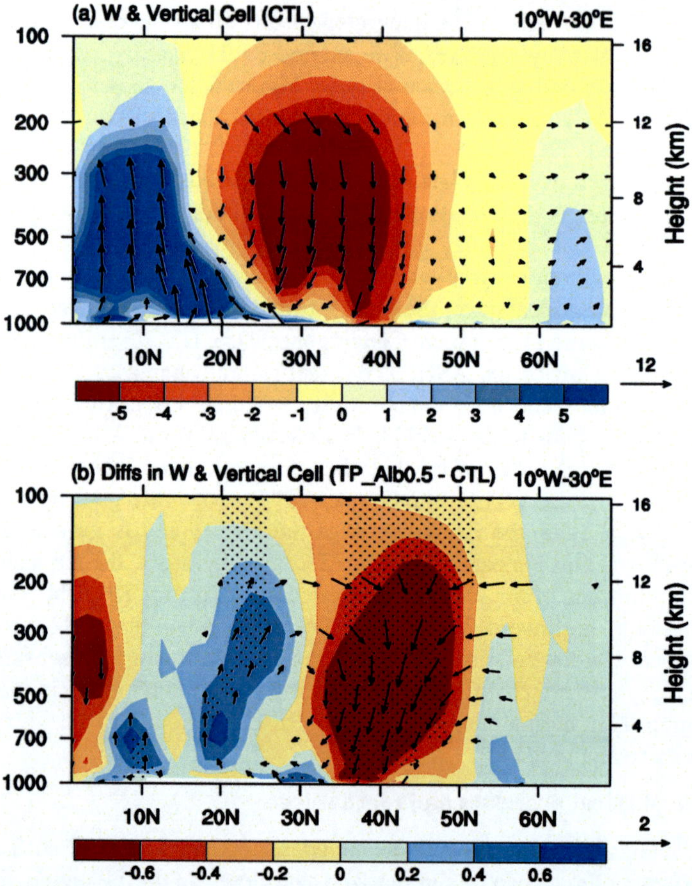

Fig. 5.22 **a** Vertical velocity (shading; units: Pa s^{-1}) and meridional circulation (vector; meridional wind in m s^{-1}; vertical velocity is in Pa s^{-1} and is multiplied by 200) averaged over 10° W–30° E in CTL, and **b** differences between TP_Alb0.5 and CTL. Stippling in **b** indicates it is above the 95% confidence level. Only the values of the vertical cell exceeding the 95% confidence level are plotted. After Lu et al. (2018)

transport from the tropical Atlantic to the Sahel and then intensifies the Sahel rainfall. In the regions to the south, especially over northern India, anomalous meridional circulation caused by plateau heating leads to decreases in rising motion and precipitation over northern India. Furthermore, the westerlies to the south of the anomalous cyclonic circulation over the TP cuts off the water vapor transport from the BoB to northern India, which also plays a role in the reduction of rainfall over northern India. In addition, in Exp TP_Alb0.5, the reversed change between the Sahel and Indian precipitation (also found in He et al. 2017) reflects the bridge role of the TP in connecting Asian–African climate to some extent.

5.4.3 Conclusions and Discussion

We use a fully coupled model to investigate the effects of TP surface heating on the upstream climate, especially over West Asia, South Europe, North Africa, and the North Atlantic. We also examine the physical processes and possible mechanisms by conducting numerical experiments. Since the TP is located in the subtropics, its thermodynamic effects are not only on the rotational component of the atmospheric circulation (i.e., Rossby wave response to anomalous diabatic heating) but also on the divergent component (i.e., thermal-driven vertical circulation). First, the apparent Rossby wave response occurs in the regions to the west of the TP, as far as the North Atlantic, causing anticyclonic circulation anomalies and subsidence motion. Correspondingly, the tropospheric temperature is warmer. In addition, the ASH is enhanced and the center shifts northwestward, resulting in the northward shifts of the Atlantic ITCZ and the local Hadley cell. Thus, the Sahel rainfall increases. The westerly anomalies to the south side of the cyclonic anomalies over the subtropical North Atlantic play an important role in the increase of Sahel precipitation. Moreover, the anomalous vertical circulations driven by the TP heating is also significant. When the TP diabatic heating is enhanced, anomalous zonal circulation appears with ascending motion over the plateau and descending motion over the Mediterranean Sea region, inducing intensified subsidence and reduced precipitation west of the plateau. Therefore, the impact of Rossby wave response to the TP heating is almost in phase with that of the zonal circulation. In the regions to the south, anomalous meridional circulation results in anomalous descending motion and decreased rainfall over northern India. It is worth noting that the reversed change in precipitation over the Sahel and India reflects the role of the TP in linking Asian-African climate.

We also conducted an Asian heating experiment (not shown; the results are similar to those in Zhao et al. 2012), which is the same as Exp TP_Alb0.5 except the surface albedo in the Asian domain (60°–120° E, 25°–50° N) was reduced by half. The purposes were to use it as a reference for Exp TP_Alb0.5 and to analyze the role of TP heating in the effect of the Asian continent heating on the upstream climate. We found that the distributions of surface temperature, precipitation, and atmospheric circulation in the Asian heating experiment were generally similar to those in Exp TP_Alb0.5, but the amplitude of the upstream climate variation in Exp TP_Alb0.5 is relatively smaller, accounting for about 40–50% of that in the Asian heating experiment. However, the TP domain is only one-third of the Asian domain in our experiments, indicating that the TP heating can largely explain the impact of the Asian continent heating on the upstream climate. Although some of the previous studies were conducted to understand the important role of Asian monsoon heating in the formation of African deserts (Rodwell and Hoskins 1996) and the effect of Asian summer tropospheric temperature on SST and atmospheric circulation over the North Atlantic (Zhao et al. 2012). However, here we focused on analyzing the effect of the plateau, describing how the atmospheric circulation and surface climate in the upstream regions respond to the TP-intensified heating and comparing the changes with those caused by the Asian continent heating.

5.5 Climate Effect of SAH Variability Over TP

The SAH is the most intensive and stable subtropical anticyclone in the upper troposphere and lower stratosphere during boreal summer (Mason and Anderson 1963; Tao and Zhu 1964). Being influenced by both sensible and latent heating over the TP and by the latent heating over the Asian monsoon region (Krishnamurti et al. 1973; Huang 1985; Liu et al. 1999, 2001), the SAH persists over the plateaus of southern Asia and spans from North Africa, Eurasian continent to the Western North Pacific. As one of the key members in the Asian monsoon systems (Krishnamurti and Bhalme 1976), the SAH plays an important role in affecting the Indian summer monsoon (ISM; Raman and Rao 1981; Krishnamurti et al. 1989; Bansod et al. 2003; Ashfaq et al. 2009) and the summer rainfall over China (Luo et al. 1982; Zhang and Wu 2001; Zhang et al. 2002). The SAH can be an important factor for predicting summer weathers and climate in China.

5.5.1 Zonal Shift of SAH and Its Relationship with Asian Summer Monsoon Rainfall

The zonal shift is a notable feature of the SAH (Mason and Anderson 1963; Tao and Zhu 1964; Zhang et al. 2002). Previous studies suggested that the summer rainfall over China can be affected by the zonal shift of the SAH (Luo et al. 1982; Chen and Liao 1990; Zhang and Wu 2001; Huang and Qian 2003). Besides, the active phase and break of the ISM are also known to be closely related to the variation of the SAH (Mason and Anderson 1963; Krishnamurti and Bhalme 1976; Bansod et al. 2003; Ashfaq et al. 2009). As the common upper-level system in the East Asian and Indian summer monsoon systems, the SAH interacts with the monsoon rainfall over these two regions. What is the relationship between the zonal shift of the SAH and the monsoon rainfall in the ISM and EASM regions? What is the physical mechanism for the interactions among the SAH, the ISM rainfall, and the EASM rainfall? In Sect. 5.5, we will discuss the zonal shift of the SAH on the interannual time scale and its role in connecting the ISM and EASM.

The zonal shift of the SAH is significant on the synoptic time scale (Tao and Zhu 1964; Krishnamurti et al. 1973; Luo et al. 1982) and on the subseasonal time scale (Zhang et al. 2002; Liu et al. 2007; Yang and Li 2016). On the interannual time scale, the zonal displacement of the SAH is also obvious in each summer with its center over the TP (TP mode; about 75%) or over the Iranian Plateau (IP mode; about 25%) (Wei et al. 2014). A SAH index I_{EW} is defined as the difference of geopotential height between the east center region (85°–105° E, 22.5°–32.5° N) and the west center region (55°–75° E, 22.5°–32.5° N) to quantify the zonal displacement of the SAH. Composites of the 12520-gpm isolines based on the I_{EW} show the locations of the SAH in both TP mode and IP mode. In the TP mode, the SAH is located eastward with its center over the TP and the eastern ridge extends to about 115°

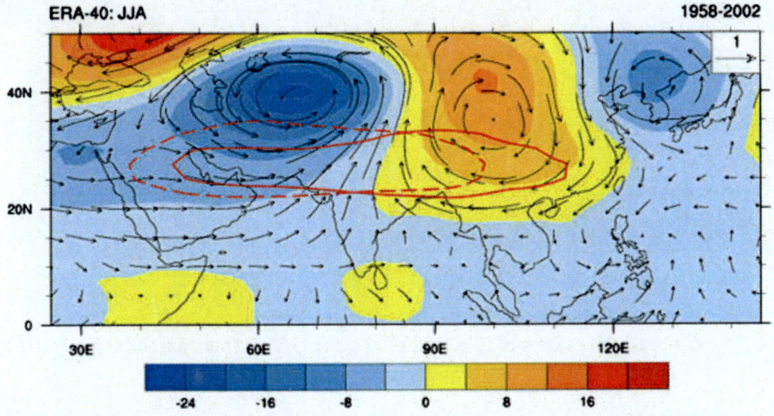

Fig. 5.23 Regressed horizontal wind (vector; units: m s^{-1}) and geopotential height anomalies (shading; units: gpm) at 200 hPa against the I_{EW}. Composites of the SAH for $I_{EW} > 1$ (red solid contour) and $I_{EW} < -1$ (red dashed contour) are indicated by the 12520-gpm contour

E (red solid contour in Fig. 5.23). In the IP mode, the SAH is located westward with its center over the IP and the western ridge extends to about 35° E (red dashed contour in Fig. 5.23). Regressed geopotential height and horizontal wind anomalies at 200 hPa against I_{EW} show an anomalous anticyclone and positive geopotential height anomalies are located over the eastern Eurasian continent with the centers in the mid-latitude, and an anomalous cyclone and negative geopotential height anomalies are located to the west (Fig. 5.23). These circulation anomalies may have an impact on the summer rainfall over China. In the following sections, we will discuss the relationship between the zonal shift of the SAH and summer monsoon rainfall. What is the physical mechanism for the zonal shift of the SAH on the interannual time scale? How does the zonal shift of the SAH influence the summer rainfall over China?

5.5.2 Relationship with the ISM and EASM Rainfall

The ISM is the most prominent tropical monsoon in the world, with abundant monsoon rainfall over the Indian subcontinent. All Indian rainfall index (AIRI) can well depict the intensity of the ISM and the convective activity over the Indian subcontinent (Parthasarathy et al. 1992; Sontakke et al. 1993; Wang and Fan 1999; Zhang et al. 1999). Time series of AIRI and I_{EW} show an opposite variation of the ISM and the SAH (Fig. 5.24). The linear correlation coefficient between the detrended AIRI and I_{EW} is −0.57, exceeding the 0.01 significance level, suggesting a significant negative relationship between the ISM and the zonal shift of the SAH. When the SAH is located eastward in the TP mode, the ISM is weak. When the SAH is located

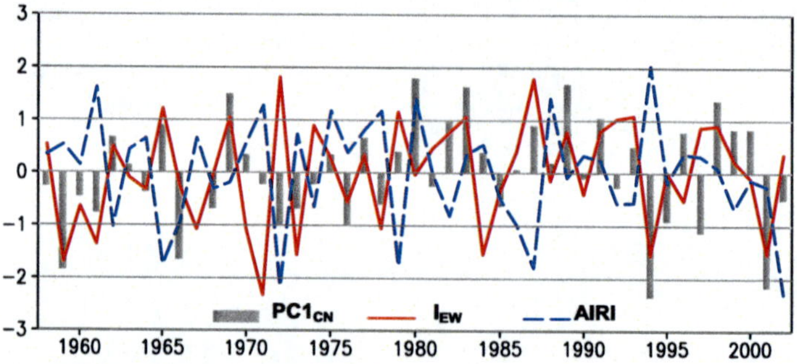

Fig. 5.24 Standardized time series of I_{EW} (red solid curve), AIRI (blue dashed curve), and PC1 of the rainfall over China (gray bar) from 1958 to 2002

westward in the IP mode, the ISM is strong. Regressed rainfall anomalies show negative rainfall anomalies are significant over the northern Indian subcontinent in the TP mode of the SAH (Fig. 5.25). It indicates that the strong (weak) ISM with intense (week) convection and more (less) rainfall over the northern Indian subcontinent is closely related to the westward (eastward) displacement of the SAH.

In East Asia, the relationship between monsoon rainfall and the zonal shift of the SAH is significant. Regressed rainfall anomalies show positive rainfall anomalies over the Yangtze River valley (YRV), South Korea, and Japan, while negative rainfall anomalies over northern and southern China (Fig. 5.25). The tripole rainfall pattern is obvious from the north to the south of eastern China, which is similar to the first spatial pattern (negative–positive–negative) of the EOF on the summer rainfall over

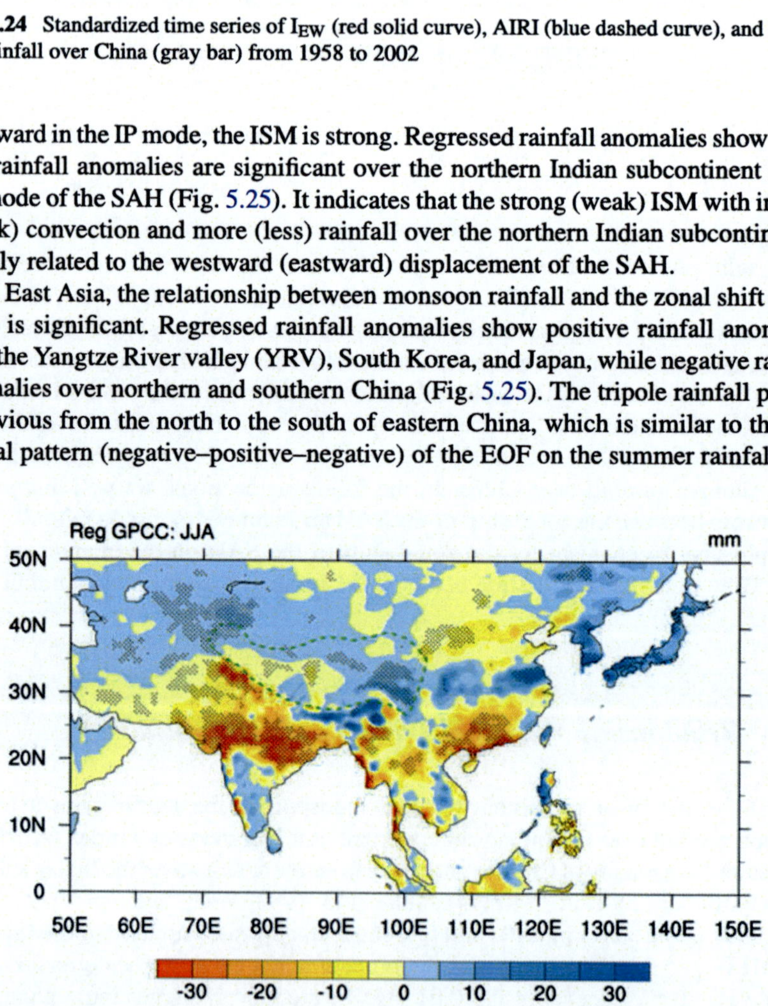

Fig. 5.25 Regressed JJA mean rainfall anomalies against I_{EW} (shading; units: mm). Stippling highlights the areas exceeding the 0.05 significance level. The green contour indicates the TP region with topography exceeding 3000 m

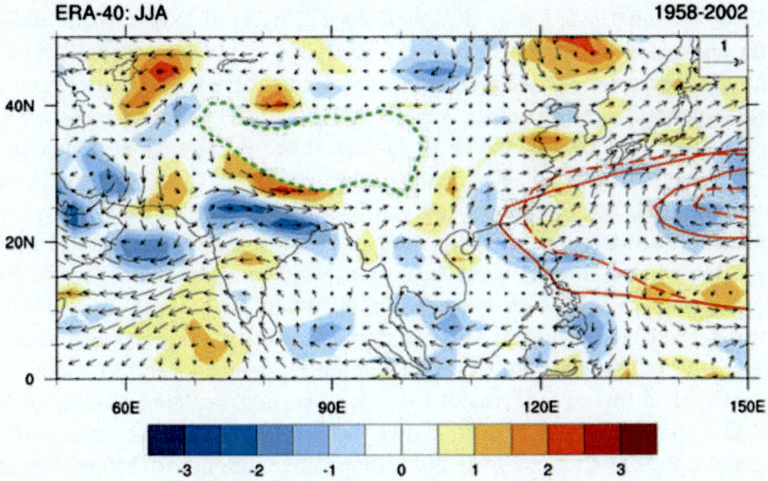

Fig. 5.26 Regressed vorticity (shading; units: 10^6 s^{-1}) and horizontal wind (vector; units: m s^{-1}) at 850 hPa against I_{EW}. Red contours indicate composite western North Pacific subtropical high (isolines 5860 and 5880 gpm) at 500 hPa for $I_{EW} > 1$ (solid contour) and $I_{EW} < -1$ (dashed contour). The green contour indicates the TP region with topography exceeding 3000 m

China (Zuo et al. 2011; Wei et al. 2015). Time series of I_{EW} and the first principal component of summer rainfall over China ($PC1_{CN}$) show a positive relationship between the zonal displacement of the SAH and the variation of the tripole rainfall pattern. The correlation coefficient between the I_{EW} and the $PC1_{CN}$ is 0.35,exceeding the 0.05 significance level. Obviously, the zonal status of the SAH is closely related to the EASM rainfall, especially the tripole rainfall pattern over eastern China.

The circulation anomalies associated with the eastward shift of the SAH show strong easterly wind anomalies over the Indian subcontinent and the Arabian Sea at 850 hPa, implying a weaker-than-normal ISM (Fig. 5.26). In East Asia, an anomalous anticyclone is located over southern China and northern South China Sea at 850 hPa, indicating a westward extension of the Western North Pacific subtropical high, which suppresses rainfall over southern China. The anomalous southerly wind on the western flank of this anticyclone converges with the northerly wind anomalies from the north over the YRV, which favor more rainfall over the YRV. These circulation anomalies associated with the zonal displacement of the SAH are consistent with the rainfall distributions. In the next section, we will discuss the physical processes of the zonal shift of the SAH connecting the ISM and EASM rainfall.

5.5.3 *Impact of ISM Rainfall on the Zonal Shift of SAH*

The formation and variation of the SAH are affected by the sensible heating from the surface of the plateau and by the latent heating in the atmosphere (Krishnamurti et al. 1973; Huang 1985; Liu et al. 1999, 2001; Zuo et al. 2011; Wu et al. 2013). Thus, the anomalous heating distribution is analyzed to find out the possible cause of the zonal displacement of the SAH on the interannual time scale. Regressions of the integrated apparent heating source $< Q_1 >$ and integrated apparent moisture sink $< Q_2 >$ against the I_{EW} show that the most significant heating anomalies appear over the northern Indian subcontinent. The nearly equivalent values of $< Q_1 >$ and $< Q_2 >$ over northern India indicate the condensational latent heating induced by convection is the dominant heating in this region. It indicates the zonal shift of the SAH is closely related to the condensational latent heating associated with the convective activity and the monsoon rainfall over northern India. When the SAH is located eastward (westward), the ISM is weak (strong) with weak (strong) convection activities and less (more) rainfall over northern India. Over the TP, deep convection is smaller than that over the ISM region in terms of frequency, vertical scale, and horizontal scale (Luo et al. 2011). Figure 5.27 also shows the heating source anomalies are much weaker and opposite to those over the Indian subcontinent. Therefore, the release of latent heat by condensation of deep convection over the Indian subcontinent may play a dominant role in affecting the zonal displacement of the SAH.

According to the simplified vorticity Eq. (5.1), the meridional wind along the ridge of the subtropical high ($u \approx 0$) is associated with the vertical distribution of diabatic heating (Liu et al. 1999; Wu and Liu 2003; Liu et al. 2001):

$$\beta v \propto \frac{f + \varsigma}{\theta_z}\frac{\partial Q}{\partial z}, \theta_z \neq 0 \tag{5.4}$$

where Q is diabatic heating, ζ is vertical vorticity, and θ_z is static stability. In the upper troposphere over the heating source, $\frac{f+\varsigma}{\theta_z}\frac{\partial Q}{\partial z} < 0$. Thus, the northerly wind anomalies will appear at the upper level accompanied by an anomalous anticyclone to the west and an anomalous cyclone to the east according to Eq. (5.4) (Liu et al. 1999, 2001). On the contrary, southerly anomalies will appear over the cooling center accompanied by an anomalous cyclone to the west and an anomalous anticyclone to the east. Thus, in the weak ISM condition, negative latent heating over northern India can excite an anomalous anticyclone to the east and an anomalous cyclone to the west, resulting in an eastward shift of the SAH (Fig. 5.23).

Numerical experiments are conducted by using anomalous AGCM (Jiang and Li 2005; Li 2006) to further test the impact of the ISM rainfall on the SAH. Two idealized numerical experiments are used: the positive heating run (PHR) and the negative heating run (NHR). Based on the observed vertical profile of Q_1 anomalies, an idealized 5-σ level vertical profile of heating anomalies is estimated to be prescribed over northern India (78° E, 30° N) in the experiments. In the two experiments, the central heating rates at levels $\sigma = 0.1, 0.3, 0.5, 0.7$, and 0.9 are 0.2/−0.2, 0.8/−0.8, 1.0/−1.0, 0.5/−0.5, and 0.1/−0.1 K day $^{-1}$, respectively. The experiments are conducted by running the anomalous AGCM for 120 days under a fixed summer (JJA) basic state. The ensemble means of the last 90 days are used to analyze the responses of upper-level atmospheric circulations to the prescribed heating anomalies associated with the ISM rainfall. The differences of horizontal wind anomalies at 200 hPa between

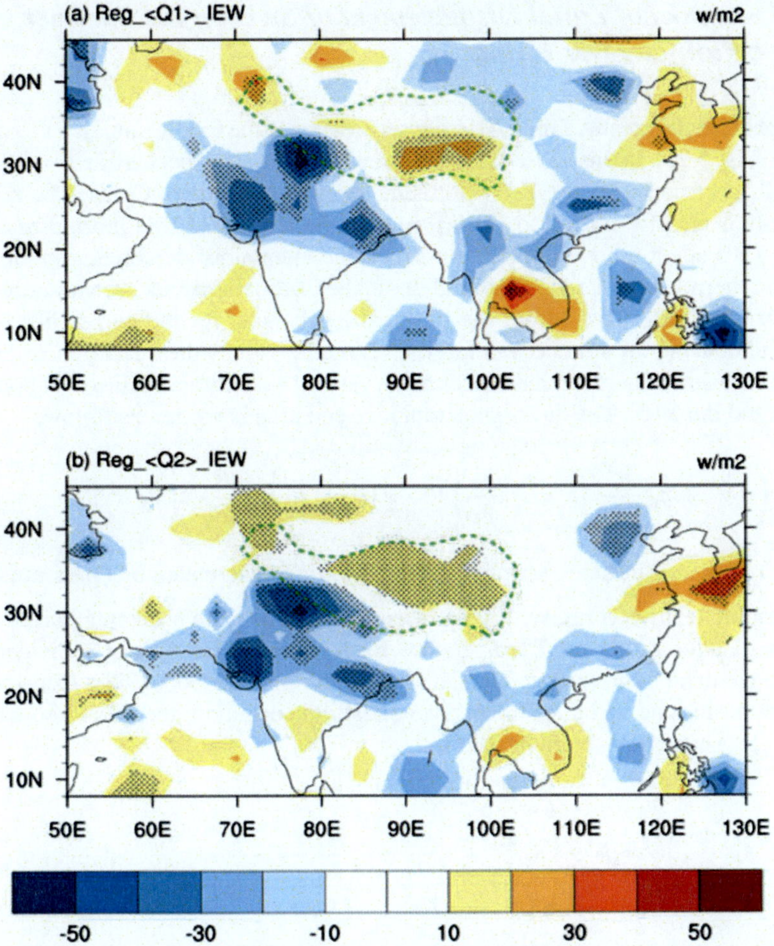

Fig. 5.27 Regressed (a) < Q_1> and (b) < Q_2> anomalies against I_{EW} (shading; units: W m^{-2}). Stippling highlights the areas exceeding the 0.05 significance level. The green contour indicates the TP region with topography exceeding 3000 m

Exp NHR and Exp PHR show obvious southerly wind anomalies over the heating source, being consistent with the proportional relationship between the meridional wind and the vertical gradient of the diabatic heating discussed in the above subsection. An anomalous anticyclone and a cyclone are excited to the east and the west of the heating source, respectively (Fig. 5.31a). This circulation pattern is similar to the observation shown in Fig. 5.23, indicating an eastward shift of the SAH. This result further demonstrates that the latent heating anomaly associated with the ISM is the main cause of the zonal displacement of the SAH.

5.5.4 Impact of Zonal Displacement of SAH on the Summer Rainfall Over China

The zonal displacement of the SAH is closely related to the tripole rainfall pattern over China (Fig. 5.25). In this subsection, we discuss the physical mechanism for different rainfall anomalies over the YRV, northern China, and southern China. The rainfall anomaly is directly related to vertical motion. When the SAH is located eastward in the TP mode, negative rainfall anomalies and anomalous descending motion are over northern China. Positive rainfall anomalies and anomalous ascending motion are over the YRV. To find out the physical mechanism for the zonal shift of the SAH influencing the summer rainfall over China, we apply the quasi-geostrophic ω equation to determine the main causes of the vertical velocity anomalies over northern China and the YRV. The quasi-geostrophic ω equation is written as follows:

$$\left(\sigma\nabla^2 + f_0^2\frac{\partial^2}{\partial p^2}\right)\omega = f_0\frac{\partial}{\partial p}\left[\vec{V}_g \cdot \nabla(\zeta_g + f)\right] - \frac{R}{p}\nabla^2\left[\vec{V}_g \cdot \nabla T\right] \tag{5.5}$$

where σ denotes the static stability, f_0 is the Coriolis parameter, p is pressure, ζ_g is geostrophic relative vorticity, $\vec{V}_g$ represents the geostrophic horizontal wind, R the gas constant for dry air, and T temperature. Each variable in Eq. 5.5 can be divided into a time-averaged basic state and its departure (Peixo´to and Oort 1984). Considering that the basic state variables themselves satisfy the equation, the quasi-geostrophic ω Eq. (5.5) can be rewritten as follows:

$$\begin{aligned}\left(\sigma\nabla^2+f_0^2\frac{\partial^2}{\partial p^2}\right)\omega' = {} & f_0\frac{\partial}{\partial p}\left(\underbrace{u'_g\frac{\partial\bar{\zeta}_g}{\partial x}}_{B1}+\underbrace{\bar{u}_g\frac{\partial\zeta'_g}{\partial x}}_{B2}+\underbrace{u'_g\frac{\partial\zeta'_g}{\partial x}}_{B3}+\underbrace{v'_g\frac{\partial\bar{\zeta}_g}{\partial y}}_{B4}+\underbrace{\bar{v}_g\frac{\partial\zeta'_g}{\partial y}}_{B5}+\underbrace{v'_g\frac{\partial\zeta'_g}{\partial y}}_{B6}+\underbrace{v'_g\frac{\partial f}{\partial y}}_{B7}\right)\\ & -\frac{R}{p}\nabla^2\left(\underbrace{u'_g\frac{\partial\bar{T}_g}{\partial x}}_{C1}+\underbrace{\bar{u}_g\frac{\partial T'_g}{\partial x}}_{C2}+\underbrace{u'_g\frac{\partial T'_g}{\partial x}}_{C3}+\underbrace{v'_g\frac{\partial\bar{T}_g}{\partial y}}_{C4}+\underbrace{\bar{v}_g\frac{\partial T'_g}{\partial y}}_{C5}+\underbrace{v'_g\frac{\partial T'_g}{\partial y}}_{C6}\right)\end{aligned} \tag{5.6}$$

where the variables with bar and prime are their basic state and perturbation, respectively. Terms B1–B7 in Eq. (5.6) represent the terms related to vorticity advection, and terms C1–C6 represent the terms related to temperature advection. These 13 terms represent 13 factors that contribute to the vertical velocity anomalies.

The diagnosis of the ω equation is applied over Northern China (110°–120° E, 35°–42.5° N) and the YRV (110°–120° E, 27.5°–35° N), respectively, at the non-divergence level 500 hPa (Palmén 1958; Serreze and Etringer 2003) to calculate the main contributors to the opposite vertical velocity anomalies over northern China and the YHR. The results show the main contributors to the descending motion anomalies over northern China are the advection of relative vorticity anomalies by the basic-state westerly wind (B2) and the cold advection by the anomalous northerly wind over eastern China (C4). Over the YRV, the main contributors to the ascending motion anomalies are the advections of the basic-state relative vorticity (B4) and the vorticity of the Earth (B7) by the anomalous northerly wind over eastern China on the east flank of the anomalous anticyclone (Fig. 10 of Wei et al. 2014). The quasi-geostrophic ω equation cannot be used to diagnose the vertical velocity anomalies over southern China, because it is close to the tropic. Figure 5.26 shows the western North Pacific subtropical high extends westward when the SAH is in the TP mode, which favors the descending motion over southern China. These circulation anomalies associated with the zonal shift of the SAH result in the tripole vertical motions and rainfall anomalies over eastern China.

5.5.5 Impact of YRV Rainfall on the Zonal Shift of SAH

In Section 5.5.2, we show the zonal variation of the SAH is closely related to the tripole rainfall pattern, which is the first EOF pattern of summer rainfall over China. Similar tripole rainfall pattern is also associated with the meridional variation of the SAH (Wei et al. 2012). When the SAH is located southward and eastward, the YRV is wetter than normal, while southern China and northern China are drier than normal. When the SAH is located northward and westward, more rainfall appears over southern China and northern China, and less rainfall is over the YRV. Comparing with the location of the anomalous anticyclone in the TP mode of the SAH (Fig. 5.23), the circulation anomalies at 200 hPa associated with the summer rainfall over China ($PC1_{CN}$) show a southward shifted anticyclone over southern China, indicating a southeastward displacement of the SAH (Fig. 5.28). It implies that the zonal movement may be accompanied by the meridional movement in the SAH.

The correlation coefficient between the detrended zonal shift index I_{EW} and the meridional shift index I_{NS} (Z200 (50°–100° E, 27.5°–32.5° N) - Z200 (50°–100° E, 22.5°–27.5° N)) is −0.44, exceeding the 0.05 significance level. It indicates the eastward (westward) movement is always accompanied by the southward (northward) movement. The result of EOF analysis on 200-hPa geopotential height also suggests that the southeast–northwest (SE–NW) movement is the dominant movement pattern on the interannual time scale (Wei et al. 2015).

An index $I_{SE\text{–}NW}$ is defined as the difference of 200-hPa geopotential height between the southeast part (85°–115° E, 20°–27.5° N) and the northwest part (50°–80° E, 27.5°–35° N) of the SAH to quantify the SE–NW movement of the SAH and its relationship with monsoon rainfall (Wei et al. 2015, 2017). Regressed summer rainfall against the $I_{SE\text{–}NW}$ shows the most significant rainfall anomalies appear over the ISM and EASM regions (Fig. 5.29). Negative rainfall anomalies over the northern Indian subcontinent, northern China, and southern China, and the positive rainfall anomalies over the YRV, South Korea, and Japan all exceed the 0.05 significance level. The rainfall anomalies over the ISM region and the YRV are opposite, as shown

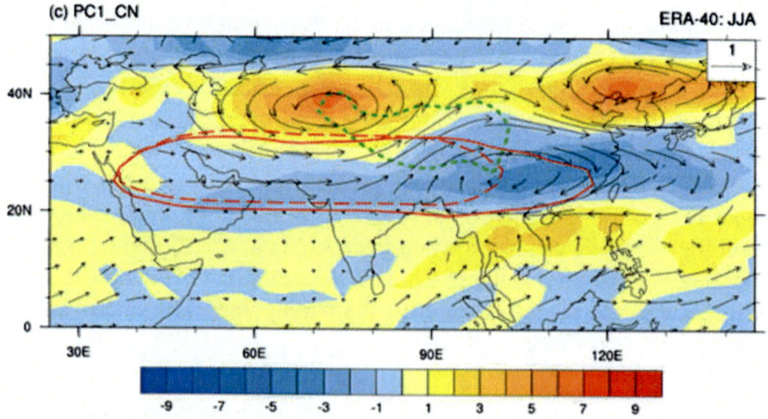

Fig. 5.28 Regressed horizontal wind (vector; units: m s^{-1}) and relative vorticity (shading; units: 10^{-6} s^{-1}) at 200 hPa against $PC1_{CN}$. Composites of 12520 gpm contours for $PC1_{CN} > 1$ (solid line) or $PC1_{CN} < -1$ (dashed line) are indicated by red contours. The green contour indicates the TP region with topography exceeding 3000 m. After Wei et al. (2015)

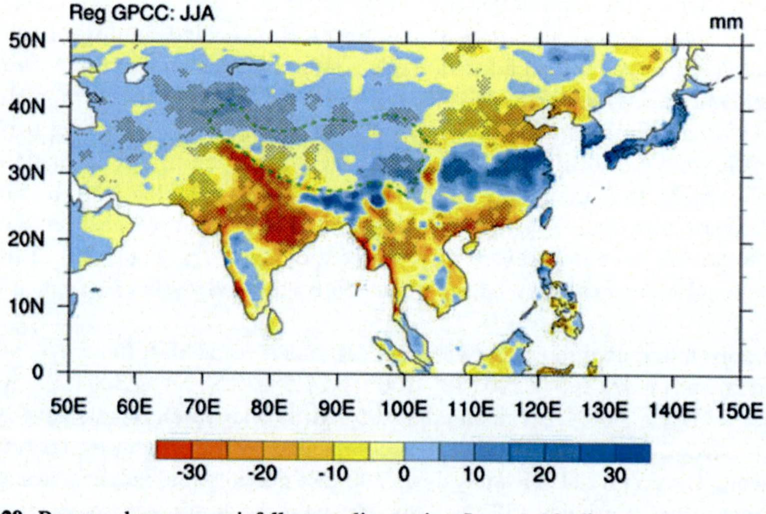

Fig. 5.29 Regressed summer rainfall anomalies against $I_{SE\text{-}NW}$ (shading; units: mm). Stippling highlights the areas exceeding the 0.05 significance level. The green contour indicates the TP region with topography exceeding 3000 m. Modified from Wei et al. (2015)

in previous studies (Zhang et al. 1999; Zhang 2001). The correlation coefficient of all Indian rainfall index AIRI with the $I_{SE\text{-}NW}$ is -0.64, exceeding the 0.01 significance level. This correlation coefficient is much higher than those with the I_{EW} (-0.49) and the I_{NS} (0.47). Besides, the correlation coefficient of the $PC1_{CN}$ with the $I_{SE\text{-}NW}$ (0.50) is also much higher than those with the I_{EW} (0.35) and the I_{NS} (-0.40). It

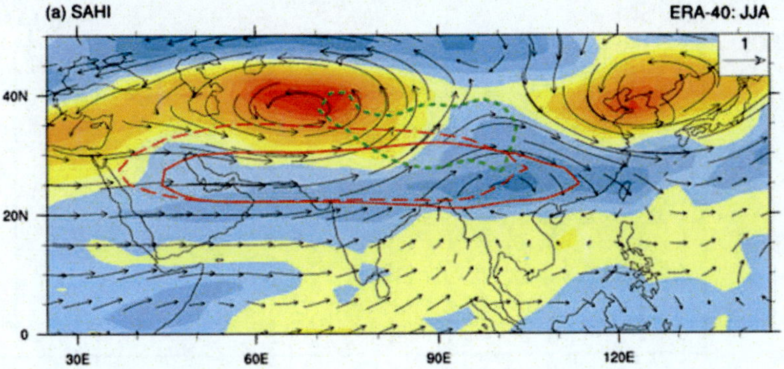

Fig. 5.30 Same as Fig. 5.28, except for the $I_{SE–NW}$ index. After Wei et al. (2015)

indicates the SE–NW variation of the SAH can better reflect the relationship among the SAH, the ISM rainfall, and the EASM rainfall.

Regressed 200-hPa horizontal wind anomalies against the $I_{SE–NW}$ show an anomalous anticyclone over southern China and two cyclones over central Asia and northeastern Asia (Fig. 5.30). This circulation pattern indicates a southeastward shifted SAH when the $I_{SE–NW}$ is positive. The composite of 12520-gpm contours based on the high/low $I_{SE–NW}$ also suggests that the SAH extends southeastward to southern China when the $I_{SE–NW}$ is larger than 1, and the SAH retreats northwestward when the $I_{SE–NW}$ is smaller than -1. This circulation pattern is similar to the circulation against $PC1_{CN}$, especially the eastern part over eastern China. The similarity implies the rainfall over eastern China may contribute to the SE–NW shift of the SAH.

Condensational latent heating associated with summer rainfall over the Western North Pacific and East Asia plays a key role in maintaining the summertime atmospheric circulation (Liu et al. 2004; Lu and Lin 2009). So, the rainfall over the EASM region can influence the upper-level circulation besides of the ISM rainfall. In Fig. 5.29, the rainfall anomalies over the YRV are opposite, but with similar amplitude, with those over the northern Indian subcontinent (Wei et al. 2015). Thus, we speculate that the southeastward extension of the anomalous anticyclone over East Asia may be caused by the rainfall anomalies over the YRV.

The anomalous AGCM (Jiang and Li 2005; Li 2006) is utilized to investigate the possible impact of the YRV rainfall on the displacement of the SAH. Three groups of experiments are conducted to compare the effects of latent heating over the ISM region and over the YRV. In Exp I, a cooling center of -1 K day^{-1} is placed at the mid-level atmosphere over the northern Indian subcontinent, indicating a weak ISM with less rainfall. An anomalous anticyclone and an anomalous cyclone are excited to the east and to the west of the TP, respectively, with two centers at the same latitude near 32.5° N, indicating an eastward displacement of the SAH (Fig. 5.31a). In Exp II, a heating center of 1 K day^{-1} is placed at the mid-level atmosphere over the YRV, indicating positive rainfall anomalies over the YRV. The heating anomalies also excite an anomalous anticyclone to the east and an anomalous cyclone to the

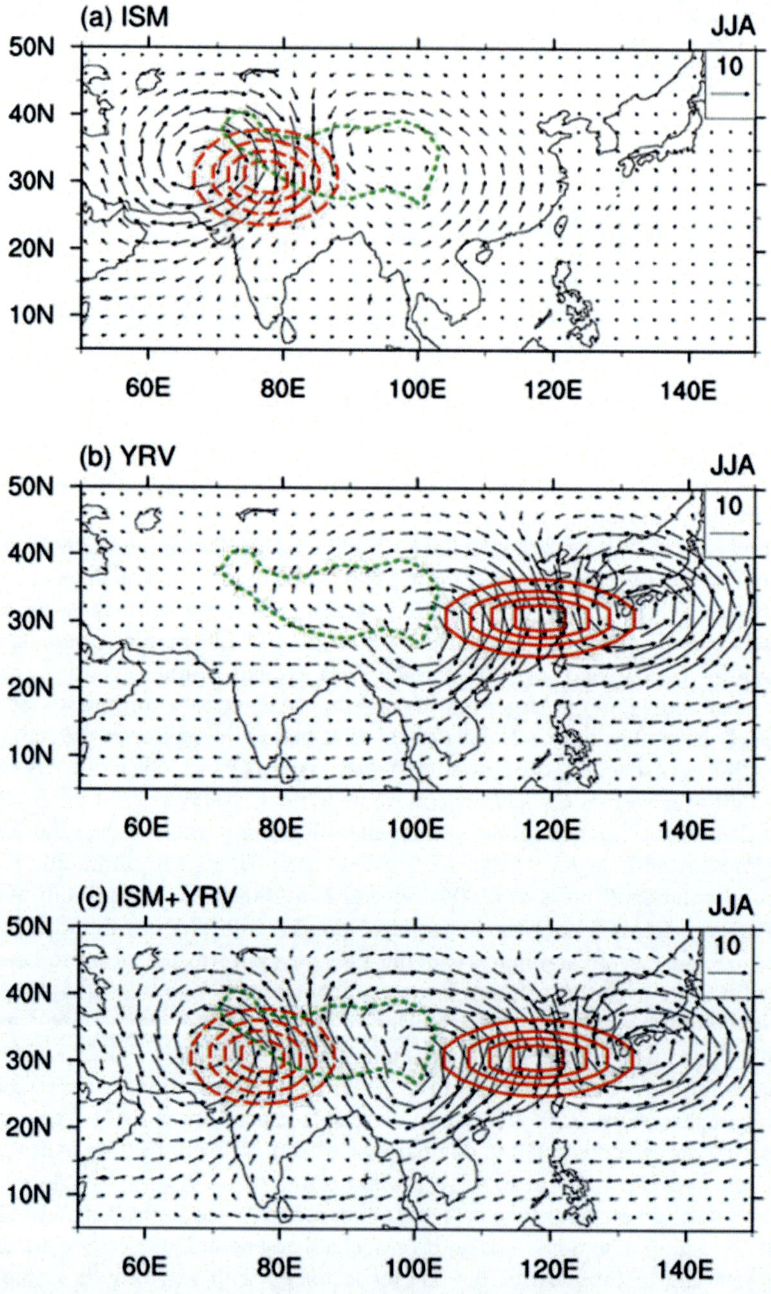

Fig. 5.31 Differences of ensemble horizontal winds at 200 hPa (vector; units: m s^{-1}) between the positive run and the negative run in **a** Exp I, **b** Exp II, and **c** Exp III. The red contours indicate the horizontal distribution of diabetic heating anomalies (units: K day^{-1}) at the height $\sigma = 0.5$. The green contour indicates the TP region with topography exceeding 3000 m. Modified from Wei et al. (2015)

west. However, the center of the anticyclone is located southward over southern China near 27.5° N, compared with that near 32.5° N in Exp I (Fig. 5.31b). When the ISM cooling and the YRV heating are both considered in Exp III, an anomalous anticyclone is excited over southern China and an anomalous cyclone is excited to the west of the TP. The axis of these two centers is oriented along the southeast-northwest direction, resulting in a SE–NW movement of the SAH (Fig. 5.31c). The results of these three experiments suggest that the latent heating associated with ISM rainfall mainly cause the zonal movement of the SAH, which can further exert impact on the summer rainfall over China. And the latent heating anomalies associated with the rainfall over the YRV excite a southward shifted anticyclone over eastern China, leading to the southeastward shifted SAH.

5.5.6 Conclusions and Discussion

The zonal displacement of the SAH is significant during boreal summer on the interannual time scale, which is closely related to the ISM rainfall and the EASM rainfall. When the SAH is located eastward in the TP mode, less rainfall and convective activity can be found over the northern Indian subcontinent, indicating a weak ISM. Over East Asia, more rainfall can be found over the YRV and less rainfall, over northern and southern China. When the SAH is located westward in the IP mode, the ISM is strong with more rainfall over the northern Indian subcontinent, and the EASM is strong with the positive–negative–positive rainfall pattern over eastern China. Diagnostic analysis and numerical experiments based on the anomalous AGCM show the latent heating associated with the ISM rainfall over the northern is the main contributor to the zonal shift of the SAH. When the ISM is weak, the negative heating anomalies excite southerly wind over the cooling center, accompanied by an anomalous anticyclone to the east and an anomalous cyclone to the west, resulting in an eastward displacement of the SAH. The circulation anomalies associated with the eastward located SAH can further lead to opposite vertical motions over the YRV and northern and southern China, resulting in the tripole rainfall pattern over China. These results indicate that SAH plays an important role in the effect of ISM on the EASM.

Further analysis shows that the zonal shift of the SAH is always accompanied by a meridional displacement. The SE–NW displacement is the dominant moving feature of the SAH on the interannual time scale, which is related to the feedback effect of the EASM rainfall. A closer relationship can be found among the SAH, ISM rainfall, and EASM rainfall after considering both zonal and meridional displacements of the SAH. The correlation coefficient of AIRI/$PC1_{CN}$ with the $I_{SE\text{–}NW}$ is −0.64/0.50, which is much higher than those with I_{EW} (−0.49/0.35) and I_{NS} (0.47/−0.40). It suggests that the SE–NW variation of the SAH can better reflect the relationship among the SAH, the ISM rainfall, and the EASM rainfall.

Physical mechanisms of the interactions among the ISM, the EASM, and the SAH are revealed by numerical experiments. Latent heating anomalies associated with ISM rainfall over the northern Indian subcontinent mainly cause the zonal shift of the SAH. The positive latent heating anomalies over the YRV excite a southward located anticyclone over eastern China, exerting a feedback effect on the SAH and leading to the SE–NW displacement of the SAH.

References

Abe M, Hori M, Yasunari T, Kitoh A (2013) Effects of the Tibetan Plateau on the onset of the summer monsoon in South Asia: the role of the air–sea interaction. J Geophys Res 118:1760–1776. https://doi.org/10.1002/jgrd.50210

Ashfaq M, Shi Y, Tung WW et al (2009) Suppression of south Asian summer monsoon precipitation in the 21st century. Geophys Res Lett 36(1):L01704. https://doi.org/10.01029/02008GL036500

Bansod SD, Yin ZY, Lin Z et al (2003) Thermal field over Tibetan Plateau and Indian summer monsoon rainfall. Int J Climatol 23(13):1589–1605

Barnett TP, Dümenil L, Schlese U, Roeckner E, Latif M (1989) The effect of Eurasian snow cover on regional and global climate variations. J Atmos Sci 46(5):661–685

Boos W, Kuang Z (2010) Dominant control of the South Asian monsoon by orographic insulation versus plateau heating. Nature 463:218–223

Chen GY, Liao QS (1990) Relationship between the location features of 100 hPa South Asia High and mid-summer rainfall in China. Plateau Meteorol 9:432–438 (in Chinese)

Duan A, Wang M, Lei Y, Cui Y (2013) Trends in summer rainfall over China associated with the Tibetan Plateau sensible heat source during 1980–2008. J Clim 26:261–275

Duan A, Wu G (2005) Role of the Tibetan Plateau thermal forcing in the summer climate pattern over subtropical Asia. Clim Dyn 24:793–807

Duan AM, Wu GX, Liang XY (2008) Influence of the Tibetan Plateau on the summer climate patterns over Asia in the IAP/LASG SAMIL model. Adv Atmos Sci 25(4):518–528. https://doi.org/10.1007/s00376-008-0518-2

Duan AM, Xiao ZX (2015) Does the climate warming hiatus exist over the Tibetan Plateau? Sci Rep 5:13711. https://doi.org/10.1038/srep13711

Flohn H (1957) Large-scale aspects of the "summer monsoon" in South and East Asia. J Meteorol Soc Jpn 75:180–186

Folland CK, Knight J, Linderholm HW, Fereday D, Ineson S, Hurrell JW (2009) The Summer North Atlantic Oscillation: Past, Present, and Future. J Clim 22:1082–1103

Hahn DG, Manabe S (1975) The role of mountains in the South Asian monsoon circulation. J Atmos Sci 32(8):1515–1541

Hansen J, Ruedy R, Sato M, Lo K (2010) Global surface temperature change. Rev Geophys 48:RG4004. https://doi.org/10.1029/2010rg000345

He S, Yang S, Li Z (2017) Influence of latent heating over the Asian and western Pacific monsoon region on Sahel summer rainfall. Sci Rep 7:7680. https://doi.org/10.1038/s41598-017-07971-6

Hu J, Duan A (2015) Relative contributions of the Tibetan Plateau thermal forcing and the Indian Ocean sea surface temperature basin mode to the interannual variability of the East Asian summer monsoon. Clim Dyn 45:2697–2711

Huang RH (1985) The influence of the heat source anomaly over Tibetan Plateau on the northern hemispheric circulation anomalies. Acta Meteorol Sin 43:208–219 (in Chinese)

Huang Y, Qian YF (2003) Relationship between South Asian High and summer rainfall in North China. Plateau Meteorol 22:602–607 (in Chinese)

Jiang X-A, Li T (2005) Reinitiation of the Boreal Summer Intraseasonal Oscillation in the Tropical Indian Ocean. J Clim 18(18):3777–3795

Jiang XW, Ting MF (2017) A dipole pattern of summertime rainfall across the Indian subcontinent and the Tibetan Plateau. J Clim. https://doi.org/10.1175/jcli-d-16-0914.1

Kitoh A (2004) Effects of mountain uplift on East Asian summer climate investigated by a coupled atmosphere–ocean GCM. J Clim 17(4):783–802

Kobayashi S, Ota Y, Harada Y, Ebita A, Moriya M, Onoda H, Onogi K, Kamahori H, Kobayashi C, Endo H, Miyaoka K, Takahashi K (2015) The JRA-55 Reanalysis: General Specifications and Basic Characteristics. J Met Soc Jap 93:5–48

Krishnamurti TN, Bedi HS, Subramaniam M (1989) The summer monsoon of 1987. J Clim 2:321–340

Krishnamurti TN, Bhalme H (1976) Oscillations of a monsoon system. Part I. Observational aspects. J Atmos Sci 33:1937–1954

Krishnamurti TN, Daggupaty SM, Fein J et al (1973) Tibetan high and upper tropospheric tropical circulation during northern summer. Bull Amer Meteor Soc 54:1234–1249

Li T (2006) Origin of the Summertime Synoptic-Scale Wave Train in the Western North Pacific. J Atmos Sci 63(3):1093–1102

Linderholm HW, Qu T, Jeong J-H, Folland C, Gong D, Liu H, Liu Y, Chen D (2011) Interannual teleconnections between the summer North Atlantic Oscillation and the East Asian summer monsoon. J Geophys Res 116:D13107. https://doi.org/10.1029/2010jd015235

Liu Y, Hoskins B, Blackburn M (2007) Impact of Tibetan orography and heating on the summer flow over Asia. J Met Soc Jap 85B:1–19

Liu Y, Wu G, Ren R (2004) Relationship between the subtropical anticyclone and diabatic heating. J Clim 17(4):682–698

Liu YM, Wu GX, Hong JL, Dong BW, Duan AM, Bao Q, Zhou LJ (2012) Revisiting Asian monsoon formation and change associated with Tibetan Plateau forcing: II. Change. Clim Dyn 39(5):1183–1195

Liu YM, Wu GX, Liu H, Liu P (1999) The effect of spatially nonuniform heating on the formation and variation of subtropical high part III: condensation heating and South Asia high and western Pacific subtropical high. Acta Meteorol Sin 57:525–538 (in Chinese)

Liu YM, Wu GX, Liu H, Liu P (2001) Condensation heating of the Asian summer monsoon and the subtropical anticyclone in Eastern Hemisphere. Clim Dyn 17:327–338L

Lu MM, Yang S, Li ZN, He B, He S, Wang ZQ (2018) Possible effect of the Tibetan Plateau on the "upstream" climate over West Asia, North Africa, South Europe and the North Atlantic. Clim Dyn 51:1485–1498

Lu R, Lin Z (2009) Role of Subtropical Precipitation Anomalies in Maintaining the Summertime Meridional Teleconnection over the Western North Pacific and East Asia. J Clim 22(8):2058–2072

Luo H, Yanai M (1984) The large-scale circulation and heat sources over the Tibetan Plateau and surrounding areas during the early summer of 1979. Part II: Heat and moisture budgets. Mon Wea Rev 112:966–989

Luo SW, Qian ZA, Wang QQ (1982) The climatic and synoptical study about the relationship between the Qinghai-Xizang High Pressure on the 100 mb surface and the flood and drought in East China in summer. Plateau Meteorol 1:1–10 (in Chinese)

Luo Y, Zhang R, Qian W et al (2011) Intercomparison of Deep Convection over the Tibetan Plateau-Asian Monsoon Region and Subtropical North America in Boreal Summer Using CloudSat/CALIPSO Data. J Clim 24(8):2164–2177

Ma D, Boos W, Kuang Z (2014) Effects of Orography and Surface Heat Fluxes on the South Asian Summer Monsoon. J Clim 27:6647–6659

Mason RB, Anderson CE (1963) The Development and Decay of the 100mb Summertime Anticyclone over Southern Asia. Mon Wea Rev 91:3–12

Neale RB, Richter J, Park S, Lauritzen PH, Vavrus SJ, Rasch PJ, Zhang M (2013) The mean climate of the community atmosphere model (CAM4) in forced SST and fully coupled experiments. J Clim 26:5150–5168

Noh Y, Kim H-J (1999) Simulations of temperature and turbulence structure of the oceanic boundary layer with the improved near surface process. J Geophys Res 104:15621–15634

Palmén E (1958) Vertical circulation and release of kinetic energy during the development of Hurricane Hazel into an extratropical storm. Tellus 10(1):1–23

Parthasarathy B, Kumar KR, Kothawale D (1992) Indian summer monsoon rainfall indices: 1871-1990. Meteor Mag 121:174–186

Peixóto JP, Oort AH (1984) Physics of climate. Rev Mod Phys 56:365–429

Plumb RA, Hou AY (1992) The response of a zonally symmetric atmosphere to subtropical thermal forcing: Threshold behavior. J Atmos Sci 49:1790–1799

Raman CRV, Rao YP (1981) Blocking highs over Asia and monsoon droughts over India. Nature 289:271–273

Rayner N, Parker D, Horton E, Folland C, Alexander L, Rowell D, Kent E, Kaplan A (2003) Global analyses of sea surface temperature, sea ice, and night marine air temperature since the late Nineteenth Century. J Geophys Res 108(D14):4407

Reynolds R, Smith T, Liu C, Chelton D, Casey K, Schlax M (2007) Daily high-resolution-blended analyses for sea surface temperature. J Clim 20:5473–5496

Rodwell MJ, Hoskins BJ (1996) Monsoon and the dynamics of deserts. Q J R Meteorol Soc 122:1385–1404

Rodwell MJ, Hoskins BJ (2001) Subtropical anticyclones and summer monsoons. J Clim 14:3192–3211

Schneider EK (1977) Axially symmetric steady-state models of the basic state for instability and climate studies. Part II. Nonlinear calculations. J Atmos Sci 34:280–296

Schneider EK (1987) A simplified model of the modified Hadley circulation. J Atmos Sci 44:3311–3328

Schneider EK, Lindzen RS (1977) Axially symmetric steady-state models of the basic state for instability and climate studies. Part I: Linearized calculations. J Atmos Sci 34:263–279

Schneider U, Becker A, Finger P, Meyer-Christoffer A, Ziese M, Rudolf B (2014) GPCC's new land surface precipitation climatology based on quality-controlled in situ data and its role in quantifying the global water cycle. Theor Appl Climatol 115:15–40

Serreze MC, Etringer AJ (2003) Precipitation characteristics of the Eurasian Arctic drainage system. Int J Climatol 23(11):1267–1291

Si D, Ding Y (2013) Decadal change in the correlation pattern between the Tibetan Plateau winter snow and the East Asian summer rainfall during 1979–2011. J Clim 26:7622–7634. https://doi.org/10.1175/jcli-d-12-00587.1

Skamarock W, Klemp J, Dudhia J, Gill D, Barker D, Duda M, Huang X, Wang W (2008) A description of the advanced research WRF version 3. NCAR Technical Note NCAR/TN-475 + STR, https://doi.org/10.5065/d68s4mvh

Sontakke N, Pant G, Singh N (1993) Construction of All-India Summer Monsoon Rainfall Series for the Period 1844-1991. J Clim 6:1807–1811

Sun J, Wang H (2012) Changes of the connection between the summer North Atlantic Oscillation and the East Asian summer rainfall. J Geophys Res 117:D08110. https://doi.org/10.1029/2012jd017482

Sun J, Wang H, Yuan W (2008) Decadal variations of the relationship between the summer North Atlantic Oscillation and middle East Asian air temperature. J Geophys Res 113:D15107. https://doi.org/10.1029/2007jd009626

Takaya K, Nakamura H (2001) A formulation of a phase-independent wave-activity flux for stationary and migratory quasigeostrophic eddies on a zonally varying basic flow. J Atmos Sci 58:608–627

Tao SY, Zhu FK (1964) The 100-mb flow patterns in southern Asia in summer and its relation to the advance and retreat of the West Pacific subtropical anticyclone over the Far East. Acta Meteorol Sin 34:385–396 (in Chinese)

Wang B, Bao Q, Hoskins B, Wu G, Liu Y (2008a) Tibetan Plateau warming and precipitation changes in East Asia. Geophys Res Lett 35:L14702. https://doi.org/10.1029/2008gl034330

Wang B, Fan Z (1999) Choice of south Asian summer monsoon indices. Bull Amer Meteor Soc 80:629–638

Wang B, Wu Z, Li J, Liu J, Chang C-P, Ding Y, Wu G (2008b) How to measure the strength of the East Asian summer monsoon. J Clim 21:4449–4463

Wang Z, Duan A (2012) A New Ocean Mixed-Layer Model Coupled into WRF. Atmos Oceanic Sci Lett 5:170–175

Wang Z, Duan A, Li M, He B (2016) Influences of thermal forcing over the slope/platform of the Tibetan Plateau on Asian summer monsoon: Numerical studies with WRF model. Chinese J Geophys 59(5):474–487

Wang Z, Duan A, Wu G (2014a) Impacts of boundary layer parameterization schemes and air-sea coupling on WRF simulation of the East Asian summer monsoon. Sci China: Earth Sci 57:1–14

Wang Z, Duan A, Wu G (2014b) Time-lagged impact of spring sensible heat over the Tibetan Plateau on the summer rainfall anomaly in East China: case studies using the WRF model. Clim Dyn 42:2885–2898

Wang Z, Duan A, Yang S (2018a) Potential regulation on the climatic effect of Tibetan Plateau heating by tropical air-sea coupling in regional models. Clim Dyn. https://doi.org/10.1007/s00382-018-4218-z

Wang Z, Duan A, Yang S, Ullah K (2017) Atmospheric moisture budget and its regulation on the variability of summer precipitation over the Tibetan Plateau. J Geophys Res 122:614–630

Wang Z, Yang S, Lau N-C, Duan A (2018b) Teleconnection between Summer NAO and East China Rainfall Variations: A Bridge Effect of the Tibetan Plateau. J Clim. https://doi.org/10.1175/jcli-d-17-0413.1

Wei W, Zhang R, Wen M (2012) Meridional variation of South Asian High and its relationship with the summer precipitation over China. J Appl Meteorol Sci 23:650–659 (in Chinese)

Wei W, Zhang R, Wen M, Kim B-J, Nam J-C (2015) Interannual Variation of the South Asian High and Its Relation with Indian and East Asian Summer Monsoon Rainfall. J Clim 28:2623–2634

Wei W, Zhang R, Wen M, Rong X (2014) Impact of Indian summer monsoon on the South Asian High and its influence on summer rainfall over China. Clim Dyn 43:1257–1269

Wei W, Zhang R, Wen M, Yang S (2017) Relationship between the Asian westerly jet stream and summer rainfall over central Asia and north China: roles of the Indian monsoon and the South Asian high. J Clim 30(2):537–552

Wu G, He B, Liu Y, Bao Q, Ren R (2015) Location and variation of the summertime upper-troposphere temperature maximum over South Asia. Clim Dyn 45:2757–2774

Wu G, Liu Y (2003) Summertime quadruplet heating pattern in the subtropics and the associated atmospheric circulation. Geophys Res Lett 30(5):1201. https://doi.org/10.1029/2002GL016209

Wu G, Liu Y, Wang T, Wan R, Liu X, Li W, Wang Z, Zhang Q, Duan A, Liang X (2007) The Influence of the Mechanical and Thermal Forcing of the Tibetan Plateau on the Asian Climate. J Hydrometeorol. 8:770–789

Wu G, Ren S, Xu J et al (2013) Impact of tropical cyclone development on the instability of South Asian High and the summer monsoon onset over Bay of Bengal. Clim Dyn 41:2603–2616

Wu G, Zhuo H, Wang Z, Liu Y (2016) Two types of summertime heating over the Asian large-scale orography and excitation of potential-vorticity forcing I. over the Tibetan Plateau. Sci China: Earth Sci 59:1996–2008

Wu GX, Duan AM, Liu YM, Mao JY, Ren RC, Bao Q, He B, Liu BQ, Hu WT (2014) Tibetan Plateau climate dynamics: recent research progress and outlook. Nat Sci Rev. https://doi.org/10.1093/nsr/nwu045

Wu GX, Li WP, Guo H, et al (1997) Sensible heat driven air-pump over the Tibetan Plateau and its impacts on the Asian summer monsoon. In: Ye DZ (ed) Collections on the Memory of Zhao Jiuzhang (in Chinese). Science Press, Beijing, pp 116–126

Wu GX, Liu Y, Zhu X, Li W, Ren R, Duan A, Liang X (2009) Multi-scale forcing and the formation of subtropical desert and monsoon. Ann Geophys 27(9):3631–3644

Wu GX, Liu YM, He B, Bao Q, Duan AM, Jin FF (2012) Thermal controls on the Asian summer monsoon. Sci Rep 2:404. https://doi.org/10.1038/srep00404

Wu Z, Li J, Jiang Z, Ma T (2012) Modulation of the Tibetan Plateau snow cover on the ENSO teleconnections: from the East Asian summer monsoon perspective. J Clim 25:2481–2489

Yanai M, Li C, Song Z (1992) Seasonal heating of the Tibetan Plateau and its effects on the evolution of the Asian summer monsoon. J Meteor Soc Japan 70:319–351

Yanai M, Wu G (2006) Effects of the Tibetan Plateau, in The Asian Monsoon, edited by B. Wang, pp. 513–549, Springer Praxis
Yang H, Wang B, Wang B (2012) Reduction of systematic biases in regional climate downscaling through ensemble forcing. Clim Dyn 38(3–4):655–665
Yang S, Lau KM (1998) Influence of sea surface temperature and ground wetness on Asian summer monsoon. J Clim 11:3230–3246
Yang S, Li T (2016) Causes of intraseasonal diabatic heating variability over and near the Tibetan Plateau in boreal summer. Clim Dyn 1–22
Yang S, Webster PJ, Dong M (1992) Longitudinal heating gradient: another possible factor influencing the intensity of the Asian summer monsoon circulation. Adv Atmos Sci 9(4):397–410
Ye D, Gao Y (1979) Tibetan Plateau Meteorology. Science Press, Beijing (in Chinese)
Ye DZ, Wu GX (1998) The role of the heat source of the Tibetan Plateau in the general circulation. Meteorol Atmos Phys 67:181–198
Yeh TC, Lo SW, Chu PC (1957) The wind structure and heat balance in the lower troposphere over Tibetan Plateau and its surrounding. Acta Meteorol Sin 28:108–121
Yeh TC, Wetherald RT, Manabe S (1984) The effect of soil moisture on the short-term climate and hydrology change: a numerical experiment. Mon Weather Rev 112:474–490
Zhang Q, Wu GX (2001) The large area flood and drought over Yangtze River Valley and its relation to the South Asia High. Acta Meteorol Sin 59:569–577 (in Chinese)
Zhang Q, Wu GX, Qian YF (2002) The bimodality of the 100 hPa South Asia High and its relationship to the climate anomaly over East Asia in summer. J Meteorol Soc Jpn 80:733–744
Zhang R (2001) Relations of water vapor transport from Indian monsoon with that over East Asia and the summer rainfall in China. Adv Atmos Sci 18:1005–1017
Zhang R, Sumi A, Kimoto M (1999) A diagnostic study of the impact of El Nino on the precipitation in China. Adv Atmos Sci 16:229–241
Zhao P, Yang S, Wu RG, Wen ZP, Chen JM, Wang HJ (2012) Asian origin of interannual variations of summer climate over the extratropical North Atlantic Ocean. J Clim 25:6594–6609
Zhou L, Bao Q, Liu Y, Wu G, Wang W-C, Wang X, He B, Yu H, Li J (2015) Global energy and water balance: characteristics from finite-volume atmospheric model of the IAP/LASG (FAMIL1). J Adv Model Earth Syst 7:1–20
Zuo Z, Zhang R, Zhao P (2011) The relation of vegetation over the Tibetan Plateau to rainfall in China during the boreal summer. Clim Dyn 36(5–6):1207–1219

Chapter 6
Feedback Attributions of Climate Changes Over the Globe and Over Southeast Asia and Its Adjacent Regions

6.1 Overview

The global mean surface temperature has experienced a rapid warming from the 1980s to the early-2000s, but a muted warming since then, referred to as the global warming hiatus in the literature (Trenberth and Fasullo 2013; Meehl 2015). During the period of accelerated warming, the spatial distribution of the trend of global surface temperature mainly had the following characteristics (Hu et al. 2017): (1) warming over most places around the globe, especially over the tropical oceans; (2) pronounced polar warming amplification (PWA), especially in the Arctic region; and (3) warmer land but colder ocean over the extratropical regions in the Northern Hemisphere. The evidence of global warming is also present in other climatic variables, such as stratospheric temperatures (Ramaswamy et al. 2001; Forster et al. 2011), water vapor (Trenberth et al. 2005), cloud cover (Wylie et al. 2005), and precipitation (Simon and Montanari 2019). Stratospheric cooling is one important evidence of the greenhouse effect caused by the increased CO_2 concentration. From the perspective of energy balance, it has been shown that CO_2 adjusts energy distribution within the climate system, trapping more energy in the troposphere and causing stratospheric cooling (Thompson et al. 2012). A large number of studies have revealed the causes of global warming in many aspects. In this chapter, we attempt to provide a more comprehensive picture on the attribution of global warming, and quantitatively analyze relative contributions by external forcing of the climate system and by internal processes to global warming.

During global warming, the internal variability of the climate system has also undergone significant changes. El Niño-Southern Oscillation (ENSO) is the most typical ocean–atmosphere coupled event in the climate system on the time scale of interannual variability, which has important impacts on global weathers and climate at various time scales, especially on the climate change in Southeast Asia. In the warming world, ENSO has shown new features. Since the 1990s, the frequency of emergence of the central Pacific (CP) El Niño (Yu and Kao 2007) has increased, with its maximum warm sea surface temperature (SST) anomaly located in the central

S. Yang et al., *Climate Change in Southeast Asia and Surrounding Areas*,
Springer Climate, https://doi.org/10.1007/978-981-15-8225-7_6

equatorial Pacific, also known as El Niño Modoki (Ashok et al. 2007) or warm-pool El Niño (Kug et al. 2009). In contrast to the traditional El Niño, also referred to as the eastern Pacific (EP) El Niño, the CP El Niño has new features in many aspects, including the spatial distribution and intensity of SST anomaly, the mechanism of development and evolution, etc. Correspondingly, the impact of ENSO on climate in Southeast Asia has also changed. Therefore, the attribution of different spatial distributions of SST anomalies of the two types of El Niño can help us have a better understanding of climate change in Southeast Asia.

Southeast Asia is located in the junction of the world's largest continent—Asia–Europe continent and the largest ocean—the Pacific. It experiences the longest period in the world affected by the monsoon. Both the East Asian winter monsoon (EAWM) and the East Asian summer monsoon are the products of the coupling between the ocean and the atmosphere. The EAWM is not only the most powerful winter monsoon in the world, but also the most active circulation system in winter in the Northern Hemisphere (Chang et al. 2004), causing changes in atmospheric circulation over the globe. However, the EAWM is also affected by the surrounding atmospheric circulation, especially the circulation in the mid- and high-latitudes, and by the thermal contrast between sea and land. The anomalous atmospheric circulation, water vapor transport, cloud cover, and aerosols all affect the evolution and intensity of the EAWM. Due to the interaction between the factors in the low-latitude and the mid-to-high- latitudes, the intensity of the EAWM has obvious interannual variation. Therefore, factors affecting the intensity of the EAWM are also one of the issues discussed in this chapter.

The Maritime Continent (MC) is one of the major terrestrial groups in Southeast Asia. It includes the Malay Peninsula, Indonesia, Borneo, New Guinea, and many other islands, including their surrounding regions. Since the MC is located in the middle of the Asian monsoon region and the Australian monsoon region, the climate over the MC is regulated by these two monsoon systems. At the same time, as the key component of the "atmospheric bridge," the climate change in the MC can affect the climate in the other regions over the globe (Hung et al. 2004; Chang et al. 2016). For example, the Madden–Julian Oscillation (MJO) passes over the MC during its mature phase, as the deep convection of the MJO positive phase weakens, and the land-separated rain bands reconverge in the South Pacific convergence zone after passing through the MC (Sui and Lau 1992; Chang et al. 2005). The MC is likened to a "boiling box" (Ramage 1968; Simpson et al. 1993), because the strong deep convection over the MC is one of the equatorial heat sources (Neale and Slingo 2003). Strong convection releases large amount of latent heat at high-altitude, exciting Rossby wave trains to transmit energy beyond the equator, which can affect the winter atmospheric circulation and surface temperature over North America (Yanai and Tomita 1998; Yang et al. 2002), Northeast Asia and Europe (Neale and Slingo 2003). Under this complex circulation, the attribution of temperature changes in the MC and the South China Sea region needs to fully consider the internal climate variability and external forcing.

The dynamical diagnostic method used in this chapter for attribution of climate change is the "climate response-feedback analysis method" (CFRAM; Cai and Lu

2009; Lu and Cai 2009). The CFRAM is a statistical diagnostic method that uses temperature to characterize the contributions of external forcing and internal climatic processes to the observed changes in energy or temperature. It is possible to quantify the partial energy or temperature changes caused by individual external forcing or internal climate processes, and to exam the relative contributions of all factors. Changes in the climate system can be represented by the differences between the two climate states, namely, the initial and final states. External forcing includes solar radiation, CO_2 concentration, ozone, and aerosols. The internal climate processes include dynamic and thermodynamic components. The dynamic process includes the energy transport by the large-scale atmospheric circulation and convection, and oceanic heat storage and energy transport. Thermodynamic processes include surface sensible and latent heat fluxes, water vapor feedback, cloud feedback, and albedo feedback. These processes interact with each other and ultimately equilibrate with the radiative cooling due to temperature change. The CFRAM has been used to quantify the contribution of the atmospheric dynamic process to the PWA (Lu and Cai 2009) and the relative contributions of climate feedback processes to global warming (Hu et al. 2017), compare the similarities and differences between climate responses to a doubling of CO_2 and a 2% increase in solar radiation (Cai and Tung 2012), analyze the relative contributions of radiation and dynamic processes to different spatial distributions of SST anomalies related to the two types of El Niño (Hu et al. 2016), among others.

In summary, global warming is an indisputable fact based on observed data. However, except for the change in temperature, the trends of other climate factors in the observed datasets and model simulations are not highly consistent. The largest uncertainty exists in the trend of cloud. Some studies have shown that since the end of the 1980s, the global mean cloud volume has decreased, but the total cloud coverage over most of the land surface has increased. However, some regions showed reduced cloudiness, such as China, Italy, and Central Europe. The variability of global mean precipitation is a nonlinear process. Since the 1980s, the global average annual precipitation had decreased, but has continued to increase after the 1990s, which means the water vapor circulation in the atmosphere has been accelerated, and more water vapor has been transported from the tropics to the mid- and high-latitudes. The uncertainties of precipitation and cloud in the model directly affect the simulation and projection of global warming. Therefore, in the last section of this chapter, the uncertainties of tropical precipitation and SST in the model will be discussed.

6.2 Climate Response-Feedback Analysis Method

All the data used by the CFRAM are obtained from the latest European Centre for Medium-Range Weather Forecasts (ECMWF) Reanalysis Interim (ERA-Interim; Dee et al. 2011), as in Deng et al. (2012) and Park et al. (2012). The variables include temperature, specific humidity, ozone mixing ratio, cloud amount, cloud liquid/ice water content, downward solar energy flux at the top of the atmosphere

(TOA), surface albedo, and surface sensible and latent heat fluxes. The time series of the annual mean CO_2 concentration from 1984 to 2013, is downloaded from the Earth System Research Laboratory website (http://www.esrl.noaa.gov/gmd/ccgg/trends/).

Every climate variable that responds to global surface temperature change and affects the Earth's radiation budget may constitute a climate feedback agent (Bony et al. 2006). However, the commonly used framework for climate feedback analysis, based on the premise that the temperature change is the response to (radiative) energy exchange with outer space, has focused on the radiation perturbation at the TOA, or at the tropopause for troposphere–surface system. Popular TOA-based climate feedback analysis methods include the "partial radiative perturbation" method (PRP; Wetherald and Manabe 1988), the cloud forcing analysis method (Cess et al. 1990), and the radiative kernel method (Soden and Held 2006; Soden et al. 2008). However, none of these TOA-based methods calculates the partial temperature changes due to individual feedback processes. In addition, the TOA-based feedback analysis framework cannot explicitly take into consideration the internal non-radiative processes (e.g., oceanic heat storage, convective and large-scale atmospheric energy transport, and surface turbulence fluxes such as evaporation and sensible heat flux). All the aforementioned methods are offline climate feedback diagnosis methods. However, there is an online feedback suppression method to calculate partial temperature change due to a specific feedback process. It requires running the same climate model twice: once with all processes on (original climate system), and the other time with one specific process turned off (a virtual climate system). The difference between these two simulations corresponds to the partial temperate change due to the process under consideration (Hall and Manabe 1999; Schneider et al. 1999). The partial temperature change due to specific feedback inferred by the online feedback suppression method also includes the difference in the other feedbacks between the original and virtual climate systems.

The CFRAM considers the energy budget for the atmosphere-surface column due to all radiative and non-radiative feedbacks. By applying the CFRAM, we can calculate partial temperature changes associated with individual feedbacks. The resultant partial temperature changes are addable, and their sum gives rise to the total temperature response to the external forcing. Therefore, the total temperature change calculated using the CFRAM can be directly compared with its counterpart in the observations or model simulations. In this sense, we isolate the contributions to the (total) temperature change from the external forcing alone, and from individual feedbacks, although physically speaking, these feedbacks are not independent of one another.

According to Lu and Cai (2009), when considering two different equilibrium climate states, the CFRAM method allows us to calculate the vertical profile of partial temperature change, $\Delta \mathbf{T}^{(X)}$, by using the corresponding vertical profile of (linearized) thermal radiative cooling perturbation, $\left(\frac{\partial \mathbf{R}}{\partial \mathbf{T}}\right)\Delta \mathbf{T}$, to be in radiative-equilibrium balance with $\Delta \mathbf{H}^{(X)}$, namely

$$\Delta\mathbf{T}^{(X)} = \left(\frac{\partial\mathbf{R}}{\partial\mathbf{T}}\right)^{-1}\Delta\mathbf{H}^{(X)} \tag{6.1}$$

where $\left(\frac{\partial\mathbf{R}}{\partial\mathbf{T}}\right)$ is the Planck feedback matrix whose jth column represents the vertical profile of (linearized) thermal radiative cooling perturbation due to 1 K warming in the jth layer alone. Note that, as in Lu and Cai (2009), all the vertical profiles are denoted in the form of a vector with the bottom component corresponding to the surface layer and the remaining components corresponding to the atmospheric layers from the layer next to the surface layer to the top layer.

Using the ERA-Interim data, we can evaluate each of the vertical profiles of non-temperature-induced radiative heating perturbations between the two periods listed on the right-hand side (RHS) of Eq. (6.2)

$$\Delta\mathbf{H}^{rad} = \Delta^{(CO_2)}(\mathbf{S}-\mathbf{R}) + \Delta^{(WV)}(\mathbf{S}-\mathbf{R}) + \Delta^{(CLD)}(\mathbf{S}-\mathbf{R}) + \Delta^{(AL)}\mathbf{S} + \Delta^{(O_3)}(\mathbf{S}-\mathbf{R}) \tag{6.2}$$

where **S** and **R** denote the vertical profiles of the solar energy absorbed and the net longwave radiation emitted (or cooling rate) in each layer, respectively. In (6.2), the superscripts "CO_2," "WV," "CLD," "AL," and "O_3" after the symbol "Δ" denote the partial radiative heating perturbations due to changes, respectively, in CO_2 concentration alone, in atmospheric water vapor alone, in atmospheric cloud properties alone (which include cloud area as well as ice and water cloud concentrations), in surface albedo alone, and in ozone alone.

In addition to these non-temperature-induced partial radiative heating perturbations, we also calculated partial radiative heating perturbation due to changes in the vertical profile of temperature, which is denoted as $\Delta^{(Temp)}\mathbf{R}$. We have verified

$$\left(\frac{\partial\mathbf{R}}{\partial\mathbf{T}}\right)\Delta\mathbf{T} \approx \Delta^{(Temp)}\mathbf{R} \tag{6.3}$$

where $\frac{\partial\mathbf{R}}{\partial\mathbf{T}}$ is evaluated using the input taken from the climate mean state, and $\Delta\mathbf{T}$ is the vertical profile of differences in atmospheric and surface temperatures between two different climate states. Equation (6.3) implies that partial radiative heating perturbation due to changes in the vertical profile of temperature can be calculated under the linear approximation. The total radiative heating perturbation, $\Delta^{total}(\mathbf{S}-\mathbf{R})$, can also be calculated as the difference of two calculations of the Fu-Liou radiative transfer model: one with all input fields are taken from one climate mean state, and the other with all input fields from the other climate mean state. We also verified that (not shown here)

$$\begin{aligned}-\left(\frac{\partial\mathbf{R}}{\partial\mathbf{T}}\right)\Delta\mathbf{T} + \Delta^{(CO_2)}(\mathbf{S}-\mathbf{R}) + \Delta^{(WV)}(\mathbf{S}-\mathbf{R}) + \Delta^{(CLD)}(\mathbf{S}-\mathbf{R})\\ +\Delta^{(AL)}\mathbf{S} + \Delta^{(O_3)}(\mathbf{S}-\mathbf{R}) \approx \Delta^{total}(\mathbf{S}-\mathbf{R})\end{aligned} \tag{6.4}$$

This is equivalent to state that the changes in radiative heating rates between the two mean states of the two periods are small enough that they can be partitioned as the sum of individual partial radiative heating perturbations due to changes in one variable/parameter alone. This is the basis of our decomposition at the process level.

Besides radiative heating perturbations, changes in atmospheric and surface temperatures are also due to non-radiative heating perturbations and heat storage terms. The ERA-Interim only has very limited non-radiative heating fields, such as surface latent and sensible heat fluxes. As in Deng et al. (2012), we estimated the sum of non-radiative heating perturbations and storage terms based on the energy balance equation below

$$\Delta \mathbf{H}^{non\_rad} = -\Delta^{total}(\mathbf{S} - \mathbf{R}) \tag{6.5}$$

$$\Delta \mathbf{H}^{non\_rad} = \Delta \mathbf{Q}^{ATM} + \Delta \mathbf{Q}^{S} \tag{6.6}$$

Note that the surface component of $\Delta \mathbf{H}^{non\_rad}$ represents the sum of changes in surface latent and sensible heat fluxes, land surface heat storage (plus some heat loss/gain due to runoff and snow/ice melting/freezing) when the grid point is land. Over a grid point in the ocean, it represents the sum of changes in surface latent and sensible heat fluxes, the net convergence of energy transport by oceanic motions into the entire oceanic column, and the oceanic heat storage term. For an easy reference, we simply denote the surface component of $\Delta \mathbf{H}^{non\_rad}$ as ΔQ_s and the remaining components of $\Delta \mathbf{H}^{non\_rad}$ for the atmospheric layers are all zero (referred to as surface processes hereafter). Similarly, we denote the atmospheric components of $\Delta \mathbf{H}^{non\_rad}$ as $\Delta \mathbf{Q}^{(ATM)}$ whose surface component is zero and the atmospheric components equal to their counterparts of $\Delta \mathbf{H}^{non\_rad}$. Because the atmospheric heat storage is very small when taking the decadal mean, $\Delta \mathbf{Q}^{(ATM)}$ represents atmospheric dynamic processes, including latent heat perturbations due to changes in convection and large-scale atmospheric motion, and to changes in dry static energy flux convergence in each atmospheric layer by vertical and horizontal atmospheric motions, as well as due to changes in sensible heat flux that enters the atmosphere from the surface below.

Using the surface latent heat (Q_{LH}) and sensible heat (Q_{SH}) fluxes provided by the ERA-Interim, we can further decompose changes in net non-radiative heating in the surface layer based on the surface energy balance equation, namely,

$$\Delta Q_{OCH} = \Delta Q_s - \Delta Q_{SH} - \Delta Q_{LH} \tag{6.7}$$

Note that the sign convention of Q_{LH} and Q_{SH} is that negative value that indicates upward energy fluxes, which causes cooling at the surface. The term ΔQ_{OCH} represents changes in land surface heat storage when the surface is land. Over the ocean, it represents the sum of the net convergence of energy transport by oceanic motions into the entire oceanic column and the storage term at a given grid point. Since the term ΔQ_{OCH} is small over land compared to that over the ocean, we simply refer to this term as the oceanic dynamic and heat storage term.

Replacing $\Delta\mathbf{H}^{(X)}$ in Eq. (6.1), with each of the five terms on the RHS of Eq. (6.2), we obtain the vertical profile of partial temperature changes, $\Delta\mathbf{T}^{(X)}$, where "X" denotes the superscript of the corresponding term. Replacing $\Delta\mathbf{H}^{(X)}$ with $\Delta\mathbf{Q}^{(ATM)}$ whose surface component is zero and the atmospheric components equal their counterparts of $\Delta\mathbf{H}^{non\_rad}$, we obtain $\Delta\mathbf{T}^{(ATM)}$ by solving (6.1). We obtain $\Delta\mathbf{T}^{(OCH)}$ by letting the surface component of $\Delta\mathbf{H}^{(X)}$ be ΔQ_{OCH} and setting its atmospheric components to zero. Through the procedures outlined above, we can obtain a total of nine vertical profiles of partial temperature changes: five for radiative processes and four for non-radiative processes. The surface components of these nine vertical profiles of $\Delta\mathbf{T}^{(X)}$ correspond to partial surface temperature changes associated with the process "X" (denoted as $\Delta T_s^{(X)}$ hereafter). Note that $\Delta T_s^{(ATM)}$ is purely associated with the change in the downward thermal radiation resulting from changes in air temperature in response to $\Delta\mathbf{Q}^{(ATM)}$ since its surface component is zero.

Note that as any other offline feedback analysis methods, our feedback analysis is a post-processing diagnostic that cannot predict changes in other fields. However, the CFRAM method allows us to explicitly calculate partial temperature changes associated with individual processes without requiring the information of the total temperature change. The sum of these partial temperature changes can be directly compared to the total temperature change in the observation. The small differences between ΔT_s^{OBS} and $\sum_X \Delta T_s^{(X)}$ are due to errors introduced in linearization of the radiative transfer model, namely, approximating the RHS of (6.4) with the terms on the left-hand side. Furthermore, numerical values of $(\sum_X \Delta T_s^{(X)} - \Delta T_s^{OBS})$ are very close to zero, much smaller than individual $\Delta T_s^{(X)}$.

6.3 Attribution of Decadal Climate Difference Between 2002–13 and 1984–95

Figure 6.1 shows the annual mean and 5-yr running mean global mean surface temperature (GMST) anomalies from 1979 to 2014, presenting a strong warming trend since the mid-1980s to the end of the twentieth century but coming to a warming hiatus

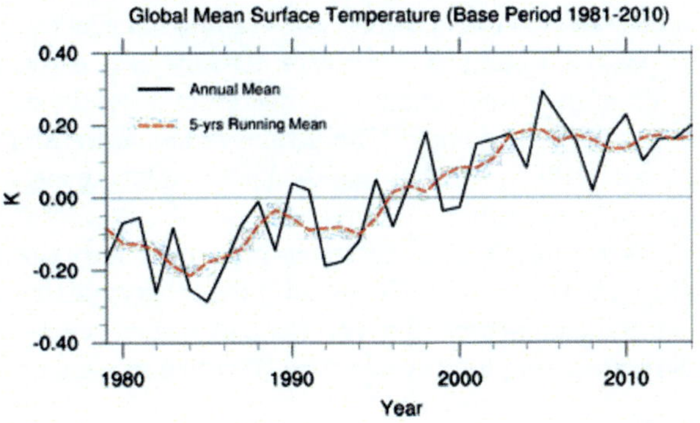

Fig. 6.1 Time series of annual global mean surface temperature anomalies (black solid line) and 5-yr running mean (red dashed line) from 1979 to 2014, based on the climatology of 1981–2010. After Hu et al. (2017)

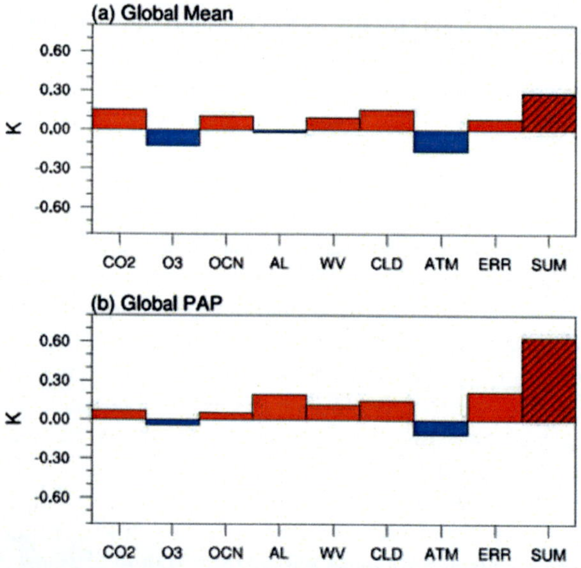

Fig. 6.2 Contributions from individual partial surface temperature changes between two periods associated with changes in "X" alone to the observed surface temperature change and their sum: **a** GMST anomalies and **b** their global pattern–amplitude projection (PAP) coefficients. The hatched portion of the bar labeled as SUM corresponds to the values derived from the ERA-Interim directly (units: K). After Hu et al. (2017)

since 2000. In Hu et al. (2017), we use the slowdown-warming decade 2002–13 as the end decade of the fast warming and 1984–95 as the beginning decade. A process-level attribution of the decadal climate difference between 2002–13 and 1984–95 is discussed in this section to find the sources of fast global warming.

Figure 6.2a shows the partial GMST anomalies associated with different processes and the observed GMST anomalies between the two periods. It indicates that the partial GMST change due to the increase of CO_2 from 1984 to 2013, alone can explain nearly half of the observed global warming between these two periods. The largest contributor to global warming is oceanic heat storage term with the most heat released into the atmosphere by the enhanced evaporation. Changes in clouds contribute 0.14 K to the observed warming. The increase in GMST is further amplified by positive water vapor feedback. The major negative feedback processes are ozone feedback and atmospheric dynamic process, totally contributing a cooling of −0.3 K to the GMST anomaly. The offline radiative transfer model calculation causes an error of 0.99 K, which is mainly due to replacing instantaneous radiation fluxes with climate mean values.

As we discussed in the overview, the changes in surface temperature over the globe are not uniform. To quantify the relative contributions of individual processes to the spatial pattern and amplitude of surface temperature anomalies, we introduce the pattern-amplitude projection (PAP) coefficient below

$$PAP_A^{(X)} = \frac{A^{-1} \int\limits_A a^2 \Delta T_s^{(X)} \Delta T_s^{OBS} \cos\phi d\lambda d\phi}{\sqrt{A^{-1} \int\limits_A a^2 (\Delta T_s^{OBS})^2 \cos\phi d\lambda d\phi}} \tag{6.8}$$

where Φ and λ are latitude and longitude, respectively, and a is the mean radius of the Earth. For an easy reference, the term $\sum_X PAP_A^{(X)} = \sqrt{A^{-1} \int_A a^2 (\Delta T_s^{OBS})^2 \cos\phi d\lambda d\phi}$ is the amplitude of the spatial pattern of observed surface temperature anomalies over the globe. Figure 6.2b shows the PAP coefficients, with large values indicating major contributors to spatial pattern and amplitude of global surface temperature anomalies. Compared to Fig. 6.2a, the contribution of the albedo feedback is larger but that of the CO_2 greenhouse effect is smaller. These differences between Figs. 6.2a, b, are due to the most significant feature of the spatial pattern of the observed surface temperature change being the PWA, which is mainly caused by the albedo feedback.

Displayed in Fig. 6.3, are partial surface temperature anomalies due to individual processes. The greenhouse effect due to a uniform increase in CO_2 tends to be stronger over the regions where climatological water vapor and cloud cover are scarce (Fig. 6.3a). The partial surface temperature changes due to ocean heat storage term (Fig. 6.3c) and surface latent heat flux over the ocean (Fig. 6.3d), largely resemble each other, indicating that most of the heat released by the ocean goes into the atmosphere in the form of latent heat. Abnormal heat entered the atmosphere is further transported from the tropics to the Arctic polar region, resulting in the PWA (Fig. 6.3j). The albedo feedback is also a major contributor to the PWA in the Arctic region (Fig. 6.3f). Moreover, an increase in water vapor content in the warmer climate results in surface warming (Fig. 6.3g), and more clouds over the tropical oceans. The net radiative effect of more clouds cools the tropical oceans (Fig. 6.3h). The reduction of the solar energy flux reaching the surface (Fig. 6.3i), is due to the upward trend of the stratosphere ozone. Because of a stronger increase of the stratosphere ozone in the tropics, as well as more available incoming solar energy, the reduction in the downward solar energy flux at the surface is stronger in the tropics. The sum of all partial temperature changes (Fig. 6.3l), is almost the same as the observed surface temperature change (Fig. 6.3k).

6.4 Feedback Attribution of Interannual Variability

6.4.1 Attribution of the Distinct Spatial Patterns of SST Anomalies for Two Types of El Niño

According to the location of the warmest SST anomalies, El Niño can be roughly divided into two types: eastern Pacific (EP) El Niño (Fig. 6.4b; citations) and central Pacific (CP) El Niño (Fig. 6.4a; Yu and Kao 2007). The CFRAM is applied to explain

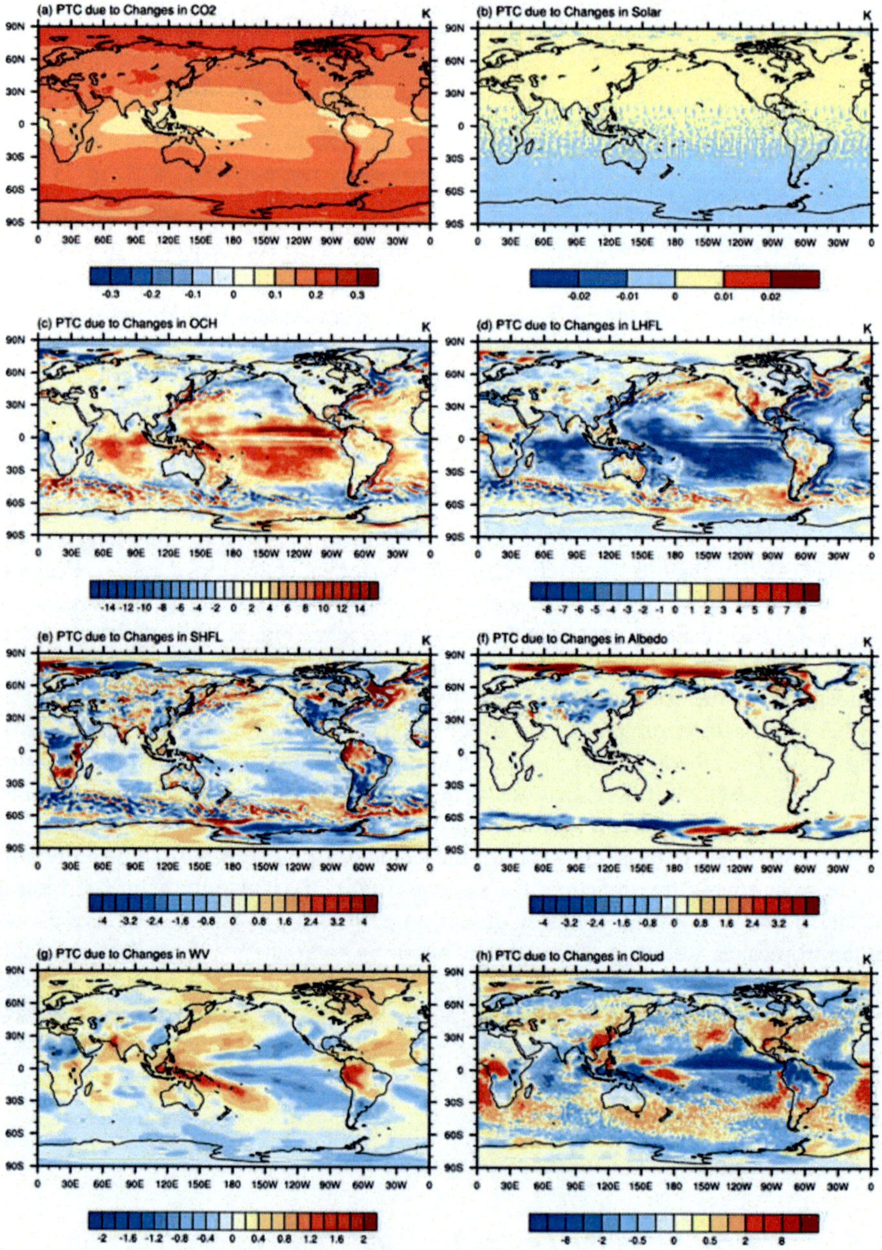

Fig. 6.3 Maps for partial surface temperature changes (units: K) due to individual processes between two periods: **a** ΔT_S^{CO2}, **b** ΔT_S^{SR}, **c** ΔT_S^{OCH}, **(d)** ΔT_S^{LH}, **e** ΔT_S^{SH}, **f** ΔT_S^{AL}, **g** ΔT_S^{WV}, **h** ΔT_S^{CLD}, **i** ΔT_S^{O3}, **j** $\Delta T_S^{atmos\_dyn}$, **k** ΔT_S^{ERR}, and **(l)** the sum of **(a)** to **(j)**

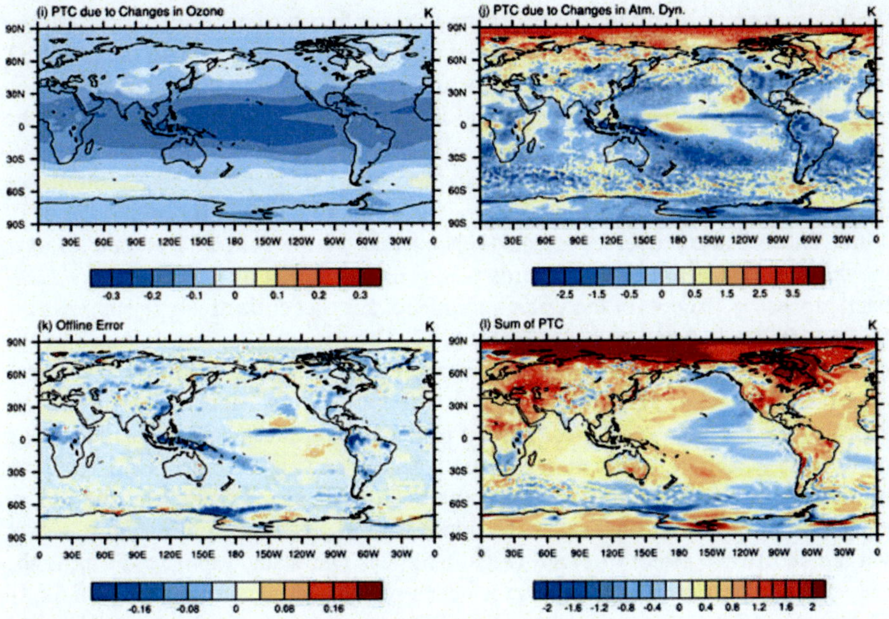

Fig. 6.3 (continued)

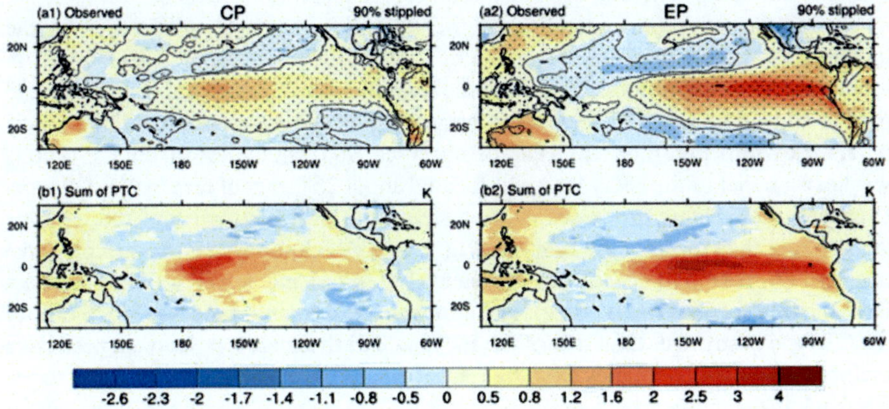

Fig. 6.4 DJF mean SST anomalies: observed SST anomalies **a**, **b** and the sum of partial SST anomalies for the CP El Niño (left panels) and EP El Niño (right panels) in winter derived via the CFRAM at the surface (**c**, **d**). Stippling in a and b indicates the 90% confidence level of statistical significance. Units: K. After Hu et al. (2016)

the different spatial distributions SST anomalies for the two types of El Niño in winter. In Figs. 6.4c, d, we can see that the partial SST anomalies do add up to the total SST anomalies reasonably well, which gives us the confidence to carry out attribution analysis. Next, we focus on the contributions of individual feedback processes to the SST anomaly patterns of the two types of El Niño.

As shown in Figs. 6.5a and b, oceanic processes lead to positive SST anomalies over the central and eastern equatorial Pacific as the root causes for both types of El Nino. The oceanic processes are also responsible for the cold SST anomalies over the western equatorial basin. Figures 6.5c–j illustrate that all other negative feedback processes, except for the water vapor feedback, act collectively to suppress the SST anomalies induced by oceanic processes. The abnormal energy released by the oceanic processes induces anomalous energy going into the atmosphere via evaporation (Figs. 6.5c, d) and sensible heat flux (Figs. 6.5e, f). More energy and water vapor in the air enhance the convection over the central equatorial Pacific, taking more energy from the lower troposphere to the mid and upper troposphere, resulting in SST cooling (Figs. 6.5i, j). At the same time, more cloud cover over the central equatorial Pacific also suppresses the initial warming caused by oceanic processes via the shortwave effect of cloud (Figs. 6.5g, h). The water vapor feedback is the only positive process that amplifies the ocean-related initial warming (Figs. 6.5k, l). The most significant difference in partial SST anomalies between EP and CP El Niño events is that the location of the maximum partial SST anomalies for CP El Niño in winter is located to the west of the location for EP El Niño in winter.

To gain a better understanding of the contributions to the difference in the longitudinal distributions of SST anomalies between the two types of El Niño from individual feedback processes, we plot the longitude profiles of total (Fig. 6.6a) and partial (Figs. 6.6b–h), SST anomalies averaged over the equatorial band of 5° S–5° N. As shown in Fig. 6.6a, the peak of observed SST anomalies of the CP type is weaker than that of the EP type, and is located about 25° west of that for the CP type. However, the oceanic processes that induce partial SST anomalies of the CP type are stronger than those of the EP type (Fig. 6.6b), which indicates that the collective negative effect (Fig. 6.6c), of surface latent (Fig. 6.6e) and sensible (Fig. 6.6f) heat fluxes, cloud feedback (Fig. 6.6g), and atmospheric dynamic process (Fig. 6.6h), of the CP type is stronger than that of the EP type. All these aforementioned processes mainly affect SST anomalies over the central equatorial Pacific, and the water vapor feedback is the major source of the different SST anomalies over the eastern equatorial Pacific between the two types of El Niño. The water vapor feedback is the sole factor that explains the feature that the warming over the eastern basin is stronger in the EP type than in the CP type (Fig. 6.6d).

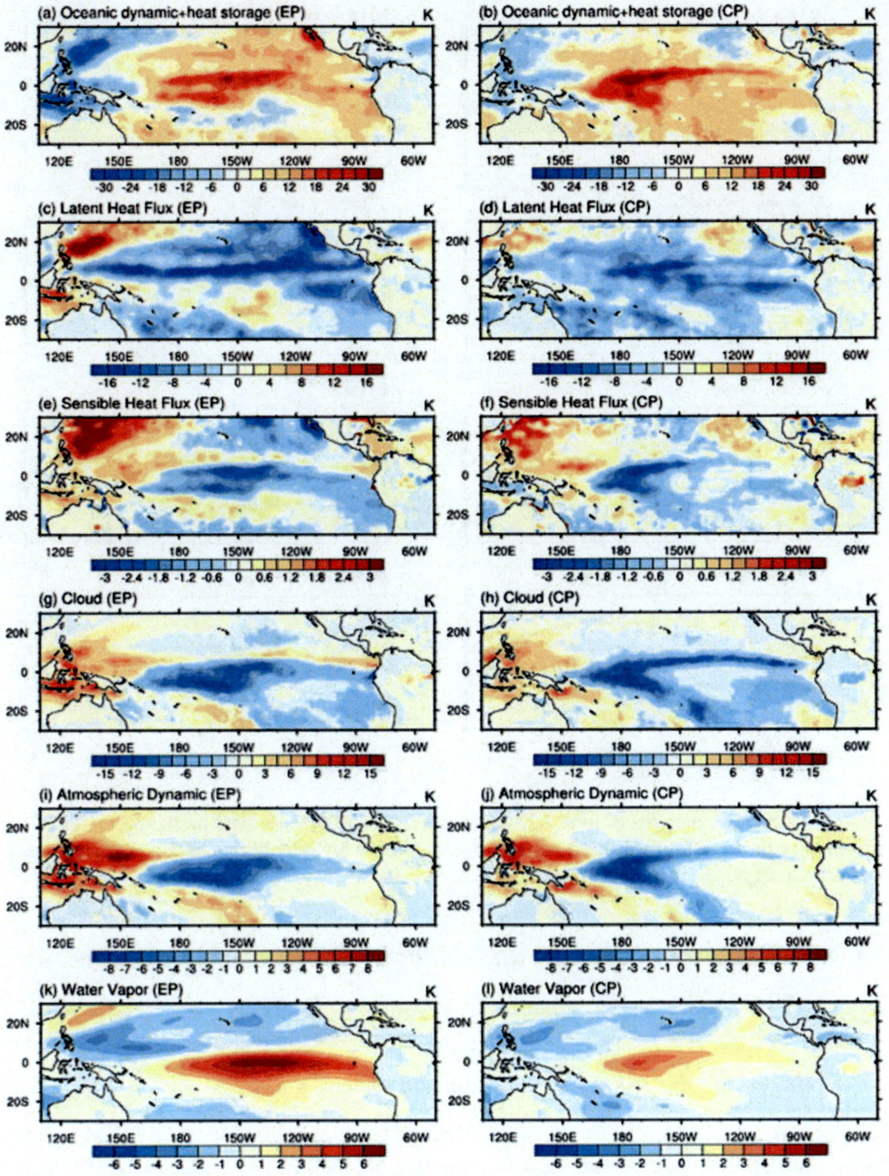

Fig. 6.5 CFRAM-derived partial SST anomalies for the EP El Niño in the boreal winter due to oceanic process (**a**), latent heat flux (**c**), sensible heat flux (**e**), cloud (**g**), atmospheric dynamic feedback (**i**), and water vapor feedback (**k**) processes at the surface. **b**, **d**, **f**, **h**, **j**, and **l** are the same as (**a**, **c**, **e**, **g**, **i**, and **k**), except for the CP El Niño. Units: K. After Hu et al. (2016)

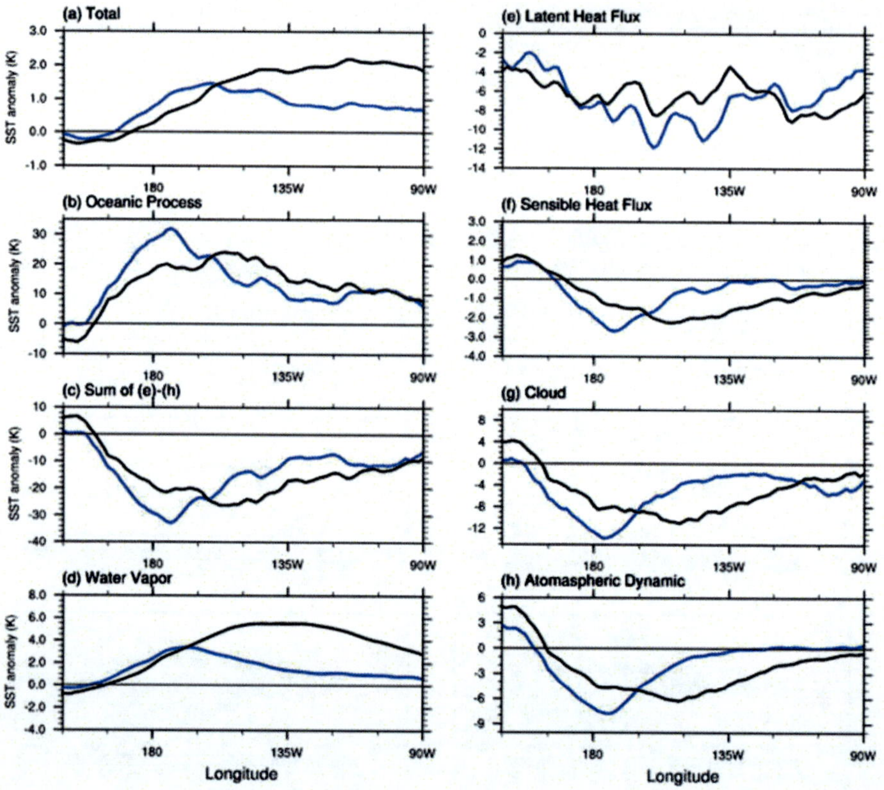

Fig. 6.6 Composite patterns of total SST anomalies and CFRAM-derived partial SST anomalies (specified in the y-coordinate) meridionally-averaged over 5° S–5° N for the EP El Niño (black line) and CP El Niño (blue line) in winter. Units: K. After Hu et al. (2016)

6.4.2 Variations of Dominant Modes of the East Asian Winter Monsoon

a. Coupling of the East Asian winter monsoon system

The movement and development of the East Asian winter monsoon (EAWM) affect global atmospheric circulation. At the same time, its intensity is controlled by changes in the atmospheric circulation, especially by the changes from the mid and high latitudes (Chang 2004; Wang et al. 2009). For example, when the Arctic Oscillation (AO) is in its negative phase, the frequency and intensity of cold surges increase, and the EAWM is stronger (Wu and Wang 2002; Gong and Wang 2003; Park et al. 2014). The emergence and enhancement of the Ural blocking can usually affect the intensity of the Siberian high, which further influences the intensity of the EAWM and the path and extent of its southward invasion (Chang and Lau 1982; Gong and Ho 2002; Wang et al. 2010; Cheung et al. 2012). Climate change in the low latitudes can also significantly affect the interannual variation of EAWM intensity. For example, in the winter of La Niña, the EAWM is stronger, while in the El Niño winter, the EAWM is weak (Li 1990). The ENSO mainly influences the strength of the EAWM through the anomalous anticyclone over the northwestern Pacific Ocean (Wang et al. 2010). In the strong years of EAWM, strong northeasterly cold air can advance southward to reach the South China Sea and the MC, inducing deep convection in the region (Chang et al. 2004). The latent heat released from deep convection at high altitudes acts on the upper atmosphere, strengthening the East Asian jet stream, affecting teleconnection patterns and thus global climate change (Lau and Chang 1987; Yang and Webster 1990; Lau and Nath 1996; Yang et al. 2002; Chang 2004).

According to previous studies, the meteorological elements that characterize the strength of the EAWM include near-surface temperature, wind speed, sea surface pressure, and 500 hPa geopotential height. The meteorological elements for the mode analysis of the EAWM mainly include near-surface air temperature (Wang et al. 2010), 850 hPa wind (Wu et al. 2006), and precipitation (Wang and Feng 2011). Taking into account the spatial distribution uniformity of the meteorological elements and the record length of the data, we chose 2-m air temperature to characterize the intensity of the EAWM.

b. Data and methods

The data used in this chapter are mainly from the monthly ERA-Interim reanalysis (Dee et al. 2011), with a spatial resolution of 1° × 1° and 37 isobaric surfaces. They cover the period of 1979–2013. The meteorological variables include CFRAM variables (consistent with the CFRAM method, see Sect. 6.1), wind field and geopotential height. Precipitation data are from the GPCP, which has a spatial resolution of 2.5° × 2.5°.

We apply the EOF analysis on air temperature 2 meters in winter. The strong and weak monsoon years are selected using one standard deviation of the principal component (PC) of each mode. Then, we use the CFRAM to analyze the temperature difference between the strong and weak EAWM years, using Eq. (6.9)

$$\Delta\vec{T} = \left(\frac{\partial\vec{R}}{\partial\vec{T}}\right)^{-1}\left[(\Delta\vec{S}^{wv} - \Delta\vec{R}^{wv}) + (\Delta\vec{S}^{c} - \Delta\vec{R}^{c}) + (\Delta\vec{S}^{O_3} - \Delta\vec{R}^{O_3}) + \Delta\vec{S}^{\alpha} + (\Delta\vec{Q} - \frac{\partial\vec{E}}{\partial t})\right] \tag{6.9}$$

where $\Delta\vec{T}$ is the temperature difference between the strong and weak EAWM years, "wv" is water vapor, "c" is cloud, "O_3" is ozone, and "α" is albedo. Since the changes in solar radiation and carbon dioxide are too small between the two EAWM states to have any effect on the EAWM interannual variation, we set the two factors as constants in the two states. According to the energy balance, we decompose the temperature change into partial temperature changes (PTCs) caused by all radiative and non-radiative processes that affect the energy balance in the Earth-atmosphere system. The magnitude and spatial distribution of each PTC determine its contribution to the overall temperature change.

c. Dominant modes of the East Asian winter monsoon

Our study domain is (100°–140° E, 0°–60° N), the same as that in Wang et al. (2010) and Sun et al. (2016). An EOF analysis indicates that in the first mode (EOF1; Fig. 6.7a), accounts for 52.2% of the total variance. The negative anomalies are located north of 40° N with its cold core over the Mongolia-Siberia region. The second mode (EOF2; Fig. 6.7b), accounts for 18.7% of the total variance. It illustrates that temperature decreases over most of southeastern China and the cold area extending to the equatorial region in strong EAWM years. In addition to the obvious interannual variation, both modes exhibit interdecadal variability. The EOF1 exhibits strong EAWM phases in the early-1980s and late-2010s, and weak phases during the 1990s and the 2000s (Fig. 6.7c), while the EOF2 shows weak phases in the late-1980s, the late-1990s and the 2000s, and a strong phase in the early-1980s (Fig. 6.7d), which is consistent with Ding et al. (2014). Due to the record length of the data analyzed in

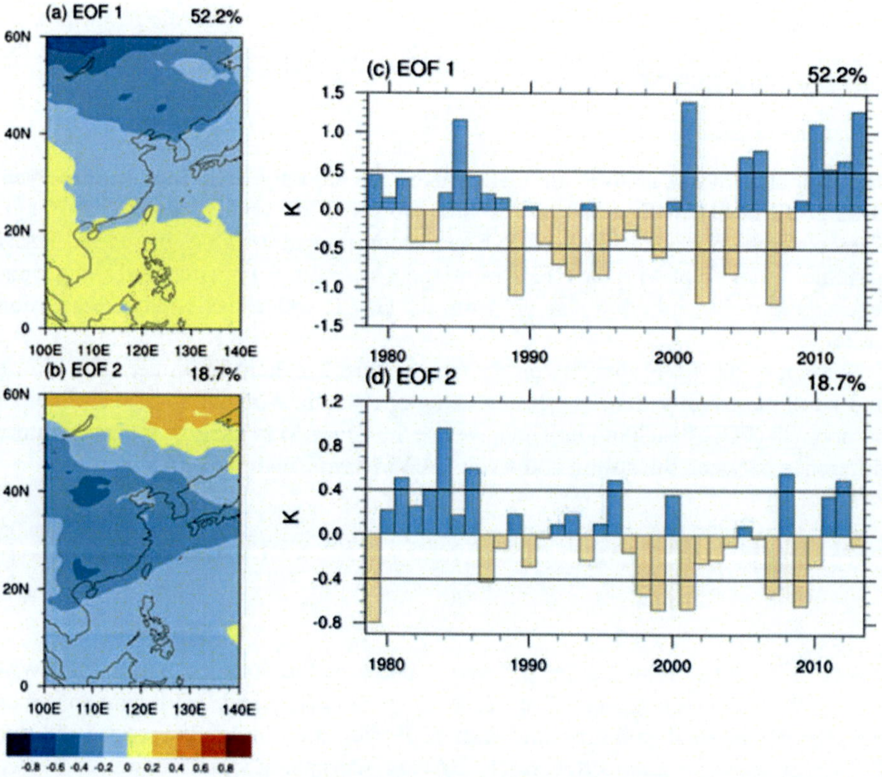

Fig. 6.7 Spatial patterns of **a** the first and **b** second EOF modes of boreal winter (December to February) mean 2-m air temperature and the PCs of **c** the first and **d** second EOF modes. The black solid lines in (**c, d**), indicate the values corresponding to one standard deviation. After Li and Yang (2017)

this study, the results of this study focus mainly on interannual features rather than on interdecadal variation of the EAWM.

From the spatial distributions of the two main modes and the related circulation (not shown), the EOF1 is closely related to the circulation in the mid-to-high latitudes. In the years of strong EAWM-EOF1, the cold air from the Arctic polar region intrudes southeast toward southeastern China, eastern China, Japan, and the Northwest Pacific, and an abnormal barotropic system exists in northern Siberia. The system extends from near surface to 500 hPa. It is much more stable and accompanied by an anomalous easterly wind. At the same time, the East Asia trough is located more to the west-northwest, and the 200 hPa jet stream strengthens over central and northern China. Therefore, the cold source of EAWM-EOF1 is located in the mid-to-high latitudes, which is intercepted by the abnormal low-level strong easterly wind along with the anomalous high-pressure system in northern Siberia, and the strong westerly wind along the East Asia trough. Such anomalous circulations do not facilitate the southward intrusion of cold air. The EOF2 is coherent with low-latitude atmospheric circulation. In the strong years of EAWM-EOF2, the cold air is mainly accumulated in the southern part of East Asia and the western North Pacific. The abnormal high-pressure over the East Asian continent and low- pressure over the western North Pacific strengthen the ocean-land pressure gradient in winter, which is favorable to the strengthening of the northeasterly monsoon along the East Asian coast. At the same time, the East Asian trough deepens to the southeast, and the high-level westerly jet weakens. Such circulations facilitate the enhancement of the EAWM, which further affects the southern part of East Asia and even the equatorial region.

d. Crucial factors of strong and weak monsoon years

To explore the main contributing factors that affect the intensity of the EAWM, we apply the CFRAM to quantitatively determine the key factors affecting the strength of the two main modes. The factors include all radiative and non-radiative processes: water vapor, cloud, atmospheric dynamics, surface latent/sensible heat flux, oceanic dynamics and ocean/land heat storage (OCH), albedo, and ozone.

We define strong (weak) EAWM years using one standard deviation of each PC. For EOF1, 1985, 2001, 2005, 2006, 2010, 2011, 2012, and 2013 are eight strong EAWM years (referred to as strong EAWM-EOF1 years), while 1989, 1992, 1993, 1995, 1999, 2002, 2004, and 2007 are eight weak EAWM years (referred to as weak EAWM-EOF1 years). For EOF2, 1981, 1984, 1986, 1996, 2008, and 2012 are six strong EAWM years (strong EAWM-EOF2 years), and 1979, 1987, 1998, 1999, 2001, 2007, and 2009 are seven weak EAWM years (weak EAWM-EOF2 years). According to Eq. (6.9), we separately decompose T2m changes $(\Delta \vec{T}\ \Delta T)$ of the two main modes between the weak and strong EAWM into the PTC due to each feedback process. Since both albedo and ozone contribute insignificantly to changes in T2m, the following analyses will focus on the other factors.

For EOF1 (Fig. 6.8), the changes in water vapor, cloud, atmospheric dynamic processes, and surface sensible heat flux are positive contributors for the EOF1 cold

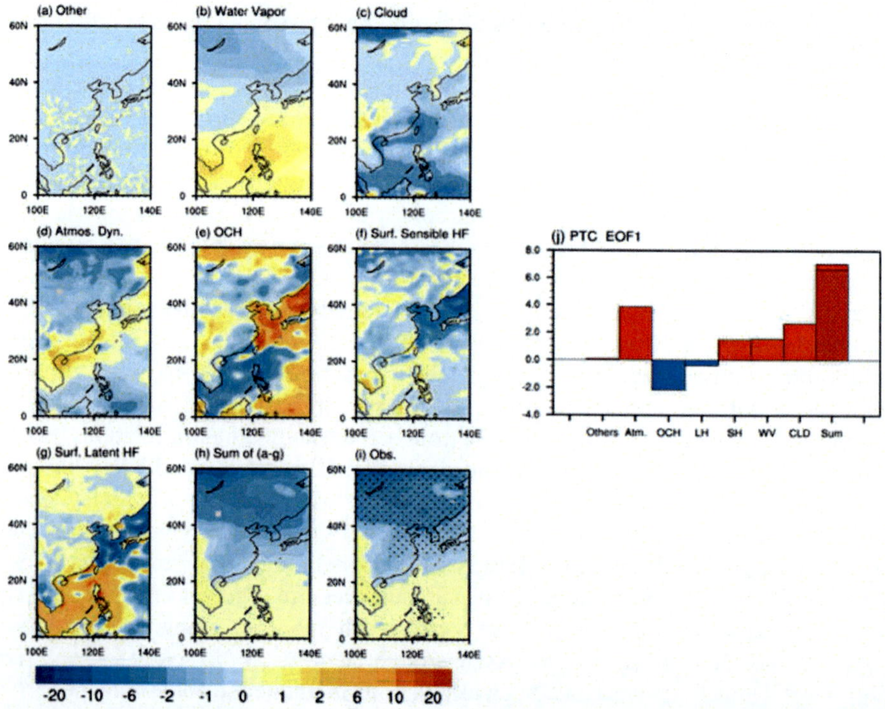

Fig. 6.8 For EOF1, the partial temperature changes near the surface due to the changes in **a** other feedback processes (albedo and ozone), **b** water vapor, **c** cloud, **d** atmospheric dynamics, **e** oceanic dynamics and ocean/land heat storage (OCH), and **f** sensible heat flux at the surface, and **g** latent heat flux at the surface. **h** is the sum of (**a**) to (**h**), and the observation is shown in (**i**). Stippling indicates it is significant at the 95% confidence level. **j** shows pattern-amplitude projections (PAPs) of eight partial temperature anomalies onto the changes in temperature near the surface at the cold core (100°–120° E, 50°–60° N). After Li and Yang (2016)

center (around 40° N), while the changes in land heat storage and surface latent heat flux are negative contributors, which is not favorable for the formation of the cold center. It is worth mentioning that change in OCH cools the land in the southern part of the cold center, but exerts heating effect on the adjacent sea. Therefore, the land is abnormally cold and the ocean is abnormally warm, strengthening the regional ocean-land thermal contrast, which facilitates the strengthening of the northeasterly wind. However, at the low-latitudes (south of 30° N) the magnitude of oceanic cooling is greater than that over the adjacent land, thus weakening the ocean-land thermal contrast, which is not conducive to the strengthening of low-latitude northeasterly wind. Although the change in surface latent heat flux in lower latitudes weakens the influence of sea-land thermodynamic processes on EAWM intensity to a certain extent, it does not completely offset that influence.

For EOF2 (Fig. 6.9), the changes in water vapor, atmospheric dynamic processes, and surface sensible heat flux during the strong and weak EAWM years cool the

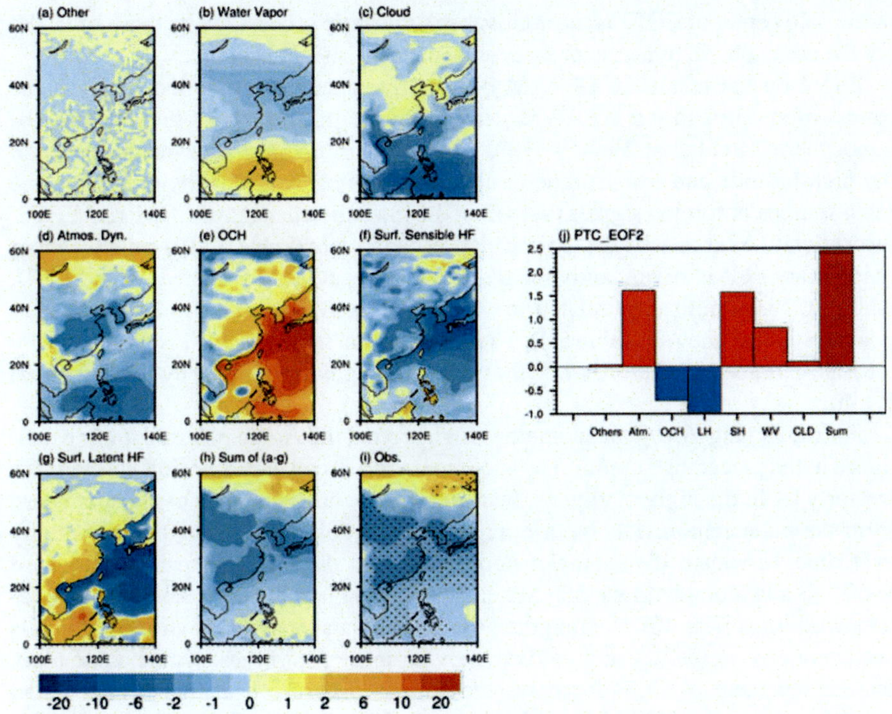

Fig. 6.9 Same as Fig. 6.8, except for EOF2 and (j) of the cold core (105°–120° E, 25°–45° N). After Li and Yang (2016)

surface, favoring the formation of a cold center (in northern and central China). The cold center of the PTC of water vapor is in northern China, while the changes in atmospheric dynamics are over central and eastern China; and the change in surface sensible heat flux is located along the coast of China, cooling the entire East Asia. The change in cloud mainly cools southern China and heats up northern and central China, which is not conducive to the formation of a cold center in this area. The change in OCH significantly warms up eastern China and its adjacent ocean surface, and the change in surface latent heat flux heats up land and cools the ocean, both of which are not conducive to the formation of terrestrial cold centers of EOF2.

In the following, we discuss the relative contribution of each factor to T2m spatial distribution and amplitude variation based on the "pattern-amplitude" projection (PAP) coefficient (Deng et al. 2012; Park et al. 2012). From the perspective of spatial distribution and contribution magnitude (Figs. 6.8j and 6.9j), the cold centers of the two main modes obtain positive contributions from the changes in water vapor, cloud, surface sensible heat flux, and atmospheric dynamics, and negative ones from the changes in surface latent heat flux and OCH, warming the surface. However, the relative magnitude of each factor varies for each main mode. The positive contributor of the cold center of EOF1 is the atmospheric dynamic process, while those of EOF2 are the atmospheric dynamic process and surface sensible heat flux.

Based on the results of CFRAM quantitative analysis, we further analyze the causes of strong and weak EAWM years from the perspective of the evolution of atmospheric circulation. First, we select the AO and ENSO indices to characterize the high-latitude and low-latitude circulation systems, respectively, to explore the main sources of forcing signals that affect the intensity changes of EOF1 and EOF2, respectively. As shown by the lag correlation coefficient (Table 6.1), from November to February PC1 is significantly related to the AO, but weakly related to the ENSO, while PC2 is significantly related to the ENSO from September to December but poorly related to the AO (Table 6.2). Therefore, the intensity of EOF1 is dominated by the high-latitude circulation, while that of EOF2 is greatly affected by the equatorial Pacific.

For EOF1 (Fig. 6.10), an anomalous high around the Arctic polar region is established in the preceding autumn. The accompanying strong easterly wind weakens the westerly jet in the higher latitudes, favoring the dry and cold air moving southward from the polar region. The barotropic anomalous cold high strengthens gradually over time. In winter, the anomalous cold high near the surface expands southward to 50° N; and along with the 500 hPa East Asia trough more northwest, the anomalous westerlies near 40° N strengthen the East Asian westerly jet stream, which is not conducive to the dry and cold air moving more southward. At the same time, the enhancement of OCH increases the ability to absorb heat in the East Asian coastal areas south of 30° N, resulting in anomalous cooling of the offshore surface, thereby weakening the regional ocean–land thermal contrast south of 30° N; this is not conducive to dry and cool air moving southward. In the spring of the following year, the atmospheric circulation signals related to EOF1 are no longer significant.

For EOF2 (Fig. 6.11), the anomalous signals in the preceding autumn circulation are not as significant as those of EOF1, except for an anomalous high at 500 hPa over the Siberian region. In winter, anomalous high exists over the East Asian continent, and an anomalous low appears over the western Pacific; thus, the enhancement of ocean-land pressure gradient over the East Asian coastal region will benefit the northeasterly trade wind. As the Siberian high rapidly strengthens, the anomalous easterly wind at its southern part weakens the westerly jet, thus facilitating the dry and cold air moving southward. The OCH in the entire east coast of East Asia weakens as its heat storage rate weakens, leading to anomalous heating over this region. Such anomalous circulations strengthen the ocean–land thermal contrast, which favors dry and cold air moving southward, strengthening the EAWM. In the following spring, there is an abnormally high near the barotropic structure over the Okhotsk Sea. The trade winds along the eastern coast of East Asia continue to exist in the following spring. Therefore, EOF2 can be used as one of the predictors for climate change in the following spring.

This section starts with the two main modes of the EAWM as EOF1 and EOF2, and we use the CFRAM to quantitatively analyze the main contributors of monsoon intensity of the two modes. EOF1 explains a variance of 52.2% with its cold center located north of 40° N, closely related to the mid-to-high-latitude circulation system,

Table 6.1 Correlation coefficients of the AO and Nino3.4 with PC1 from the previous September to the following May. Statistical significant values at 95% confidence level are shown in bold

PC1	Sep	Oct	Nov	Dec	Jan	Feb	Mar	Apr	May
AO	−0.0846	−0.0928	−0.3326	−0.3832	−0.5265	−0.4817	−0.3122	0.0855	−0.1466
Nino34	−0.1085	−0.0835	−0.1062	−0.1378	−0.1058	−0.0855	−0.0805	−0.0858	−0.0760

Table 6.2 Same as Table 6.1, except for PC2

PC2	Sep	Oct	Nov	Dec	Jan	Feb	Mar	Apr	May
AO	0.1781	0.1514	−0.0470	−0.0010	0.2220	0.0005	0.0323	−0.0015	−0.0517
Nino34	−0.3702	−0.3732	−0.3709	−0.3291	−0.3028	−0.3092	−0.3146	−0.2722	−0.2052

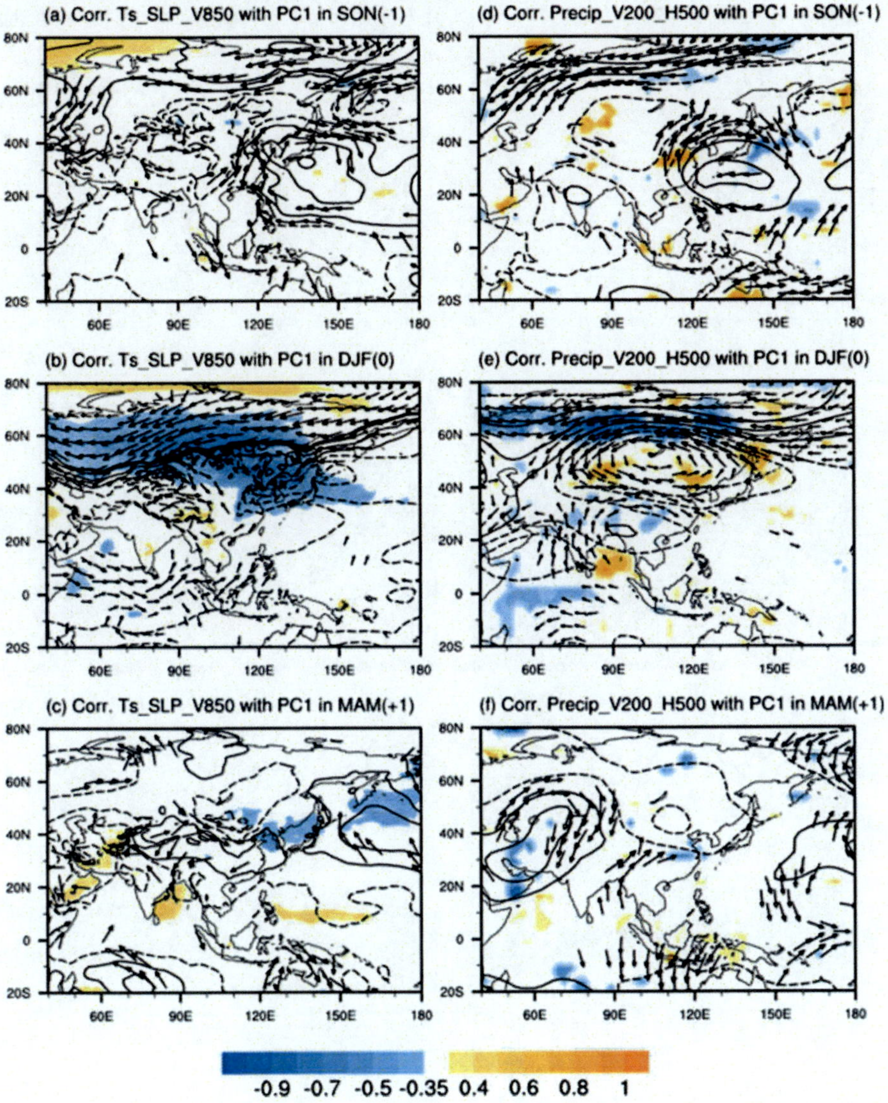

Fig. 6.10 Spatial patterns of correlation coefficients between PC1 and **a** Ts (shading), sea-level pressure (SLP; contour with interval of 0.15 hPa/K) and 850 hPa wind (vector), and **d** precipitation (shading), 200 hPa wind (vector) and 500 hPa geopotential height (contour with interval of 0.2 gpm/K) with PC1 in the preceding autumn (SON(-1)). **b** and **e** are for winter (DJF(0)); **c** and **f** for the following spring (MAM(+1)). The magnitude of vectors in each panel varies. Note that 0.1 cm refers to 1 m/s in (**a**), (**c**), (**d**), and (**f**), while 0.1 cm refers to 2 m/s in (**b**) and (**e**). The correlation coefficients in the areas with shading or vector are all above the 95% confidence level

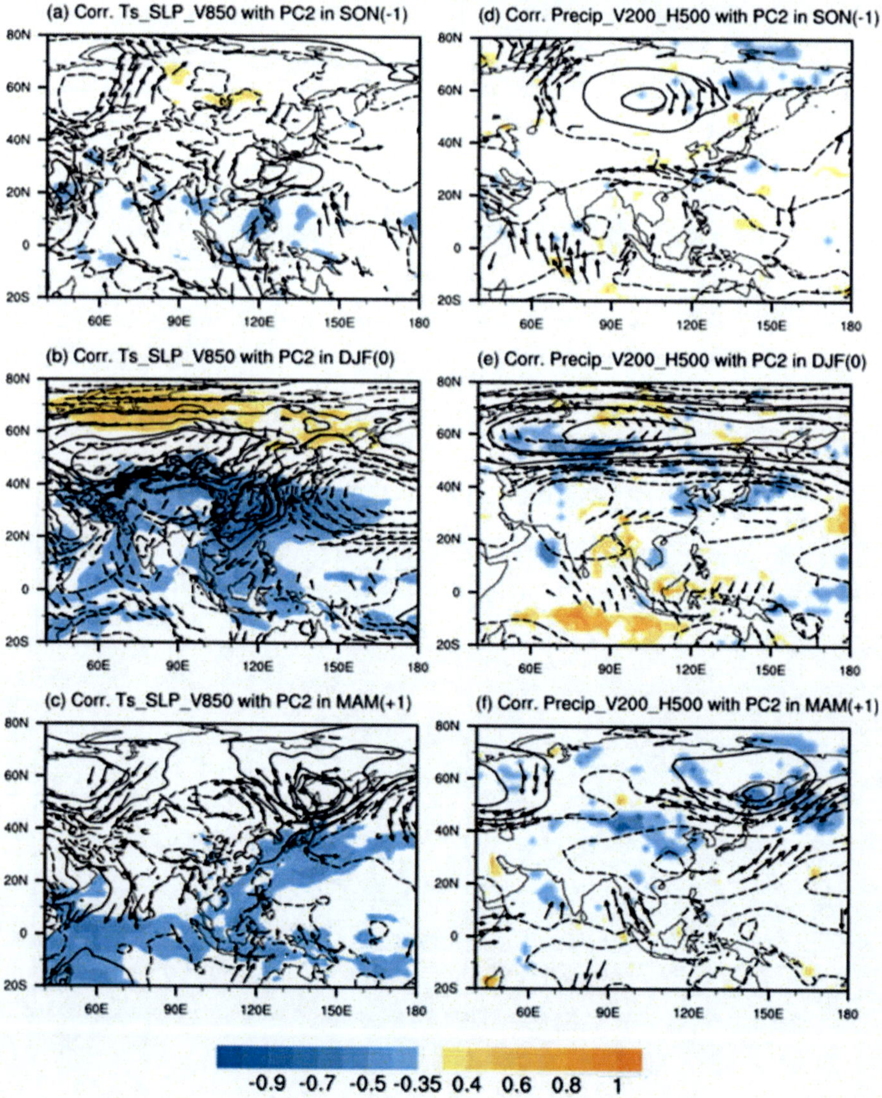

Fig. 6.11 Same as Fig. 6.10, except for PC2

and the correlation coefficient with the AO is as high as -0.53. The variance of EOF2 is 18.7%, and the cold center is located south of 40° N, which is closely related to the low-latitude circulation, with a correlation coefficient of −0.37 with the ENSO.

The main contributors to the two EAWM modes are the same, but the relative contribution strengths vary. The changes in water vapor, clouds, surface sensible heat flux, and atmospheric dynamic processes cause cooling of the cold center, namely, as positive contribution factors. However, the changes in surface latent heat flux, oceanic dynamic processes, and ocean–land heat storage are not conducive to the formation of a cold center due to their heating impact; they are negative contribution factors. Atmospheric dynamic process is the main positive contribution factor of EOF1, while the atmospheric dynamic process and surface sensible heat flux are the main positive contribution factors of EOF2.

6.5 Feedback Attributions of Climate Changes Over the South China Sea and Its Adjacent Regions

6.5.1 Annual Cycles of Surface Temperature

Compared with deep convective precipitation, less research has been carried out on variation of temperature over the MC and its surrounding regions, although it is an integral part of the regional energy and moisture circulations (Sui and Lau 1992). Nicholls et al. (1996) pointed out that the spatial distribution of MC convective precipitation is closely related to the MC near-surface temperature and its adjacent SST, because changes in near-surface temperature can be used to predict sea-level pressure change; and sea-level pressure anomalies indicate anomalous atmospheric divergence, which can trigger deep convection. Therefore, the change of MC near-surface temperature plays an important role in regulating the large-scale circulation. Because the MC sea-land thermal contrast varies with season, the deep convection zone jumps northward during winter, while deep convection gradually moves southward during summer (Sukanto 1969; Hung et al. 2004; Chang et al. 2005). "Dry" biases over the MC land occur in most models, mainly due to models' inability to mimic the sea-land thermodynamic difference (Simpson et al. 1993; Van den Dool and Saha 1993; Chang et al. 2005). Webster and Yang (1992) pointed out that the weather forecast obstacles in spring over the western tropical Pacific were mainly due to the rapid development of errors in spring. Therefore, it is extremely important to understand the MC temperature change and to deepen our understanding of the mechanism involved. Next, we first analyze the climatic annual cycle of thermal characteristics over the MC, and then use the CFRAM to analyze the attribution of the annual cycle, and finally explain the underlying mechanism.

The data used in this subsection are mainly from the monthly ERA-Interim reanalysis (Dee et al. 2011), which has a spatial resolution of $1° \times 1°$ and 37 isobaric surfaces. The data cover the period of 1981–2010. Meteorological elements include the CFRAM variables (consistent with the CFRAM method described in Sect. 6.2) and outgoing longwave radiation.

The monthly climate state was obtained over the period of 1981–2010. The temperature difference between two adjacent months is defined as the temperature change between the two months. For example, temperature in January minus that in February is the monthly temperature change between January and February, and the time is recorded as January; and the "December" of the monthly temperature difference indicates the temperature change from December to January. We use the CFRAM to analyze the MC near-surface temperature change, as shown in Eq. (6.10)

$$\Delta T = \left(\frac{\partial R}{\partial T}\right)^{-1}\left[\Delta S^{solar} + (\Delta S - \Delta R)^{wv} + (\Delta S - \Delta R)^{c} + \Delta S^{\alpha} + \Delta Q^{atmos\_dyn} + \Delta Q^{surface}\right] \quad (6.10)$$

where ΔT is the temperature change in each month, "solar" is the solar radiation at the TOA, "wv" is water vapor, "c" is cloud, α is albedo, "atmos_dyn" is atmospheric dynamic process, and "surface" indicates the surface process including surface sensible/latent heat fluxes and OCH. According to the energy balance, we decompose the temperature change into PTCs caused by all factors that affect the energy budget in the climate system. The contribution of each process to the overall temperature change and its magnitude is represented by PAP coefficient.

PAP is used for quantitative analysis of each month. On this basis, to obtain the contribution of each factor to the annual cycle of surface temperature, we extend it to "temporal-PAP" (TPAP), as shown in Eq. (6.11):

$$TPAP_i = \frac{\sum_1^{12} (PAP_{in} * \Delta T_n)}{\sum_1^{12} (\Delta T_n)^2} \quad (6.11)$$

where i and n denote the i-th factor and the nth month (interval), respectively. ΔTn denotes the average temperature during 12 months in area A. Therefore, the combination of PAP and TPAP can be used to quantitatively analyze the relative contribution of each factor in space and time.

a. Observations

The MC is located in the warmest part of the world ocean. Although the near-surface temperatures are above 300 K throughout the year, obvious seasonal variation in temperature exists (Fig. 6.12). The northern MC (north of the equator) gradually warms up from February to May, and gradually cools from June to the following January. The southern MC (south of the equator) gradually warms up from February to April and from August to November, and gradually cools from May to July and from December to the following January. Due to the small thermal inertia on land, the land surface temperature changes significantly annually, compared with the SST in the adjacent sea. As we all know, solar radiation is one of the major forces that drive seasonal change. Comparing the annual cycle of solar radiation at the TOA with that of near-surface temperature (Fig. 6.12i), both have double peaks, and the temperature peaks lag behind the peaks of solar radiation for 1–2 months.

The MC gradually warms up during February–May and September–November, and becomes colder during June–August and December–January (January of the following year). Rapid warming occurs in March–April and September–October, while rapid cooling is in June–July. The peak of month-to-month variation in near-surface temperature is later than the peak of solar radiation by two months (Fig. 6.12j).

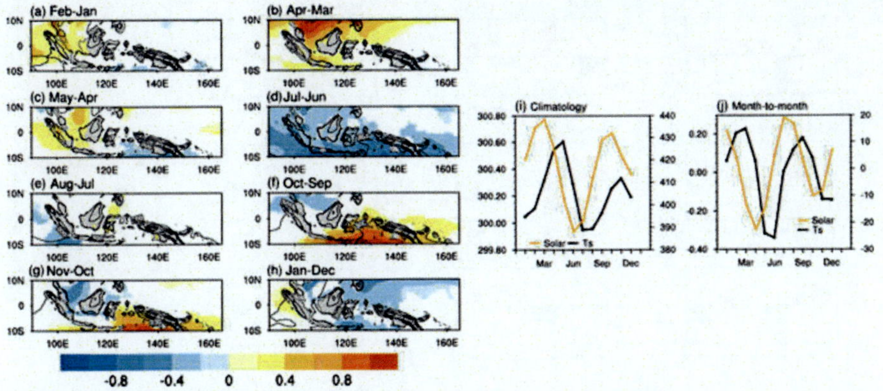

Fig. 6.12 Left (eight panels): Month-to-month variation in near-surface temperature during 1981–2010 (shading; K), significant at the 95% confidence level. The contours indicate the climatology for each month. For example, the contours in **a** represent the climatology in January, and those in **h** is the climatology in December. Right (two panels): **i** Monthly mean TOA solar radiation (W/m^2; yellow curve that uses the y-coordinate on the left) and near-surface temperature (K; black curve that uses the y-coordinate on the right) over (90°–165° E, 10° S–10° N) during 1981–2010. **j** is the same as (**i**), except for month-to-month variation. After Li et al. (2017)

b. Attribution analysis

To explore the main factors affecting the annual cycle of temperature, we apply the CFRAM to decompose the temperature evolution between two neighboring months into PTCs to obtain the spatial distribution of temperature change due to each factor in each month from January to December (not shown). Then, we use the PAP method to obtain the contribution of each factor to the monthly temperature change.

According to the results of PAP (Fig. 6.13) and from the perspective of contribution magnitude, the contributions of solar radiation and OCH are the largest, about ±5 K. The contributions of surface latent heat flux and cloud are the second major factors, about ±2 K; those of water vapor, atmospheric dynamics and surface sensible heat flux are about ±0.6 K. As a whole, the PAP of solar radiation is in the same phase as that of the observed monthly temperature change (both being positive or negative at the same time) except for March–May, June–July, and December–January. The OCH is out of phase with temperature change. In the warm pool region, the intensity of surface sensible and latent heat fluxes over the MC is mainly determined by the thermal condition of the ocean surface. The near-surface heat fluxes are in phase with the temperature change, but their contribution magnitude is much smaller than the OCH. Since the surface heat fluxes act as a compensator to the negative contribution of the OCH, the net surface process including the three processes (Fig. 6.13c), is still out of phase with the observed temperature evolution. Water vapor and clouds are in phase with near-surface temperature changes. Atmospheric dynamic processes are out of phase with temperature changes in April–May and August–October, and are in phase in the other months.

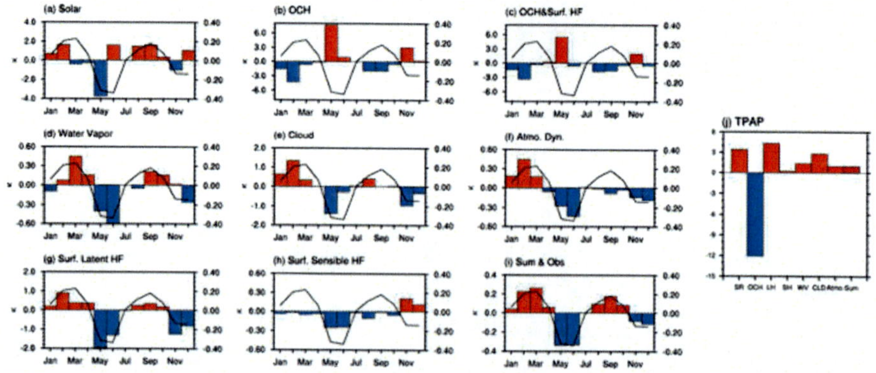

Fig. 6.13 Annual cycle of PAP coefficient of surface temperature averaged over (90°–165° E, 10° S–10° N) due to changes in **a** solar radiation at the TOA, **b** oceanic dynamic and ocean/land heat storage (OCH), **c** OCH and surface heat flux, **d** water vapor, **e** cloud, **f** atmospheric dynamics, **g** surface latent heat flux, **h** surface sensible heat flux, and **i** the sum of all individual feedback processes. The black curve in **a**–**i** refers to observations. **j** Temporal PAP (TPAP) coefficients associated with each radiative and non-radiative forcing of the areal averaged surface anomalies over (90°–165° E, 10° S–10° N). The stippled bar (labeled as Sum) in **j** refers to observations

To obtain the effect of each factor on the annual cycle of near-surface temperature, we calculated TPAP according to Eq. (6.11). For the entire MC (Fig. 6.13j), all radiative and non-radiative terms except for OCH positively contribute to the annual cycle of near-surface temperature. Because of the strong heat storage capacity, the ocean first stores some of the shortwave radiation like a buffer and then influences the surface temperature change through surface heat fluxes. This is why the peak of solar radiation is ahead of the peak of near-surface temperature by 1–2 months (Fig. 6.12i, j). The contribution of each factor to the north and south parts of MC varies in magnitude (not shown). For example, the largest positive contribution factors in the north and south of the MC are the surface latent heat flux and solar radiation, respectively. Due to the low heat storage capacity of land, shortwave solar radiation can immediately affect near-surface temperature change, while oceans can absorb a large amount of solar shortwave radiation and store it. Therefore, the ratio of land and sea can regulate the influence of solar radiation on near-surface temperature. The ratio of sea to land south of the MC is lower than that north of the MC, so the influence of solar radiation on the near-surface temperature south of the MC is greater than that north of the MC.

From March to April (Fig. 6.14), the MC is warming up overall, and the temperature increase in the north is greater than that in the south. As the Sun gradually moves northward, the PTC of solar radiation is positive north of 5° N and negative south of it. The variation in OCH cools the western tropical Pacific and heats up the southern MC. The PTCs of surface latent heat flux and OCH are asymmetric. The magnitude of PTC of surface sensible heat flux is far smaller than that of surface latent heat flux and OCH. Warming over land and cooling over the ocean increases the sea–land thermal contrast. The PTC patterns of atmospheric dynamic process

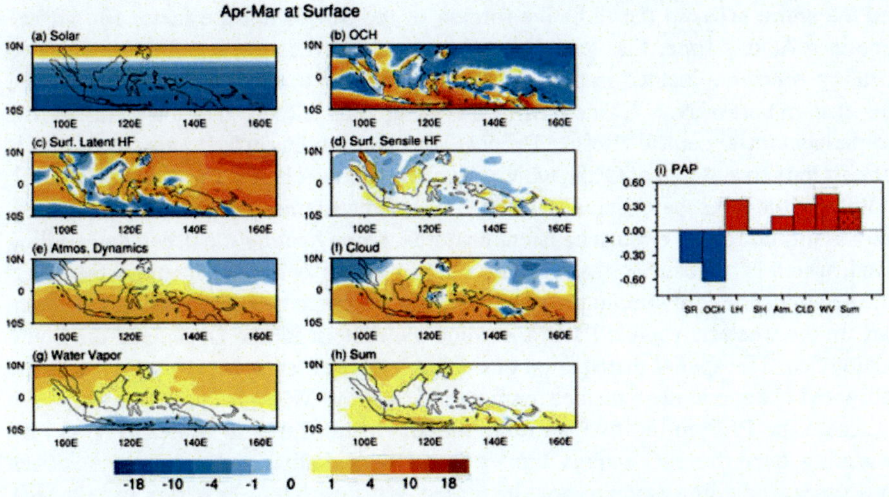

Fig. 6.14 Partial temperature changes (units: K) at the surface from March to April due to changes in **a** solar insolation, **b** surface processes including OCH and surface latent/sensible heat fluxes, **c** atmospheric dynamics, **d** cloud, and **e** water vapor. **f** is the sum of all individual feedback processes. **i** PAPs of **a**–**g** at the surface over (90°–165° E, 10° S–10° N). The stippled bar (labeled as Sum) in **i** refers to observations

and cloud are similar: heating in the southern MC, cooling east of the Bay of Bengal and the western tropical Pacific. The water vapor mainly heats up the southern MC.

From June to July (Fig. 6.15), the total surface temperature of the MC decreases, and the temperature in the south is larger than that in the north. Solar radiation heats

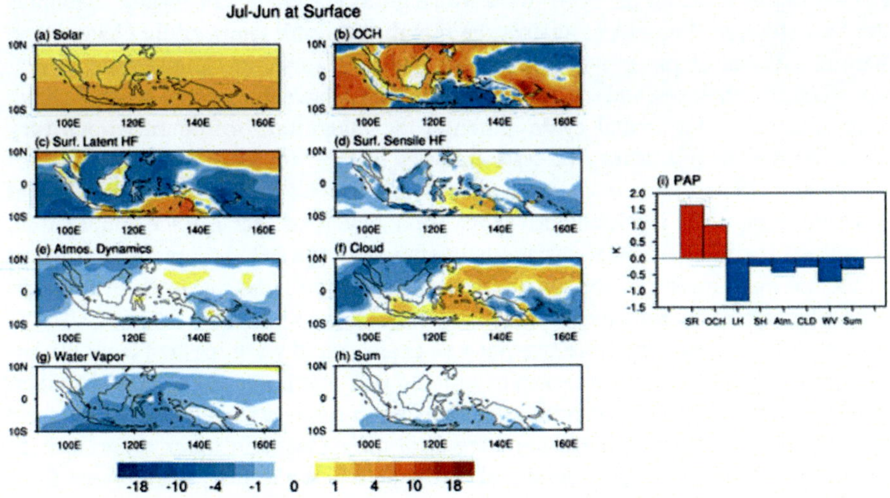

Fig. 6.15 Same as Fig. 6.14, except for changes in T2m between June and July

up the entire area, so the radiative forcing of the Sun is not the factor for surface cooling. At this time, the spatial distribution of OCH is quite different. The two heating zones are located in the eastern Indian Ocean to the Northwest Pacific and the eastern part of New Guinea, while a cooling zone is located in the southern MC to the equatorial central Pacific. The PTC distribution of surface latent heat flux is almost the same as that of OCH, which is the same as the change from March to April. Although the PTC scale is small, the PTC spatial distribution of surface sensible heat flux is similar to that of surface latent heat flux, when heating the ocean and cooling land, which is opposite to that in March–April. Same as in March–April, the PTCs of the atmospheric dynamic processes and clouds are similar: heating the southern MC to the western tropical Pacific, cooling the eastern Indian Ocean and the South China Sea. The spatial distribution of cooling effect of water vapor is similar to the observed change in near-surface temperature over the MC.

From the PAP, in the two warming months from March to March (Fig. 6.14i), warming over the MC mainly comes from the radiative and dynamic feedback processes, including water vapor (increased), surface latent heat flux (weakened), cloud volume (decreased), and atmospheric dynamic processes (energy convergence); from September to October (not shown), the warming over the MC mainly comes from solar radiation. In the rapid cooling month from June to July (Fig. 6.15i), the main factors that cause the cooling are the same as those in March–April, as water vapor (decreased), cloudiness (increased), surface sensible heat flux (enhanced), and atmospheric dynamic process (energy divergence). The atmospheric circulation changes in March–April and June–July are almost asymmetric.

Based on the energy balance in individual atmosphere-surface columns, we apply the CFRAM to quantity the contributions of various radiative and non-radiative (dynamic) processes to the annual cycle of surface temperature of the MC. Solar forcing and surface latent flux changes have the largest contributions to the near-surface temperature annual cycle in the MC region. The change in oceanic dynamics and heat storage often works against the month-to-month temperature change, and therefore, is out of phase with the annual cycle. Changes in clouds, water vapor, atmospheric dynamics, and surface sensible heat flux have the secondary but positive contributions to the annual cycle although the magnitudes of contributions vary across processes and differ over various parts of the MC (e.g., south vs. north of the equator). Therefore, in contrast to most extratropical regions where seasonal variation in the solar insolation drives the annual cycle of near-surface temperature, the annual cycle of the near-surface temperature in the MC is jointly determined by the seasonal march of the maximum solar insolation and the work of multiple atmospheric/oceanic radiative and dynamic processes.

Throughout a year, there are two positive peaks (rapid warming) in the month-to-month temperature increment, occurring in March–April and September–October, and one negative peak (rapid cooling), occurring in June–July, in the MC. The rapid warming in March–April is mainly a result of the changes in water vapor, surface latent heat flux, clouds, and atmospheric dynamics, while the contributions from direct solar forcing and OCH are negative. This is in contrast to the warming in September–October, which is driven directly by the changes in solar forcing with considerable contributions from water vapor and latent heat flux change. The main contributors to the rapid cooling in June–July are the same as those for the rapid warming in March–April, and the cooling also has negative contributions from direct solar forcing and OCH. As the largest contributor of the quick atmospheric processes, changes in water vapor contribute positively to the total temperature change in all three periods, and they are associated with the changes in the location of the center of large-scale moisture convergence during the onset and demise stages of the East Asian summer monsoon.

The contributions from individual processes differ between the two periods of rapid warming months, both of which are in the seasons of Asian and Australian monsoon transition. We provide a quantitative analysis of all radiative and non-radiative feedback processes in the two rapid warming months in monsoon transition seasons. Differences in quick atmospheric processes (i.e., the feedback processes of water vapor, clouds, surface latent heat flux, and atmospheric dynamics) and ocean–atmosphere coupling between the two periods are accompanied by temporal asymmetric movement of the deep convection belt. With almost symmetric solar insolation into the surface of the MC in the two periods, more contributions from quick atmospheric processes in March–April coexist with the sudden establishment of the East Asian summer monsoon over the northern MC and its adjacent regions, while there is relatively smaller effect by quick atmospheric processes in September–October, due to the stepwise retreat of the deep convection belt. Chang et al. (2005) proposed a "mass redistribution" hypothesis based on the difference in land-ocean thermal memories that the quick atmospheric processes and ocean–atmosphere coupling are linked to land–ocean distributions, which needs further investigation. The results allow us to look further into the asymmetric seasonal transition in precipitation by directly linking near-surface temperature and related precipitation-driving circulation anomalies to individual radiative and non-radiative processes, which will be one of the main issues to address in our future research.

6.5.2 Interdecadal Change in Troposphere Temperature

As one of the main energy sources in the equatorial region, the effects of the South China Sea and its surrounding areas on global climate anomalies, especially tropical–subtropical interaction and teleconnection patterns, are mainly achieved through latent heat released from deep convections over this region (Ramage 1968; Yanai and Tomita 1998; Hendon 2003; Chang et al. 2016; McBribe et al. 2003; Neale and Slingo 2003). Under the context of global warming, the tropospheric warming in the equatorial region in recent years is much more significant than the surface warming, and the equatorial belt has widened (Seidel et al. 2008). Based on satellite data, Fu and Johanson (2005), showed that the upper-tropospheric warming in the equatorial region was greater than the near-surface warming and gradually increased with height. The IPCC AR4 GCMs also predict that the upper troposphere warming will gradually increase with height, reach the maximum around 200 hPa and then gradually weaken. In this subsection, the interdecadal variation of the troposphere mid-upper temperature and its causes are discussed using monthly reanalysis data.

The data used in this subsection mainly comes from the monthly ERA-Interim and NCEP reanalysis products. The data from the ERA-Interim include temperature, wind, vertical velocity (spatial resolution 1° × 1°, 37 isobaric layers, time period 1979–2016) and geopotential height (spatial resolution 2.5° × 2.5°, 37 isobaric surfaces, time period 1979–2016). The NCEP data used include geopotential height, diabatic heating rate Q1/Cp (spatial resolution 2.5° × 2.5°, 10 isobaric layers, time period 1979–2016), and tropospheric height (horizontal resolution of 2.5° × 2.5°, 10 isobaric layers, time period 1979–2016).

a. Interdecadal variation

According to the pressure-height formula, we use the thickness of atmosphere referred to as "air thickness," as an indicator to characterize the temperature change of the troposphere. The 700–200 hPa air thickness is obtained from 200 hPa geopotential height minus 700 hPa geopotential height, which represents the temperature of the entire troposphere between 700 and 200 hPa. Figure 6.16 shows the warmest region of 700–200 hPa is in the western Pacific warm pool region, and the warming in the mid and high latitudes is faster than that in the low latitudes. In the equatorial region, with the exception of the central-eastern Pacific, the 700–200 hPa troposphere shows a significant warming trend, especially in the Atlantic, Sahara, and the western Pacific warm pool regions. In terms of monthly mean (not shown), the 700–200 hPa air thickness over the South China Sea and its surrounding areas is significantly warmer except for June. To improve the objectivity of the analysis results, we compare the results of two sets of reanalysis data from ERA-Interim and NCEP. The analysis results of the two data sets have similar spatial patterns, and the result obtained from ERA-Interim shows stronger magnitude than that from the NCEP. Since the upper-tropospheric warming over the South China Sea and its surrounding areas is the most significant in the boreal winter, we mainly discuss the causes of the interdecadal changes in December-February next.

The significant upper-tropospheric warming in the boreal winter is over (100°–150° E, 20° S–20° N). From the vertical-latitude distribution of zonal average over 100°–150° E of this area (Figs. 6.17a–c), the warming in December is the greatest, covering the whole area between 1000 and 150 hPa in the low latitudes (30° S–30° N), with a strong center at 400–200 hPa. In January and February, the troposphere is warmer than the lower troposphere, mainly at 400–200 hPa. In January, there are two strong centers located near 30° S and 30° N, respectively, and the strong center in the Southern Hemisphere is stronger than that in the Northern Hemisphere. The strong center in February is mainly around 30° N. For the vertical distribution, the 400–200 hPa troposphere is the strongest and most significant area of warming in the Northern Hemisphere in winter. Therefore, we focus on the warming at 400–200 hPa in the South China Sea and its surrounding areas (100°–150° E, 20° S–20° N) in the following analyses.

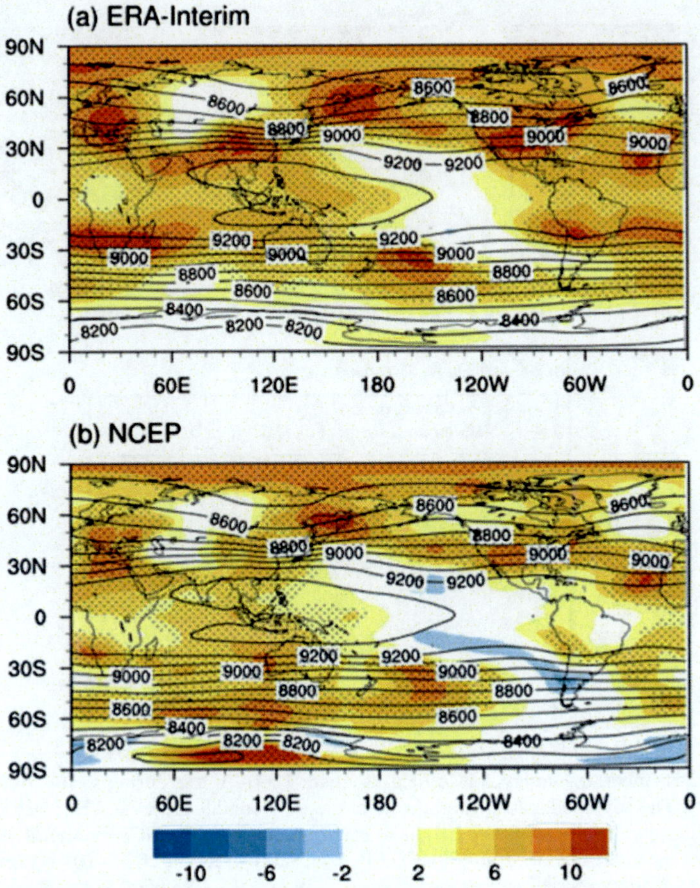

Fig. 6.16 Trends in 700–200 hPa air thickness over the period 1979–2015 (shading; m/10 years) and its climatology (contour; m) in 1981–2010 from **a** ERA-Interim and **b** NCEP reanalysis. The stippled areas are significant at the 95% confidence level. After Li et al. (2019)

b. Interdecadal attribution analyses

From 1979, to the present (Fig. 6.17d), the upper-tropospheric temperature shows a significant warming trend, with an interdecadal abrupt change in the late-1990s. The air thickness at 400–200 hPa increased by about 150 m^2/s^2. The interannual variation before the late-1990s was strong, and that after the late-1990s, was weak. That means the troposphere over this region has been in a more stable state since the 2000s. We select the two periods of 1986–1995 and 2004–2013 to represent the two anomalously cooling and warming periods, and to explore the underlying mechanism for the formation of the interdecadal upper-tropospheric warming.

Local temperature changes are determined by temperature advection $-\vec{V}_h * \nabla T$, temperature vertical transport $-\omega * \frac{T}{\theta} * \frac{\partial \theta}{\partial p}$ and diabatic heating rate Q1/Cp. The

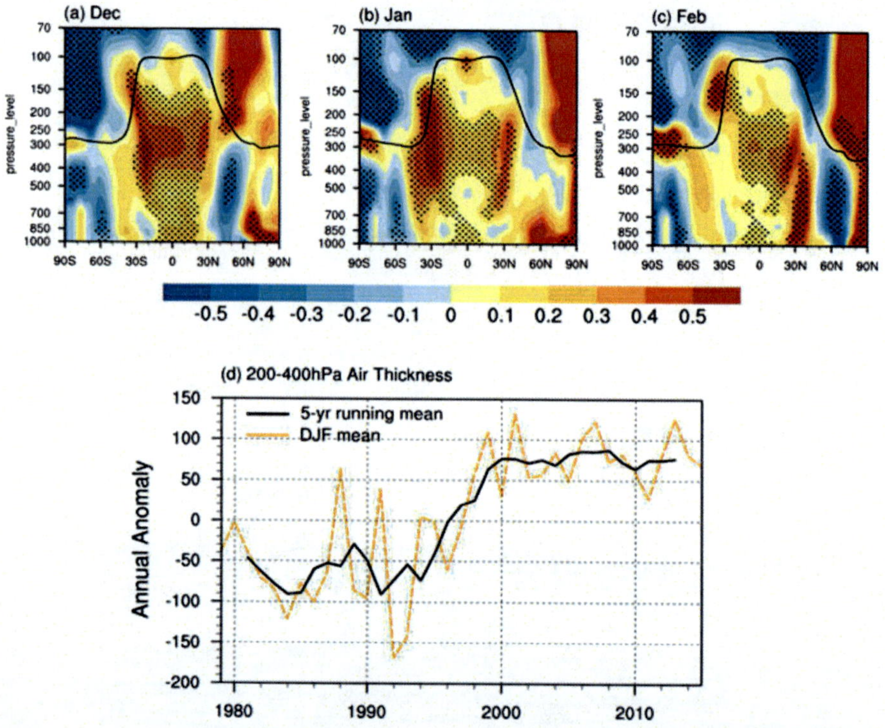

Fig. 6.17 Vertical-latitude cross section of trends of 100°–150° E zonal mean troposphere temperature in **a** December, **b** January, and **c** February over 1000–70 hPa during 1979–2015 (shading; K/10 years). The stippled areas are significant at the 90% confidence level. The black curve indicates the climatological tropopause height in each month (from the NCEP data). **d** Variation of the seasonal mean air thickness at 400–200 hPa over (100°–150° E, 20° S–20° N) (m^2/s^2) from December to January (dotted yellow curve) from 1979 to 2015. The black curve shows the result of 5-year running mean

interdecadal changes in these three items are significant and have decisive effects on the decadal change in temperature. The decadal strengthening of $-\vec{V}_h * \nabla T$ is mainly contributed by the enhancement of $-\vec{u}_h * \frac{\partial T}{\partial x}$ (not shown).

Compared to $-\vec{V}_h * \nabla T$, $-\omega\frac{T}{\theta}\frac{\partial\theta}{\partial p}$ is one order of magnitude larger (Figs. 6.18a–c), which negatively contributes to local temperature changes. In winter, $-\omega\frac{T}{\theta}\frac{\partial\theta}{\partial p}$ significantly strengthens over the South China Sea and its surrounding areas. More explicitly, in December, the enhanced cooling zone due to strengthened $-\omega\frac{T}{\theta}\frac{\partial\theta}{\partial p}$ is mainly located at (80°–120° E, 10° S–10° N); in January, the enhanced cooling zone is mainly located at (120°–160° E, 5° S–10° N), and a weakening trend is over (80°–90° E, 10°–5° S); that is to say, there are a significant cooling anomaly north of the climatological center and a warming anomaly on the southern side, indicating the climatological cold center of $-\omega\frac{T}{\theta}\frac{\partial\theta}{\partial p}$ moves from the Southern Hemisphere to the

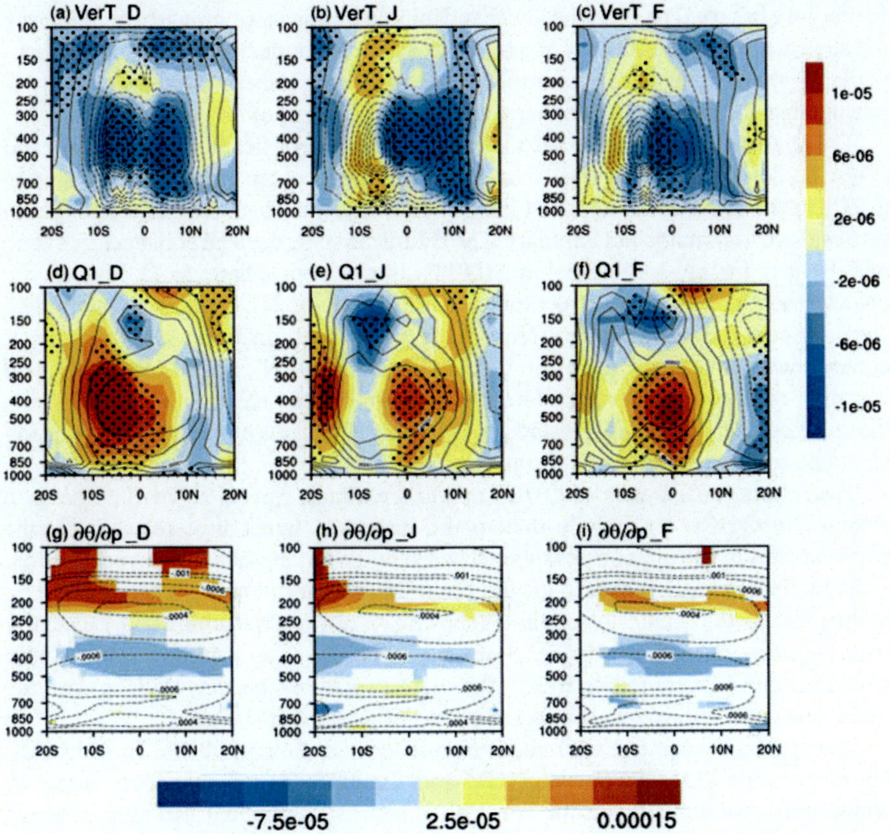

Fig. 6.18 Differences in **a–c** temperature vertical transport term $-\omega\frac{T}{\theta}\frac{\partial\theta}{\partial p}$, **d–f** diabatic heating rate Q1/Cp, and **g–i** lapse rate $\frac{\partial\theta}{\partial p}$ averaged over 100°–150° E in December (left), January (middle), and February (right) between 1986–1995 and 2004–2013 (shading; K/s). Contours indicate the climatology in each month during 1981–2010. The stippled areas are significant at the 90% confidence level

equator in January. In February, the enhanced $-\omega\frac{T}{\theta}\frac{\partial\theta}{\partial p}$ is mainly located at (120°–150° E, 10° S–10° N), which is similar to that of January. The cooling center moves toward the equator.

Moreover, $-\omega\frac{T}{\theta}\frac{\partial\theta}{\partial p}$ is determined by two parts, vertical velocity ω and lapse rate $\frac{\partial\theta}{\partial p}$. From the interdecadal variation of ω (not shown), the decadal enhancement of $-\omega\frac{T}{\theta}\frac{\partial\theta}{\partial p}$ in winter is mainly due to the enhancement of ω in the region. Therefore, the cold center moving toward the equator in January and February is caused by the enhanced ω near the equator and by weakened ω south of the cold center. Meanwhile, $\frac{\partial\theta}{\partial p}$ (Figs. 6.18g–i) shows an interdecadal weakening anomaly in the upper troposphere and an interdecadal strengthening anomaly in the lower troposphere in December–January. Note that the intensity in December is stronger than that in

January–February. The $\frac{\partial\theta}{\partial p}$ in the lower and middle troposphere strengthens, implying the stratification instability is strengthened, which is conducive to deep convection, while $\frac{\partial\theta}{\partial p}$ weakens in the upper troposphere, indicating the stratification stability is strengthened, which is not conducive to the enhancement of ω.

The diabatic heating rate Q1/Cp also shows a significant interdecadal trend (Figs. 6.18d–f). In winter, the anomalous warming center caused by changes in Q1/Cp spans from 10° S to 10° N in December, and shifts slightly to the Southern Hemisphere in January and February. The significant interdecadal enhancement zone of Q1/Cp is mainly located below 200 hPa in the troposphere. In December, the enhancement area is mainly over (60°–120° E, 15° S–5° N), and the climatological warming center in the Southern Hemisphere is enhanced. In January, the significant enhancement zone is mainly over (120°–150° E, 0°–10° N), and the climatological warming center in the Northern Hemisphere significantly strengthens. In February, the enhanced zone is mainly located near 10° S–5° N, and the climatological warming center near the equator is significantly enhanced.

The interdecadal trends of Q1/Cp and ω are almost opposite. The interdecadal changes in Q1/Cp are mainly due to the increased latent heat released by the condensation in the upper troposphere caused by the strengthened deep convection.

From 1979 to 2015, except for the central and eastern Pacific, the tropospheric temperature in the equatorial region has shown a significant warming trend, especially over the western tropical Pacific. Since the troposphere temperature in the South China Sea and its surrounding areas is the warmest in winter and has obvious warming trend, our analyses are focused on the period of December–January.

The upper-tropospheric warming is mainly located at 400–200 hPa. In winter, the air thickness of 400–200 hPa over (100°–150° E, 20° S–20° N) experiences a significant interdecadal change in the late-1990s, and its interannual variation weakens since the 2000s. In this subsection, we further explore the causes of the interdecadal variation in upper-tropospheric warming. Two periods of 1986–1995 and 2004–2013 are selected. The difference between these two periods represents the interdecadal variation. The factors affecting local temperature change include horizontal temperature advection term $-\vec{V}_h * \nabla T$, vertical temperature transport term $-\omega\frac{T}{\theta}\frac{\partial\theta}{\partial p}$ and diabatic heating rate Q1/Cp. The interdecadal enhancement of $-\vec{V}_h * \nabla T$ is mainly contributed by the enhancement of zonal temperature transport term $-\vec{u}_h * \frac{\partial T}{\partial x}$, which positively contributes to the formation of a tropospheric warming center. However, the magnitude of $-\vec{V}_h * \nabla T$ is smaller than that of the remaining two terms. Due to the enhancement of vertical velocity ω, the cooling effect of $-\omega\frac{T}{\theta}\frac{\partial\theta}{\partial p}$ presents a significant enhancement. At the same time, the significant enhancement of $\frac{\partial\theta}{\partial p}$ in the middle and lower troposphere enhances the convective instability, which facilitates the enhancement of vertical velocity. Thus, the strengthened upward motion results in enhanced release of latent heat of condensation in the upper and middle troposphere. This anomalous circulation is the direct cause of the formation of significant upper-tropospheric warming in this region.

6.6 Uncertainty in Tropical Climate Change Under Global Warming

The tropical SST and rainfall changes are critical to global climate disasters, agriculture, ecology, and economy. As an indispensable part of climate change in the western tropical Pacific, the climate over the South China Sea is influenced by the whole tropical climate change. Therefore, understanding the tropical climate change under global warming is substantial to further comprehend the changes in global moisture and energy balances and in hydrological cycle (Kummerow et al. 1998; Adler et al. 2000; IPCC 2013a).

Nowadays, the projection of tropical climate changes under global warming primarily depends on the results simulated by numerical models, such as the models participating in the Coupled Model Intercomparison Projection (CMIP). In the fifth report of the Intergovernmental Panel on Climate Change (IPCC), more than 60 models from different nations and regions joined the Phase 5 of the CMIP (CMIP5). The abundant model outputs offer us multifarious ways to study the projections of future tropical climate change. A conventional way to project the tropical climate change is based on the multimodel ensemble mean (MME), which regards intermodel spreads as noise and removes them from the MME to improve the reliability of projection (IPCC 2013b). However, the MME result lacks persuasiveness and credibility when there are large intermodel spreads among models. Previous studies have revealed that the projections of the future tropical climate, such as changes in tropical precipitation and SST, both involve pronounced inconsistency among CMIP5 models (Hawkins and Sutton 2009; Huang and Ying 2015; Kent et al. 2015; Long et al. 2016; Ying and Huang 2016). Therefore, an investigation for the uncertainty in the projections of tropical climate change and the underlying causes is an effective way to improve the credibility of model projections for tropical climate change under global warming.

In this subsection, we discuss the intermodel uncertainty in tropical climate change and the corresponding sources of uncertainty by mainly focusing on the changes in tropical precipitation and SST changes, the two factors that are critical to the climate change in response to global warming.

6.6.1 Uncertainty in Tropical Rainfall Change

The monthly outputs in the historical and Representative Concentration Pathway (RCP) 4.5 runs from 18 CMIP5 models are used here. The models are BCC-CSM1.1, CanESM2, CCSM4, CNRM-CM5, CSIRO Mk3.6.0, GFDL CM3, GFDL-ESM2G, GISS-E2-R, GISS-E2-H, HadGEM2-ES, INM-CM4, IPSL-CM5A-LR, IPSL-CM5A-MR, MIROC5, MIROC-ESM, MIROC-ESM-CHEM, MRI-CGCM3, and NorESM1-M. The historical climatology is defined as the long-term mean from

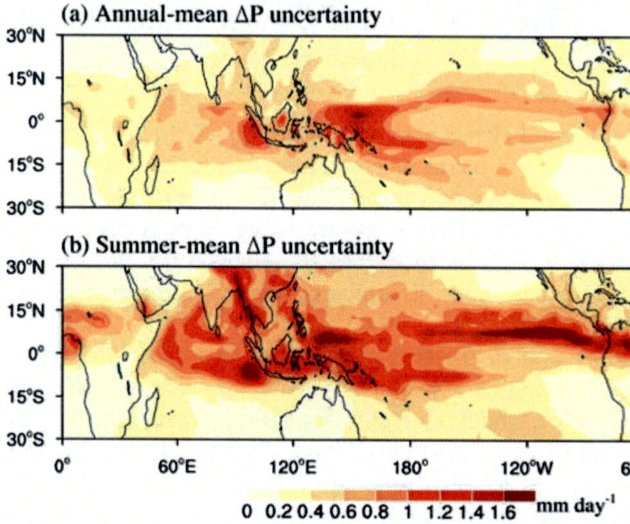

Fig. 6.19 The annual mean **a** and summer mean **b** intermodal standard deviation of tropical rainfall change

1981 to 2000, in the historical run, and the future climatology is defined as the long-term mean from 2079 to 2098, in RCP4.5. The future change (denoted as Δ) is defined as the difference between the historical and future climatology. The summer season is defined as the average of June, July, and August. The MME is the average of outputs from all 18 models. All the model outputs are interpolated onto 2.5° × 2.5° grid using bilinear interpolation before analysis.

Figure 6.19 shows the intermodel standard deviation (SD) of future rainfall change among the 18 CMIP5 models. The intermodel SD is commonly used to examine intermodel discrepancies of the projection. The intermodel uncertainty of rainfall change is large in the tropics, which means low reliability.

Moisture budget decomposition (MBD) is widely used to study the mechanism of tropical rainfall change (Chou and Neelin 2004; Chou et al. 2009; Seager et al. 2010; Huang et al. 2013; Huang 2014; Long and Xie 2015). Here, we use the MBD given in Huang et al. (2013)

$$\widehat{\Delta P}_H \sim -\frac{1}{\rho_w g}(\Delta\omega \cdot q_{20} + \omega_{20} \cdot \Delta q), \tag{6.12}$$

Here, SD and signal-to-noise ratio (SNR) is used to measure intermodel uncertainty in rainfall change. The SNR is the ratio of the MME to the SD of rainfall change. Figure 6.20 shows the intermodel SD and SNR of the thermodynamic and dynamic components in Eq. (6.12). The SD of the thermodynamic component is less than 0.6 in most tropical regions, which is much smaller than the SD of the dynamic component. The SNR of the thermodynamic component is greater than 1

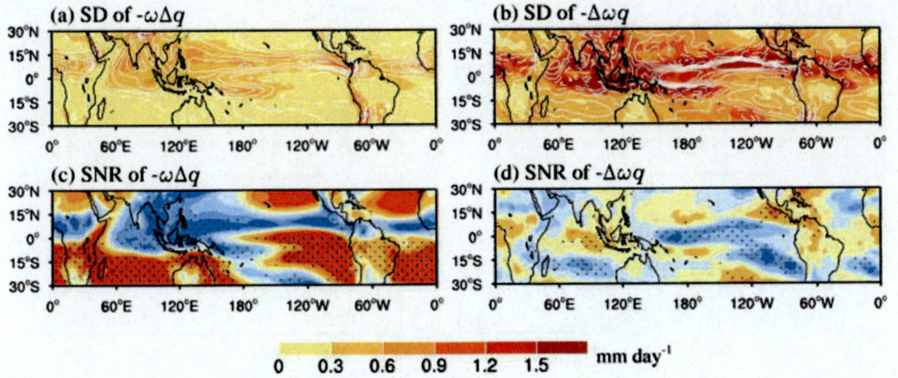

Fig. 6.20 **a**, **b** SD and **c**, **d** SNR of the thermodynamic ($-\omega\Delta q$) and dynamic ($-\Delta\omega q$) components. Contours in (**a**) and (**b**), are the MME changes in thermodynamic and dynamic components (contour interval 0.3 mm day^{-1} and negative contours dashed), respectively. Stippling in (**c**) and (**d**), indicates that more than 70% models agree on the sign of the MME change

and passes the sign agreement test, revealing that the thermodynamic component is robust among the models. In addition, the dynamic component contributes most to the intermodel uncertainty in tropical rainfall change.

Figure 6.21 presents the intermodel SD of each term in Eq. (6.12). The SD of the term related to the circulation change ($\Delta\omega$) is the largest, and the SD of the term related to specific humidity (q) is the smallest among these four terms. For the two terms in the thermodynamic component, the SD of the term related to ω is greater than the SD of the term related to Δq, indicating ω is the major cause of the uncertainty in the thermodynamic component. By comparing SD and SNR (Fig. 6.22) of each term, we can conclude that $\Delta\omega$ is the major source of the intermodel uncertainty

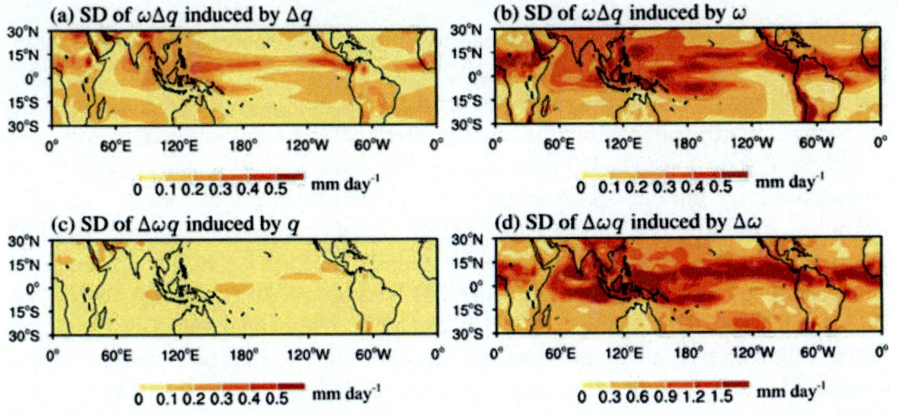

Fig. 6.21 The SD of the thermodynamic ($-\omega\Delta q$) components contributed by the SD of **a** Δq and **b** ω. The SD of the dynamic ($-\Delta\omega q$) components contributed by the SD of **c** q and **d** $\Delta\omega$. Note that the color scale used in **d** differs from the rest

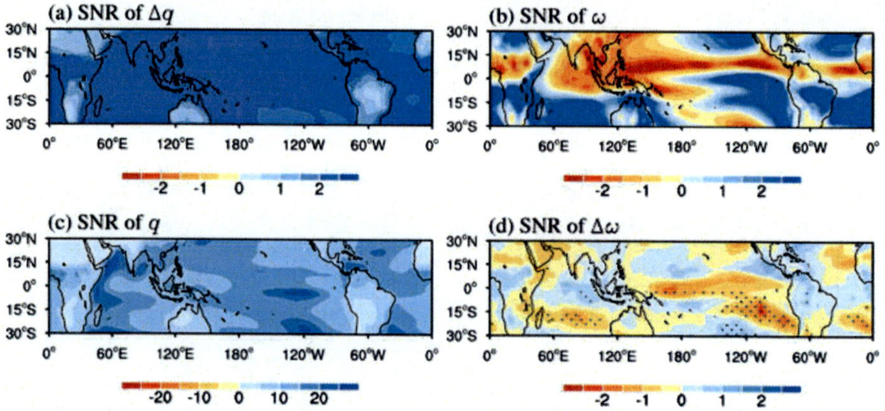

Fig. 6.22 SNR of **a** Δq, **b** ω, **c** q, and **d** $\Delta\omega$. Stippling in **d** indicates that more than 70% models agree on the sign of the MME change

of tropical rainfall change. For the other terms in Eq. (6.12), q is robust enough to be neglected, Δq is almost consistent within the projections of all models, and ω is also an important source of intermodel uncertainty of tropical rainfall change. The improvement of tropical rainfall projection is a crucial way to make the projection of tropical rainfall change more robust.

In this subsection, we attribute the key factor of intermodel uncertainty of tropical rainfall change to circulation change. However, the causes of large uncertainty in the circulation change remain unclear (Shepherd 2014).

6.6.2 Uncertainty in Tropical Pacific SST Change

Previous studies indicated that the ocean warming in the tropics is spatially nonuniform under global warming, with some regions warming fast. The nonuniform ocean warming, which is found mainly in the tropical Pacific, is important for regulating the changes in tropical precipitation and circulation. However, projecting the tropical Pacific SST warming (TPSW) pattern based on current coupled ocean–atmosphere models has large intermodel uncertainty, especially for the projections of zonal SST warming. Although the majority of CMIP5 models project an El Niño-like SST warming pattern, a few models project a zonally uniform or even a La Niña-like SST warming pattern (Clement et al. 1996; DiNezio et al. 2009; Zhang and Li 2014; Huang and Ying 2015; Zheng et al. 2016). Such intermodel uncertainty is a dominant source of uncertainty in the projection of regional climate changes, such as changes in tropical precipitation and atmospheric circulation (Long and Xie 2015).

In climate models, there are several distinct mechanisms affecting future TPSW pattern, such as the changes in the shortwave radiation-SST feedback (Ramanathan and Collins 1991), the ocean dynamic thermostat (Clement et al. 1996), and the

Walker circulation (Vecchi and Soden 2007). Any intermodel differences in these mechanisms could be the sources of intermodel uncertainty in predicting the TPSW pattern. Among these physical mechanisms, the cloud-radiation feedback is an outstanding one due to the fact that it involves both parameterized cloud processes and the climatological distribution of clouds, the two aspects that are known to carry large model biases (Cess et al. 1990; Arakawa 2004; Stephens 2005; Soden and Held 2006; Calisto et al. 2014). Therefore, in this subsection, we focus on the role of cloud-radiation feedback on the intermodel uncertainty in the TPSW pattern and elaborate on the corresponding mechanism of its climate impact.

To explore the spatial coupled relationship between the TPSW pattern and the cloud-radiation feedback, an intermodel singular value decomposition (SVD) analysis is performed on multimodel TPSW patterns and the cloud-radiation feedback. The left singular vector field is the TPSW pattern, while the right singular vector field is the cloud-radiation feedback, defined by regressing the monthly surface net shortwave radiation anomalies onto the monthly SST anomalies in the historical run (cloud-SST feedback index, or CSFI). The result shows that the first intermodel SVD modes of the TPSW pattern and CSFI explains 24% and 32% of the total intermodel variance, respectively. In addition, the spatial pattern of the first SVD model of the TPSW pattern is very close to the first intermodel EOF of the TPSW pattern (not shown), with the correlation between the two PCs up to 0.59 (t-test, 99% confidence level). This indicates the first SVD model of the TPSW pattern shown in Fig. 6.23, is the dominant mode for the intermodel spread in the TPSW pattern. Moreover, the linear correlation between the PCs associated with the modes of the TPSW pattern and CSFI is up to 0.65, which is also significant at the 99% confidence level based on

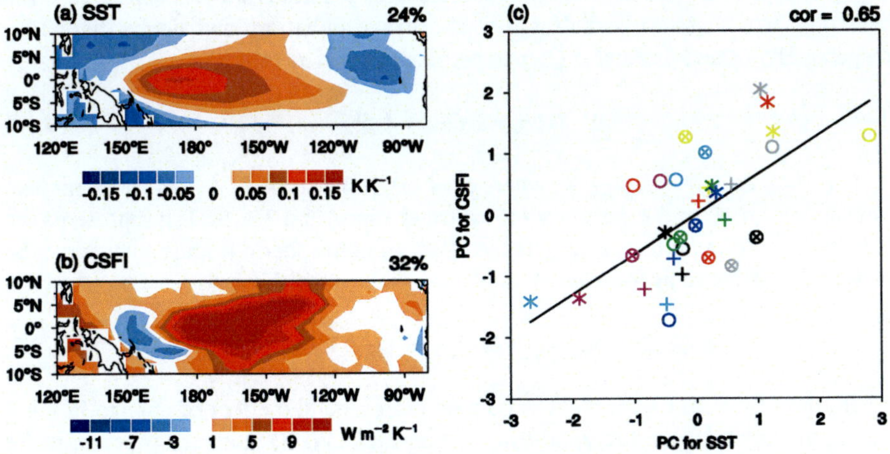

Fig. 6.23 The first intermodel SVD modes of **a** TPSW pattern and **b** CSFI. The percentage of intermodel variance explained by the respective vector field is shown in the upper right corner of each panel. **c** Scatter plot of normalized PC1s associated with the TPSW pattern and CSFI of the first SVD mode, with their correlation coefficient shown in the upper right of the panel

the χ^2 test. These results indicate that the intermodel spread in the cloud-radiation feedback must be a leading source of intermodel uncertainty in the TPSW pattern.

Figure 6.23 shows that the first SVD mode of the TPSW pattern has a broad positive pattern in the central-western equatorial Pacific, whereas the first SVD mode of CSFI features a positive area that is confined to the central equatorial Pacific. The two coupled spatial patterns could be explained as follows: A model with a positive CSFI deviation relative to the negative MME CSFI indicates a weak negative cloud-radiation feedback over the central Pacific when compared with the MME result, the relatively weak negative cloud-radiation feedback cannot suppress the SST warming as sufficiently as that in the MME, thus inducing a positive SST warming deviation relative to the MME result. However, the spatial patterns of the first SVD modes of TPSW and CSFI possess some apparent discrepancies, even though both are located around the central Pacific. The positive center of the deviation of the TPSW pattern is located in the western Pacific around 170° E (Fig. 6.23a), whereas that of the CSFI is much closer to the central Pacific (around 160° W; Fig. 6.23b). Such discrepancy implies there could be a dynamic process in the cloud-radiation feedback that affects the TPSW pattern, which is not as simple as stated above.

To figure out the mechanism via which the cloud-radiation feedback impacts the intermodel uncertainty in the TPSW pattern, the changes in the surface ocean heat budget are analyzed. For the long-term change under global warming, the balance of the surface ocean heat budget can be written as

$$\Delta Q_{\mathrm{E}} + \Delta Q_{\mathrm{H}} + \Delta Q_{\mathrm{LW}} + \Delta Q_{\mathrm{SW}} + \Delta D_{\mathrm{O}} = 0 \tag{6.13}$$

where Δ denotes the change in the RCP8.5 run relative to the historical run, ΔQ_{E}, ΔQ_{H}, ΔQ_{LW}, ΔQ_{SW}, and ΔD_{O} represent changes in surface latent heat flux, sensible heat flux, net longwave radiation, net shortwave radiation, and ocean dynamics, respectively. In these terms, ΔQ_{E} can be decomposed as follows:

$$\Delta Q_{\mathrm{E}}{=}\Delta Q_{\mathrm{EO}}{+}\Delta Q_{\mathrm{EA}} \tag{6.14}$$

In (6.14), $\Delta Q_{\mathrm{EO}} = \alpha Q_{\mathrm{E}} \Delta T$, where $\alpha{=}L(R_{\mathrm{v}}T^2) \approx 0.06K^{-1}$, L is the latent heat of evaporation, R_{v} is the gas constant for water vapor, and T is SST; it represents the Newtonian cooling effect. ΔQ_{EA} is the effect of atmospheric forcing, which can be further decomposed as follows:

$$\Delta Q_{\mathrm{EA}} = \Delta Q_{\mathrm{EW}} + \Delta Q_{\mathrm{ER}} = Q_{\mathrm{E}} \Delta V/V + \Delta Q_{\mathrm{ER}} \tag{6.15}$$

where the first term on the right-hand side represents the effect of change in wind speed, the key factor in the wind-evaporation-SST (WES) feedback process, while the second term is regarded as a residual term, representing the combined effect of changes in relative humidity and surface stability.

Figure 6.24 shows the intermodel regression pattern of the changes in surface heat fluxes onto the normalized PC1 associated with the first SVD mode of the TPSW pattern. Based on such spatial regression patterns, we propose a dynamic process via

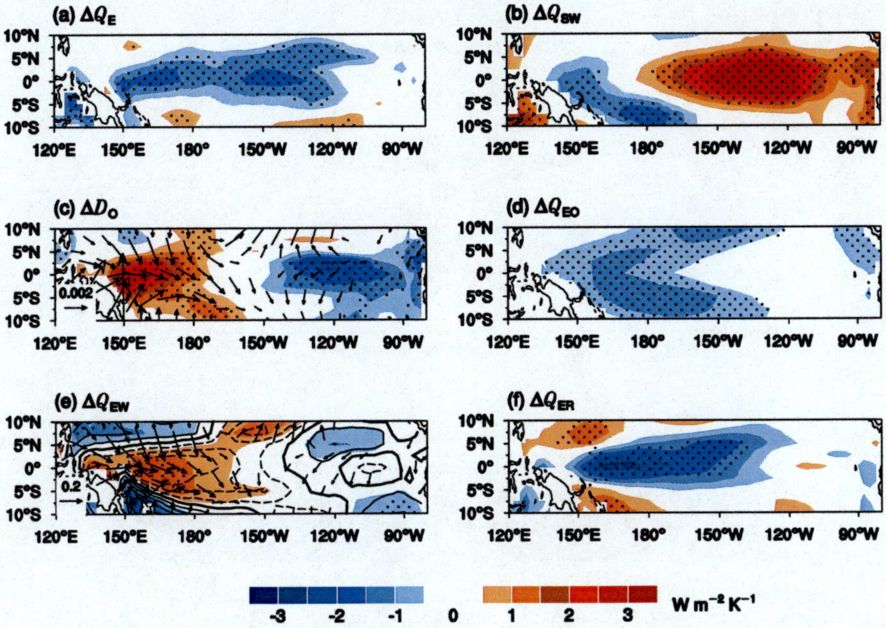

Fig. 6.24 Intermodel regression patterns of changes in surface energy budget onto normalized PC associated with the first SVD mode of the TPSW pattern. **a** Latent heat flux (ΔQ_{E}), **b** shortwave radiation (ΔQ_{SW}), **c** ocean dynamics (ΔD_{O}), **d** the Newtonian cooling effect (ΔQ_{EO}), **e** the effect of changes in surface wind speed (ΔQ_{EW}; shading), and **f** the effect of changes in relative humidity and surface stability (ΔQ_{ER}). Stippling indicates that regression is significant at the 95% confidence level based on the Student's *t*-test. The vectors in **c** and **e** are the regression patterns of changes in surface wind stress (Pa K^{-1}; values smaller than 5×10^{-5} Pa s^{-1} omitted) and the surface wind vector (m s^{-1} K^{-1}; values smaller than 5×10^{-2} m s^{-1} K^{-1} omitted). The contours in **e** are the regression patterns of changes in the scalar speed of surface wind (m s^{-1} K^{-1})

which the cloud-radiation feedback influences intermodel uncertainty in the TPSW pattern, which is depicted in Fig. 6.24. Given a positive deviation of CSFI relative to the negative CSFI in the MME result as shown in Fig. 6.23b, in a particular model, the relatively weak negative cloud-radiation feedback over the central Pacific cannot suppress the local SST warming as efficiently as the MME cloud-radiation feedback, thus leading to a positive local ΔQ_{SW} deviation and a positive SST warming deviation (Fig. 6.24a). The induced positive SST warming deviation will further force a low-level convergent deviation over the central Pacific (Fig. 6.24e, vector; Fig. 6.25a, blue arrow). Against the background of prevailing easterlies over the equatorial Pacific, the low-level convergence will increase (decrease) the surface wind speed and the evaporative cooling over the eastern (western) Pacific through the WES feedback (Fig. 6.24e, contour; Fig. 6.25b). On the other hand, the low-level convergent deviation will produce a similar deviation pattern in surface wind stress (Fig. 6.24c, vector), which will enhance (suppress) the cold equatorial ocean advection in the eastern (western) Pacific through the Bjerknes feedback (Fig. 6.24c, vector;

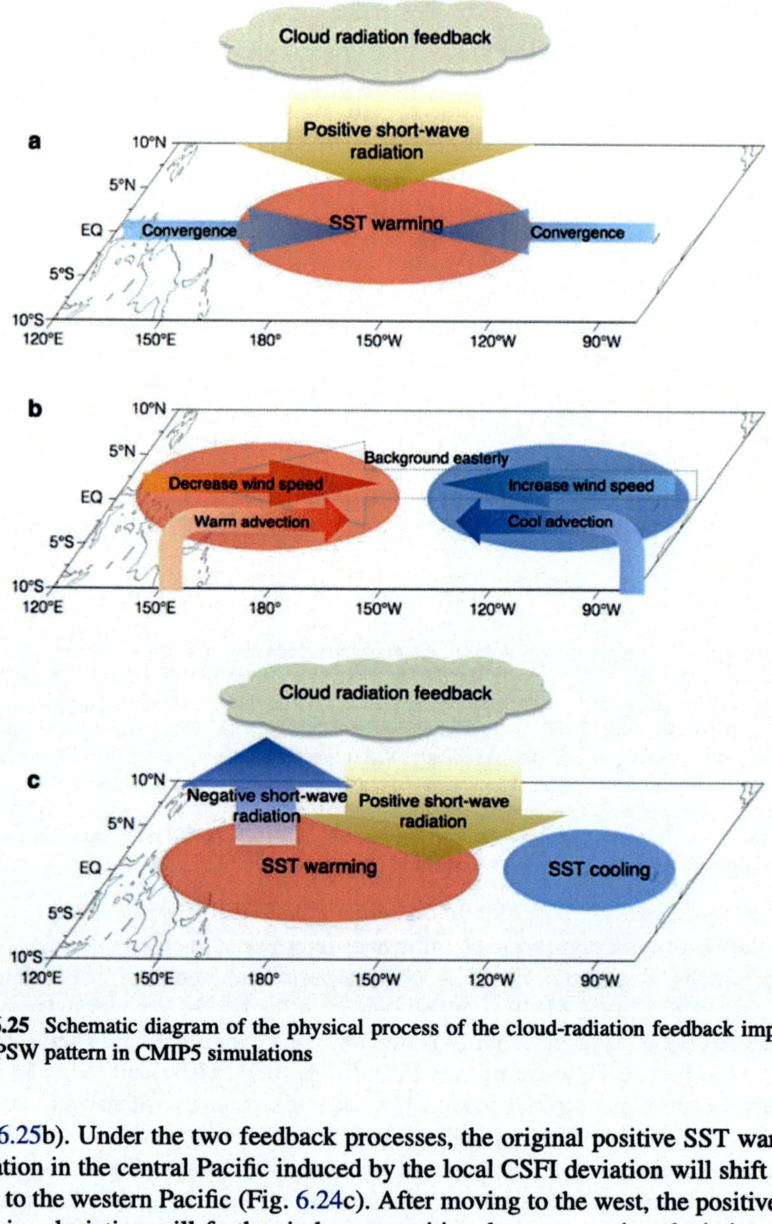

Fig. 6.25 Schematic diagram of the physical process of the cloud-radiation feedback impacting the TPSW pattern in CMIP5 simulations

Fig. 6.25b). Under the two feedback processes, the original positive SST warming deviation in the central Pacific induced by the local CSFI deviation will shift westward to the western Pacific (Fig. 6.24c). After moving to the west, the positive SST warming deviation will further induce a positive deep convection deviation and a resultant negative shortwave radiation deviation over the western Pacific, leading to a suppression of local ΔQ_{SW} and forming the final pattern of ΔQ_{SW} (Figs. 6.24b, 6.25c). During the westward moving of the positive SST warming deviation, the deviation patterns of surface wind, ΔD_O and ΔQ_{EW} also move to the west.

In this subsection, we investigate the uncertainty in the projections of changes in tropical precipitation and SST under global warming in CMIP5 models. The results show the thermodynamic component of the changes in tropical precipitation projected by CMIP5 models—the enhanced precipitation induced by the increased moisture—is quite reliable, while the dominant source of intermodel uncertainty in tropical precipitation change comes from the projections for the changes in the atmospheric circulation. Moreover, as previous studies have revealed, the uncertainty in the tropical atmospheric circulation originates from that in tropical ocean warming pattern; here, we further indicate that the intermodel difference in the cloud-radiation feedback is the leading source of intermodel uncertainty in the TPSW pattern based on our surface heat budget analysis. These studies reveal the features of uncertainty in future tropical climate change and their associated formation mechanisms, improving the credibility on the projections of future tropical climate change.

References

Adler RF, Huffman GJ, Bolvin DT, Curtis S, Nelkin E (2000) Tropical Rainfall Distributions Determined Using TRMM Combined with Other Satellite and Rain Gauge Information. J Appl Meteorol 39(12):2007–2023

Ashok K, Behera SK, Rao SA, Weng H, Yamagata T (2007) El Niño Modoki and its possible teleconnection. J Geophys Res 112:C11007

Arakawa A (2004) The cumulus parameterization problem: Past, present, and future. J Clim 17(13):2493–2525

Bony S, Colman R, Kattsov VM et al (2006) How Well Do We Understand and Evaluate Climate Change Feedback Processes? J Clim 19:3445–3482. https://doi.org/10.1175/jcli3819.1

Cai M, Lu J (2009) A new framework for isolating individual feedback processes in coupled general circulation climate models. Part II: Method demonstrations and comparisons. Clim Dyn 32:887–900. https://doi.org/10.1007/s00382-008-0424-4

Cai M, Tung K-K (2012) Robustness of dynamical feedbacks from radiative forcing: 2% solar versus 2 × CO_2 experiments in an idealized GCM. J Atmos Sci 69:2256–2271. https://doi.org/10.1175/JAS-D-11-0117.1

Calisto M, Folini D, Wild M, Bengtsson L (2014) Cloud radiative forcing intercomparison between fully coupled CMIP5 models and CERES satellite data. Ann Geophys 32(7):793–807

Cess RD, Potter G, Blanchet J, Boer G, Ghan S, Kiehl J, Le Treut H, Li Z-X, Liang X-Z, Mitchell J (1990) Interpretation of cloud-climate feedback as produced by 14 atmospheric general circulation models. Science 245(4917):513–516

Chang CP (2004) East Asian Monsoon. World Scientific Publishing Co. Ptc. Ltd

Chang CP, Harr PA, Chen HJ (2005) Synoptic disturbances over the equatorial South China Sea and western Maritime Continent during boreal winter. Mon Weather Rev 133:489–503

Chang CP, Lau KM (1982) Short-term planetary-scale interactions over the tropics and midlatitude during northern winter. Part I: Contrasts between active and inactive periods. Mon Weather Rev 110:933–946

Chang CP, Lu MM, Lim H (2016) Monsoon convection in the Maritime Continent: Interaction of large-scale motion and complex terrain. Meteorological Monographs 56, chapter 6

Chang CP, Wang Z, Ju J, Li T (2004) On the relationship between western maritime continent monsoon rainfall and ENSO during northern winter. J Clim 17:665–672

Cheung HN, Zhou W, Mok HY, Wu MC (2012) Relationship between Ural-Siberian blocking and the east Asian winter monsoon in relation to the Arctic Oscillation and the El Niño-Southern Oscillation. J Clim 25:4242–4257

Chou C, Neelin JD (2004) Mechanisms of global warming impacts on regional tropical precipitation. J Climate 17:2688–2701

Chou C, Neelin JD, Chen C-A, Tu J-Y (2009) Evaluating the "Rich-Get-Richer" Mechanism in Tropical Precipitation Change under Global Warming. J Climate 22(8):1982–2005

Clement AC, Seager R, Cane MA, Zebiak SE (1996) An ocean dynamical thermostat. J Clim 9:2190–2196

Dee DP, and Coauthors (2011) The ERA-Interim reanalysis: configuration and performance of the data assimilation system. Quart J Roy Meteor Soc 137:553–597

Deng Y, Park TW, Cai M (2012) Process-based decomposition of the global surface temperature response to El Niño in boreal winter. J Atmos Sci 69:1706–1712. https://doi.org/10.1175/jas-d-12-023.1

DiNezio PN, Clement AC, Vecchi GA, Soden BJ, Kirtman BP, Lee S-K (2009) Climate Response of the Equatorial Pacific to Global Warming. J Clim 22(18):4873–4892

Ding Y, and Coauthors (2014) Interdecadal variability of the East Asian winter monsoon and its possible links to global climate change. Chin Meteor Soc 90th Annivers Spec Collect 28:693–713

Forster PM, Fomichev VI, Rozanov E, et al (2011) Evaluation of radiation scheme performance within chemistry climate models. J Geophys Res Atmos 116:. https://doi.org/10.1029/2010jd015361

Fu Q, Johanson CM (2005) Satellite-derived vertical dependence of tropical tropospheric temperature trends. Geophys Res Lett 32:L10703

Gong DY, Ho CH (2002) The Siberian High and climate change over middle to high latitude Asia. Theor Appl Climatol 72:1–9

Gong DY, Wang S (2003) Influence of Arctic Oscillation on winter climate over China. ACTA Geographica Sin 58:559–568

Hall A, Manabe S (1999) The role of water vapor feedback in unperturbed climate variability and global warming. J Clim 12:2327–2346.

Hawkins E, Sutton R (2009) The Potential to Narrow Uncertainty in Regional Climate Predictions. Bull Amer Meteor Soc 90(8):1095–1107

Hendon HH (2003) Indonesian rainfall variability: Impacts of ENSO and local air-sea interaction. J Clim 16:1775–1790

Hu X, Yang S, Cai M (2016) Contrasting the eastern Pacific El Niño and the central Pacific El Niño: process-based feedback attribution. Clim Dyn 47:2413–2424

Hu X, Li Y, Yang S et al (2017) Process-Based Decomposition of the Decadal Climate Difference between 2002–13 and 1984–95. J Clim 30:4373–4393. https://doi.org/10.1175/jcli-d-15-0742.1

Huang P (2014) Regional response of annual-mean tropical rainfall to global warming. Atmos Sci Lett 15(2):103–109

Huang P, Xie S-P, Hu K, Huang G, Huang R (2013) Patterns of the seasonal response of tropical rainfall to global warming. Nat Geosci 6(5):357–361

Huang P, Ying J (2015) A Multimodel Ensemble Pattern Regression Method to Correct the Tropical Pacific SST Change Patterns under Global Warming. J Clim 28(12):4706–4723

Hung CW, Liu X, Yanai M (2004) Symmetry and asymmetry of the Asian and Australian summer monsoons. J Clim 17:2413–2426

IPCC (2013a) IPCC, 2013: Summary for policymakers. In: Stocker TF, Qin D, Plattner G-K et al. (eds) Climate change 2013: The physical science basis Contribution of working group I to the fifth assessment report of the intergovernmental panel on climate change. Cambridge University Press

IPCC (2013b) The physical science basis. Climate change 2013. Cambridge University Press

Intergovernmental Panel on Climate Change (IPCC) (2013), Climate change 2013: The physical science basis. Contribution of Working Group I to AR5. Eds. IPCC, Cambridge: Cambridge University Press

Kent C, Chadwick R, Rowell DP (2015) Understanding Uncertainties in Future Projections of Seasonal Tropical Precipitation. J Climate 28(11):4390–4413

Kug JS, Jin FF, An SI (2009) Two types of El Niño events: cold tongue El Niño and warm pool El Niño. J Clim 22:1499–1515

Kummerow CD, Barnes WL, Kozu T, Shiue J, Simpson J (1998) The Tropical Rainfall Measuring Mission (TRMM) Sensor Package. Journal of Atmospheric and Oceanic Technology 15(3):809–817

Lau KM, Chang CP (1987) Planetary scale aspects of winter monsoon and teleconnections. Monsoon Meteorology, C.-P. Chang and T. N. Krishnamurti, Eds., Oxford University Press, 161–202

Lau NC, Nath MJ (1996) The role of the "atmospheric bridge" in linking tropical Pacific ENSO events to extratropical SST anomalies. J Clim 9:2036–2057

Li C (1990) Interaction between anomalous winter monsoon in East Asia and El Nino events. Adv Atmos Sci 7:36–46

Li Y, Yang S (2017) Feedback attributions to the dominant modes of East Asian winter monsoon variations. J Clim 30:905–920

Li Y, Yang S, Deng Y, Hu X, Cai M (2017) A process-level attribution of the annual cycle of surface temperature over the Maritime Continent. Clim Dyn 51:2759–2772

Li Y, Yang S, Deng Y, Hu X, Cai M, Zhou W (2019) Detection and attribution of upper-tropospheric warming over the tropical western Pacific. Clim Dyn 53:3057–3068. https://doi.org/10.1007/s00382-019-04681-9

Lu J, Cai M (2009) A new framework for isolating individual feedback process in coupled general circulation climate models. Part I: Formulation Clim Dyn 32:873–885. https://doi.org/10.1007/s00382-008-0425-3

Long S-M, Xie S-P (2015) Intermodel variations in projected precipitation change over the North Atlantic: Sea surface temperature effect. Geophys Res Lett 42(10):4158–4165

Long S-M, Xie S-P, Liu W (2016) Uncertainty in tropical rainfall projections: atmospheric circulation effect and the ocean coupling. J Climate 29(7):2671–2687

McBribe JL, Haylock MR, Nicholls N (2003) Relationships between the Maritime Continent heat source and the El Niño-Southern Oscillation phenomenon. J Clim 16:2906–2914

Meehl G (2015) Decadal climate variability and he early-2000s hiatus. US CLIVAR Var 13:6. https://doi.org/10.1175/2010bams2962.1.u

Neale R, Slingo J (2003) The Maritime continent and its role in the global climate: A GCM study. J Clim 16:834–848

Nicholls N, Lavery B, Frederiksen C, Drosdowsky W, Torok S (1996) Recent apparent changes in relationships between the El Niño- Southern oscillation and Australian rainfall and temperature. Geophys Res Lett 23:3357–3360

Park TW, Deng Y, Cai M (2012) Feedback attribution of the El Niño-Southern oscillation–related atmospheric and surfacetemperature anomalies. J Geophys Res 117:D23101. https://doi.org/10.1029/2012jd018468

Park TW, Ho CH, Deng Y (2014) A synoptic and dynamical characterization of wave-train and blocking cold surge over East Asia. Clim Dyn 43:753–770

Ramage CS (1968) Role of a tropical "maritime continent" in the atmospheric circulation. Mon Wea Rev 96:365–370

Ramanathan V, Collins W (1991) Thermodynamic regulation of ocean warming by cirrus clouds deduced from observations of the 1987 El Niño. Nature 351(6321):27–32

Ramaswamy V, Chanin ML, Angell J et al (2001) Stratospheric Temperature Trends: Observations and Model Simulations. Rev Geophys 39:71–122. https://doi.org/10.1029/1999rg000065

Schneider EK, Kirtman BP, Lindzen RS (1999) Tropospheric Water Vapor and Climate Sensitivity. J Atmos Sci 56:1649–1658. https://doi.org/10.1175/1520-0469(1999)056%3c1649:twvacs%3e2.0.co;2

Seager R, Naik N, Vecchi GA (2010) Thermodynamic and Dynamic Mechanisms for Large-Scale Changes in the Hydrological Cycle in Response to Global Warming. J Climate 23(17):4651–4668

Seidel DJ, Fu Q, Randel WJ et al (2008) Widening of the tropical belt in a changing climate. Nat Geosci 1:21–24

Shepherd TG (2014) Atmospheric circulation as a source of uncertainty in climate change projections. Nat Geosci 7(10):703–708

Simon P, Montanari A (2019) Global and regional increase of precipitation extremes under global warming. Water Resour Res 55: 4901–4914.

Simpson J, Keenan TD, Ferrier B, Simpson RH, Holland GJ (1993) Cumulus merger in the Maritime Continent region. Meteor Atmos Phys 51:73–99

Soden BJ, Held IM (2006) An assessment of climate feedbacks in coupled ocean-atmosphere models. J Clim 19(14):3354–3360

Soden BJ, Held IM, Colman RC et al (2008) Quantifying climate feedbacks using radiative kernels. J Clim 21:3504–3520. https://doi.org/10.1175/2007jcli2110.1

Stephens GL (2005) Cloud Feedbacks in the Climate System: A Critical Review. J Clim 18(2):237–273

Sui CH, Lau KM (1992) Multiscale phenomena in the tropical atmosphere over the western Pacific. Mon Wea Rev 120:407–430

Sukanto M (1969) Climate of Indonesia. Climates of Northern and Eastern Asia, H. Arakawa, Ed., World Survey of Climatology, Vol 8, Elsevier, 215–229

Sun CH, Yang S, Li WJ, Zhang RN, Wu RG (2016) Interannual variations of the dominant modes of East Asian winter monsoon and possible links to Arctic sea ice. Clim Dyn. https://doi.org/10.1007/s00382-015-2851-3

Thompson DJ, and Co-authors (2012) The mystery of recent stratospheric temperature trends. Nature 491:692–697. https://doi.org/10.1038/nature11579.

Trenberth KE, Fasullo J, Smith L (2005) Trends and variability in column-integrated atmospheric water vapor. Clim Dyn 24:741–758. https://doi.org/10.1007/s00382-005-0017-4

Trenberth KE, Fasullo JT (2013) An apparent hiatus in global warming? Earth' s Futur 1:19–32. https://doi.org/10.1002/2013ef000165.received

Van den Dool HM, Saha S (1993) Seasonal redistribution and conservation of atmospheric mass in a general circulation model. J Clim 6:22–30

Vecchi GA, Soden BJ (2007) Global Warming and the Weakening of the Tropical Circulation. J Clim 20(17):4316–4340

Wang B, Wu Z, Chang CP, Liu J, Li J, Zhou T (2010) Another look at interannual-to-interdecadal variations of the East Asian winter monsoon: The northern and southern temperature modes. J Clim 23:1495–1512

Wang L, Chen W, Zhou W, Huang RH (2009) Interannual variations of East Asian Trough axis at 500 hPa and its association with the East Asian winter monsoon pathway. J Clim 22:600–614

Wang L, Feng J (2011) Two major modes of the wintertime precipitation over China. Chinese J Atmos Sci 35:1105–1116

Webster PJ, Yang S (1992) Monsoon and ENSO: Selectively interactive systems. Quart J R Meteor Soc 118:877–926

Wetherald RT, Manabe S (1988) Cloud feedback processes in a general circulation model. J Atmos Sci 45:1397–1416

Wu B, Wang J (2002) Possible impact of winter Arctic Oscillation on Siberian High, the East Asian winter monsoon and sea-ice extent. Adv Atmos Sci 19:300–320

Wu B, Zhang R, D'Arrigo R (2006) Distinct modes of the East Asian winter monsoon. Mon Wea Rev 134:2165–2179

Wylie D, Jackson DL, Menzel WP, Bates JJ (2005) Trends in global cloud cover in two decades of HIRS observations. J Clim 18:3021–3031. https://doi.org/10.1175/jcli3461.1

Yanai M, Tomita T (1998) Seasonal and interannual variability of atmospheric heat sources and moisture sinks as determined from NCEP–NCAR reanalysis. J Clim 11:463–482

Yang S, Lau KM, Kim KM (2002) Variations of the east Asian jet stream and Asian–Pacific–American winter climate anomalies. J Clim 15:306–325

Yang S, Webster PJ (1990) The effect of summer tropical heating on the location and intensity of the extratropical westerly jet streams. J Geophys Res 95:18705–18721

Ying J, Huang P (2016) Cloud-Radiation Feedback as a Leading Source of Uncertainty in the Tropical Pacific SST Warming Pattern in CMIP5 Models. J Climate 29(10):3867–3881

Yu J-Y, Kao H-Y (2007) Decadal changes of ENSO persistence barrier in SST and ocean heat content indices: 1958–2001. J Geo-phys Res 112:D13106. https://doi.org/10.1029/2006JD007654

Zhang L, Li T (2014) A Simple Analytical Model for Understanding the Formation of Sea Surface Temperature Patterns under Global Warming. J Clim 27(22):8413–8421

Zheng X-T, Xie S-P, Lv L-H, Zhou Z-Q (2016) Intermodel Uncertainty in ENSO Amplitude Change Tied to Pacific Ocean Warming Pattern. J Clim 29(20):7265–7279

Chapter 7
Impact of Climate Change Over Southeast Asia and Its Adjacent Regions on Global Climate

7.1 Long-Term Trend and Interannual Variation of Springtime Precipitation

The blue contours in Fig. 7.1a, show that in southern China, Taiwan, and southern Japan, the ratio of climatological springtime (March–April–May, or MAM) rainfall to the annual rainfall is more than 25%. In the mountain areas of Guangdong and Fujian provinces, the ration even approaches 40%. Meanwhile, the spring rainfall in southern China and southern Japan in the past 35 years (1979–2013) had clear declining trends. In the most significant precipitation reduction area of the coastal areas in southern China, precipitation decreased by 2.0 mm/day over the past 35 years. This reduction is equivalent to ~50% of the climatological MAM precipitation in this area. At the same time, in the southern Indochinese Peninsula, the southern South China Sea (SCS) and the Philippines Sea (PhiSea), precipitation showed a significant increasing trend in spring. The increased precipitation rate was about 2.5 mm/day over the past 35 years. This increased rate is considerable, reaching more than 40% of the 35-year springtime climatological precipitation in the region.

Changes in regional precipitation are accompanied by changes in the circulation system. To reveal the changes in atmospheric circulation associated with the precipitation change, we use Fig. 7.1b, to show the trends of spring sea level pressure (SLP), the mid-level tropospheric vertical velocity and the low-level wind at 850 hPa. We can see that the SLP in East Asia, the Japan Sea, and the western Pacific decreases significantly, along with increased SLP in the inland regions of the Asian continent. Low-level northerly anomalies appear along the East Asian coast. The linear trend of the vertical velocity at 500 hPa is located over northern Vietnam, and there are subsidence movements in southern China and southern Japan (see the red hatched regions). These features indicate that the convective activity of the atmosphere is weakened in this region. Note that the 850 hPa wind field shows an unusual wave-like structure, with anomalous cyclones over the SCS and the PhiSea. This circulation–convection coupling feature is close to the Gill-type atmospheric circulation response due to the deep convection enhancement in the tropics (Gill 1980).

S. Yang et al., *Climate Change in Southeast Asia and Surrounding Areas*, Springer Climate, https://doi.org/10.1007/978-981-15-8225-7_7

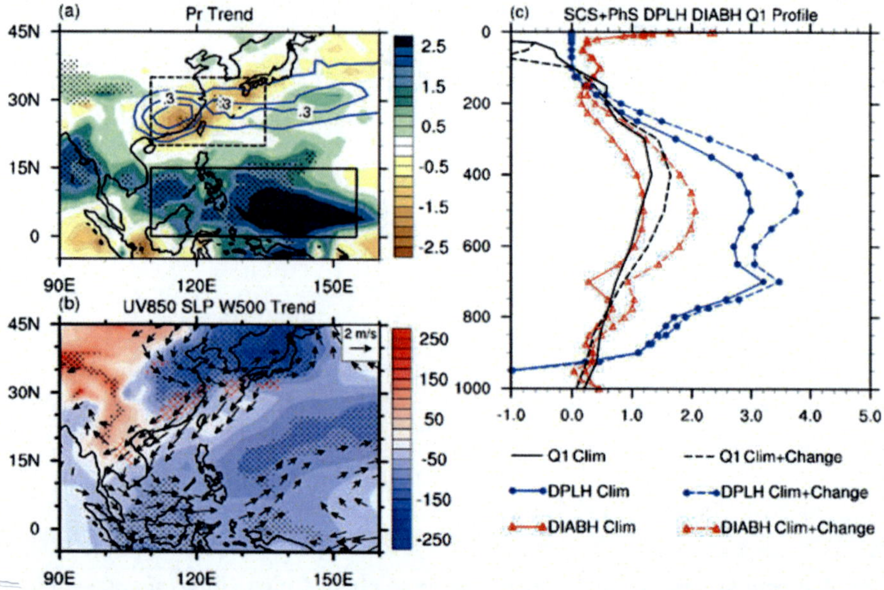

Fig. 7.1 **a** Shaded areas showing the linear trend of MAM precipitation [mm day^{-1} (35 yr)$^{-1}$] during 1979–2013. Stippling indicates a 90% confidence level. Blue contours at 5% interval show the areas where the ratio of climatological MAM precipitation to climatological annual precipitation exceeds 25%. **b** Shading and stippling are used the same way as in (**a**), but for sea level pressure [Pa (35 yr)$^{-1}$]. Vectors denote the linear trend of 850 hPa wind, and hatching indicates where the linear trend of 500 hPa vertical velocity is above 0.025 [Pa s^{-1} (35 yr)$^{-1}$]. Both are statistically significant at the 90% confidence level. **c** MAM climatology, and climatology plus 35-yr linear trend, of heating profile averaged over the black box in **a**. Black curves are for Q1 derived from the NCEP-DOE reanalysis data. Red and blue curves are for diabetic heating and deep convection latent heating, respectively, from the 1-h NCEP CFS forecasting data. After Li et al. (2016)

We further estimated the changes in the latent heating rate of the atmosphere in the SCS and the PhiSea over the past 35 years. In Fig. 7.1c, the solid curves depict the springtime climatological profiles for deep convective heating, atmospheric diabatic heating, and apparent heat (Q1), respectively. The dashed curves show the climatological states of these variables plus a 35-year linear trend. We can see that there is a significant increase in both diabatic heating and deep convective heating in the troposphere from 800 to 200 hPa. Among them, the most intense change of latent heating occurs from 600 to 300 hPa, and there is a daily maximum heating rate of about +0.9 K at 450 hPa. Compared to the latent heating profile of the climatological state, the increase in heating in the mid and upper troposphere is greater than the increase in shallow convective heating in the lower layers. In addition, the increase in diabatic heating is similar to the change in the latent heating profile, indicating that the change in deep convective heating dominates the change in diabatic heating in the SCS and the PhiSea. Similar enhancement can also be found through the diagnosis of Q1, which verifies the robustness of the strengthened convective heating over the SCS and the PhiSea.

7.2 Effect on East Asian Climate

Studies have shown that spring (MAM) rainfall in southern China is related to the vortex activity in southwestern China and to the activity of the western North Pacific subtropical high (WNPSH): The frequent southwest vortex activity often contributes to the water vapor transport from the warm, moist ocean surface of the SCS to mainland China. As a result, continuous precipitation occurs; and the southerly wind on the western rim of the WNPSH also contributes to water vapor transport from the warm ocean to southern China, inducing continuous precipitation (Yang and Lau 1998). In addition, the sea surface temperature (SST) anomaly in the western Pacific (Qiang and Yang 2013) and the thermal and dynamic processes from the Tibetan Plateau (TP; Wan and Wu 2007), also affect spring precipitation in southern China. We can see that these studies mainly focused on the interannual variability and prediction of spring precipitation in southern China. The trend of spring precipitation in southern China, under the background of long-term climate change, however, has not received much attention. One of our aims here is to use observations and reanalysis data from the SCS and the PhiSea (through cross-validation of various data sets) to study the changes in convective heating sources and reveal their roles in driving atmospheric circulation, which affects precipitation changes in southern China.

7.2.1 *Impact of Convection Over the South China Sea and the Philippine Sea on Southern China Precipitation*

Figure 7.1 shows the atmospheric convective heating over the SCS and the PhiSea had significantly increased in the past 35 years. Since deep convective heating is difficult to observe via instruments, especially over the oceans, it can only be diagnosed through other means, for example, using reanalysis data that contain observations. Adequate cross-validation of the heating trend should then be conducted. First, the trends and magnitudes of the Climate Prediction Center merged analysis of precipitation (CMAP) and Global Precipitation Climatology Project (GPCP) data sets are generally consistent across the SCS and the PhilSea (see Fig. 7.2a, b). In addition, the significant reduction in outgoing longwave radiation (OLR; Fig. 7.2c) also indicates the enhancement of deep convection. However, since longwave radiation is only an indicator of the strength of convection, a lower OLR may also be caused by observational biases in satellite products due to increased probability of more cloud overlap over the tropical regions (Huang et al. 2005, 2006). Further analysis showed that the SLP decreases significantly. The cyclonic response of the lower troposphere

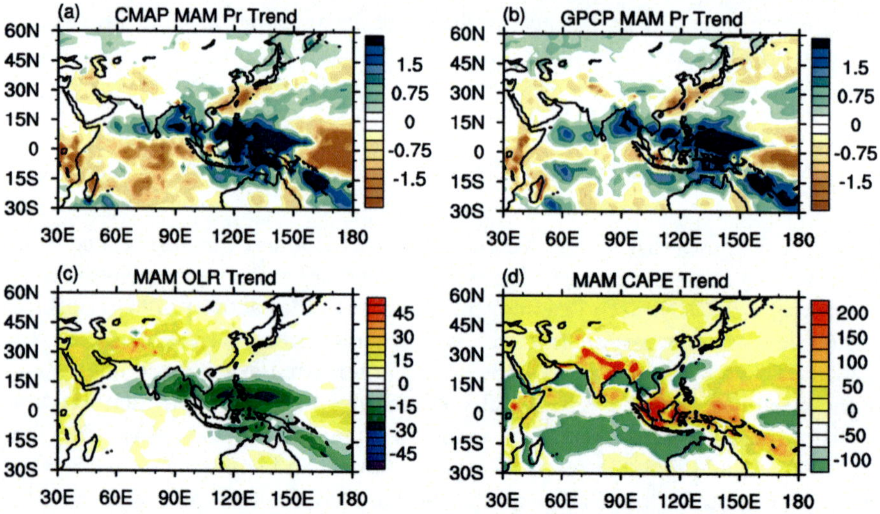

Fig. 7.2 **a** and **b**: Long-term trends of CMAP and GPCP spring precipitation in 1979–2013, respectively. **c** Long-term trend of NOAA spring OLR in 1979–2013. **d** Long-term trend in convective available potential energy in the ERA-Interim reanalysis data. After Li et al. (2016)

wind (Fig. 7.2b) and the increasing trend of the convective available potential energy (Fig. 7.2d), are consistent with the precipitation change over the SCS and the PhiSea. In summary, reanalysis and observational data consistently indicate that changes in the atmospheric circulation in spring have been accompanied by increased rainfall and convection over the SCS and the PhiSea. At the same time, rainfall in southern China decreased significantly over the 35 years.

To describe the potential link between precipitation changes over the SCS and the PhiSea and changes in springtime (MAM) rainfall over southern China, we analyze the precipitation time series averaged over these two regions. Figure 7.3 shows the area-weighted spring precipitation in the two boxes shown in Fig. 7.1a, calculated based on multiple data sets. During the period from 1979 to 2013, the average precipitation in southern China decreased year by year, and the three-year moving average decreased steadily (Fig. 7.3a). Three different data sets show a consistent drying trend, and even the gauge-based precipitation data (PREC/L) show the same trend. On the other hand, the spring rainfall over the SCS and the PhiSea increased significantly (Fig. 7.3b). Although the CMAP and GPCP data sets show some systematic differences, their trends and magnitudes of the changes are consistent. On the interannual timescale, the amount of spring precipitation over the SCS and the PhiSea is highly correlated with the rainfall in southern China (the correlation between 1979 and 2013 was as high as −0.56, with the 99.9% confidence level). Based on these results, we wonder whether there is a causal link between this high correlation: Was it because of the enhancement of convection over the SCS and the PhiSea that caused a long-term drying trend in southern China? If so, is the physical mechanism similar to that for the Pacific–Japan teleconnection in the summertime mean flow?

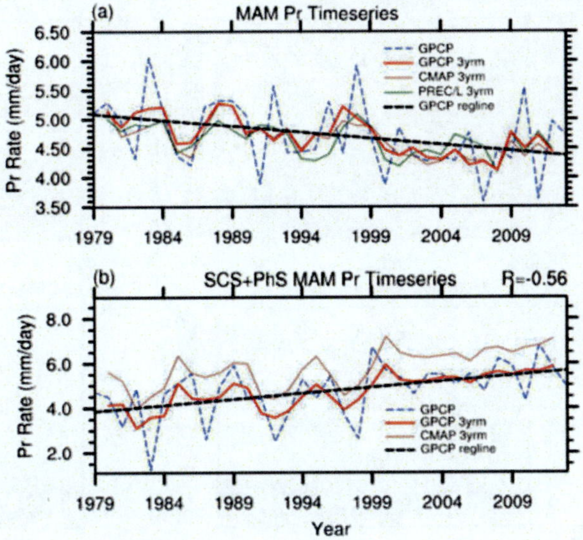

Fig. 7.3 **a** Time series of precipitation change from 1979 to 2013 over southern China. **b** Time series of precipitation change over the SCS and the PhiSea from 1979 to 2013. After Li et al. (2016)

7.2.2 Results from Numerical Model Simulations

We conducted several numerical experiments using the Community Earth System Model Version 1.2.2 (CESM 1.2.2). Specifically, we enhanced the convective heating over the SCS and the PhiSea in an air–sea–land fully-coupled system. A parallel atmospheric dynamic core simulation was also conducted to investigate the thermodynamic process. In the CESM, we use a horizontal resolution of 1.9° (lat) × 2.5° (lon) in the atmosphere component model, with 26 layers in the vertical; and we use an ocean component model of approximately 1° horizontal resolution, with xx layers in the vertical. We designed a fully-coupled control experiment (CTRL_CP). It was based on the cold start initial condition, and continuously integrated for 200 years as the spin-up for the ocean circulation; it was then integrated for 80 years. Years 201–280 outputs are used as CTRL_CP for investigation. In a sensitivity experiment, we restarted from Year 151 of CTRL_CP, and enhanced the deep convective condensation heating over the SCS and the PhiSea. It was integrated for another 130 years to reach Year 280. This experiment is called HEAT_CP. By analyzing the differences between the outputs of the two experiments in the period of Years 201–280, we can reveal the climate responses produced by the enhancement of deep convective heating over the SCS and the PhiSea.

First, from the differences between HEAT_CP and CTRL_CP outputs, we can see that in the springtime of the Northern Hemisphere, the convective heating over the SCS and the PhiSea is enhanced (Fig. 7.4b. The characteristics of "enhancement of convection over the SCS-PhiSea–and reduction of precipitation in southern China" can be well captured. Compared to the observations (Fig. 7.4a), although there are some systematic biases, they do not affect our conclusions.

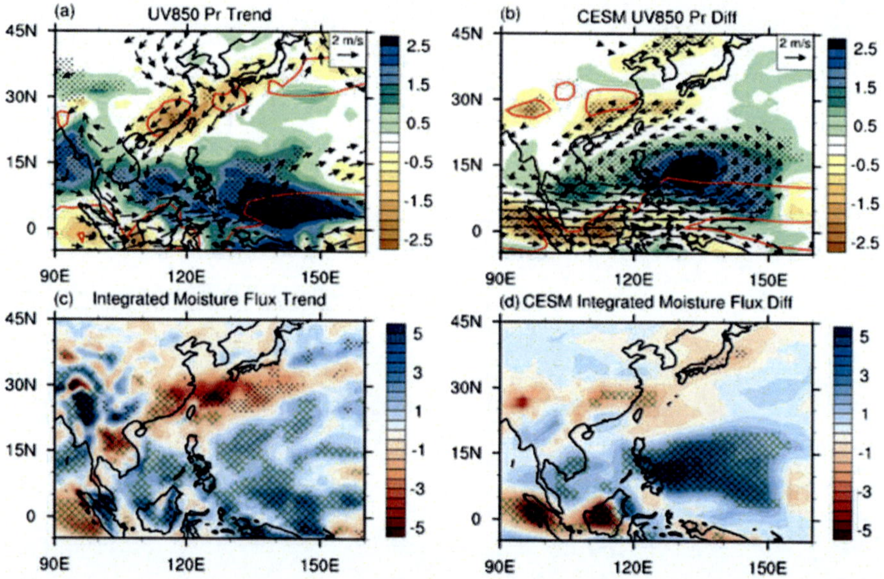

Fig. 7.4 **a** Thirty-five years trend of springtime precipitation and low-level wind at 850 hPa. Stippling represents the linear trend exceeds the 90% confidence level. Vectors denote the linear trend of 850 hPa wind, masking out the areas <90% confidence level. Red contours show the locations of rain-belts, defined by the grids where precipitation rate >6.0 mm day^{-1} in CTRL_CP **b** Same as (**a**), except for CESM-simulated differences between HEAT_CP and the CTRL_CP. **c** Shading areas show the linear trend of convergence of vertical integrated water vapor flux (CVIMF) [mm day^{-1} (35 yr)$^{-1}$]. Green hatching denotes where the convergence term takes up more than 50% of CVIMF change, and stippling indicates the advection term takes up more than 50% of CVIMF change. **d** Same as (**c**), except for CESM-simulated differences between HEAT_CP and CTRL_CP. After Li et al. (2016)

First, the drought trend in southern China is smaller than that in the observations. This is acceptable because the process that contributes to this observational trend is not only the low-latitude thermodynamic forcing process. The observations show that the atmospheric circulation had undergone through significant changes in the mid-latitudes over the past 35 years, and anomalous northerly wind over the mid-latitude continent had also contributed to the transport of dry air advection to southern China, thereby aggravating the drought trend (Fig. 7.4a). In HEAT_CP, we only changed the convective heating in the low-latitudes, so the suppressed spring precipitation in southern China should be attributed only to the convective heating change, and did not include the effects of latitude changes in the mid- latitudes.

Second, the simulated precipitation enhancement in the PhiSea (Fig. 7.4b), has a northwestward shift compared to the observation (Fig. 7.4a). This difference may be related to two factors. (1) In CTRL_CP, the main rain belt over the western tropical Pacific shifts approximately 5° northward and 10° westward. Thus, the location of rainfall response also shifts due to this bias. (2) According to the Gill-type response theory and the heating profile added to HEAT_CP, the anomalous cyclone, which is essential for the convergence in the planetary boundary layer and the ascending motion in the troposphere, is located to the northwest of the anomalous heating center (Gill 1980). Therefore, the anomalous precipitation center corresponding to the water vapor convergence is also located in the northwest of the heating center. Despite the above-mentioned biases, the consistency between observations and CESM fully-coupled experiments is clear, which means that the enhanced convective heating over the SCS and the PhiSea is conducive to the reduction of precipitation in southern China.

To understand the mechanism through which convective changes over the SCS and the PhiSea affect precipitation in southern China, we conducted a diagnostic analysis of the overall water vapor budget. In general, the convergence or divergence of vertical integrated water vapor flux (CVIMF) is balanced by precipitation minus evaporation (P-E; Rodwell and Hoskins 2001). In addition, CVIMF can be further divided into water vapor convergence term caused by convergence or divergence of the mean wind field, and moisture advection term dominated by nonuniform moisture field. To understand the effect of convective heating over the SCS and the PhiSea on springtime precipitation in southern China, we calculated changes in CVIMF and its two components. In the observations and model simulations (Fig. 7.4c, d), the CVIMF changes correspond well with the changes in precipitation over East Asia and the western Pacific, where the increase (decrease) in precipitation is consistent with positive (negative) CVIMF. A closer examination of changes in CVIMF suggests some inconsistencies compared to the observed rainfall trend, which is mainly due to the change in the residual term, namely, evaporation (E).

When focusing on the relative contributions of the two terms in the convergence of vertically integrated moisture flux (CVIMF), several significant features can be discovered from the observations. Specifically, regardless of positive or negative values, changes in CVIMF are dominated by convergence term in the low-latitude regions because of the warm SST there, abundant water vapor supply, and relatively low moisture gradient in the boundary layer (Fig. 7.4c). Note that in southern China, CVIMF is also mainly attributed to the convergence or divergence of the wind. However, over the East China Sea near the southeastern coast of China, the contribution mainly comes from the advection term. The difference between the two regions indicates that there are different mechanisms for controlling changes in precipitation on land and at sea (Fig. 7.4c). Although the simulated CVIMF shifted northward by about 5° in southern China, it is mainly limited to the southern inland area (Fig. 7.4d). In the model simulation, CVIMF changes are dominated by the convergence term in East Asia and the western Pacific (Fig. 7.4d). Indeed, water vapor divergence contributes more than 80% of the reduction of CVIMF in southern China. This remarkable dynamic effect leads us to speculate that the rapid and direct atmospheric dynamic response caused by increased convective heating over the SCS and the PhiSea inhibits springtime precipitation in southern China.

To verify the above inference, we used the prescribed heating profile over the SCS and the PhiSea in the idealized simplified atmosphere model to investigate the atmospheric response. The idealized model is the Community Atmospheric Model version 5 (CAM5) dynamic core embedded with the idealized physical package. Figure 7.5

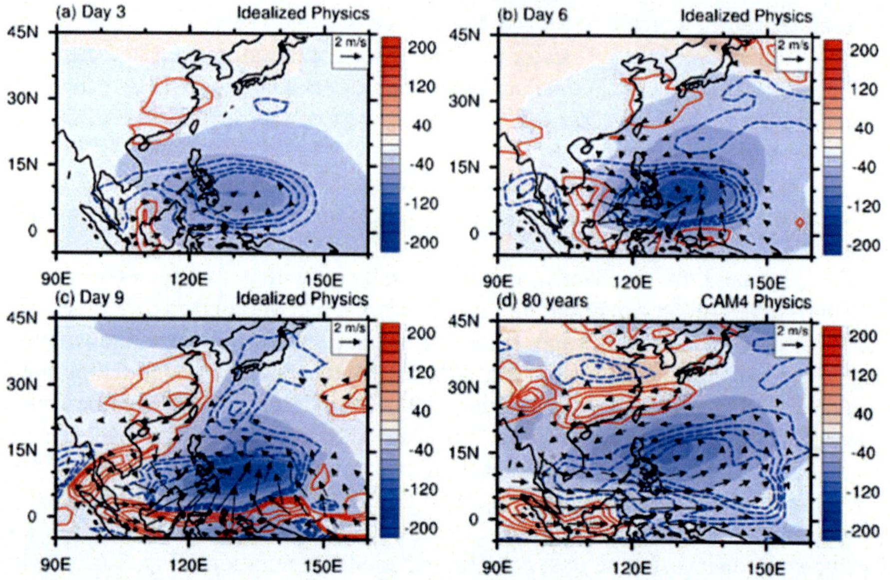

Fig. 7.5 Sea level pressure anomalies (shading; Pa). Blue (or red) contours at 0.004 interval show 500 hPa negative (or positive) anomalies (Pa s^{-1}) of vertical velocity. Vectors show 850 hPa wind anomalies, masking out wind speed smaller than 0.5 m s^{-1} . **a–c** Three different snapshots: differences on day 3, day 6, and day 9 between HEAT_IDEAL and CTRL_IDEAL of the CAM dynamic core integration with idealized physics. **d** Eighty-year-mean differences between HEAT_CP and CTRL_CP of the coupled climate model CESM with CAM4 full physics packages. After Li et al. (2016)

shows the 80-year-averaged results for the fully-coupled CESM experiment and 3-day-averaged snapshots for the idealized model simulation. The given heating over the SCS and the PhiSea results in local updraft associated with the southerly wind in the lower troposphere and a decrease in SLP (Fig. 7.5a). Note that after three days of integration, there is a sinking motion over mainland China. On the 6th day, anomalous ascending motion around the heating center intensifies, accompanied by a predominantly lower-tropospheric anomalous cyclone and significant SLP drop. At the same time, the sinking motion in southern China spreads eastward (Fig. 7.5b). On the 9th day, the low-level anomalous cyclone marches to the northwest of the heating center (Fig. 7.5c). The ascending branch is located around the heating center, and extends over the ocean. Importantly, another more powerful sinking center emerges near the southeastern coast of China. In most parts of East Asia, large-scale sinking motion occurs after the anomalous circulation matures. The six ensemble members from perturbed initial conditions also capture this response pattern (Fig. 7.6), which means that the anomalous overturning circulation response is robust in the context of spring atmospheric conditions. The results of 80-year-averaged fully-coupled run using the CAM4 physical scheme also produced a response similar to that of the idealized experiment, although the result was smoother (Fig. 7.5d).

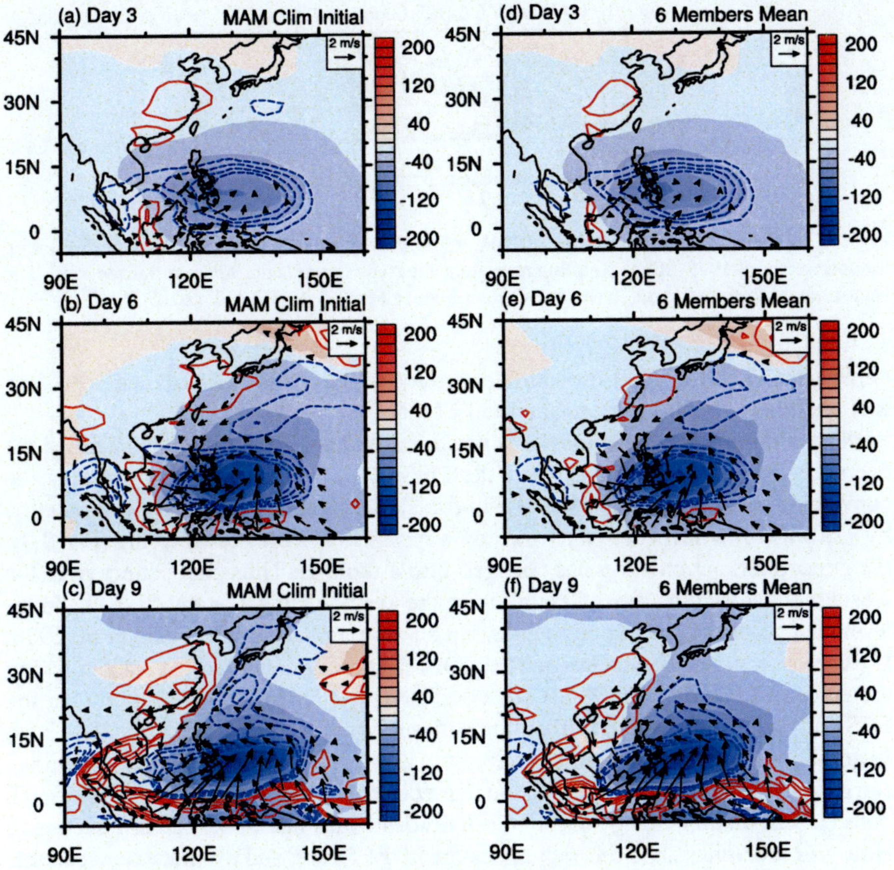

Fig. 7.6 **a–c** Sea level pressure (shading; Pa), 500 hPa vertical velocity (contour; Pa s^{-1}), and 850 hPa wind (vector; m s^{-1}) from idealized heating experiment driven by springtime mean flow on day 3, day 6, and day 9. **d–f** Same as (**a–c**), except for the results of six ensemble members with different perturbed initial conditions. After Li et al. (2016)

The similarity between the idealized model simulation (Fig. 7.5a, c) and the fully-coupled experiment (Fig. 7.5d), illustrates that the thermodynamic circulation caused by convective condensation latent heating over the SCS and the PhiSea affects the springtime rainfall over southern China directly. More specifically, the enhanced convective heating over the SCS and the PhiSea in spring results in a decrease in local SLP, a cyclonic response in the lower troposphere, and an anomalous rising motion. At the same time, subsidence appears in southern China. Sinking motion and low-level divergence together reduce the moisture convergence in southern China, and thus reduce precipitation. Although this thermal response process is regulated by synoptic-scale fluctuations under the mean flow in spring, the subsidence in southern China continues to emerge with the dynamic effects of convection over the SCS and

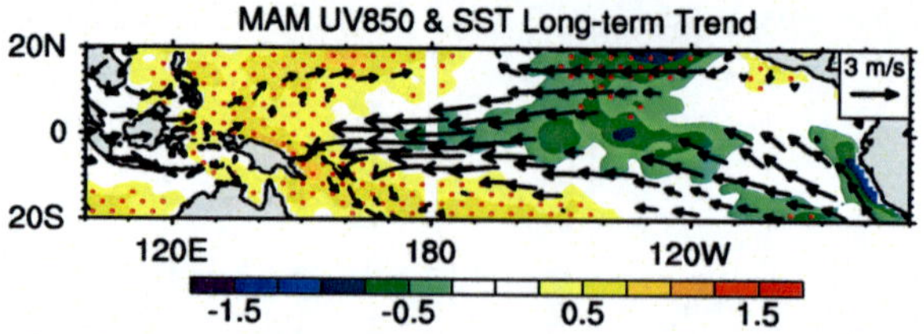

Fig. 7.7 **a** Thirty-five years trend of 850 hPa wind (vector; m s^{-1}) and SST (shading; K) during the springtime of 1979–2013. Stippling represents the trend exceeds the 90% confidence level. The vectors show the wind trend above the 90% confidence level. After Li et al. (2016)

the PhiSea. Therefore, this low-latitude convective activity is a rapid dynamic process that modulates precipitation in southern China.

Note although the spring drought trend in southern China is modulated by the deep convection over the SCS and the PhiSea, the mechanism that controls the convection enhancement over the SCS and the PhiSea remains unclear. Accompanied by a stronger low-level easterly wind and warmer SST in the boreal spring (Fig. 7.7), the atmospheric circulation has changed into a more La Niña-like pattern over the western equatorial Pacific. Although from the average annual perspective, the long-term change of the Walker circulation is still an open question (Vecchi et al. 2006; L'Heureux et al. 2013; McGregor et al. 2014; Ma and Zhou 2016), at least in spring the enhanced Pacific trade winds are associated with enhanced convection over the SCS and the PhiSea, which further intensifies the release of atmospheric latent heat. In addition, the intensification of convection activities over the SCS and the PhiSea can only partially explain the reduction in precipitation in southern China. How much does the circulation change in the mid-latitudes contribute to the rainfall decrease? How will precipitation in this region change in the future, and to what extent can this change be attributed to human activities? For example, recent studies have shown that limited amount of aerosol pollution may have caused an increase in local rainfall in eastern China in summer, whereas local precipitation decreases in the case of an excessive aerosol concentration (Guo et al. 2017). Heavy air pollution events sometimes prevail in spring in southern China. What effect does it have on spring precipitation in the area? These issues deserve more comprehensive research.

7.3 Influence of Climate Change Over Southeast Asia on Sahel Summer Monsoon Rainfall

In Sect. 7.3, we discuss enhanced summer latent heating over the South Asian and western Pacific summer monsoon (SAWPSM) region on the decadal timescale (He et al. 2017). We also illustrate the dynamic effect of the heating on the decreased Sahel summer rainfall (He et al. 2017).

The Sahel is located in Afro-Eurasia. Its summer rainfall exerts a significant influence on both local ecology and local economy. Its rainfall variation ranges apparently from synoptic timescale to decadal timescale. During the second half of the twentieth century under global warming, an obvious drying trend was seen in the Sahel in summer (Nicholson 1980; Ma and Fu 2007). The Sahel rainfall shifts from a humid state in the 1950s and 1960s to a lesser humid one in the 1970s and 1980s, with an abrupt change around 1969 (L'Hôte et al. 2002; Ozer et al. 2003; Greene et al. 2009). Furthermore, the Sahel droughts can influence the transport of desert dust and thus global climate by means of the radiative properties of dust and its climatic effects (Prospero and Lamb 2003).

In addition to the consequences of Sahel rainfall, the causes of droughts are researched. The drying trend in the Sahel is primarily related to global SST variability (Folland et al. 1986; Palmer 1986), such as the Pacific Decadal Oscillation (PDO; Ma and Fu 2007), El Niño-Southern Oscillation (ENSO), and SST fluctuations in the South Atlantic. These SST patterns contribute to different rainfall variation through several mechanisms, for instance, the Pacific– North America Oscillation (Ward 1998; Hunt 2000; Camberlin et al. 2001; Giannini et al. 2003). The reduction in Sahel rainfall can be attributed to both the internal variability of SST and the externally forced change in SST (Held et al. 2005). Apart from SST variation, previous studies discovered that the rainfall variation on different timescales is influenced by different factors. First, the synoptic-scale Sahel rainfall is mainly regulated by cloud clusters, which are modulated by the mid-tropospheric African easterly jet stream and the tropical easterly jet stream (Nicholson and Grist 2003). Therefore, the variations of the West African westerly jet stream and the African easterly waves intimately alter the variation of the rainfall (Pu and Cook 2012; Skinner and Diffenbaugh 2014). This feature indicates the remote connection to the Sahel rainfall through the jet stream. Second, an increase in local albedo due to desertification can decrease precipitation on the interannual to decadal timescales (Charney et al. 1975, 1977; Courel et al. 1984; Tucker et al. 1991; Yongkang and Shukla 1993), and local vegetation interaction intensifies the rainfall variability (Zeng et al. 1999; Giannini et al. 2003). Third, increased greenhouse gasses (GHGs) and changes in anthropogenic aerosols affect the Sahel rainfall on the decadal timescale. However, these effects vary among models, with some model results indicating droughts and while some others indicating normal rainfall condition (Held et al. 2005; Biasutti and Giannini 2006; Dong and Sutton 2015).

Afro-Eurasia contains influential climate systems that interact with each other. For instance, there are major summer heat sources in Afro-Eurasia (Yanai and Tomita 1998; Li et al. 2001), and monsoon systems with substantial latent heating (Webster and Yang 1992; Webster et al. 1998a, b). Particularly, the Asian summer monsoon is part of the influential systems. Since both mean diabatic heating and its fluctuation affect the mean climate (Sardeshmukh and Sura 2007), the climate of the Sahel region can be affected remotely by the changes in the Asian monsoon. Two factors are involved in spreading the influence of the Asian monsoon to the Sahel region. One is a zonal cell of the Asian summer monsoon circulation, which is associated with the diabatic heating gradient over Afro-Eurasia (Song et al. 1992; Li et al.

2001). The other is the Rossby wave response to the west of subtropical monsoon heating. They can influence the Sahel summer rainfall directly or affect it indirectly by modulating local meridional circulation (Liu et al. 2001a, b; Rodwell and Hoskins 2001). In addition, the Asian summer monsoon (Yang and Lau 1998), and Sahel summer rainfall as a whole are influenced by the same factors and are demonstrated to coexist as twin features of multi-scale forcing (Wu et al. 2009). In short, there is a close connection between the Asian summer monsoon and Sahel summer rainfall.

Hence, what is the role of the monsoon–desert mechanism provided by Rodwell and Hoskins (2001), in Sahel rainfall reduction? In this subsection, we analyze the NOAA's Precipitation Reconstruction over Land and the NCEP/NCAR Reanalysis, and conduct numerical model simulations. These results demonstrate that the interdecadal change in the Asian summer monsoon plays a vital role in changing Sahel summer rainfall.

7.3.1 Observed Features

The Sahel is a narrow semiarid region. Its precipitation mainly occurs during summer (from July to September), ranging up to over 14 mm/day. Therefore, a small change in the summer rainfall makes a big difference in its annual precipitation. The early-1980s witnesses the driest years of the Sahel. The difference between 1960–1969 and 1980–1989 indicates a considerable decrease in Sahel summer rainfall (by 20.5%) and a slight increase to the south (Figs. 7.8a and 7.9a). Meanwhile, as indicated by the observed changes in divergent/convergent wind, anomalous ascents emerge over southern Asia and the western Pacific while anomalous descents appear over the Sahel (Fig. 7.10). Moreover, an interdecadal change in precipitation (of 5.2% increase in the NCEP/NCAR Reanalysis and of 23.3% increase in the ERA40) occurred in the SAWPSM region between 1960–1969 and 1980–1989 (Fig. 7.8b–c). The coincidence of the decrease in Sahel rainfall and the increase in SAWPSM precipitation suggests an anti-correlation between the two during 1960–1989, and thus a hypothesis that the remote latent heating of the SAWPSM region modulates the Sahel rainfall. Figure 7.9b–c show the change in precipitation mainly emerged in the western Pacific and around the Strait of Malacca, aside from a slight decrease near the equator. The precipitation increase in the SAWPSM region is also verified by the increase in local cloudiness (Fig. 7.8d). In addition, the changes in Asian and African precipitation over land show coherent variability (Fig. 7.9a) (Feudale and Kucharski 2013; Wang et al. 2014).

In this subsection, we focus on the role of the monsoon heating over the SAWPSM region (Rodwell and Hoskins 2001), since this region includes vast tropical-subtropical oceans with considerable diabatic heating (Fig. 7.11). However, when precipitation over the oceans is considered along with the connection between the Sahel and the SAWPSM region, changes in Sahel rainfall and SAWPSM precipitation may be out of phase (Feudale and Kucharski 2013).

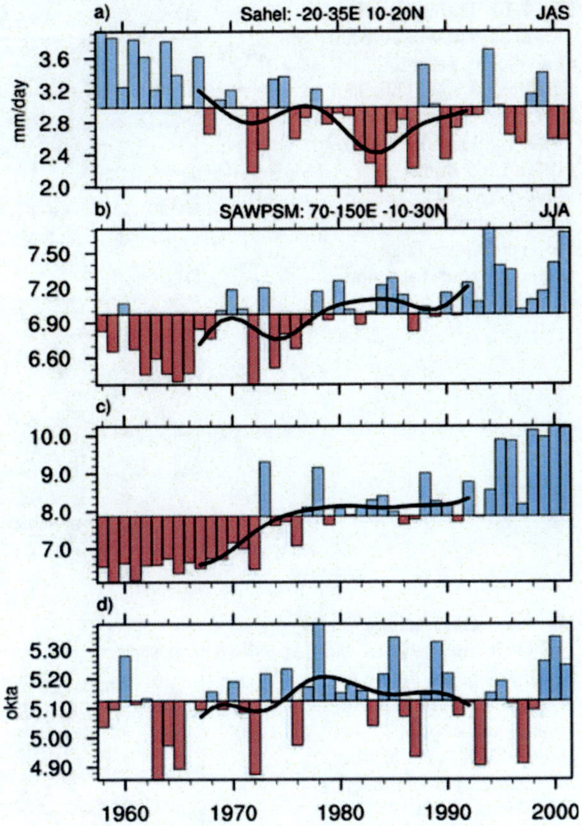

Fig. 7.8 Time series (bar) of **a** JAS rainfall over the Sahel in the PRECL precipitation data, **b** June-August (JJA) SAWPSM precipitation in the NCEP reanalysis data, **c** JJA SAWPSM precipitation in the ERA40, and **d** JJA SAWPSM cloudiness. The black curve indicates the weighted running average with 9-year Lanczos low-pass filter of each time series. After He et al. (2017)

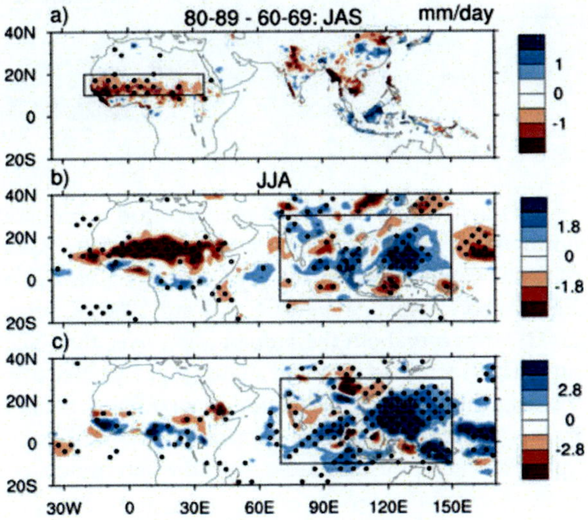

Fig. 7.9 Differences between 1980–1989 and 1960–1969 (1980–1989 minus 1960–1969) in **a** JAS Sahel rainfall in the PRECL precipitation data, **b** JJA SAWPAM precipitation in the NCEP reanalysis data, and **c** JJA SAWPAM precipitation in the ERA40 reanalysis data. The boxes define the Sahel (**a**) and SAWPSM (**b** and **c**) areas. Stippling denotes the 95% confidence level based on Student's *t*-test. After He et al. (2017)

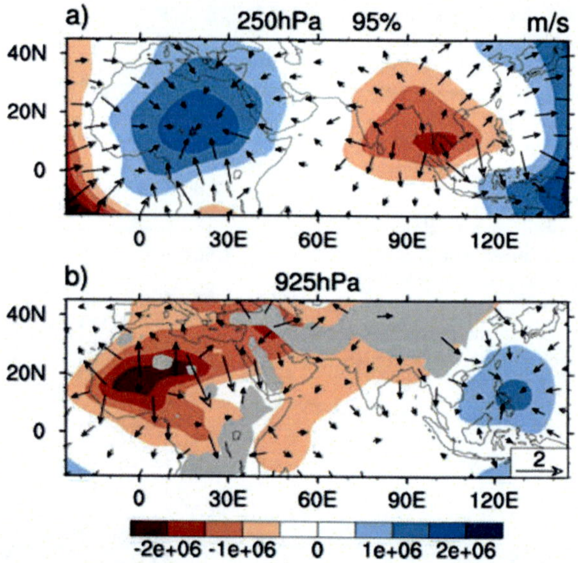

Fig. 7.10 Differences in JAS divergent wind (vector) and velocity potential (shading) at **a** 250 hPa and **b** 925 hPa between 1980–1989 and 1960–1969 (1980–1989 minus 1960–1969). Only the vectors significant at the 95% confidence level (Student's *t*-test) are plotted. After He et al. (2017)

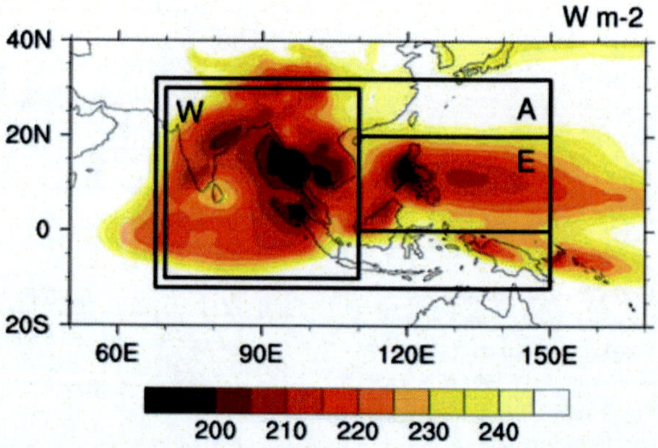

Fig. 7.11 Climatology of JAS OLR from 1979 to 2010. The boxes define the regions (A, W, and E) of anomalous deep convectional latent heating for sensitivity experiments. After He et al. (2017)

This increase in SAWPSM precipitation is probably linked to the increase in the northern Indian Ocean and western Pacific SSTs (Xie et al. 2010), and to the increase in their heating, whereas the high-frequency variation of Sahel rainfall results from ENSO and tropical Atlantic SST (Ward 1998; Camberlin et al. 2001; Giannini et al. 2003). However, both the precipitation over the Sahel and that over the SAWPSM region has increased since the 1990s (Fig. 7.8). Because of the much missing data over the tropical oceans, the decrease in cloudiness since the 1980s, shown in Fig. 7.8d, mainly reveals the features over the subtropical oceans, which indeed indicates an opposite trend of precipitation over the subtropical oceans since the 1980s.

7.3.2 Numerical Experiments with SAWPSM Latent Heating Anomalies

To explore the possible influence of the decadal change in SAWPSM latent heating on Sahel summer rainfall, the CESM 1.2.2 was employed (Kay et al. 2015). The component set of F_2000_CAM5 was chosen. It features the CAM5 on 1.9° × 2.5° grid and 26 levels. Experiments were forced by monthly climatology of SST and sea ice from 1950 to 2010; other external forcing includes CO_2, ozone, and aerosol with their values set at those in 2000. Each model integration was 18 years from January 1st, 2000, and the mean values of the last 15 years are analyzed. Predicted heating of deep convection was multiplied by an invariant coefficient over different parts of the SAWPSM region in several sensitivity experiments from June to September when the predicted heating of deep convection in the model was positive for each air parcel at each integration step. Specific coefficient is chosen according to the interdecadal change in the Sahel or SAWPSM precipitation (Table 7.1).

The first experiment, referred to as Exp SEN_A, was performed to identify the effect of the heating anomaly over the whole SAWPSM region on Sahel rainfall. In this experiment (Fig. 7.12a), precipitation increases substantially over the northern SAWPSM region but diminishes slightly over the southern SAWPSM. These features of precipitation indicate that precipitation increases due to the increase in SAWPSM latent heating. The local responses of precipitation to the heating anomaly in the model match the general features of the observed precipitation anomaly. In accordance with the observed Sahel rainfall, the precipitation over the Sahel region declines with enhanced SAWPSM heating (Fig. 7.8a). At the same time, there exist anomalous rising motion over the SAWPSM region and anomalous sinking motion over the Sahel region (Fig. 7.13a). Over the SAWPSM region, anomalous moist convective heating is balanced by the adiabatic cooling associated with the ascent (Rodwell and Hoskins 1996).

Table 7.1 List of model experiments and experimental design

Exp	Region applied	Heating coefficient	Integration (year)	Spin up (year)
CON/CON_60/CON_80	None	\	18	3
SEN_A	70°–150° E, 10° S–30° N	1.1		
SEN_W	70°–110° E, 10° S–30° N	1.1		
SEN_E	110°–150° E, 0°–20° N	1.1		
SEN_COOL1	70°–150° E, 10° S–30° N	0.8		
SEN_COOL2	20° W–35° E, 10°–20° N	0.8		

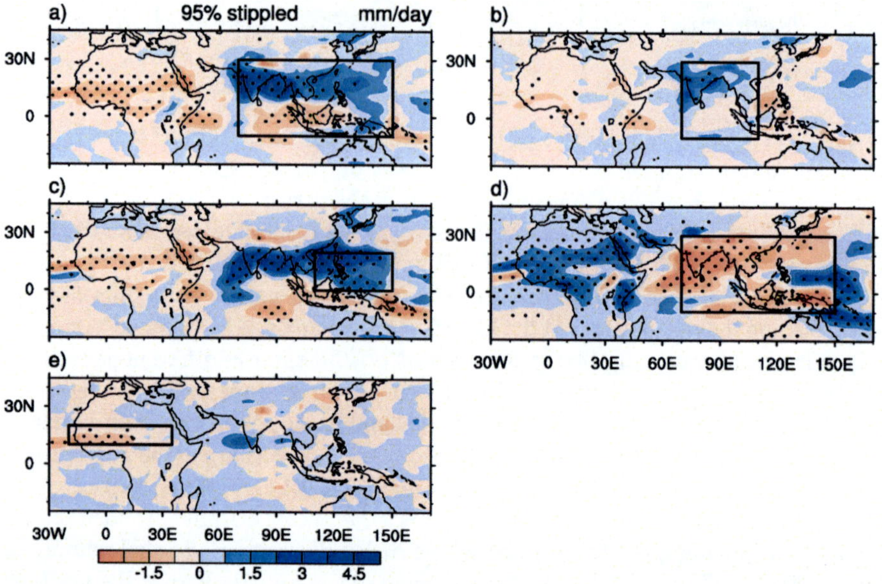

Fig. 7.12 **a** Differences in JAS precipitation between SEN_A and CON, **b** between SEN_W and CON, **c** between SEN_E and CON, **d** between SEN_COOL1 and CON, and **e** between SEN_COOL2 and CON. The boxes indicate the anomalous heating regions for sensitivity experiments. Stippling denotes the 95% confidence level based on Student's *t*-test. After He et al. (2017)

Fig. 7.13 Differences in JAS **a** omega and zonal wind (vector), and meridional wind (shading) at 14.2° N between SEN_A and CON, and **b** water vapor flux (vector) and its divergence (shading) at 900 hPa between SEN_E and CON. The meridional winds and vectors denote the changes are statistically significant at the 95% confidence level based on Student's *t*-test. Omega is scaled so that the vertical motion is visible. After He et al. (2017)

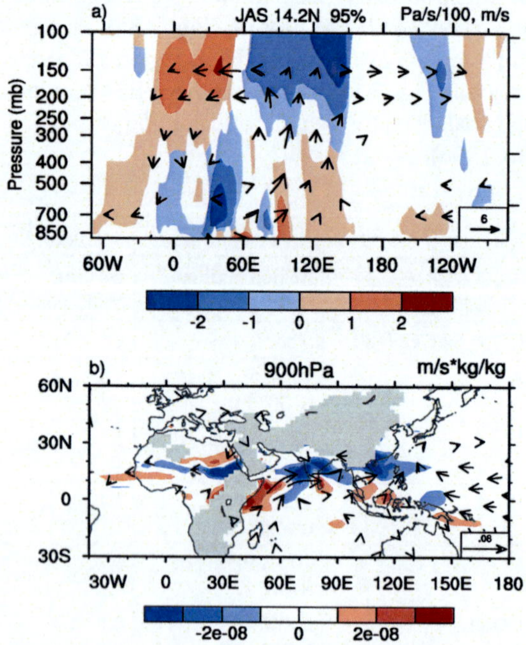

The features of the anomalous circulation indicate an enhancement of the zonal cell of the Asian summer monsoon circulation that links the Sahel and tropical Asia (Yang et al. 1992). Additionally, there are anomalous southerlies below the maximum rising motion over the SAWPSM region and anomalous northerlies below the maximum sinking motion over the Sahel. The anomalous northerlies are caused by the vortex shrinking associated with the SAWPSM-induced sinking motion, in agreement with the Sverdrup relation that a vertical column of air is squashed when it moves toward the equator (Rodwell and Hoskins 2001). Thus, more water vapor is transported from the Indian Ocean and the SCS to maintain stronger Asian monsoon, whereas the Sahel region receives less moisture supply from the Gulf of Guinea.

We conducted additional experiments by increasing the heating of deep convection over Southeast Asia and the western North Pacific (WNP), respectively (Fig. 7.11), referred to as Exps SEN_W and SEN_E, to identify the atmospheric responses over surrounding regions to these heating anomalies. The features of the summer monsoon over Southeast Asia are unlike those over the WNP. These two monsoons exhibit different life cycles and are characterized by different moisture influxes and modes with different symmetric and asymmetric components (Murakami et al. 1999). Therefore, they should be investigated separately. Moreover, we reduced the heating over the SAWPSM region in Exp SEN_COOL1 and reduced the heating over the Sahel region in Exp SEN_COOL2. Exp SEN_COOL1 was conducted to make clear the linear atmospheric response, which implies anomalies on the condition of decreased SAWPSM heating, and such anomalies are the inverse of those when SAWPSM heating increases. If that is the case, an increase in Sahel rainfall is then expected. Exp SEN_COOL2 is aimed to discern the feedback effect of decreased Sahel rainfall. In Exp SEN_W (Fig. 7.12b), precipitation increases over South Asia and decreases over the Sahel. Meanwhile, there is an anomalous atmospheric circulation similar to that in Exp SEN_A. In Exp SEN_E (Figs. 7.12c and 7.13b), precipitation increases over the SAWPSM region and diminishes over the Sahel. Similar to Exp SEN_A, there exists an anomalous vertical motion, which connects the Sahel and tropical Asia. Furthermore, an intensification of the WNP monsoon heating is accompanied by an increase in South Asian monsoon precipitation, which overlaps with a cyclonic anomaly in the lower levels of the atmosphere (Fig. 7.13b). The cyclonic anomaly may be caused by the increasing heating to its east, in agreement with the Gill-type response (Gill 1980).

In Exp SEN_COOL1 (Fig. 7.12d), the precipitation reduction over the SAWPSM region results in an increase in Sahel rainfall. Within this limit, the Sahel summer rainfall changes monotonously along with the change in the precipitation over the SAWPSM region (Fig. 7.12a, d). In Exp SEN_COOL2 (Fig. 7.12e), a decrease in Sahel rainfall results in a slight increase in precipitation over the SAWPSM region through the enhancement of the thermally-driven zonally asymmetric circulation. This feature indicates a positive feedback between Sahel rainfall and the precipitation over the SAWPSM region. In other words, the enhanced SAWPSM monsoon heating diminishes Sahel rainfall and local latent heating. In return, the diminished rainfall and latent heating slightly increase the originally enhanced SAWPSM precipitation.

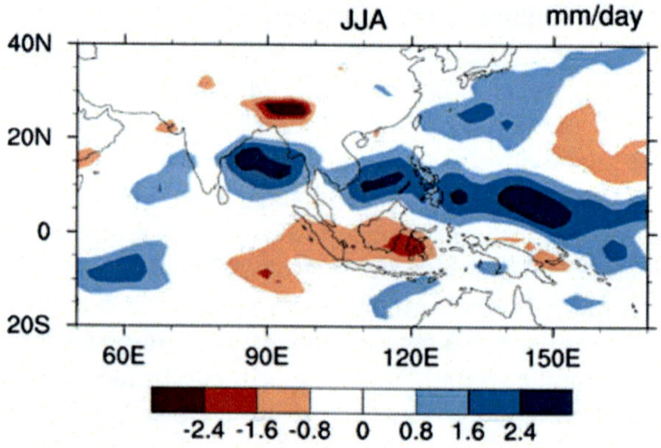

Fig. 7.14 Differences in JJA precipitation between Exps CON_80 and CON_60 (Exp CON_80 minus Exp CON_60). After He et al. (2017)

In addition, Exps CON_60 and CON_80 were performed to distinguish whether the global SST changes are the main cause of the SAWPSM heating anomaly. All the settings in Exps CON_60 and CON_80 are the same as those in Exp CON, except for the monthly climatology of SST and sea ice: for Exp CON_60, the period is from 1960 to 1969, whereas for Exp CON_80, the period is from 1980 to 1989. Although there exist some disparities between the results from the model and those from the observation, compared with Exp CON_60, the precipitation in Exp CON_80 increases substantially over the western Pacific and the Bay of Bengal (BOB) instead of the Strait of Malacca and decreases slightly near the equator (Fig. 7.14). The model responses to the SST anomalies match the general features of observed precipitation anomaly. The results from Exps CON_60 and CON_80 imply that the global SST change can be a primary cause of the SAWPSM heating anomaly (Xie et al. 2010). Further research is needed to reveal the mechanism at work.

7.3.3 Summary and Discussion

From the 1960s, the Sahel region has experienced summer rainfall deficit. The rainfall variations from seasonal to interdecadal timescales have generated substantial research interests due to the drought. Note that an interdecadal change in the SAWPSM latent heating emerged along with the change in Sahel rainfall (Fig. 7.8). In this subsection, we pay close attention to the role of the monsoon–desert mechanism in Sahel rainfall reduction. Meanwhile, there exists an enhancement of the zonal cell of the Asian summer monsoon circulation. The circulation connects the Sahel with the SAWPSM region (Fig. 7.10). Also, a negative correlation is found between Asian heating and Sahel rainfall on the interdecadal timescale (with a correlation coefficient of −0.58, which significantly exceeds the 99% confidence level). On the contrary, the increased heating does not reduce the rainfall variability on the interannual timescale (with a correlation coefficient of 0.22, significant at the 90% confidence level).

The absence of a negative correlation between the two since the 1990s, is another interesting feature (Fig. 7.8). Note that the Atlantic multidecadal oscillation switched to its warm phase in the early-1990s, and the warm SST may weaken the longitudinal heating gradient over Afro-Eurasia. Consequently, weakening occurs in the thermally-driven zonal cell of the Asian summer monsoon circulation (Song et al. 1992), and in the Rossby wave propagation from Asia to the Atlantic along with anomalous westerlies. The weakening in these two factors implies a decline in the monsoon–desert coupling. In addition to the weakening in the circulation, the warm Atlantic multidecadal oscillation phase itself can generate anomalous westerlies with moist air to increase Sahel rainfall (Knight et al. 2006). To reveal the dominant mechanism, further investigation is needed.

The results of several numerical experiments designed to examine this phenomenon also showed that the enhanced heating over different parts of the SAWPSM region exerts a remote dynamic effect on Sahel rainfall. The Asian monsoon conveys signals to the west through two mechanisms. One is the Rossby wave propagation to the west, which is induced by diabatic heating of the Asian monsoon (Rodwell and Hoskins 2001). The propagation of the wave produces sinking motion over Africa. The other is the transverse circulation connecting the Sahel region and SAWPSM region. The circulation is modulated by the longitudinal heating gradient over Afro-Eurasia, for instance, between the Sahel and SAWPSM region (Fig. 7.13a) (Song et al. 1992). The results from the numerical experiments in this subsection support the two mechanisms stated above. Thus, the role of the monsoon-desert mechanism in Sahel rainfall reduction is confirmed. Further research is needed to explore the role of the monsoon-desert mechanism under the influence of oceans. In Exp SEN_COOL2, a change in convection over the Sahel affects the Asian monsoon by enhancing the longitudinal heating gradient. Then, reinforcement occurs in the transverse Asian summer monsoon circulation, along with the increase in SAWPSM precipitation.

Note that although it may not be optimal to multiply the predicted heating of deep convection by an invariant coefficient, this method should be sufficient to confirm the linkage between Sahel rainfall and SAWPSM precipitation. The fact that the local responses over the Sahel and SAWPSM region to the heating anomaly match the features of the observed precipitation anomalies also confirms the effectiveness of the method (Figs. 7.9 and 7.12). Nonetheless, it is suggested to replace the predicted heating in the model with a model-derived latent heating climatology in further investigation. Apart from applying model-derived heating, coupled experiments are also proposed to explore the role of air–sea interaction induced by the heating anomaly over the SAWPSM region in future. The reason is that the along-shore anomalous northerlies over the Sahel induced by the heating anomaly (Fig. 7.13a), may result in offshore Ekman transport and upwelling, which can lead to cold SSTs. Consequently, the intertropical convergence zone (ITCZ) may shift to the south (Philander et al. 1996), with less Sahel summer rainfall. In addition, the causes of increased heating and its relationship with the warming in the northern Indian Ocean await future studies.

7.4 Impact of Precipitation and Associated Latent Heat Release Over South China Sea and Western Pacific on Global Climate During Boreal Spring and Summer

The latent heat release in the troposphere undoubtedly is important for the large-scale atmospheric circulation changes (Matsuno 1966; Webster 1972). Previous studies showed that the east-west asymmetric circulation in the tropics can be illustrated by solutions due to heating placed in specific areas (Gill 1980). The location and magnitude of the westerly jet stream is also modulated by the tropical heating in the boreal summer (Yang and Webster 1990). Rodwell and Hoskins (2001) pointed out that the equatorward portion of each subtropical anticyclone may be viewed as the Kelvin wave response to monsoon heating over the continent to the west. In the upper troposphere, the decadal intensification of upper-troposphere temperature maximum over the Asian continent in the boreal summer may be attributed to the increasing latent heating over East Asia (Wu et al. 2015). Therefore, it is worth investigating the changes of latent heating over the SCS and its surrounding regions under the global warming background, and their climate effects. These studies can provide us a new understanding of global climate change.

7.4.1 Recent Change in Precipitation

Global warming is one of the most significant climate change signals at the Earth's surface. According to the recent Fifth Assessment Report (AR5) of the Intergovernmental Panel on Climate Change (IPCC 2013), the GHGs increased rapidly after the 1950s, and the rise of surface temperature from the 1950s to the present almost contributed 85% of the total warming from the 1850s to the present. To identify the changes in surface temperature and precipitation under recent global warming, we first define two climate mean states to distinguish climate background between relatively low and relatively high global warming states. The climate mean of 1949–1958 is regarded as the relatively low global warming climate state, and the climate mean of 1999–2008 is regarded as the relatively high global warming climate state. Their difference represents the total climate response to recent global warming effect.

Based on the above definitions, we examine the spatial differences in surface temperature from the Goddard Institute for Space Studies (GISS) data set, SST from the Hadley Centre Sea Ice and Sea Surface Temperature data set (HadISST), and precipitation from the NOAA Precipitation Reconstruction data set (PREC) between the two states (Fig. 7.15). For the observed June–July–August (JJA) surface temperature (Fig. 7.15a), the strongest warming over land covers the central and northeastern Asian continent, and the strongest warming over the ocean occurs in the tropical Indian Ocean. A similar SST warming pattern can be found in the HadISST (Fig. 7.15b), but the warming in the tropical oceans is somewhat weaker. Different from the consistent warming pattern of SST, the changes in JJA precipitation show

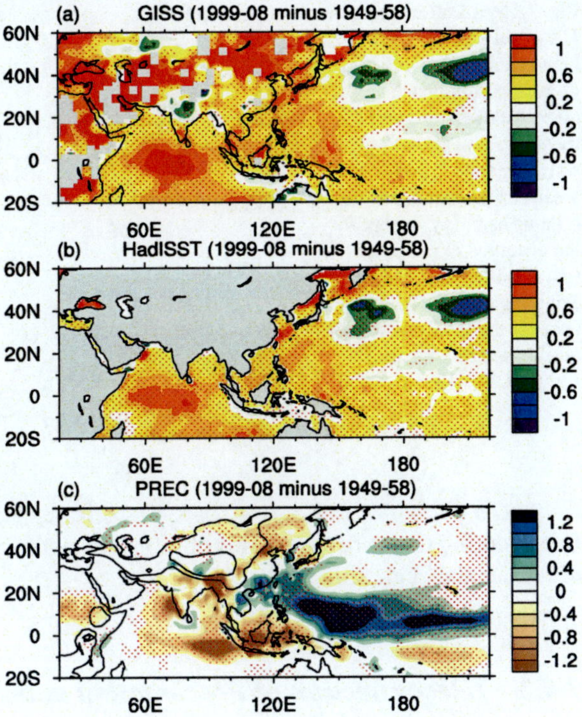

Fig. 7.15 Differences in JJA mean fields between the relatively high global warming state (1999–2008 mean) and the relatively low global warming state (1949–1958 mean) for **a** GISS SST (°C) and **b** HadISST (°C). **c** Same as (**a**), except for precipitation (mm day^{-1}) in PREC. The black contours in (c) denote the 1000 and 3000-m topography. Stippling denotes the value exceeds the 95% confidence level based on Student's *t*-test. (After He et al. 2016)

large regional features (Fig. 7.15c). Precipitation increases significantly over the SCS, the western Pacific Ocean, and southern China, while it decreases significantly over the BOB, the India subcontinent, the Arabian Sea, the tropical Indian Ocean, and northern China. Theoretically, these changes in precipitation also represent the spatial pattern changes of latent heat release in the middle troposphere.

The changes in precipitation are connected with the changes in atmospheric circulation anomaly. Here, we show some observed evidence from the SLP and clouds in the International Comprehensive Ocean–Atmosphere Data Set (ICOADS). Figure 7.16 shows the difference in JJA mean between the high and low global warming states defined above. SLP drops significantly over the SCS and WNP but increases over the entire Indian Ocean. This pattern is highly consistent with that of precipitation (Fig. 7.15c). From the viewpoint of atmospheric dynamics, the decreased SLP suggests that convergence and ascending atmospheric motion may be intensified over the SCS and WNP, while the increased SLP implies that divergence and descending motion may be dominant over the Indian Ocean. Moreover, Fig. 7.16b, shows increased cloud cover over the SCS and the WNP, while a clear reduction in cloud cover over the BOB and the tropical Indian Ocean. Therefore, the differences in the changes of SLP, clouds, and precipitation suggest latent heat release has been strengthened over the SCS and WNP, but weakened over the Indian

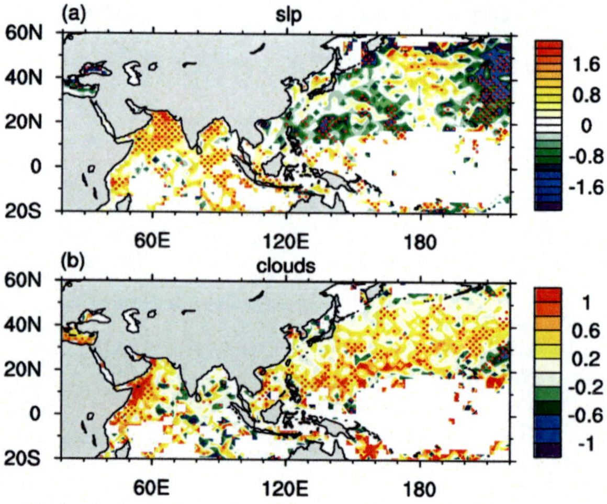

Fig. 7.16 **a** Difference in JJA mean SLP (hPa) between the relatively high global warming state (1999–2008 mean) and the relatively low global warming state (1949–1958 mean) using the ICOADS. **b** Same as in (**a**), except for the observed cloud cover (okta) using the ICOADS. Stippling denotes the value exceeds the 95% confidence level based on Student's *t*-test. (After He et al. 2016)

Ocean during recent decades. This observation evidence is consistent with the study of Annamalai et al. (2013).

7.4.2 Teleconnection of Precipitation and Atmospheric Circulation Anomaly

It is well known that teleconnection patterns link the variability of extratropical large-scale circulation with tropical heating (Wallace and Gutzler 1981; Simmons 1982; Simmons et al. 1983; Kasahara and Silva Dias 1986). Extensive studies (Ferranti et al. 1994; Graham et al. 1994; Kumar et al. 1994; Lau and Nath 1994; Kharin 1995) showed that the tropical SST and associated latent heating can affect extratropical atmospheric circulation via Rossby wave train in the upper troposphere. As summarized in Trenberth et al. (1998), the distribution of deep convection varies along with the changes in heating, low-level convergence, and upper-level divergence, thereby altering the generation of atmospheric vorticity and forcing large-scale atmospheric Rossby waves that can propagate into higher latitudes.

To investigate possible connections of latent heating over the SCS and its surrounding regions with extratropical atmospheric circulation anomaly during the boreal spring, we first identify the major modes of the latent heating in these areas (40°–160° E, 20° S–30° N), which mainly cover the Indian Ocean, SCS, the western Pacific, the Maritime Continent, and South Asia. Due to consistency in spatial and temporal variation, precipitation is often used to infer condensational heating by simply multiplying the latent heating coefficient (Li et al. 2012). For simplicity, we examine the major modes of the empirical orthogonal function (EOF) in the observed precipitation in the following, instead of studying condensational heating. The monthly GPCP data during MAM from 1979 to 2014 is used.

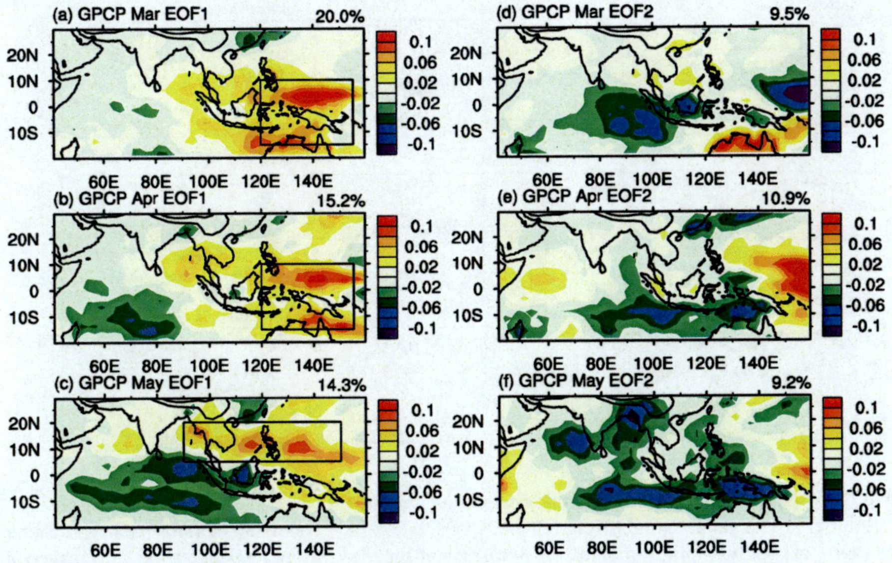

Fig. 7.17 Spatial patterns of the first EOF mode (EOF1) of precipitation in **a** March, **b** April, and **c** May during 1979–2014 from the GPCP data. **d**, **e**, and **f** are the same as (**a**), (**b**), and (**c**), respectively, except for EOF2. The explained variance is shown on the top right of each panel. The black rectangle denotes the region (120–155° E, 15° S–10° N) in (**a**) and (**b**), and the region (90–150° E, 5°–20° N) in (**c**). (After He and Yang 2018)

The spatial pattern of the first two modes is shown in Fig. 7.17. During March, the first EOF mode (EOF1) accounts for 20% of the total variance (Fig. 7.17a), while the second EOF model (EOF2) only accounts for 9.5% (Fig. 7.17d). The EOF1 pattern shows a positive phase over the Maritime Continent and western tropical Pacific (WTP), with a maximum over the southern PhiSea (Fig. 7.17a). The EOF2 pattern shows two obvious negative centers over the tropical Indian Ocean and WTP (Fig. 7.17d). The spatial patterns of EOF1 and EOF2 in April (Fig. 7.17b, e), show similar patterns as those in March (Fig. 7.17a, d), except that the intensity over the PhiSea is somewhat weaker. However, the EOF1 pattern in May (Fig. 7.17c) is substantially different from those in March and April, due to the Asian summer monsoon onset. The positive phase in May covers the BOB, the SCS, and the PhiSea, while the negative phase covers the entire southern Indian Ocean. The EOF2 (Fig. 7.17f), shows an overall negative phase in South Asia, the Indian Ocean, and the tropics, which accounts for 9.2% of the total variance. The above analysis suggests that the major mode of precipitation appears in the WTP region in recent decades. To find the associated teleconnection patterns of the precipitation major modes in the WTP during the boreal spring, we show correlation, regression, and composite analysis next.

Figure 7.18 shows the correlation coefficients between the time series of EOF1 (PC1) and the geopotential height field in the upper, middle, and lower troposphere, respectively. At 200 hPa (Fig. 7.18a, d, g), large negative values appear over the

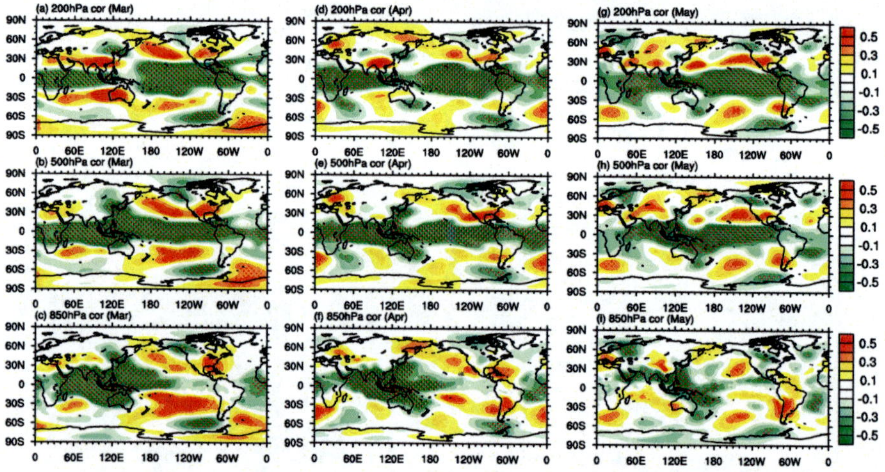

Fig. 7.18 Spatial patterns of Pearson's correlation coefficient between the time series of precipitation EOF1 and geopotential height for **a** 200 hPa, **b** 500 hPa, and **c** 850 hPa in March. **d–f** Same as (**a–c**), except for April. **g–i** Same as (**a–c**), except for May. Stippling denotes the value exceeds the 95% confidence level of Student's *t*-test. (After He and Yang 2018)

tropics in all spring months, which indicates that increase in tropical heating lowers the pressure associated with the change in the Walker circulation. In the extratropics, two Rossby wave trains occur in both hemispheres in March (Fig. 7.18a), but the wave train pattern is clearer in the Northern Hemisphere, which is characterized by three positive and two negative geopotential height anomalies across the Asian and North American continents (Fig. 7.18a). The correlation becomes weaker in April and May (Fig. 7.18d, g). At 500 hPa, the correlation patterns of MAM (Fig. 7.18b, e, h) are all similar to those at 200 hPa, except that the correlations are weaker in the extratropical regions. However, in the lower troposphere (850 hPa), the correlation patterns show substantially different features from those at 200 and 500 hPa. The negative correlation over Asia-Pacific is the dominant signal in all spring months (Fig. 7.18c, f, i), and the strongest signal presents in March.

The above analysis implies that the teleconnection between WTP precipitation and extratropical atmospheric circulation anomaly is more solid in March than in April and May. We also obtain the regressions of geopotential height, SLP, and air temperature to the precipitation PC1 in March (Fig. 7.19). The regression pattern for 200 hPa geopotential height shows a series of anticyclone-cyclone circulation anomalies in the high latitudes (Fig. 7.19a), and resembles the correlation pattern (Fig. 7.18a). The extratropical response of geopotential height is more apparent than the tropical response at 500 hPa (Fig. 7.19b), which is in phase with that at 200 hPa and is consistent with the regressions of SLP (Fig. 7.19c). In the subtropics, under the constraints of quasi-geostrophic balance, hydrostatic balance, and thermal wind balance, the air temperature anomaly is often in phase with the upper-level geopotential height anomaly (Wu et al. 2015). Therefore, we show the regressions

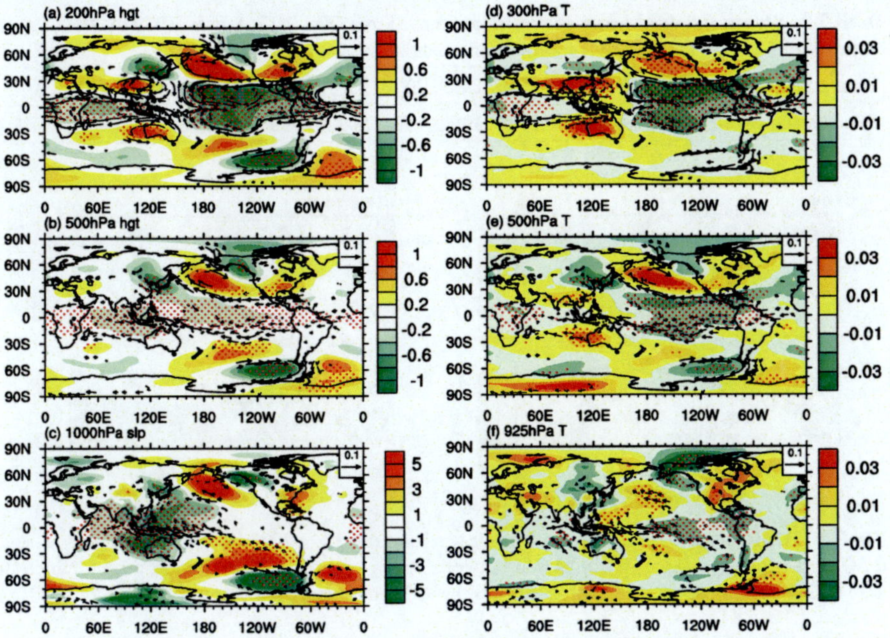

Fig. 7.19 Regression patterns of standardized time series of precipitation EOF1 in March on geopotential height and wind at **a** 200 hPa and **b** 500 hPa. **c** Same as (**a**), except for SLP (1000 hPa for wind). **d**–**f** Same as (**a**), except for air temperature at **d** 200 hPa, **e** 500 hPa, and **f** 925 hPa. Stippling denotes the value exceeds the 95% confidence level based on Student's *t*-test. (After He and Yang 2018)

of air temperature at 300, 500, and 925 hPa in Fig. 7.19d–f, respectively. These results indicate that the spatial distributions of temperature and geopotential height regressions in the upper troposphere are highly consistent with those in the lower troposphere, known as equivalent barotropic structure. This is helpful for us to infer that the variation of near-surface temperature over the extratropical regions can be attributed to tropical heating and the associated teleconnection pattern. In summary, the above analyses indicate that the intensified WTP precipitation in March leads to an equivalent barotropic Rossby wave train from South Asia to North America in the upper troposphere, which causes surface warming over the North American continent.

Next, we investigate the associated changes in SST and surface air temperature using composite analysis. Based on the precipitation PC1 in March, we list the composite years in Table 7.2. The years when the PC1 is above (below) its 0.5 (−0.5) standard deviation are recognized as the years of strong (weak) WTP precipitation.

The SST differences (Fig. 7.20a), indicate a La Niña pattern, with strong negative anomalies over the central and eastern tropical Pacific and positive anomalies over the WTP. This result suggests that the warmer SST could trigger stronger moist convection and intensifies latent heat release in the middle troposphere in the

Table 7.2 Strong and weak years of precipitation in March (Fig. 7.17a), from 1979 to 2014, based on the criterion of 0.5 standard deviation of the standardized PC1

Strong years	1982, 1986, 1988, 1989, 1999, 2000, 2001, 2004, 2006, 2008, 2011, 2012
Weak years	1981, 1983, 1987, 1991, 1992, 1993, 1997, 1998, 2002, 2010, 2014

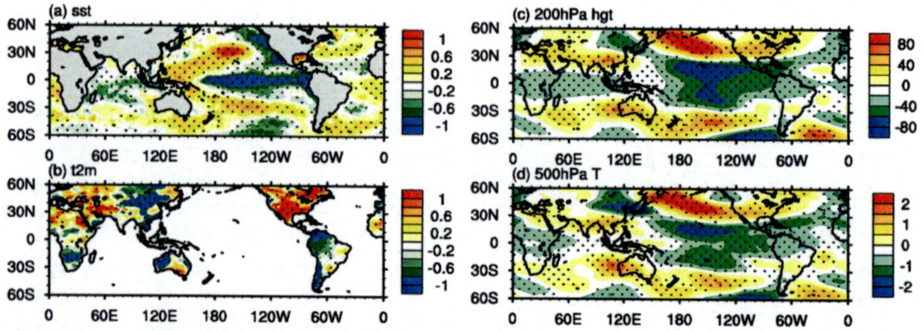

Fig. 7.20 Differences in **a** SST (K), **b** 2-m air temperature (K), **c** 200 hPa geopotential height, and **d** 500 hPa air temperature (K) in March between strong and weak precipitation years defined in Table 7.2. Stippling denotes the value exceeds the 95% confidence level. (After He and Yang 2018)

WTP region. The composite 2-m air temperature (t2m) overland (Fig. 7.20b), illustrates strong surface warming mainly appearing over the North American continent. The composite 200 hPa geopotential height (Fig. 7.20c) and 500 hPa temperature (Fig. 7.20d), present similar patterns, as shown in the correlation map (Fig. 7.18a) and regression map (Fig. 7.19a). Therefore, correlation, regression, and composite analysis together provide a clear image that the SST warming in the WTP triggers local moist convection; this causes strong precipitation and associated condensational heating in the middle troposphere, and induces Rossby wave propagation in the upper troposphere from the WTP to North America and warm the surface climate. However, it is not clear why this physical process is only seen in March during the boreal spring.

To understand the Rossby wave propagation features in the boreal spring, we investigate the key elements that control the propagate in the upper troposphere. Previous theoretical studies (Hoskins et al. 1977; Hoskins and Karoly 1981; Webster 1981, 1982; Hoskins and Ambrizzi 1993; Ting and Sardeshmukh 1993), proposed that the westerly jet is the waveguide of the Rossby wave propagation. The extratropical Rossby wave propagation related to tropical heating can be understood using the vorticity equation. For simplicity, the linearized vorticity equation about a basic state consisted of a zonal mean flow on the β plane (Trenberth et al. 1998) is used here

$$\left(\frac{\partial}{\partial t} + [\bar{u}]\frac{\partial}{\partial x}\right)\nabla^2\psi' + \frac{\partial \psi'}{\partial x}\left(\beta - \frac{\partial^2[\bar{u}]}{\partial y^2}\right) = Q \tag{7.1}$$

where $[\bar{u}]$ denotes climatological zonal flow, ψ' is perturbation stream function, and Q is vorticity forcing, i.e., the heating anomaly. A Wentzel–Kramers–Brillouin (WKB) solution to Eq. (7.1), exists in the form of $\psi' = A\exp[i(kx + ly - \sigma t)]$, with a dispersion relation of

$$\sigma = [\bar{u}]k - \frac{\beta_* k}{K^2} \tag{7.2}$$

where

$$\beta_* = \beta - \frac{\partial^2[\bar{u}]}{\partial y^2} \tag{7.3}$$

Equation (7.3) is the meridional gradient of absolute vorticity, and $K = (k^2+l^2)^{1/2}$ is the total wavenumber. For stationary Rossby waves ($\sigma = 0$),

$$K = \left(\frac{\beta_*}{[\bar{u}]}\right)^{1/2} \tag{7.4}$$

Therefore, Rossby wave propagation is largely affected by the mean zonal flow, the meridional gradient of vorticity, and the β effect. This helps us to understand why the teleconnection patterns in March, April, and May are different.

Based on Eq. (7.4), we show the climatological mean 200 hPa zonal wind, divergence, and vorticity, respectively, in Fig. 7.21. The strong divergence (blue contour) in the upper troposphere is in the same location as the strong precipitation in the WTP region. Although the distributions of divergence in March (Fig. 7.21a), April (Fig. 7.21b), and May (Fig. 7.21c) are similar, the intensity of the westerly jet stream,

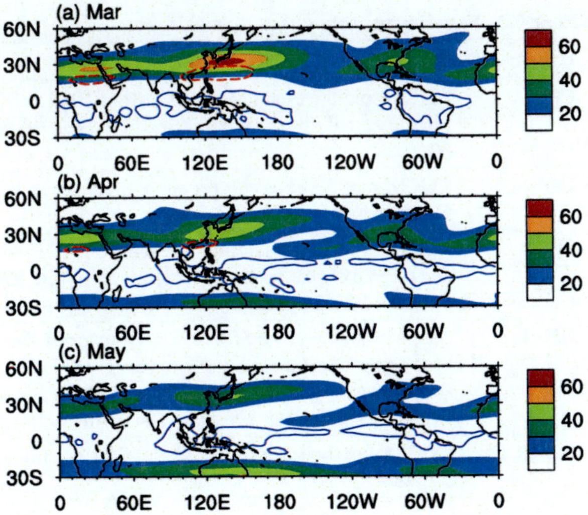

Fig. 7.21 Climatological mean (1979–2014) 200 hPa zonal wind (shading; m s^{-1}), divergence (blue contour; 2 $\times 10^{-6}$ s^{-1}) and vorticity (red dashed contour; -3×10^{-5} s^{-1}) in **a** March, **b** April, and **c** May. (After He and Yang 2018)

especially that over East Asia, is strongest in March than in the other two months. Specifically, $[\bar{u}]$ exceeds 50 m s^{-1} over southern Japan in March, associated with strong vorticity (red dashed contour) to the south of the westerly jet stream where the meridional gradient of $[\bar{u}]$ is strong (Fig. 7.21a). In April (Fig. 7.21b), both $[\bar{u}]$ and vorticity are weakened above Japan. In May (Fig. 7.21c), the westerly jet stream breaks down over the eastern subtropical Pacific and vorticity forcing is not apparent. The above results indicate that the teleconnection between the major mode of latent heating over the WTP and surface temperature change over North America is sensitive to the change in the intensity of westerly jet stream during the boreal spring.

7.4.3 Heating Effect During Boreal Summer

The local and remote effects of latent heat release in the WTP region documented in Sect. 7.4.2, are further studied by a series of numerical simulations next. To investigate the influences of heating over the WTP on Asian summer climate, five experiments are carried out (Table 7.3). Two types of control runs are based on atmospheric general circulation model (AGCM) SAMIL2 and coupled GCM (CGCM) FGOALS-s2, respectively. SAMIL2 (version 2 of the spectral atmospheric model) and FGOALS-s2 (Flexible Global Ocean-Atmosphere-Land System Model spectral version 2) are developed at the State Key Laboratory of Numerical Modeling for Atmospheric Sciences and Geophysical Fluid Dynamics, Institute of Atmospheric Physics (LASG/IAP). The first control experiment (CON) is performed by using SAMIL2 with the climatological mean SST from 1949 to 1958. The model is integrated for 30 years, and the mean fields of the last 20 years are analyzed. The second

Table 7.3 Experiment design

Exp	Description
CON	AGCM run, the climatology SST from 1949 to 1958 is used as low boundary, and other external forcing is prescribed by the 1949 values. The model is integrated for 30 years
CON_CP	CGCM run, the external forcing is the same as in CON but SST is simulated. The model is integrated for 50 years
SEN	AGCM run, the climatological SST from 1999 to 2008 is used as low boundary, and other external forcing is prescribed by the 1999 values. The model is integrated for 30 years
HEAT	AGCM run, the same SST and other external forcing as in CON, but a specific heating profile is added in the region (110°–150° E, 0°–30° N) during JJA. This specific heating is derived from the monthly mean outputs from SEN minus those from CON in the same region. The model is integrated for 30 years
HEAT_CP	CGCM run, the same as HEAT but SST is simulated. The model is integrated for 50 years

control experiment (CON_CP) is performed by using FGOALS-s2. The model setup is the same as CON, except that SST is simulated in CON_CP. The model is integrated for 50 years, and the mean fields of the last 20 years are analyzed. Three sensitivity runs are then performed to identify the atmospheric response to the WTP heating. The first sensitivity experiment is referred to as SEN, using SAMIL2 with the climatological mean SST from 1999 to 2008, while other external forcing is set to their values in 1999. The difference between SEN and CON denotes climate changes in response to global warming. To distinguish the effect of heating over the WTP, another sensitivity experiment (Exp HEAT) is performed, in which external forcing is the same as in CON, but heating is added over the region (110°–150° E, 0°–30° N) during JJA. This specific heating is the monthly mean difference between SEN and CON outputs in the same region. Finally, to study the air–sea interaction induced by the WTP heating, a coupled experiment (HEAT_CP) is performed using FGOALS-s2. All the settings are exactly the same as in CON_CP, except for adding the same heating as in Exp HEAT. The model integrates 50 years and the results of the last 20 years are analyzed.

The basic model response to global warming is shown in Fig. 7.22a. The JJA mean precipitation difference between SEN and CON shows a substantial increase over the WTP region and decreases over the southern slope of the TP and the Arabian Sea. This pattern is similar to the observed (Fig. 7.15c). The associated total column heating change (Fig. 7.22b), is similar to the precipitation pattern (Fig. 7.22a), which implies the change of the total atmospheric heating is dominated by the condensation heating in the free troposphere.

We derive the heating difference profile from SEN and CON (SEN minus CON), and show the regional (110°–150° E, 0°–30° N) mean vertical diabatic heating profiles of JJA in Fig. 7.22c. The total diabatic heating reaches its maximum over 400–500 hPa in both experiments as in the ERA40 Reanalysis. The heating profiles of CON, CON_CP, and ERA40_B (1958–1967 mean in ERA40 Reanalysis), which represent the relatively low global warming state, closely match each other. Note that the heating profiles of both Exp HEAT (red line) and HEAT_CP (blue line) are close to that of SEN, indicating that our sensitivity experiments successfully capture the heating characteristics over the WTP region under the global warming.

The responses of summer monsoon precipitation and 850 hPa winds to the WTP heating are derived by the difference of Exp HEAT and CON (Exp HEAT minus CON; Fig. 7.23a) and of HEAT_CP minus CON_CP (HEAT_CP minus CON_CP; Fig. 7.23b). It is not surprising that the local precipitation, where heating is added, increases significantly. However, precipitation decreases over the southern slope of the TP in South Asia, consistent with the feature over the continental South Asia in the observation (Fig. 7.15c). At 850 hPa, a cyclone appears over the SCS and western Pacific, and the cross-equatorial flow also intensifies over the tropical Indian Ocean. When the air–sea interaction is considered (Fig. 7.23b), the magnitude of the increased precipitation over the WTP decreases, and the positive precipitation anomaly over the BOB also weakens to a large extent. Additionally, the easterly anomaly at 850 hPa near the equator in the AGCM run (HEAT minus CON) (Fig. 7.23a) vanishes in the coupled experiment of (HEAT_CP minus CON_CP)

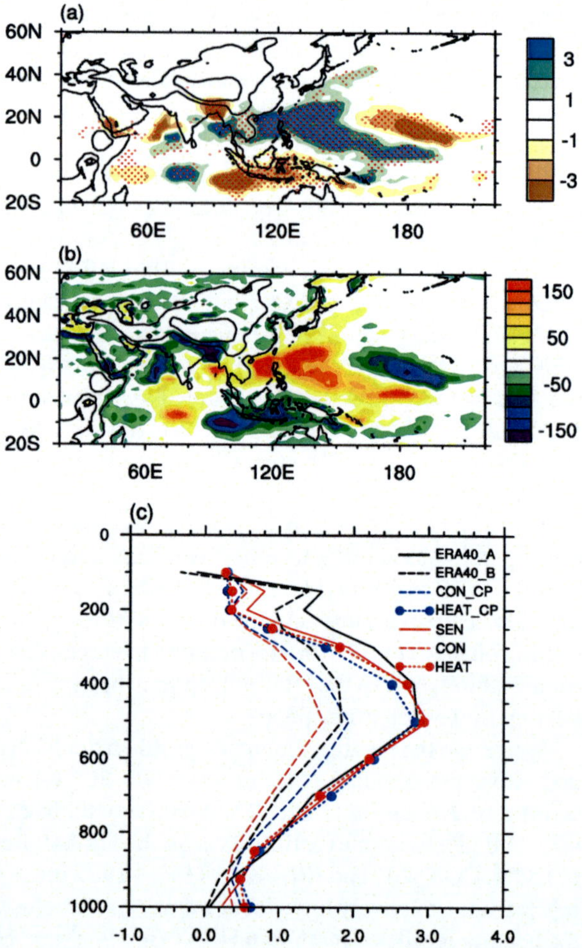

Fig. 7.22 **a** Difference in JJA precipitation (mm day^{-1}) between CON and SEN (SEN minus CON). Stippling denotes the value exceeds the 95% confidence level based on Student's *t*-test. The black contours denote the 1000- and 3000-m topography. **b** Difference in total column latent heating ($W\ m^{-2}$) between SEN and CON. **c** Total diabatic heating profile ($K\ d^{-1}$) for the region (110°–150° E, 0°–30° N) in each experiment. (After He et al. 2016)

(Fig. 7.23b). Moreover, the precipitation pattern in the CGCM run (HEAT_CP minus CON_CP) (Fig. 7.23b), is closer to the observed one (Fig. 7.15c), than that in the AGCM run (HEAT minus CON) (Fig. 7.23a), which indicates that the active air-sea coupling is important in regulating the Asian summer monsoon.

To understand the decrease in monsoon precipitation over the continental South Asia, we analyze the subtropical dynamics according to the vertically inhomogeneous diabatic heating theory (Liu et al. 2001a, b; Wu and Liu 2003). An increase in diabatic heating forces local ascending motion based on the thermodynamical constraint, and the intensity of vertical velocity is proportional to the magnitude of diabatic heating (Matsuno 1966; Gill 1980). Here, the changes of vertical velocity in the upper and lower troposphere are proportional to the heating added. At the surface, the vertical velocity anomalies are weak in both AGCM run of name (Fig. 7.24e) and CGCM run of name (Fig. 7.24f), since the added heating at the surface is weak (Fig. 7.22c).

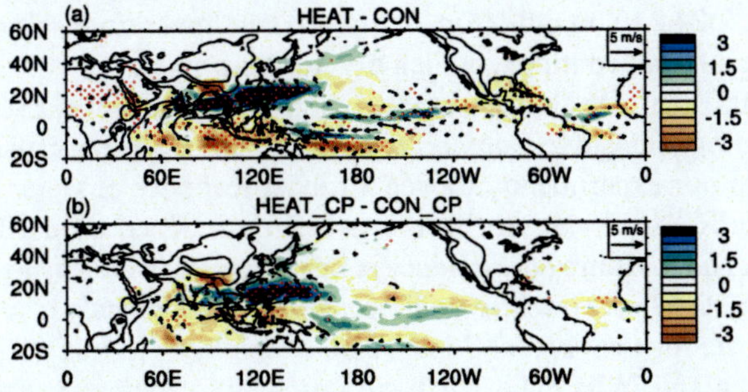

Fig. 7.23 **a** Differences in JJA precipitation (shading; mm day^{-1}) and 850 hPa wind (vector; m s^{-1}) between Exp HEAT and CON (Exp HEAT minus CON). The black contours denote the 1000 and 3000-m topography. Stippling denotes the value exceeds the 95% confidence level based on Student's *t*-test. **b** Same as in (**a**), except for the differences between HEAT_CP and CON_CP. (After He et al. 2016)

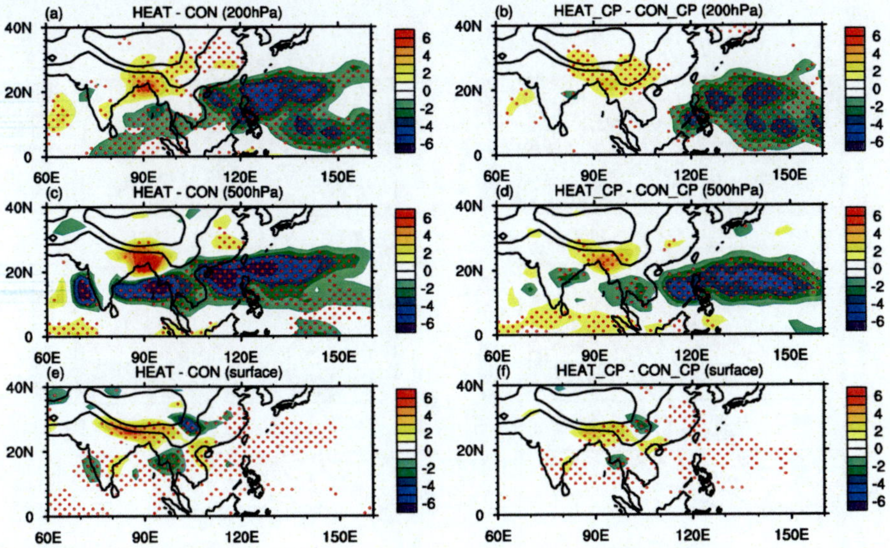

Fig. 7.24 **a** Difference in JJA 200 hPa vertical velocity (hPa s^{-1}) between Exp HEAT and CON (Exp HEAT minus CON). The negative (positive) values indicate ascending (descending) motion. Stippling denotes the value exceeds the 95% confidence level based on Student's *t*-test. The black contours denote the 1000 and 3000-m topography. **c** Same as (**a**), except for the difference at 500 hPa. **e** Same as (**a**), except for the difference at surface. **b**, **d**, and **f** are the same as (**a**), (**c**), and (**e**), respectively, except for the difference between HEAT_CP and CON_CP. (After He et al. 2016)

However, maximum ascending motion appears at 500 hPa, where the maximum heating is added, and the ascending motion further extends to the upper troposphere (Fig. 7.24a, b). Under the quasi-geostrophic balance and thermal wind balance, the adiabatic descending motion is forced just to the west of the WTP heating throughout the upper (Fig. 7.24a, b), middle (Fig. 7.24c, d), and lower troposphere (Fig. 7.24e, f) over the South Asian continent, which dry northern India.

Liu et al. (2001a, b) revealed that in the region where the upward vertical gradient of diabatic heating is positive, southerly wind will be triggered, causing a cyclonic convergence. In our experiments, we see a robust response of water vapor convergence over the SCS and western Pacific, which is induced by the heating added (Fig. 7.25a, b). The moisture convergence is instrumental in increasing the precipitation over the WTP. Meanwhile, the cyclonic circulation anomaly also causes northeasterly wind over the continental South Asia, which leads to less water vapor being transported to inland India. The adiabatic descending motion in the troposphere provides a less cloudy condition over the South Asian continent (Fig. 7.25c, d), associated with less precipitation and increased surface temperature (Fig. 7.25e, f). In the coupled experiment (HEAT_CP minus CON_CP), the results are similar to those in the AGCM run (HEAT minus CON) but SST decreases in the SCS and increases in the tropical Indian Ocean in the coupled experiment (Fig. 7.25f), dynamically coordinated with changes of precipitation (Fig. 7.23b).

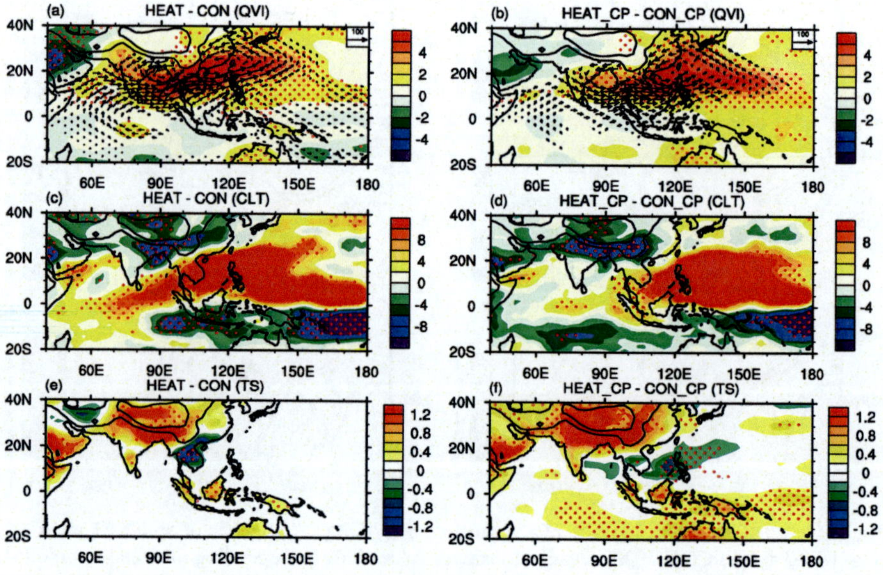

Fig. 7.25 **a** Difference in JJA total column water vapor content (shading; kg m^{-2}) and 850 hPa water vapor flux (vector; kg m^{-1} s^{-1}) between Exp HEAT and the CON (Exp HEAT minus CON). Stippling denotes the value exceeds the 95% confidence level based on Student's *t*-test. The black contours denote the 1000 and 3000-m topography. **b** Same as (**a**), except for the difference between HEAT_CP and CON_CP. **c** Differences in JJA total cloud cover (%) between Exp HEAT and CON. **d** Same as (**c**), except for the difference between HEAT_CP and CON_CP. **e** Difference in JJA surface temperature (°C) between Exp HEAT and CON. **f** Same as (**e**), except for the difference between HEAT_CP and CON_CP. (After He et al. 2016)

7.4.4 Heating Effect During Boreal Spring

As documented in Sect. 7.4.2, the teleconnection between the WTP heating and climate change over North America is sensitive to the location of the heating and the intensity of the westerly jet stream during the boreal spring. To better understand the related physical processes, we conduct a series of sensitive experiments using both Linear Baroclinic Model (LBM) and AGCM here.

Nine sensitive experiments are carried out (Table 7.4). A heating source of idealized ellipse located over the WTP region, with its center over (140° E, 5° N), is prescribed in the model. The vertical structure of the heating source is an idealized latent heating, with a maximum heating rate of 9 K day^{-1} at 400 hPa. Three basic backgrounds in the boreal spring are tested, which are named Exps TWP_Mar, TWP_Apr, and TWP_May for March, April, and May, respectively. To compare the Rossby wave propagation induced by the WTP heating with the heating over other tropical regions, six additional experiments for the heating source over the central tropical Pacific (Exps CP_Mar, CP_Apr, and CP_May) and over the western Indian Ocean (Exps WIO_Mar, WIO_Apr, and WIO_May) are also performed for March, April, and May, respectively.

For the three Exps TWP (Fig. 7.26), the 200 hPa geopotential height (HGT) anomalies from different basic flows all show a positive anomaly in the forcing region on day 1 (Fig. 7.26a, d, g), which is attributed to the local latent heating. On day 3 (Fig. 7.26b, e, h), the HGT anomalies extend to the surrounding regions, similar to the Gill-type pattern (Gill 1980). Additionally, weak Rossby wave propagating to the high latitudes can be found in all three cases. On day 7 (Fig. 7.26c, f, i), the Rossby wave propagation becomes different under these three basic flows. In March (Fig. 7.26c), a Rossby wave path is clearly seen from East Asia to North America with three pairs of positive/negative HGT anomaly centers, in which the anomalies over the northwestern Pacific are the strongest. In April (Fig. 7.26f), the Rossby wave propagation becomes weaker in the Northern Hemisphere, but stronger in the Southern Hemisphere. In May (Fig. 7.26i), the HGT anomalies in the Southern Hemisphere are stronger, while the anomalies over the Northern Hemisphere are

Table 7.4 Experimental design for LBM simulations

Exp	Center of heating	Background field
TWP_Mar	140° E, 5° N	March
TWP_Apr	140° E, 5° N	April
TWP_May	140° E, 5° N	May
CP_Mar	160° W, 5° N	March
CP_Apr	160° W 5° N	April
CP_May	160° W, 5° N	May
WIO_Mar	60° E, 5° N	March
WIO_Apr	60° E, 5° N	April
WIO_May	60° E, 5° N	May

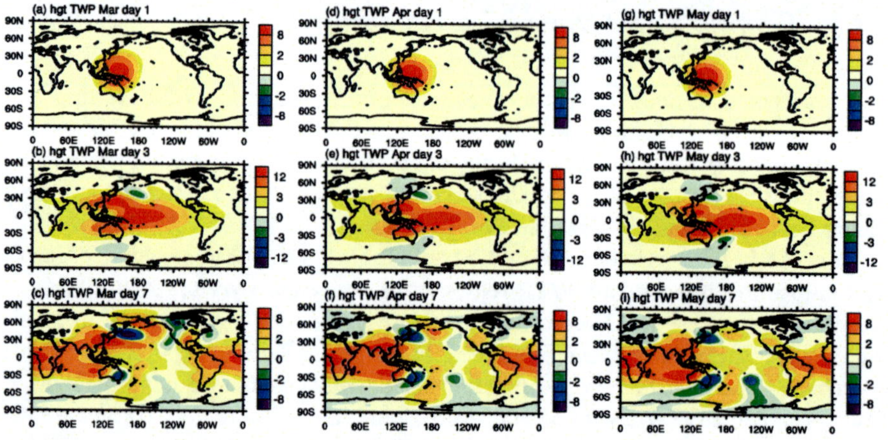

Fig. 7.26 Geopotential height anomaly (units: gpm) at 200 hPa in Exp TWP_Mar for **a** day 1, **b** day 3, and **c** day 7. **d–e** Same as (**a–c**), except for Exp TWP_Apr. **g–i** Same as (**a–c**), except for Exp TWP_May. (After He and Yang 2018)

weaker than those in April. In the above experiments, the heating forcing is identical; therefore, the different Rossby wave propagation features are due to the change of the basic flow. It is thus testified that the location and intensity of the westerly jet stream are crucial for the teleconnection pattern in the Northern Hemisphere.

To understand the importance of heating location in producing Rossby wave, two other groups of the experiments with the heating over the central Pacific and western Indian Ocean are compared next. For the heating over the three Exps CP (Fig. 7.27), the 200 hPa HGT anomaly patterns on day 1 (Fig. 7.27a, d, g) and day

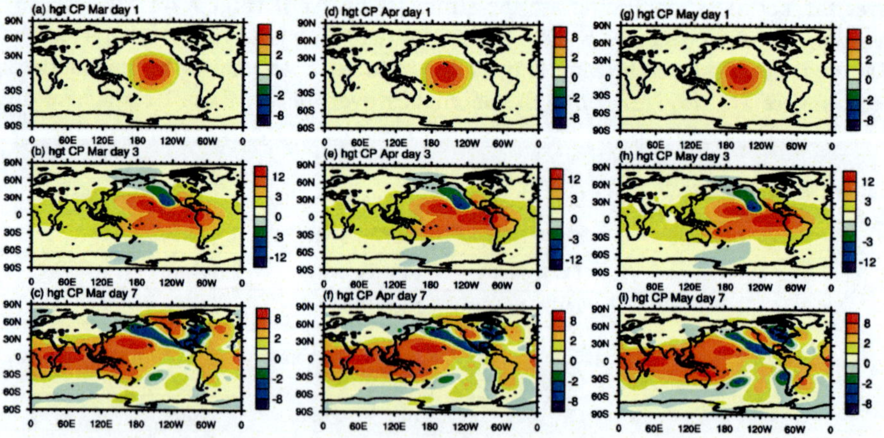

Fig. 7.27 Geopotential height anomaly (gpm) at 200 hPa in Exp CP_Mar for **a** day 1, **b** day 3, and **c** day 7. **d–e** Same as (**a–c**), except for Exp CP_Apr. **g–i** Same as (**a–c**), except for Exp CP_May. (After He and Yang 2018)

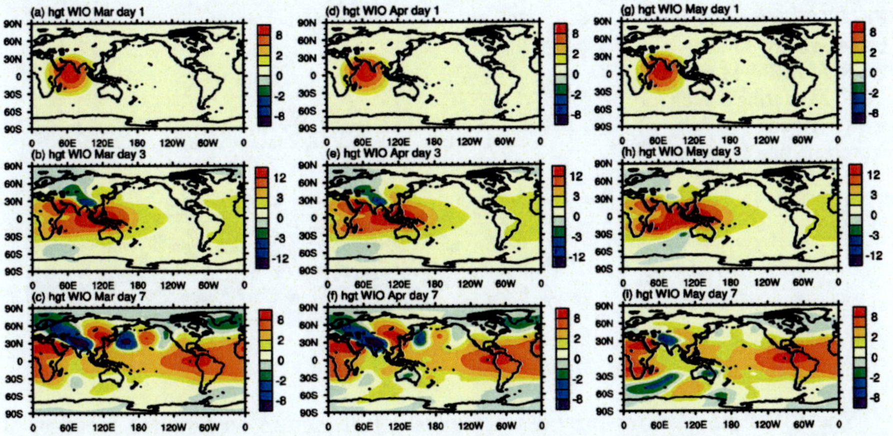

Fig. 7.28 Geopotential height anomaly (gpm) at 200 hPa in Exp WIO_Mar for **a** day 1, **b** day 3, and **c** day 7. **d–e** Same as (**a–c**), except for Exp WIO_Apr. **g–i** Same as (**a–c**), except for Exp WIO_May. After He and Yang (2018)

3 (Fig. 7.27b, e, f), are similar as those in Exps TWP. However, on day 7, although the basic backgrounds are different in CP_Mar (Fig. 7.27c), CP_Apr (Fig. 7.27f), and CP_May (Fig. 7.27i), the HGT anomalies show a solid Pacific–North American teleconnection pattern.

For the results from the three Exps WIO (Fig. 7.28). The HGT anomaly patterns on day 1 (Fig. 7.28a, d, g) and day 3 (Fig. 7.28b, e, h), are similar to those in Exps TWP and Exps CP. However, on day 7, a Rossby wave train is triggered from tropical Africa to the North Pacific during March (Fig. 7.28c) and April (Fig. 7.28f). Meanwhile, Rossby wave propagation becomes weaker in the Northern Hemisphere and stronger in the Southern Hemisphere and is confined in the eastern hemisphere in May (Fig. 7.28i).

The different Rossby wave propagations induced by these three heating sources can be understood by the changes in heating location and westerly jet stream intensity. The westerly jet streams have two cores over the northwestern Pacific and North America, respectively, and they weaken gradually from March to May. In Exps CP, the heating source is just located at the entrance of the North American jet stream; therefore, it can trigger a similar Rossby wave pattern from March to May under all the basic flows. Similarity, in Exps WIO, the Rossby wave is able to propagate along the East Asian jet stream from Central Asia to the western Pacific, but cannot propagate east to North America. In Exps TWP, the westerly jet stream is strong, with its maximum exceeding 60 m s^{-1} in March; but it weakens obviously from April to May. Therefore, the Rossby wave propagation is only apparent in March; it weakens and even ceases on the edge of the westerly jet stream in April and May. In summary, whether the WTP heating can exert a remote influence on North America depends on both the location of the heating source and the strength of the westerly jet stream.

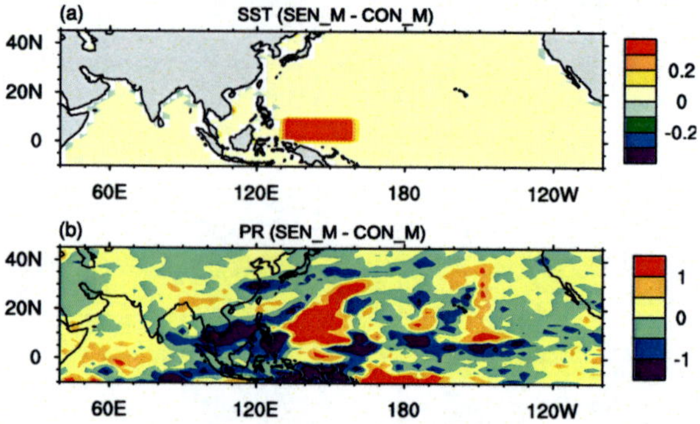

Fig. 7.29 Difference between SEN_M and CON_M (SEN_M minus CON_M) during March for **a** SST (K) and **b** precipitation (mm day^{-1})

The LBM simulations provide a clear physical picture on how the location of atmospheric heating and intensity of the westerly jet could influence the Rossby wave propagation in the upper troposphere. However, the observed connections between the precipitation over the WTP and the surface temperature over North America have not been verified yet. In this subsection, we perform additional AGCM experiments to test the connection of SST warming in the WTP and surface temperature change in North America during March. The control experiment is named CON_M, which is forced by climatological SST. The model is integrated for 35 years, and the outputs from the last 30 years are used for analysis. The sensitivity run is named SEN_M, which is the same as CON_M, except with prescribed SST anomaly in March only in the WTP (130°–160° E, 0°–10° N) using the climatological values, as shown in Fig. 7.29a.

The precipitation differences between SEN_M and CON_M are shown in Fig. 7.29b. The precipitation not only increases over the WTP but also extends northward to the mid-latitudes. Meanwhile, the deep convection also causes a decrease in precipitation over the SCS and Maritime Continent. These simulation results are consistent with those of He et al. (2016), which showed the summer latent heating in the middle troposphere over the WTP generated local anticyclonic anomaly that led to a strong divergence in the upper troposphere. Consequently, this divergence caused descending anomalies on the west, leading to a decrease in precipitation on the south slope of the TP.

In response to the increase of WTP precipitation and the associated condensational heating, the 200 hPa HGT anomaly (Fig. 7.30a), exhibits an apparent Rossby wave train from southern China to North America. This pattern is similar to the result from the LBM experiment TWP_Mar and the composite analysis, showing negative and positive anomalies over western and central-eastern North America, respectively. At the surface, the response of surface temperature (Fig. 7.30b), also shows a similar pattern as the upper-tropospheric HGT, with cooling over northern North America and warming over central-eastern North America, which is also close to the observations. These results of AGCM experiments testify our hypothesis that the atmospheric heating over the WTP triggered by local SST warming contributes to the remote surface warming over North America through Rossby wave propagation.

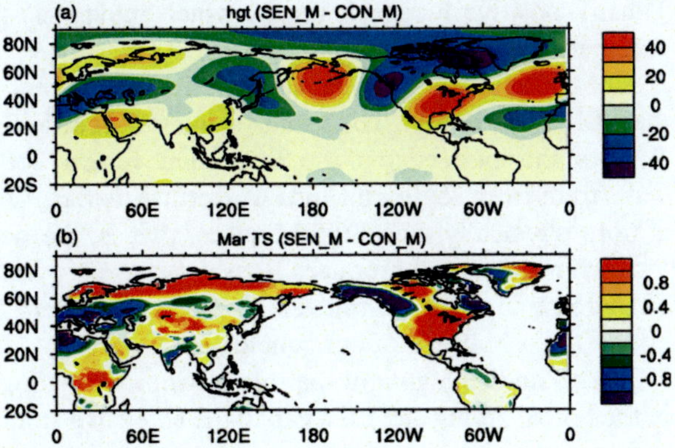

Fig. 7.30 Difference between SEN_M and CON_M (SEN_M minus CON_M) during March for **a** 200 hPa geopotential height (gpm) and **b** surface temperature (K) over land

7.4.5 Summary and Discussion

The main results of Sect. 7.4, can be concluded as follows. The observed changes in JJA SST indicate a large increase in the SST of the Indian Ocean, the SCS, and the western Pacific, with a maximum in the tropical Indian Ocean, under global warming. Different from the SST changes, the precipitation changes show a distinct east–west spatial pattern, with increased precipitation over the WTP but decreased precipitation over South Asia and the tropical Indian Ocean. The changes of the observed SLP, clouds, and atmospheric circulations all indicate that moist convection intensifies over the WTP and associated condensational heating increases in the troposphere.

During the boreal spring in recent decades, precipitation mainly occurs in the WTP in March and April, and the main rain-belt shifts northward to the Indochinese Peninsula and the SCS in May after the Asian summer monsoon onset. The EOF1 of precipitation shows a strong positive center over the WTP, explains about 20% of the total variance in March, which is the largest among the three spring months. Meanwhile, precipitation is strongly positive correlated with local SST in March, implying that SST is an important trigger for the occurrence of local convection. Further analysis suggests that the WTP heating is an important Rossby wave source in the upper troposphere. The strength of the westerly jet stream, which acts as a waveguide, decreases substantially from March to May, and the strength of the Rossby wave train decreases accordingly. The surface temperature over North America increases in phase with the HGT anomalies aloft, as the vertical structure of the Rossby wave train is equivalent barotropic.

The influences of increased condensational heating in the TWP on the Asian summer climate are investigated by a series of sensitivity experiments using both AGCM and CGCM. Results indicate that the WTP heating strengthens the upper-tropospheric South Asian high and forms a cyclonic circulation below the lower-tropospheric heating center. An adiabatic descending motion occurs in the continental South Asia, and moist convergence appears over the SCS and western Pacific regions.

Thus, a positive feedback forms, which intensifies the dry and warm climate in continental South Asia.

The Rossby wave propagation during the boreal spring is also testified using the LBM and AGCM. The results suggest that only the heating prescribed over the WTP on the south edge of the East Asian westerly jet core can produce an apparent wave train from southern China to North America. Our AGCM experiments testify a more physically meaningful feature, that is, the warming of WTP SST triggers local convection and produces Rossby wave train in March, which, in turn, causes an increase in surface temperature in central-eastern North America. These results ensure the robustness of our conclusions.

These studies highlight the role of tropical heating on regulating global climate in the boreal spring and fill a gap in the extensive studies that focused on the Rossby wave propagation in winter and summer seasons. Previous studies (Nitta 1987; Huang and Sun 1992; Lau 1992), emphasized the WTP heating could produce Rossby waves that propagate northeastward in the boreal summer. The basic PJ pattern (Fig. 1 in Nitta 1987), is a sandwich-type teleconnection with a positive correlation center east of the SCS and two negative centers over Japan and New Guinea, respectively. The correlation of heating with 500 hPa HGT (Fig. 17 in Nitta 1987), showed two pairs of HGT anomalies across the Pacific from the east to the west. Therefore, an increase in heating over the WTP leads to a decrease in surface air temperature over North America in June. Lau (1992) roughly repeated the experiments in Nitta (1987). He found that the Rossby wave train over the Pacific is sensitive to the heating over 95°–140° E. Meanwhile, the stream function showed a pair of negative and positive anomalies over North America, which was somehow different from that in Nitta (1987), but supported by our current study. These findings suggest that the location and strength of heating and westerly jet stream are crucial in determining Rossby wave pattern on the global scale.

7.5 Impacts of Climate Change in the Central-to-Western Tropical Pacific on Arctic Climate

7.5.1 *Expansion of the Central-Western Pacific Warm Pool Under Global Warming*

The SST in most of the central-western tropical Pacific (110° E–160° W) is higher than 28 °C (Fig. 7.31), which is referred to as the central-western Pacific Warm Pool. The SST in the eastern tropical Pacific (145° W–80° W) is relatively cooler and is often referred to as the cold tongue. This SST distribution contributes to the locations of the rising branch and the sinking branch of the Walker circulation; the rising branch largely determines the convection activity over the tropical Pacific Ocean.

In the past 60 years, the globally-averaged temperature has increased significantly. Recent studies have shown that the SST, ocean heat content, and sea surface height

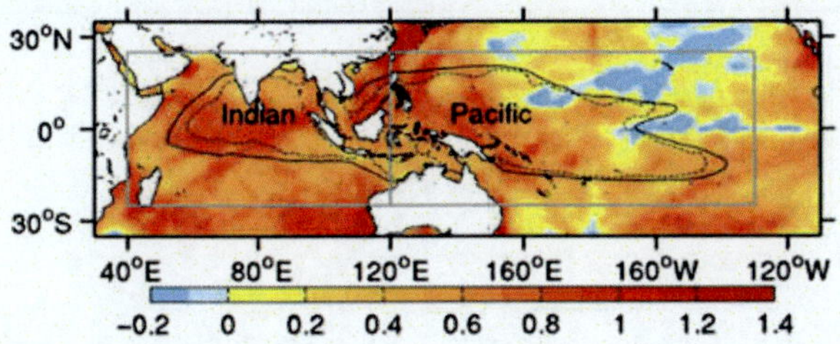

Fig. 7.31 Indo-Pacific SST trend (°C 60 yr^{-1}) and Warm Pool expansion (dashed contour for 1953–1959 average, and solid contour for 2000–2012 average). Figure is cited from Weller et al. (2016)

of the central-western Pacific have significantly increased due to human activities, resulting in an expansion of the central-western Pacific Warm Pool (Weller et al. 2016). Besides, the natural variability represented by the PDO also plays a role in the expansion of the Warm Pool, but its contribution is less than one-tenth of the GHG forcing caused by human activities (see Fig. 1c–d of Weller et al. 2016). In the context of global change, the expansion of the Warm Pool area will inevitably lead to changes in ENSO and convection in the tropical Pacific, and ultimately resulting in changes of the large-scale climate modes via teleconnections of atmospheric bridges (e.g., Wallace and Gutzler 1981; Hu et al. 2016b).

7.5.2 *Shifting Mode of El Niño in the Contexts of Global Warming and Warm Pool Expansion*

With the global warming and the Warm Pool expansion, "Warm-Pool (WP) El Niño" (Kug et al. 2009), has appeared more and more frequently in the past three decades comparing to the traditional eastern Pacific (EP) El Niño (Fig. 7.32, red line), with the maximum SST anomalies in the central tropical Pacific (Fig. 7.32, green line). Different names have been used for this type of El Niño, such as the "western Pacific El Niño" (Fu et al. 1986), "Dateline El Niño" (Larkin and Harrison 2005), "El Niño Modoki" (Ashok et al. 2007), and "CP warming (CPW)" (Kim et al. 2009). We use the name of central Pacific (CP) El Niño here, as in many studies (e.g., Graf 1986; Kao and Yu 2009; Yu and Kim 2010; Graf and Zanchetti 2012).

Most of the warm SST anomalies related to CP El Niño are located in the central-to-eastern part of the Pacific Warm Pool (Fig. 7.31 vs. Fig. 7.32b), where the SST background is significantly higher than that in the eastern Pacific cold tongue region. Therefore, even if the warm SST anomaly is milder in CP El Niño than in EP El Niño, it is still more likely to cause significant convective activity anomalies (Hoerling et al. 2001). ENSO and tropical convective activity anomalies are important to atmospheric teleconnection modes and global climate (e.g., Shukla and Paolina 1983; Webster et al. 1998a, b; Jiang et al. 2013; Yang and Jiang 2014). Many studies have shown that

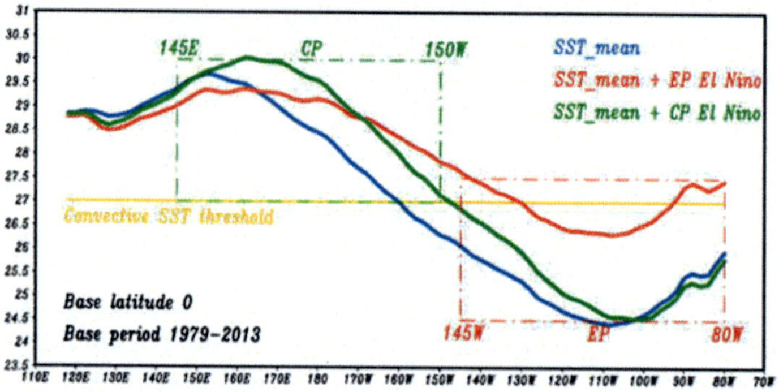

Fig. 7.32 Equatorial Pacific (110° E–70° W, 0°) SST profile. The blue curve is the annual mean SST in 1979–2013. The red and green curves are the annual mean SST plus the double SST anomalies related to EP El Niño and CP El Niño, respectively. The orange line shows the convective SST threshold at 27 °C. The red and green boxes indicate the longitude ranges of Niño3b and Niño4b, respectively. Figure is cited from Hu et al. (2016a)

the leading modes of tropical Pacific SST interannual variability (i.e., EP El Niño and CP El Niño) are important drivers of global climate "teleconnection" (Ashok et al. 2007; Yeh et al. 2009). However, atmospheric teleconnection modes are sensitive to the characteristics of El Niño diversity. Therefore, the shifting mode of El Nino under the background of global warming and Warm Pool expansion has attracted much attention.

Many studies have suggested that El Niño features show considerable changes in the context of global warming, including its frequency of occurrence, location, intensity, and meridional dimension (e.g., An and Wang 2000; McPhaden et al. 2006; Tang et al. 2008; Yeh et al. 2009; Zhang et al. 2009; Li et al. 2011). The CP El Niño occurs more and more frequently and strongly (e.g., Yeh et al. 2009; Lee and McPhaden 2010; Yu and Kim 2013). Meanwhile, the strongest Arctic sea ice ablation has been observed in summer, reaching the lowest value of annual sea ice volume in September (Serreze et al. 2009). However, in the season (i.e., summer) of severe Arctic sea ice melting, there is no significant warming trend in the Arctic Ocean (Ding et al. 2014), suggesting the domination of interannual variability. In addition, Arctic warming is not uniform (Bader 2014; Ding et al. 2014)—such seemingly irrelevant two phenomena remind us of an intriguing problem: Does the almost concurrent changes in the shifting mode of El Nino have some connection with the interannual variability of Arctic summer climate anomalies?

7.5.3 Teleconnection Between Shifting El Niño and Arctic Summer Climate Anomalies

The Arctic climate has undergone tremendous changes since the satellite observations started in the late-1970s (Duarte et al. 2012; Jeffries et al. 2013), such as rapid warming and a sharp decline in sea ice area. Many studies have focused on Arctic sea ice's responses to climate change caused by external forcing and found that the Arctic warming rate is significantly faster than the rest of the world, the so-called "Arctic amplification effect" (Graversen et al. 2008; Screen and Simmonds 2010). However, the ratio of surface air temperature and atmospheric circulation tends to the detrended interannual variability is only about 10% (Fig. 7.33), suggesting that even in the context of global warming, natural interannual variability is still dominant.

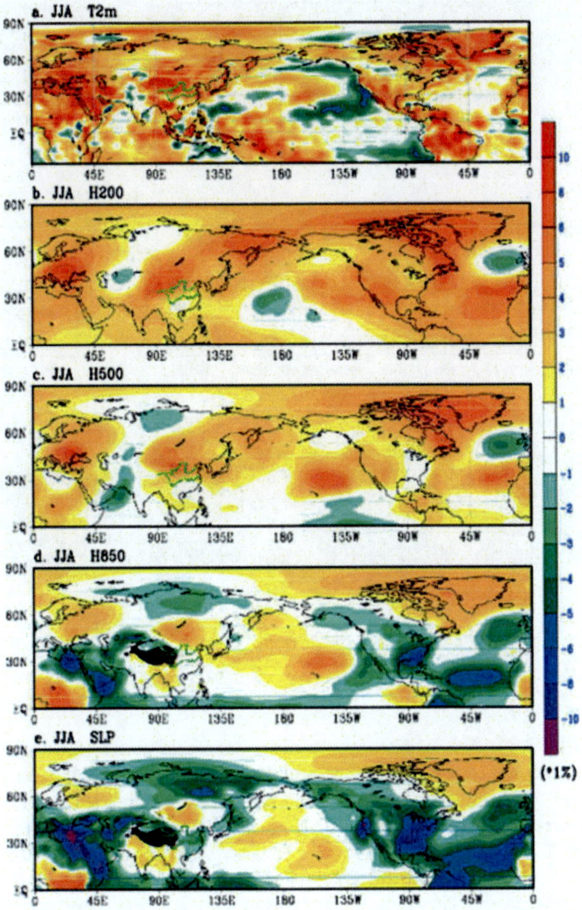

Fig. 7.33 Ratios of linear trend to standard deviation of interannual variability during 1979–2013. **a–e** show T2m, H200, H500, H850, and SLP (ERA-interim), respectively. Figure is cited from the supporting information of Hu et al. (2016b)

It is well established that the Arctic climatic trend is attributed to human activities (e.g., Gillett et al. 2008; Polyakov et al. 2012) and the tropical Pacific SST (e.g., Lee 2012; Ding et al. 2014). The interannual variability of the Arctic climate, however, may respond more strongly to the interannual variability of the leading SST modes over the tropical Pacific.

The temporal and spatial characteristics of the EP and CP El Niños during the boreal summer and the corresponding precipitation anomalies are shown in Fig. 7.34a–d. Both two modes exhibit strong interannual variation, emphasizing

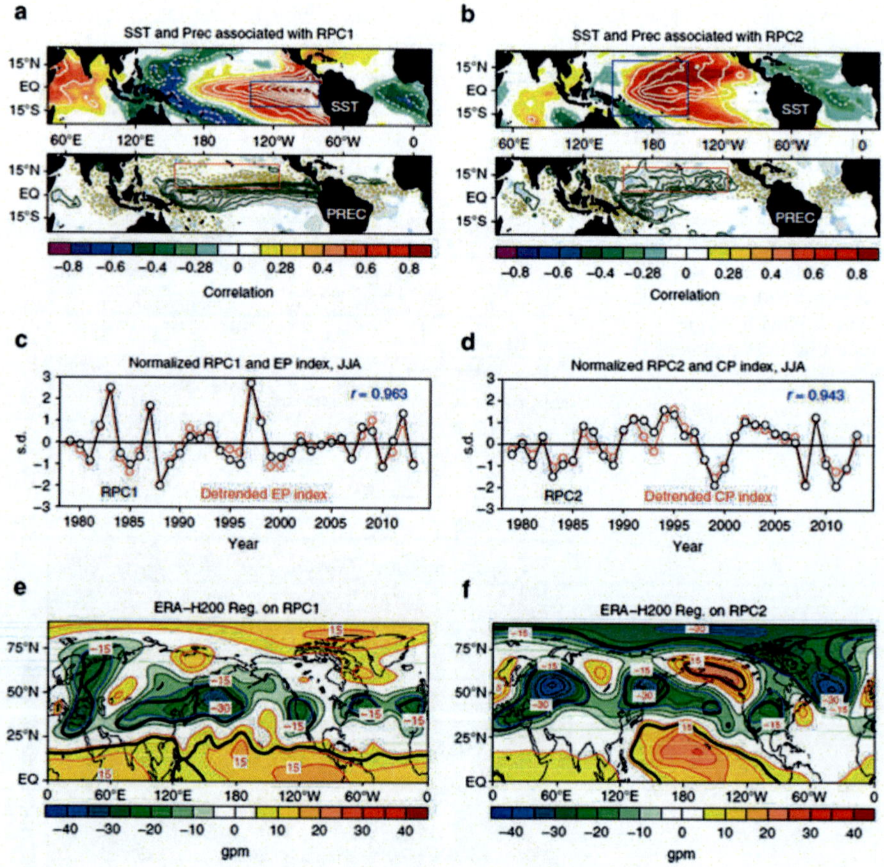

Fig. 7.34 The leading modes of the tropical Pacific SST anomalies and the corresponding H200 anomalies during the summers of 1979–2013. **a–b** SST [shadings and contours are the correlations and regressions (interval 0.2 K)] and precipitation (regression with an interval of 0.4 mm day^{-1}) correlated/regressed with RPC1 and RPC2, respectively. **c–d** Black curves show a time series of RPC1 and RPC2, respectively. Red sequence is the normalized and detrended EP index and CP index, respectively. **e–f** ERA H200 regressions (contour interval 5 gpm) onto RPC1 and RPC2, respectively. The thick black contour indicates the correlation passing the 90% confidence level. Figure is cited from Hu et al. (2016b)

the location and magnitude of the two types of ENSO-related SST warming, respectively. Changes in the SST associated with El Niño can cause changes in tropical convective activity, thereby stimulating large-scale Rossby wave activity through convergent and divergent disturbances in the upper troposphere (Sardeshmukh and Hoskins 1988). The 200 hPa geopotential height (H200) anomalies associated with the two types of El Niño exhibit planetary-scale long wave characteristics, which subsequently cause mid-latitude and Arctic climate anomalies via teleconnections. Compared to Fig. 7.34e, Fig. 7.34f reveals a significant teleconnection between the CP El Niño and the Arctic upper-tropospheric circulation, namely, the deepening Arctic polar vortex, especially over the Canadian Basin, which is closely linked to the warm SST anomalies in the tropical Pacific.

The deepening of the tropospheric Arctic polar vortex can result in Arctic climate anomalies in the lower troposphere. Figure 7.35a–c show that 925 hPa temperature (T925), land surface temperature (LST), SST, and Arctic sea ice concentration (SIC) are all significantly associated with CP El Niño, with the strongest correlations

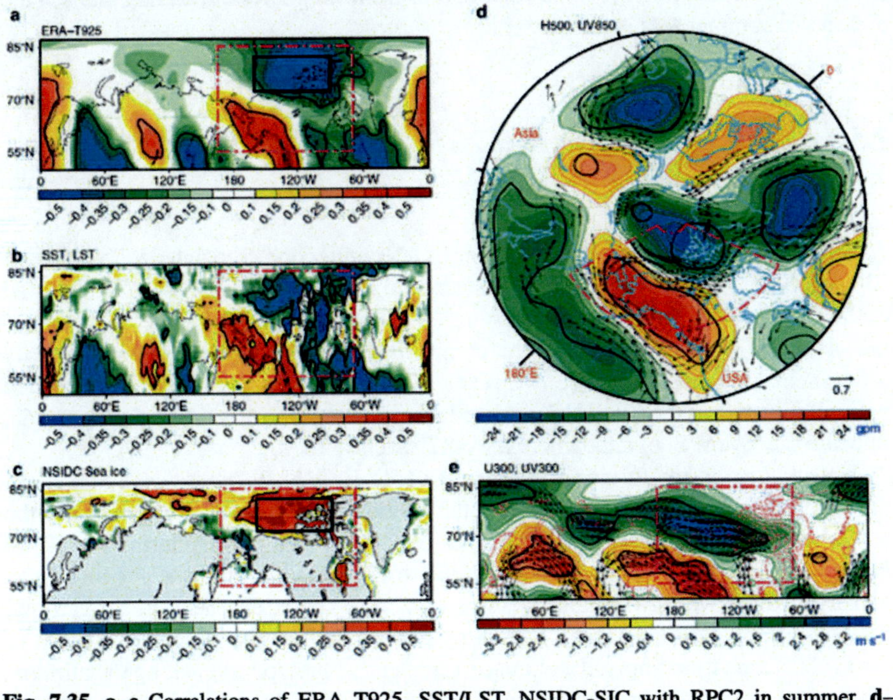

Fig. 7.35 a–c Correlations of ERA T925, SST/LST, NSIDC-SIC with RPC2 in summer. **d–e** Shading indicates the regressions of ERA H500 and ERA U300 onto RPC2, respectively. Arrows are the correlations of ERA UV850 and ERA UV300 with RPC2, respectively, showing only the vector with absolute value >0.28. The thick black contour indicates the correlation passing the 90% confidence level. **a, c** The solid box outlines the area (160°–90° W, 72°–82° N), which is used to define the T925 index and SIC index for later use. The area outlined by the purple dashed box indicates the same region in all five panels. Figure is cited from Hu et al. (2016b)

appearing in the Canada Basin. Here, the northwesterly wind anomalies (Fig. 7.35d, arrows) and the resulting cold advection are conducive to local cooling and sea ice growth. Moreover, such responses may also be magnified by some complex local positive feedback mechanisms among atmosphere-ocean-sea ice, such as thermal advection, ice-temperature feedback, cloud-radiation feedback, surface heat flux exchange, etc. (e.g., Kay et al. 2008; Serreze et al. 2009; Screen and Simmonds 2010; Ogi and Wallace 2012; Stammerjohn et al. 2012; Goosse and Zunz 2013). The 500 hPa geopotential height (H500) anomalies (Fig. 7.35d) also exhibits a wavelike character similar to that in Fig. 7.34f, accompanied by the deepening of the upper and lower polar vortexes and the strengthening of the circumpolar westerly wind (Fig. 7.35d–e). Such high- and low-level consistent wind anomalies (Fig. 7.34d–e, arrows), indicate that the CP El Niño-induced Arctic circulation anomaly is quasi-barotropic. In contrast, the EP El Niño does not show any significant contribution to the Arctic summer climate anomaly (Fig. 7.34e). Therefore, the CP El Niño can stimulate the summer tropical-Arctic teleconnection, resulting in the deepening of the Arctic polar vortex and the enhancing of circumpolar westerly wind, which helps to keep more cold air in the polar region, thus inhibiting Arctic warming and sea ice melting in summer to a certain extent.

7.5.4 Convection Differences Associated with Shifting El Niño

To further illustrate different mechanisms of teleconnection associated with different El Niño types, we show the spatial distribution characteristics of total SST related to the two types of El Niño (i.e., the summer SST climatology plus the SST anomalies shown in Fig. 7.34a–b), and the composites of tropical precipitation anomaly differences associated with the two types of El Niño (i.e., the CP El Niño-related precipitation minus the EP El Niño-related precipitation). The isotherms in Fig. 7.36a, indicate the Warm Pool area where the SST reaches 28 °C.

Note that in the key forcing area (i.e., area 155° E–115° W, 5°–20° N in Fig. 7.36), the dashed black line (corresponding to the EP El Nino) is always surrounded by the red solid line (corresponding to the CP El Nino). In other words, during the CP El Niño, the 28 °C isotherm to the north of the tropical Pacific expands, suggesting an expansion of the western tropical Pacific Warm Pool. This strengthens the background SST over the key forcing area, facilitating the enhancement and northward shift of the ITCZ. Correspondingly, the enhanced convective precipitation along the climatic ITCZ location and its northern critical location (Fig. 7.36b) confirm the enhancement and northward shift of the ITCZ. Linearly, the enhancement and northward shift of the ITCZ represent the enhancement and northward shift of tropical convection, which will inevitably lead to high-level convergence and divergence (Sardeshmukh and Hoskins 1988), and to the northward shift of the north subsidence of the Hadley cell and subtropical westerly jet stream (Hoskins and Karoly 1981; Jin and Hoskins

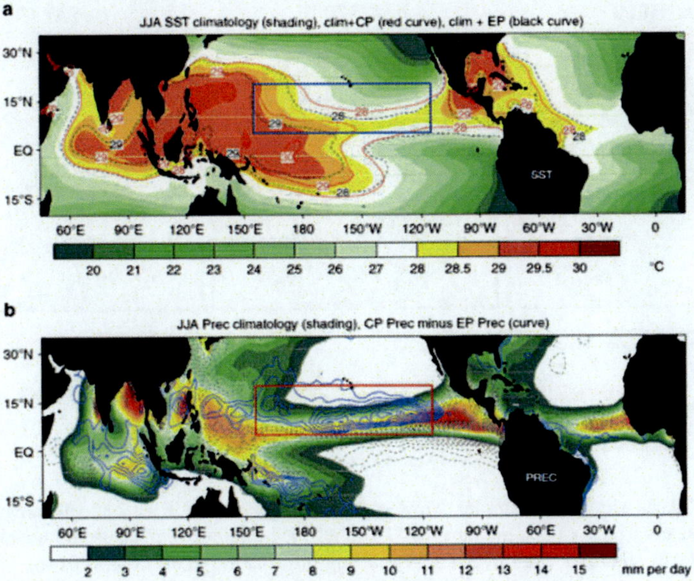

Fig. 7.36 **a–c** Differences in SST and precipitation between the two types of El Niño. **a** The shading shows SST climatology in summer. The red and black contours are the total SST (only showing SST >28 °C with 1 °C interval) related to CP El Niño and EP El Niño, respectively. **b** The shading shows precipitation climatology in summer. The contours indicate the CP El Niño-related precipitation minus the EP El Niño-related precipitation (interval is 0.4 mm d^{-1}). The rectangle outlines the key forcing area (155° E–115° W, 5°–20° N). Figure is cited from Hu et al. (2016b)

1995), thus favoring the poleward propagation of Rossby waves (Hoskins and Karoly 1981; Lee and Yoo 2014) and the large-scale teleconnection from the tropics to the polar region.

The Arctic H200 response obtained from the CAM4 simulation (Neale et al. 2013; Hu et al. 2016a, c), is consistent with the observed H200 anomalies, both being significant negative anomalies (Fig. 7.37), with comparable magnitude, although there are some differences in terms of location and strength. This suggests that the CAM4 can simulate the deepening of the Arctic vortex in summer by the CP El Niño. The response of CAM4 simulation can also capture the enhancement of the circumpolar westerly wind and Arctic cooling (Fig. 7.37b–c). This suggests that the deepening of the Arctic vortex and the cooling of the polar region can be considered as the linear response to the forcing of the given SST anomalies in the tropics. It also confirms that the Arctic cooling and sea ice increase are indeed related to the change of upper atmosphere circulation induced by the CP El Niño SST anomalies in summer.

In the summers of 2014 and 2015, there were anomalous CP El Niño and EP El Niño, respectively (Fig. 7.38a–b). As shown in Fig. 7.38c–f, the corresponding low-level temperature anomalies and sea ice anomalies in the Arctic Canada Basin are indeed consistent with the statistical results: in the summer of 2014, the CP El

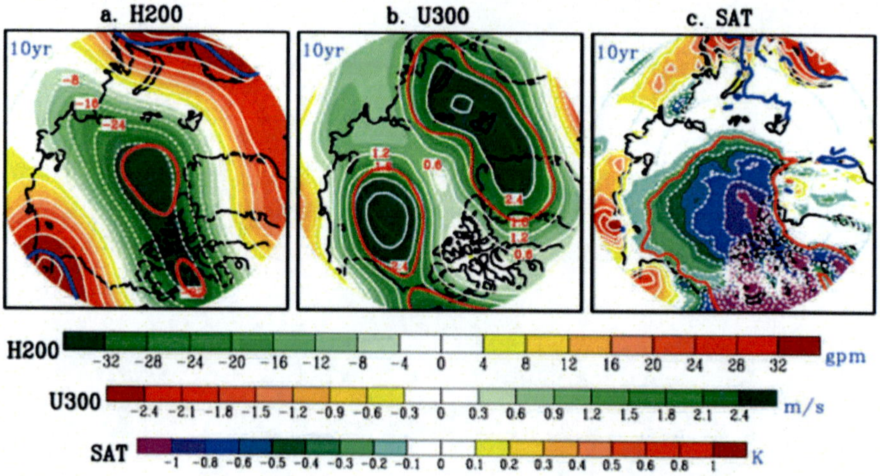

Fig. 7.37 CAM4-simulated summer Arctic climate response to CP El Niño forcing: **a** H200, **b** U300, and **c** surface air temperature (SAT). The thick contour represents the estimated 90% confidence level. Figure is cited from the supporting information of Hu et al. (2016b)

Niño corresponded to the anomalous cold in the Arctic Canada Basin and the relative increase in sea ice; and in the summer of 2015, the EP El Niño corresponded to the warmer anomalies in the Arctic Canada Basin and the relative decrease in sea ice. The clear examples of these two years confirmed that the shifting El Niño has an inhibitory effect on the summer Arctic warming and sea ice melting, especially over the Arctic Canada Basin.

7.5.5 Summary and Discussion

The above results suggest that there is a significant large-range teleconnection between tropical convective activity anomalies caused by CP El Niño SST warming and the Arctic summer climate variability on the interannual timescale. First, the linear response of upper troposphere circulation over the Canadian Basin to the CP El Niño was strong, while there was no obvious linear response to the EP El Niño, although the intensity of the SST anomaly corresponding to the EP El Niño was almost twice that of the CP El Niño. Second, the different effects of the two types of El Niño on Arctic atmospheric circulation can be explained as follows. The background SST of the central tropical Pacific was higher than that of the eastern tropical Pacific, so the CP SST warming was more conducive to the deep convection over the ITCZ and its northern waters. The enhanced and northwardly shifted ITCZ was conducive to large-scale teleconnected wave trains that could propagate to higher latitudes. Third, the teleconnection related to the CP El Nino induced the deepening of the Arctic vortex and the enhancement of circumpolar westerly. Such abnormal Arctic

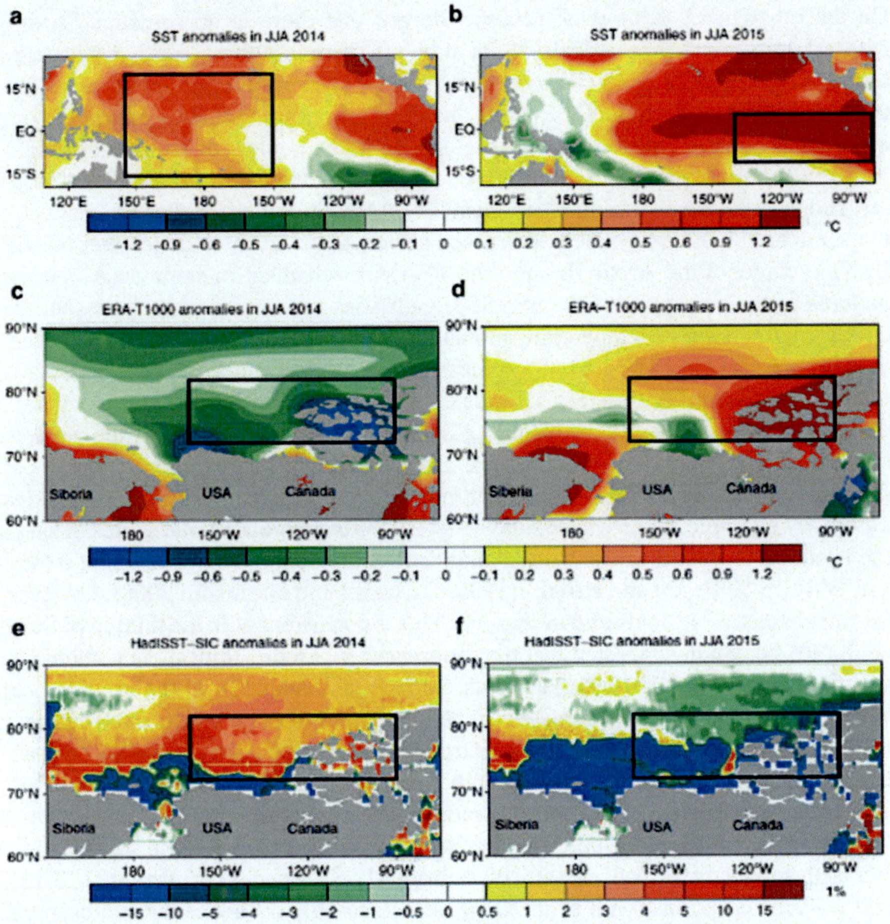

Fig. 7.38 **a, c, e** Tropical Pacific SST, T1000, and SIC anomalies in the summer of 2014, respectively. **b, d, f** Same as (**a, c, e**), except for the summer of 2015. The boxes in **a, b**) outline the CP region (140° W–80° W, 12° S–5° N) and the EP region (145° E–150° W, 16° S–20° N), respectively. The boxes in (**c–f**) indicate the same region (160° W–90° W, 72–82° N). Figure is cited from Hu et al. (2016b)

atmospheric circulation change provided favorable large-scale dynamic conditions for the cold air mass to stay in the polar region. As a result, the Arctic was relatively colder and sea ice increased, especially in the Canadian Basin.

Therefore, on the interannual variability scale, the CP El Niño has an inhibitory effect on the Arctic summer warming and sea ice melting. In other words, if there is no CP El Niño effect in the context of global warming, summer Arctic sea ice loss will be more dramatic than the current state. However, this may not last, because the continuous increases of GHGs and aerosols by human activities would cause the external forcing dominant in the Arctic warming sooner or later (Bader 2014).

On the other hand, Kim et al. (2009), showed that there is no so-called "spring forecasting barrier" for the CP El Niño, although there is a clear "spring forecasting barrier" for the EP El Niño (Webster and Yang 1992; Webster 1995). Therefore, the teleconnection between the CP El Niño and the Arctic is undoubtedly enhancing our ability to predict Arctic summer climate change in the context of global warming.

Lee (2012) suggested that the EP El Niño cools the Arctic through a "Tropical Induced Arctic Warming Mechanism (TEAM)" in winter. Although this is not the same as the summer results discussed above, there is no evidence that the EP El Nino can cool the Arctic through the TEAM mechanism in summer. Albeit the patterns of the two types of El Niño SST anomalies are similar in each season, the corresponding surface temperature anomalies over the Canadian Basin are different in different seasons. The reason for these different teleconnection mechanisms is mainly the seasonal differences in climate background and different air-sea interactions (Alexander et al. 2002), including the location and expansion of the western tropical Pacific Warm Pool (Wang et al. 2000), strength and location of the ITCZ (convection heating), strength and location of the subtropical jet (Jin and Hoskins 1995). For example, studies have shown that ENSO can regulate polar vortex through the Aleutian low and Pacific–North America teleconnection (PNA) in winter (Horel and Wallace 1981; Jin and Hoskins 1995; Garfinkel and Hartmann 2008); however, this mechanism was not found in summer. This is consistent with the finding of Sung et al. (2014), which suggested that the differences in Arctic atmosphere response to the two types of El Niño originate from the seasonal evolution of the extratropical wave trains, which is closely related to the location of tropical convective heating.

Note that the response of the winter stratospheric Arctic vortex to ENSO is sensitive to the phase of Quasi-biannual Oscillation (e.g., Garfinkel and Hartmann 2007, 2008; Garfinkel et al. 2010) and the sudden stratospheric warming (e.g., Iza and Calvo 2015). In addition, other factors such as Arctic sea ice (Kim et al. 2014), Eurasian snow cover (Cohen et al. 2007), tropical SST trend (Garfinkel et al. 2015), and volcanic eruptions (Graf et al. 2014) can all strongly influence the stratospheric polar vortex in winter. Whether or not the response of summer tropospheric polar vortex to CP El Niño depends on other factors (such as sea ice, snow cover, MJO, AMO, etc.) also needs investigation. Although these factors may be interrelated, the roles played by each factor needs to be systematically studied.

References

Alexander MA et al (2002) The atmospheric bridge: The influence of ENSO teleconnections on air–sea interaction over the global oceans. J Clim 15:2205–2231

An S-I, Wang B (2000) Interdecadal change of the structure of the ENSO mode and its impact on the ENSO frequency. J Clim 13:2044–2055

Annamalai H, Hafner J et al (2013) Global warming shifts the monsoon circulation, drying South Asia. J Clim 26:2701–2718

Ashok K, Behera SK, Rao SA et al (2007) El Niño Modoki and its possible teleconnection. J Geophys Res 112:C11007

Bader J (2014) Climate science: The origin of regional Arctic warming. Nature 509:167–168
Biasutti M, Giannini A (2006) Robust Sahel drying in response to late 20th century forcings. Geophys Res Lett 33:L11706
Camberlin P, Janicot S, Poccard I (2001) Seasonality and atmospheric dynamics of the teleconnection between African rainfall and tropical sea-surface temperature: Atlantic vs. ENSO. Int J Climatol 21:973–1005
Charney J, Stone PH, Quirk WJ (1975) Drought in the Sahara: A biogeophysical feedback mechanism. Science 187:434–435
Charney J, Quirk WJ, Chow SH, Kornfield J (1977) Comparative study of the effects of albedo change on drought in semi-arid regions. J Atmos Sci 34:1366–1385
Cohen J, Barlow M, Kushner PJ et al (2007) Stratosphere–troposphere coupling and links with Eurasian land surface variability. J Clim 20:5335–5343
Courel MF, Kandel RS, Rasool SI (1984) Surface albedo and the sahel drought. Nature 307:528–531
Ding Q et al (2014) Tropical forcing of the recent rapid Arctic warming in northeastern Canada and Greenland. Nature 509:209–212
Dong B, Sutton R (2015) Dominant role of greenhouse-gas forcing in the recovery of Sahel rainfall. Nature Clim Change 5:757–760
Duarte C, Lenton T, Wadhams P et al (2012) Abrupt climate change in the Arctic. Nature Clim Change 2:60–62
Ferranti L, Molteni F, Palmer TN (1994) Impact of localized tropical and extratropical SST anomalies in ensembles of seasonal GCM integrations. Quart J Roy Meteor Soc 120:1613–1645
Feudale L, Kucharski F (2013) A common mode of variability of African and Indian monsoon rainfall at decadal timescale. Clim Dyn 41:243–254
Folland CK, Palmer TN, Parker DE (1986) Sahel rainfall and worldwide sea temperatures, 1901-85. Nature 320:602–607
Fu C, Diaz HF, Fletcher JO (1986) Characteristics of the response of sea surface temperature in the central Pacific associated with warm episodes of the Southern Oscillation. Mon Wea Rev 114:1716–1739
Guo J, Su T, Li Z, Miao Y, Li J, Liu H, Xu H, Cribb M, Zhai P (2017) Declining frequency of summertime local-scale precipitation over eastern China from 1970 to 2010 and its potential link to aerosols. Geophys Res Lett 44(11):5700–5708
Garfinkel CI, Hartmann DL (2007) Effects of the El Niño Southern Oscillation and the Quasi-Biennial Oscillation on polar temperatures in the stratosphere. J Geophys Res 112:D19112
Garfinkel CI, Hartmann DL (2008) Different ENSO teleconnections and their effects on the stratospheric polar vortex. J Geophys Res 113:D18114
Garfinkel CI, Hartmann DL, Sassi F (2010) Tropospheric precursors of anomalous Northern Hemisphere stratospheric polar vortices. J Clim 23:3282–3299
Garfinkel CI, Hurwitz MM, Oman LD (2015) Effect of recent sea surface temperature trends on the Arctic stratospheric vortex. J Geophys Res 120:5404–5416
Giannini A, Saravanan R, Chang P (2003) Oceanic forcing of Sahel rainfall on interannual to interdecadal time scales. Science 302:1027–1030
Gill AE (1980) Some simple solutions for heat induced tropical circulation. Quart J Roy Meteor Soc 106:447–462
Gillett NP et al (2008) Attribution of polar warming to human influence. Nat Geosci 1:750–754
Goosse H, Zunz V (2013) Decadal trends in the Antarctic sea ice extent ultimately controlled by ice-ocean feedback. Cryosphere Discuss 7:4585–4632
Graf H-F (1986) On El Niño/Southern Oscillation and Northern Hemispheric temperature. Gerlands Beitr Geophys 95:63–75
Graf H-F, Zanchettin D (2012) Central Pacific El Niño, the “subtropical bridge”, and Eurasian climate. J Geophys Res 117:D01102
Graf H-F, Zanchettin D, Timmreck C et al (2014) Observational constraints on the tropospheric and near-surface winter signature of the Northern Hemisphere stratospheric polar vortex. Clim Dyn 43:3245–3266

Graham NE, Barnett TP, Wilde R et al (1994) On the roles of tropical and midlatitude SSTs in forcing interannual to interdecadal variability in the winter Northern Hemisphere circulation. J Clim 7:1416–1441

Graversen RG, Mauritsen T, Tjernström M et al (2008) Vertical structure of recent Arctic warming. Nature 451:53–56

Greene AM, Giannini A, Zebiak SE (2009) Drought return times in the Sahel: A question of attribution. Geophys Res Lett 36:L12701

He B, Yang S, Li ZN (2016) Role of atmospheric heating over the South China Sea and western Pacific regions in modulating Asian summer climate under the global warming background. Clim Dyn 46:2897–2908

He B, Yang S (2018) Role of Latent Heating over the South China Sea and Adjacent Regions on the Change in Climate over North America during Boreal Spring. J Clim 31:2169–2184

He S, Yang S, Li ZN (2017) Influence of Latent Heating over the Asian and Western Pacific Monsoon Region on Sahel Summer Rainfall. Sci Rep 7:1–8

Held IM, Delworth TL, Lu J, Findell KL, Knutson TR (2005) Simulation of Sahel drought in the 20th and 21st centuries. Proc Natl Acad Sci USA 102:17891–17896

Hoerling MP, Hurrell JW, Xu TY (2001) Tropical origins for recent North Atlantic climate change. Science 292:90–92

Horel JD, Wallace JM (1981) Planetary-scale atmospheric phenomena associated with the Southern Oscillation. Mon Wea Rev 109:813–829

Hoskins BJ, Ambrizzi T (1993) Rossby wave propagation on a realistic longitudinally varying flow. J Atmos Sci 50:1661–1671

Hoskins BJ, Karoly DJ (1981) The steady linear response of a spherical atmosphere to thermal and orographic forcing. J Atmos Sci 38:1179–1196

Hoskins BJ, Simmons AJ, Andrews DG (1977) Energy dispersion in a barotropic atmosphere. Quart J Roy Meteor Soc 103:553–568

Hu C, Yang S, Wu Q et al (2016a) Reinspecting Two Types of ENSO: A New Pair of Niño Indices for Improving Real-Time ENSO Monitoring. Clim Dyn 47:4031–4049

Hu C, Yang S, Wu Q et al (2016b) Shifting El Niño Inhibits Summer Arctic Warming and Sea Ice Melting over the Canada Basin. Nature Communications 7:11721

Huang J, Minnis P, Lin B, Yi Y, Fan TF, Sun-Mack S, Ayers JK (2006) Determination of ice water path in ice-over-water cloud systems using combined MODIS and AMSR-E measurements. Geophys Res Lett 33(21)

Huang J, Minnis P, Lin B, Yi Y, Khaiyer MM, Arduini RF, Fan A, Mace GG (2005) Advanced retrievals of multilayered cloud properties using multispectral measurements. J Geophys Res: Atmos 110:(D15)

Huang RH, Sun F (1992) Impact of the tropical western Pacific on the East Asian summer monsoon. J Meteor Soc Japan 70:213–256

Hunt BG (2000) Natural climatic variability and Sahelian rainfall trends. Global Planet Change 24:107–131

IPCC (2013) Climate Change 2013: The Physical Science Basis. Contribution of Working Group I to the Fifth Assessment Report of the Intergovernmental Panel on Climate Change, Stocker TF, Qin D, Plattner GK, Tignor M, Allen SK, Boschung J, Nauels A, Xia Y, Bex V, Midgley PM Cambridge University Press, Cambridge, United Kingdom and New York, NY, USA, 3–32

Iza M, Calvo N (2015) Role of stratospheric sudden warmings on the response to Central Pacific El Niño. Geophys Res Lett 42:2482–2489

Jeffries M, Overland J, Perovich D (2013) The Arctic shifts to a new normal. Phys Today 66:35–40

Jiang X, Yang S, Li J et al (2013) Variability of the Indian Ocean SST and its possible impact on summer western North Pacific anticyclone in the NCEP Climate Forecast System. Clim Dyn 41:2199–2212

Jin F, Hoskins BJ (1995) The direct response to tropical heating in a baroclinic atmosphere. J Atmos Sci 52:307–319

Kasahara A, Silva Dias PL (1986) Response of planetary waves to stationary tropical heating in a global atmosphere with meridional and vertical shear. J Atmos Sci J3:1893–1911
Kao H-Y, Yu J-Y (2009) Contrasting eastern-Pacific and central-Pacific types of ENSO. J Clim 22:615–632
Kay JE, L'Ecuyer T, Gettelman A et al (2008) The contribution of cloud and radiation anomalies to the 2007 Arctic sea ice extent minimum. Geophys Res Lett 35:L08503
Kay JE et al (2015) The community earth system model (CESM) large ensemble project: A community resource for studying climate change in the presence of internal climate variability. Bull Am Meteorol Soc 96:1333–1349
Kharin VV (1995) The relationship between sea surface temperature anomalies and atmospheric circulation in GCM experiments. Clim Dyn 11:359–375
Kim B-M et al (2014) Weakening of the stratospheric polar vortex by Arctic sea-ice loss. Nat Commun 5:4646
Kim H-M, Webster PJ, Curry JA (2009) Impact of shifting patterns of Pacific Ocean warming on North Atlantic tropical cyclones. Science 325:77–80
Knight JR, Folland CF, Scaife AA (2006) Climate impacts of the Atlantic multidecadal oscillation. Geophys Res Lett 33:L17706
Kug J-S, Jin F-F, An S-I (2009) Two types of El Niño events: Cold tongue El Niño and warm pool El Niño. J Clim 22:1499–1515
Kumar A, Leetmaa A, Ji M (1994) Simulations of atmospheric variability induced by sea surface temperatures and implications for global warming. Science 266:632–634
Larkin NK, Harrison DE (2005) On the definition of El Niño and associated seasonal average U.S. weather anomalies. Geophys Res Lett 32:L13705
Lau KM (1992) East Asian summer monsoon rainfall variability and climate teleconnection. J Meteor Soc Japan 70:211–242
Lee S (2012) Testing of the tropically excited Arctic warming mechanism (TEAM) with traditional El Niño and La Niña. J Clim 25:4015–4022
Lee S, Yoo C (2014) On the causal relationship between poleward heat flux and the Equator-to-Pole temperature gradient: A cautionary tale. J Clim 27:6519–6525
Lee T, McPhaden MJ (2010) Increasing intensity of El Niño in the central-equatorial Pacific. Geophys Res Lett 37:L14603
L'Hôte Y, Mahé G, Somé B, Triboulet JP (2002) Analysis of a Sahelian annual rainfall index from 1896 to 2000; the drought continues. Hydrol Sci J 47:563–572
Li Z, Yang S, He B, Hu C (2016) Intensified springtime deep convection over the South China Sea and the Philippine Sea dries southern China. Sci Rep 6: 30470
Li J, Xie S-P, Cook ER et al (2011) Interdecadal modulation of El Niño amplitude during the past millennium. Nature Clim Change 1:114–118
Li WH, Li LF, Ting MF, Liu YM (2012) Intensification of Northern Hemisphere subtropical highs in a warming climate. Nat Geosci 5:830–834
Li WT, Chineke C, Xin L, Wu GX (2001) Atmospheric diabatic heating and summertime circulation in Asia-Africa area. Adv Atmos Sci 18:256–268
Liu P, Wu GX, Sun S (2001a) Local meridional circulation and deserts. Adv Atmos Sci 18:864–871
Liu YM, Wu GX, Liu H, Liu P (2001b) Condensation heating of the Asian summer monsoon and the subtropical anticyclones in the Eastern Hemisphere. Clim Dyn 17:327–338
Ma ZG, Fu CB (2007) Global aridification in the second half of the 20th century and its relationship to large-scale climate background. Sci China, Ser D Earth Sci 50:776–788
Ma S, Zhou T (2016) Robust strengthening and westward shift of the tropical Pacific Walker circulation during 1979–2012: a comparison of 7 sets of reanalysis data and 26 CMIP5 models. J Clim 29(9):3097–3118
Matsuno T (1966) Quasi-Geostrophic Motions in the Equatorial Area. J Meteor Soc Japan 44:25–43
McGregor S, Timmermann A, Stuecker MF, England MH, Merrifield M, Jin FF, Chikamoto Y (2014) Recent Walker circulation strengthening and Pacific cooling amplified by Atlantic warming. Nature Clim Change 4(10):888

McPhaden MJ, Zebiak SE, Glantz MH (2006) ENSO as an integrating concept in Earth science. Science 314:1740–1745
Murakami T, Matsumoto J, Yalagai A (1999) Similarities as well as differences between summer monsoons over Southeast Asia and the western North Pacific. J Meteorol Soc Jpn 77:887–906
Neale RB et al (2013) The mean climate of the Community Atmosphere Model (CAM4) in forced SST and fully coupled experiments. J Clim 26:5150–5168
Nicholson SE (1980) The nature of rainfall fluctuations in subtropical West Africa (Guinea Sahel Soudan). Mon Weather Rev 108:473–487
Nicholson SE, Grist JP (2003) The seasonal evolution of the atmospheric circulation over West Africa and equatorial Africa. J Clim 16:1013–1030
Nitta T (1987) Convective activities in the tropical western Pacific and their impact on the Northern Hemisphere summer circulation. J Meteor Soc Japan 65:373–390
Ogi M, Wallace JM (2012) The role of summer surface wind anomalies in the summer Arctic sea ice extent in 2010 and 2011. Geophys Res Lett 39:L09704
Ozer P, Erpicum M, Demarée G, Vandiepenbeeck M (2003) The Sahelian drought may have ended during the 1990s. Hydrol Sci J 48:489–492
Palmer TN (1986) Influence of the Atlantic, Pacific and Indian Oceans on Sahel rainfall. Nature 322:251–253
Philander SGH et al (1996) Why the ITCZ is mostly north of the equator. J Clim 9:2958–2972
Polyakov I, Walsh JE, Kwok R (2012) Recent changes of Arctic multiyear sea ice coverage and the likely causes. Bull Am Meteorol Soc 93:145–151
Prospero JM, Lamb PJ (2003) African droughts and dust transport to the Caribbean: Climate change implications. Science 302:1024–1027
Pu B, Cook KH (2012) Role of the west African westerly jet in sahel rainfall variations. J Clim 25:2880–2896
Qiang XM, Yang XQ (2013) Relationship between the first rainy season precipitation anomaly in South China and the sea surface temperature anomaly in the Pacific. Chin J Geophys Chin Edn 56(8):2583–2593
Rodwell MJ, Hoskins BJ (1996) Monsoons and the dynamics of deserts. Quart J R Meteorol Soc 122:1385–1404
Rodwell M, Hoskins BJ (2001) Subtropical anticyclones and summer monsoons. J Clim 14(15):3192–3211
Sardeshmukh PD, Hoskins BJ (1988) The generation of global rotational flow by steady idealized tropical divergence. J Atmos Sci 45:1228–1251
Sardeshmukh PD, Sura P (2007) Multiscale impacts of variable heating in climate. J Clim 20:5677–5695
Screen JA, Simmonds I (2010) The central role of diminishing sea ice in recent Arctic temperature amplification. Nature 464:1334–1337
Serreze MC, Barrett AP, Stroeve JC et al (2009) The emergence of surface-based Arctic amplification. Cryosphere 3:11–19
Shukla J, Paolina DA (1983) The Southern Oscillation and long-range forecasting of the summer monsoon rainfall over India. Mon Wea Rev 111:1830–1837
Simmons AJ (1982) The forcing of stationary wave motion by tropical diabatic heating. Quart J Roy Meteor Soc 108:503–534
Simmons AJ, Wallace JM, Branstator GW (1983) Barotropic wave propagation and instability, and atmospheric teleconnection patterns. J Atmos Sci 40:1363–1392
Skinner CB, Diffenbaugh NS (2014) Projected changes in African easterly wave intensity and track in response to greenhouse forcing. Proc Natl Acad Sci USA 111:6882–6887
Song Y, Webster PJ, Min D (1992) Longitudinal heating gradient: Another possible factor influencing the intensity of the Asian summer monsoon circulation. Adv Atmos Sci 9: 397–410
Stammerjohn S, Massom R, Rind D et al (2012) Regions of rapid sea ice change: An inter-hemispheric seasonal comparison. Geophys Res Lett 39:L06501

Sung M-K, Kim B-M, An S-I (2014) Altered atmospheric responses to eastern Pacific and central Pacific El Niños over the North Atlantic region due to stratospheric interference. Clim Dyn 42:159–170
Tang Y, Deng Z, Zhou X et al (2008) Interdecadal variation of ENSO predictability in multiple models. J Clim 21:4811–4833
Ting M, Sardeshmukh PD (1993) Factors determining the extratropical response to equatorial diabatic heating anomalies. J Atmos Sci 50:907–918
Trenberth KE, Branstator GW et al (1998) Progress during TOGA in understanding and modeling global teleconnections associated with tropical sea surface temperatures. J Geophys Res Oceans 103:14291–14324
Tucker CJ, Dregne HE, Newcomb WW (1991) Expansion and contraction of the Sahara desert from 1980 to 1990. Science 253:299–301
Vecchi GA, Soden BJ, Wittenberg AT, Held IM, Leetmaa A, Harrison MJ (2006) Weakening of tropical Pacific atmospheric circulation due to anthropogenic forcing. Nature 441(7089):73–76
Wallace JM, Gutzler DS (1981) Teleconnections in the geopotential height field during the Northern Hemisphere winter. Mon Wea Rev 109:784–812
Wan R, Wu G (2007) Mechanism of the spring persistent rains over southeastern China. Sci China, Ser D Earth Sci 50(1):130–144
Wang B, Wu R, Fu X (2000) Pacific-East Asian teleconnection: How does ENSO affect East Asian climate? J Clim 13:1517–1536
Wang PX, Wang B, Cheng H, Fasullo J, Guo ZT, Kiefer T, Liu ZY (2014) The global monsoon across timescales: Coherent variability of regional monsoons. Clim Past 10:2007–2052
Ward MN (1998) Diagnosis and short-lead time prediction of summer rainfall in tropical North Africa at interannual and multidecadal timescales. J Clim 11:3167–3191
Webster PJ (1972) Response of the tropical atmosphere to local, steady forcing. Mon Wea Rev 100:518–541
Webster PJ (1981) Mechanisms determining the atmospheric response to sea surface anomalies. J Atmos Sci 38:554–571
Webster PJ (1982) Seasonality in the local and remote atmospheric response to sea surface temperature anomalies. J Atmos Sci 39:41–52
Webster PJ (1995) The annual cycle and the predictability of the tropical coupled ocean-atmosphere system. Meteorol Atmos Phys 56:33–55
Webster PJ, Magaña VO, Palmer TN et al (1998a) Monsoons: Processes, predictability, and the prospects for prediction. J Geophys Res 103:14451–14510
Webster PJ, Yang S (1992) Monsoon and ENSO: Selectively interactive systems. Quart J Roy Meteor Soc 118:877–926
Webster PJ, Magaña VO, Palmer TN, Shukla J, Tomas RA, Yanai M, Yasunari T (1998b) Monsoons: Processes, predictability, and the prospects for prediction. J Geophys Res Oceans 103:14451–14510
Weller E, Min S-K, Cai W et al (2016) Human-caused Indo-Pacific warm pool expansion. Sci Adv 2(7):e1501719
Wu GX, Liu Y (2003) Summertime quadruplet heating pattern in the subtropics and the associated atmospheric circulation. Geophys Res Lett 30:1201
Wu GX, Liu YM, Zhu X, Li W, Ren R, Duan A, Liang X (2009) Multi-scale forcing and the formation of subtropical desert and monsoon. Ann Geophys 27:3631–3644
Wu GX, He B, Liu YM, Bao Q, Ren RC (2015) Location and variation of the summertime upper-troposphere temperature maximum over South Asia. Clim Dyn 45:2757–2774
Xie SP, Deser C, Vecchi GA, Ma J, Teng H, Wittenberg AT (2010) Global warming pattern formation: Sea surface temperature and rainfall. J Clim 23:966–986
Yanai M, Tomita T (1998) Seasonal and interannual variability of atmospheric heat sources and moisture sinks as determined from NCEP-NCAR reanalysis. J Clim 11:463–482
Yang S, Webster PJ (1990) The effect of summer tropical heating on the location and intensity of the extratropical westerly jet streams. J Geophys Res 95:18705–18721

Yang S, Jiang X (2014) Prediction of Eastern and Central Pacific ENSO Events and Their Impacts on East Asian Climate by the NCEP Climate Forecast System. J Clim 27:4451–4472

Yang S, Lau KM (1998) Influences of sea surface temperature and ground wetness on Asian summer monsoon. J Clim 11:3230–3246

Yang S, Webster PJ, Min D (1992) Longitudinal heating gradient: Another possible factor influencing the intensity of the Asian summer monsoon circulation. Adv Atmos Sci 9:397–410

Yeh S-W, Kug J-S, Dewitte B et al (2009) El Niño in a changing climate. Nature 461:511–514

Yongkang X, Shukla J (1993) The influence of land surface properties on Sahel climate. Part I: desertification. J of Clim 6: 2232–2245

Yu J-Y, Kim ST (2010) Identification of central-Pacific and eastern-Pacific types of ENSO in CMIP3 models. Geophys Res Lett 37:L15705

Yu J-Y, Kim ST (2013) Identifying the types of major El Niño events since 1870. Int J Climatol 33:2105–2112

Zeng N, Neelin JD, Lau KM, Tucker CJ (1999) Enhancement of interdecadal climate variability in the Sahel by vegetation interaction. Science 286:1537–1540

Zhang W, Li J, Jin F-F (2009) Spatial and temporal features of ENSO meridional scales. Geophys Res Lett 36:L15605

Chapter 8
Subseasonal to Seasonal Prediction of Atmospheric Circulation and Rainfall Over Southeast Asia

8.1 Introduction

Southeast Asia (SEA) is bordered by East Asia to the north, by South Asia and the Bay of Bengal to the west, by the Pacific Ocean to the east, and by Australia and the Indian Ocean to the south. Due to complex distribution of land and sea in SEA and surrounded by the warmest sea surface temperature (SST) in the world, there exist strong land–ocean–atmosphere interactions in the region. SEA is the key region where Australian monsoon interacts with East Asia monsoon, and is the transition region of South Asia monsoon to East Asia monsoon. The Indian Ocean and the Pacific Ocean also interact with each other through SEA. The weather and climate over SEA show strong variability on a wide range of time and space scales because this region is affected by atmospheric and oceanic circulation originated from various regions. The atmospheric processes over SEA and their interactions with local and remote oceans and lands are important for subseasonal to seasonal (S2S) prediction of atmospheric circulation and rainfall in the region and beyond.

SEA is one of the atmospheric convection centers, and the latent heating released by convective condensation over the maritime continent (MC) fuels the global atmospheric circulation (Yang and Webster 1990; Keenan et al. 2000; Neale and Slingo 2003). The latent heating over SEA can directly affect Asian climate, and it has a close relationship with El Niño–Southern Oscillation (ENSO) through the Walker circulation. Thus, SEA is a key region linking ENSO and Asian climate. In addition, it plays an important role in the processes by which the Indian Ocean (IO) affects East Asian climate. Overall, the atmospheric circulation and rainfall over SEA interact strongly with large-scale rainfall anomalies in the nearby regions. S2S prediction of atmospheric circulation and rainfall over SEA is important not only for local social and economic development, but also for climate prediction in its neighboring regions.

The MC is a typical region of complex distribution of land and sea in SEA, where rainfall and temperature are affected by weather systems from both the tropics and extratropics. For example, the cold surges, originated from tropospheric shortwave trough over the Lake Baikal (Chen et al. 2004), can grow and travel southward

S. Yang et al., *Climate Change in Southeast Asia and Surrounding Areas*,
Springer Climate, https://doi.org/10.1007/978-981-15-8225-7_8

to SEA, and even to the Southern Hemisphere. Because of the complex land–sea distribution and geographical location in SEA, there are strong multi-scale variabilities in atmospheric circulation and rainfall, which are associated with diurnal cycle, synoptic weather systems, the Madden–Julian oscillation (MJO), and monsoons.

Scientists from various countries have been investigating characteristics and mechanisms of the formation and variabilities of atmospheric circulation and rainfall over SEA, and studying their impacts on weathers and climate in its neighboring regions. The convection over the MC is associated with localized thunderstorms, and thus experiments for thunderstorm observations have been conducted (Keenan et al. 1989, 2000). The islands play an important role in the formation and variation of regional atmospheric circulation and rainfall. Large-scale organization of thunderstorm activity is affected by regional orography and sea-breeze circulation (Holland and Keenan 1980). The convection related to sea-breeze convergence can aggregate into mesoscale convective complexes. Because of the surface inhomogeneity of the MC, the diurnal amplitude of convective rainfall over the islands can be three times as great as that over the adjacent ocean (Yang and Slingo 2001). Thus, the complex distribution of islands and seas plays an important part in the meteorology over SEA.

The synoptic-scale variation of convection over SEA is associated with the activity of cold surges. On the one hand, cold surges can strengthen convection; on the other hand, convective heating over the MC can strengthen the local Hadley circulation and thereby the East Asian jet stream, which in turn favors for more southward intrusion of cold surges into SEA (Chang and Lau 1980, 1982; Lau et al. 1983; Zhang et al. 1997; Park et al. 2011). Besides, the convection over SEA can deepen the monsoon trough over Australia (Holland 1984; McBride 1995).

The MJO often approaches the MC during its mature phase, and has strong interaction with local convection (Sui and Lau 1992). The MJO becomes weak and moves slowly over the MC (Zhu and Wang 1993). The enhanced convection over the MC induced by the MJO can reduce rainfall and temperature over a large portion of East Asia north of 20° N (He et al. 2011). The MC is a center of convection in spring in the Asian monsoon region, and the convection plays an important role in monsoon onset. The convective heating in spring over the northeastern MC can excite the South Asian high, which then moves westward and triggers the monsoon onset over the Bay of Bengal (Liu et al. 2012).

ENSO begins to develop or delay in spring. The dominator of the interaction between ENSO and monsoon in spring determines the development of summer climate. In contrast, monsoon in autumn is weak, and ENSO develops toward its mature phase. Overall, the characteristics of interaction between ENSO and monsoon exhibit strong seasonality and are important for seasonal prediction of both local and remote rainfall and atmospheric circulation. The seasonality of rainfall over the MC is different from that in the adjacent regions. The boreal winter is the wet season in the MC, while the boreal summer is the dry season there. In addition, the climate in the MC exhibits semiannual cycle rather than annual cycle (Chang et al. 2005a, b).

On the interannual time scale, rainfall generally decreases over the MC, the resultant diabatic cooling excites two anticyclones over South Asia and the IO. The anticyclone over South Asia weakens the South Asian monsoon in summer, while the anticyclone over the IO favors the development of the Indian Ocean dipole (IOD; Saji et al. 1999; Webster et al. 1999; Lau and Wang 2006). Besides, ENSO impacts the climate over East Asia by modulating the convection over the MC. Note that the rainfall over the western MC is not significantly correlated to the ENSO, which may be due to the orography of the MC (Chang et al. 2004).

The convection over the Philippine Sea is suppressed during ENSO events, but the regions with suppressed convection and the underlying physical processes are different for different phases of the ENSO. The suppressed convection over the Philippine Sea in winter and spring is mostly caused by cold local SST, while it is mostly excited remotely by warm SST in the IO in summer (e.g., Wu et al. 2010). The suppressed convection over the Philippine Sea can excite an anomalous anticyclone over the western Pacific, which generates significant rainfall anomalies over East Asia. Thus, the variability of convection over the Philippine Sea is vital for predicting East Asian climate (e.g., Jiang et al. 2013a, b, c).

S2S prediction of atmospheric circulation and rainfall is generally based on physical–statistical model and global climate model. Current global climate models can predict some major global climate modes. For example, they can predict ENSO several months in advance. It has become an important tool for operational climate prediction. In spite of the advances in climate modeling, some apparent deficiencies exist. For example, climate models cannot well predict the Arctic oscillation (AO), which is the most dominant climate variability in the extratropics. They also have apparent biases in simulating the teleconnection associated with ENSO. Thus, it is imperative that how to use the useful information provided by climate models and how to improve climate models at the same time.

State-of-the-art climate models show large biases over SEA, which are important sources for biases in other regions. The simulation of Met Office model HadAM3 (Neal and Slingo 2003) had a significant dry bias over the MC, which was associated with the model's deficiency in simulating the diurnal cycle of convection. The dry bias could lead to systematic errors over the tropical IO and Pacific Ocean, as well as extratropical North America and Eurasia. The Climate Forecast System version 2 (CFSv2; Saha et al. 2013) from the National Centers for Environmental Prediction has apparent dry bias in the western MC, which is partly responsible for the weak EAWM in the model. Thus, it is necessary to investigate the characteristics and mechanisms of the biases over SEA in state-of-the-art climate models.

Climate prediction systems have been established by major climate prediction centers worldwide. The assessment of prediction of rainfall and atmospheric circulation showed that although models' skills in predicting major climate phenomena have considerable differences, the models always have better skills for S2S prediction of rainfall and atmospheric circulation over the tropics than over the extratropics. As for the tropics, the predicted rainfall and atmospheric circulation over the Pacific are more realistic compared to those over the IO. As such, in this chapter, we mostly analyze the S2S prediction of atmospheric circulation and rainfall over SEA and its adjacent regions by the CFSv2 from the NCEP and by the climate system model from the Beijing Climate Center (BCC_CSM; Liu et al. 2017).

In this chapter, we first analyze seasonal prediction of atmospheric circulation and rainfall over SEA and its adjacent regions, and examine their possible impact factors, including the Asian monsoon circulation and associated rainfall, dominant mode of SST anomalies in the IO and the Pacific and its climate impacts, the western Pacific subtropical high (WPSH), and the atmospheric circulation and rainfall over the MC. Second, we discuss the subseasonal prediction of the North Atlantic oscillation (NAO), the blocking, and could surges over the Northern Hemisphere, WPSH, tropical intraseasonal oscillation (ISO), as well as the seasonal transition of rainfall over the MC and the early-season rainfall over South China. Third, we explore the predictability source of the S2S prediction of rainfall and atmospheric circulation over SEA and its adjacent regions. Finally, we present a summary and an outlook for S2S prediction.

8.2 Seasonal Prediction of Atmospheric Circulation and Rainfall Over SEA and Its Adjacent Regions

8.2.1 Prediction of the Asian Summer Monsoon

The Asian summer monsoon (ASM) exerts significant impacts on climate in and outside Asia (Zhao et al. 2003, 2007; Yang et al. 2008a, b). Jiang et al. (2013a) indicated that many major features of the ASM are well predicted by the CFSv2, which includes the heavy rainfall centers, large-scale monsoon circulation pattern, as well as monsoon onset and retreat. Figure 8.1 shows the CFSv2's skill in predicting the three popular monsoon indices: the Webster-Yang monsoon index (the WY index; Webster and Yang 1992), the South Asian monsoon index (the SA index; Goswami et al. 1999), and the Southeast Asian monsoon index (the SEA index; Wang and Fan 1999). The model predicted the variability of large-scale monsoon circulation very well, including the WY index and the SEA index, when lead time was less than five months. However, it shows much lower skill in predicting the variability of the meridional circulation over South Asia, with a correlation coefficient of about 0.2 for the 2-month lead prediction. The CFSv2's prediction shows an overall low skill in the 3- and 4-month lead for the WY and SEA indices, corresponding to the predictions initiated in January and February. This low prediction skill was likely related to the so-called "Spring Prediction Barrier" phenomenon (Webster and Yang 1992). The patterns of rainfall and circulation related to these indices could also be predicted several months in advance. Similar to the prediction of the indices, rainfall and circulation associated with the WY and SEA indices were better predicted compared to those related to the SA index.

Some CFSv2 biases in predicting the ASM were also reported. Weaker-than-observed WPSH and exaggerated strong link of the ASM to ENSO were produced by the model. The CFSv2 simulated an unrealistic zonal SST gradient in the tropical IO during the ENSO developing phase, with a large warm SST bias over the tropical

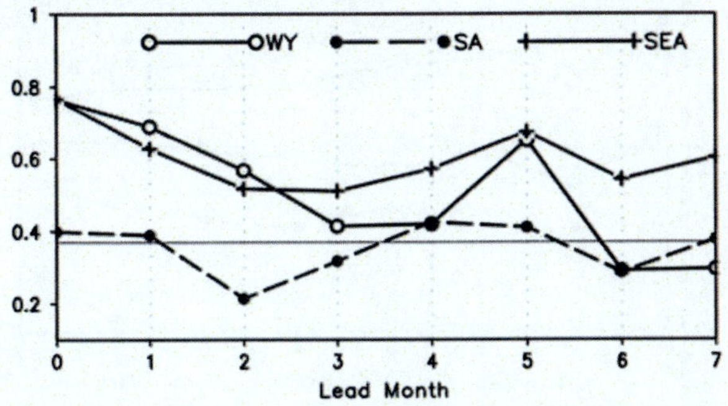

Fig. 8.1 Correlation coefficients between observed monsoon indices and hindcast monsoon indices of different lead months. Values are shown for different dynamic monsoon indices. Solid straight line denotes the 95% significance level. The solid curves with open circles indicate the WY index, defined as the vertical shear of zonal winds between 850 and 200 hPa averaged over (40°–110° E, 0°–20° S). The solid curves with pluses indicate the SEA index, defined as horizontal zonal wind shear at 850 hPa between (90°–130° E, 5°–15° S) and (110°–140° E, 22.5°–32.5° S). The dashed curves with solid dots indicate the SA index, defined as the vertical shear of meridional winds between 850 and 200 hPa averaged over (70°–110° E, 10°–30° S). After Jiang et al. (2013a)

western IO, accompanied by anomalous easterly flow over the equatorial IO. The easterly flow weakened the WY index, which might contribute to a strong-than-observed link of the WY index to ENSO, since the observed WY index is weak during the ENSO developing phase. As for the weaker-than-observed WPSH in the model, Liu et al. (2013) attributed it to the bias of internal atmospheric variability, since the development of this bias was often found in the prediction with lead time less than two weeks.

Comparison of the ASM predictions by the CFSv2 and European Centre for Medium-Range Weather Forecasts (ECMWF) System 4 indicated that although the SST, rainfall, and atmospheric circulation predicted by the two systems showed some differences, they shared some common features (Kim et al. 2012). Both models produced a cold bias of SST in the equatorial Pacific, North Atlantic, and IO, but a warm bias over the northern part of the North Pacific and North Atlantic. They had a high skill in predicting tropical SST, rainfall, and low-level atmospheric circulation. They captured the variability of large-scale circulation, but exhibited low skill for monsoon rainfall prediction. The models' skills in predicting monsoon increased with ENSO amplitude, although both overestimated the link between ENSO and monsoon.

Different from the ASM, the East Asian winter monsoon (EAWM) has a much closer relationship with atmospheric circulation systems over the mid and high latitudes. The CFSv2 can predict the major features of the EAWM, including the Siberian high, the East Asia trough, the East Asian jet stream, the low-tropospheric wind, surface temperature, and precipitation over East Asia (Jiang et al. 2013b). Figure 8.2 shows correlation coefficients between observed and predicted indices for the EAWM, AO, and Nino3.4. The model had a considerable skill in predicting

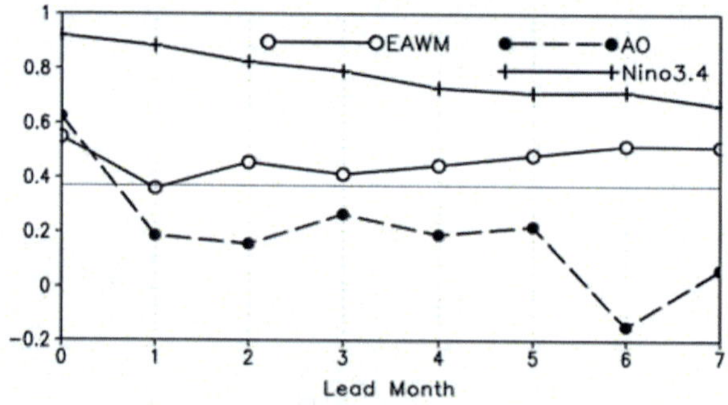

Fig. 8.2 Correlation coefficients between observed indices and hindcast indices of different lead months. Values are shown for the EAWM, AO, and Niño3.4. Solid straight line denotes the 95% confidence level. The EAWM is defined as meridional shear of zonal wind in three bands. IEAWM = (U1-U2 + U1-U3)/2, where U1 = U200(30°–35° S, 90°–160° E), U2 = U200(50°–60° S, 70°–170° E), and U3 = U200(5° S–10° S, 90°–160° E). After Jiang et al. (2013b)

a recently defined dynamic EAWM index (Li and Yang 2010; see figure caption for more details) for most lead months. As the interannual variability of EAWM is affected by the AO and ENSO, the model's ability in predicting both of them was examined. The CFSv2 can predict ENSO very well, but it cannot predict the AO even one month in advance. In addition, the CFSv2 can predict the circulation associated with the ENSO. Therefore, the CFSv2's skill in predicting EAWM came from its high skill in predicting ENSO, although it overestimated the link between the EAWM and ENSO (Jiang et al. 2013b).

The EAWM-related atmospheric circulation, 2-m temperature, and precipitation over the oceans can be predicted several months in advance. However, the model showed low skill predicting the EAWM-related extratropical atmospheric circulation over land, which can be partly attributed to its inability in reproducing the AO and its impact on atmospheric circulation. The model's higher skill in predicting the EAWM's components over the oceans is attributed to its better performance in ENSO prediction. However, the CFSv2 still had apparent bias in simulating the response of the EAWM to ENSO.

8.2.2 *Prediction of the Dominant SST Variability in the Pacific and Indian Oceans*

ENSO and related teleconnection patterns are important bases for seasonal climate prediction in many regions (e.g., Shukla and Paolina 1983; Webster et al. 1998; Jiang et al. 2013a, b). ENSO exerts a significant impact on climate in SEA and its adjacent regions. The EAWM is weak and Southeast China is warmer and wetter than normal in ENSO years (Tomita and Yasunari 1996; Zhang et al. 1996). On the other hand, the East Asian summer monsoon (EASM) strengthens during the ENSO decaying summer, because it can excite an anomalous anticyclone over the western North Pacific (WNP; Wang et al. 2000). The canonical El Niño is associated with maximum SST anomalies in the eastern equatorial Pacific. In the recent decade, a new type of tropical Pacific SST warming pattern, with maximum warm anomalies in the central equatorial Pacific, has also been discussed widely (Larkin and Harrison 2005; Ashok et al. 2007; Yu and Kao 2007). It is alternatively referred to as dateline El Niño (Larkin and Harrison 2005), El Niño Modoki (Ashok et al. 2007; Weng et al. 2007), central Pacific (CP) El Niño (Kao and Yu 2009), and warm pool El Niño (Kug et al. 2009). The two types of ENSO have different impacts on atmospheric circulation and rainfall over SEA and its adjacent regions (Weng et al. 2007; Yuan et al. 2012).

The SST variability in the IO has two dominant modes: the Indian Ocean basin warming (IOBW; e.g., Xie et al. 2009) and the IOD (Saji et al. 1999; Webster et al. 1999); both are linked to ENSO. The IOBW usually occurs in the ENSO decaying summer, which suppresses rainfall over the WNP and thereby excites an anomalous anticyclone over the region. The anomalous anticyclone can affect the rainfall over East Asia by modulating the western North Pacific subtropical high (WNPSH; Chang et al. 2000; Li and Wang 2005; Lau and Wang 2006; Xie et al. 2009; Yuan et al. 2012). Some IOD events co-occurred with ENSO, which could affect the Indian summer monsoon independently and modulate the relationship between Indian summer monsoon and ENSO (Saji et al. 1999; Ashok et al. 2001; Saji and Yamagata 2007; Behera et al. 2005).

The three tropical anomalous SST modes have significant climate impacts. The studies of Jiang et al. (2013a, b, c) indicated that model's skill in predicting the climate over SEA is linked to its skills in predicting the four anomalous SST modes. In this chapter, we discuss CFSv2's skills in predicting the three anomalous SST modes.

1. Prediction of ENSO

ENSO usually first appears in May, reaches its peak in winter, and terminates in the following summer. The CFSv2 prediction of 0-month lead can capture the life cycle of ENSO (Fig. 8.3). The prediction skill for the Niño3.4 SST index decreases as lead time increases. The prediction skill for ENSO decaying phase is higher than that for ENSO developing phase. The predicted onset and decay of ENSO are later than the observations by 1–2 months. The magnitude of interannual variability of Niño3.4 SST is strong in winter but weak in summer. The prediction of 0-month lead can capture the magnitude during ENSO decaying phase, but overestimates the magnitude during ENSO developing phase. The magnitude of predicted ENSO decreases with increasing lead time, and it is weaker than observations when lead time is longer than five months.

2. Prediction of the Indian Ocean basin warming

Figure 8.4a and c presents the spatial patterns and principal components of the first mode of an EOF analysis of the temporal covariance matrix of JJA-averaged SST for observation and 0-month lead hindcast, which account for 50% and 52% of the

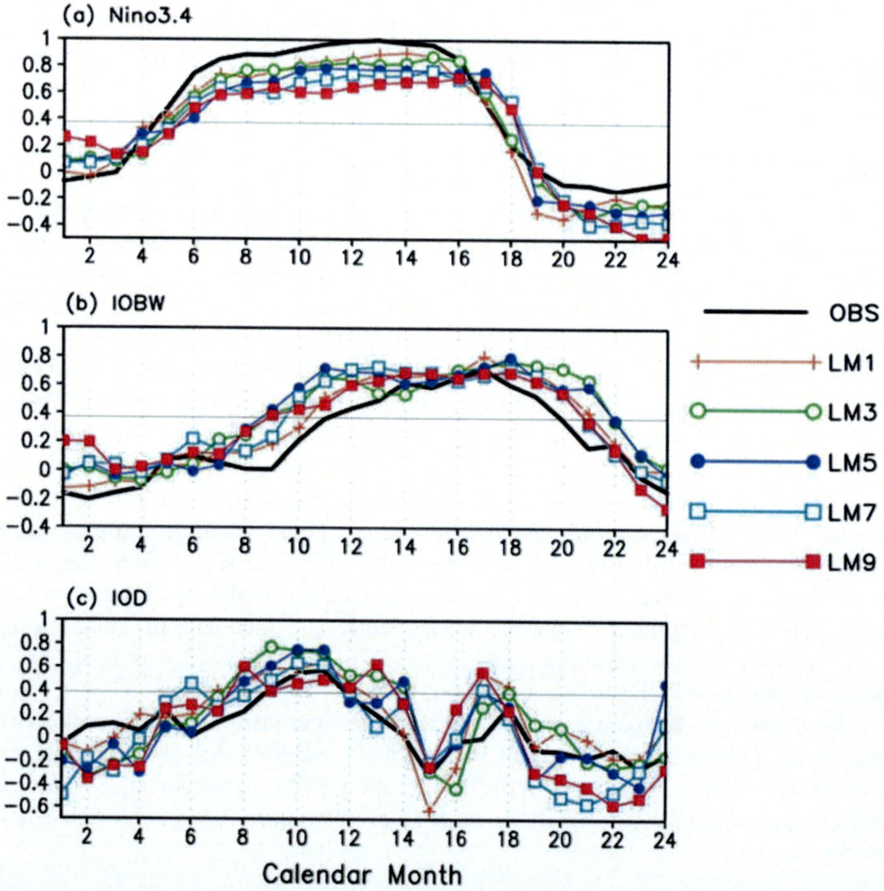

Fig. 8.3 Correlations of observed DJF (0/1) Niño3.4 SST index with **a** monthly Niño3.4 SST index, **b** monthly IOBW index, and **c** monthly IOD index during the ENSO developing and decay years for the OI SST (black solid line and hindcasts of different time leads (color lines). After Jiang et al. (2013c)

total variance, respectively. This mode shows basin-wide warming, which is called the IOBW. The CFSv2 cannot well capture the spatial pattern of SST anomalies even for the prediction of 0-month lead, with a spatial correlation coefficient of only 0.31. However, the prediction of 0-month lead can capture the temporal variability very well, with a correlation coefficient of 0.93.

As the IOBW is linked to ENSO, we checked their relationship in the CFSv2 predictions. The 0-month-lead prediction captures the correlation between ENSO and IOBW very well (Fig. 8.3b). It also captures the seasonal variation of the IOBW magnitude: strong in summer and weak in winter, but overestimates its magnitude in some seasons. The CFSv2-predicted duration of IOBW associated with ENSO is longer than the observation when lead time is longer than one month (Fig. 8.3b).

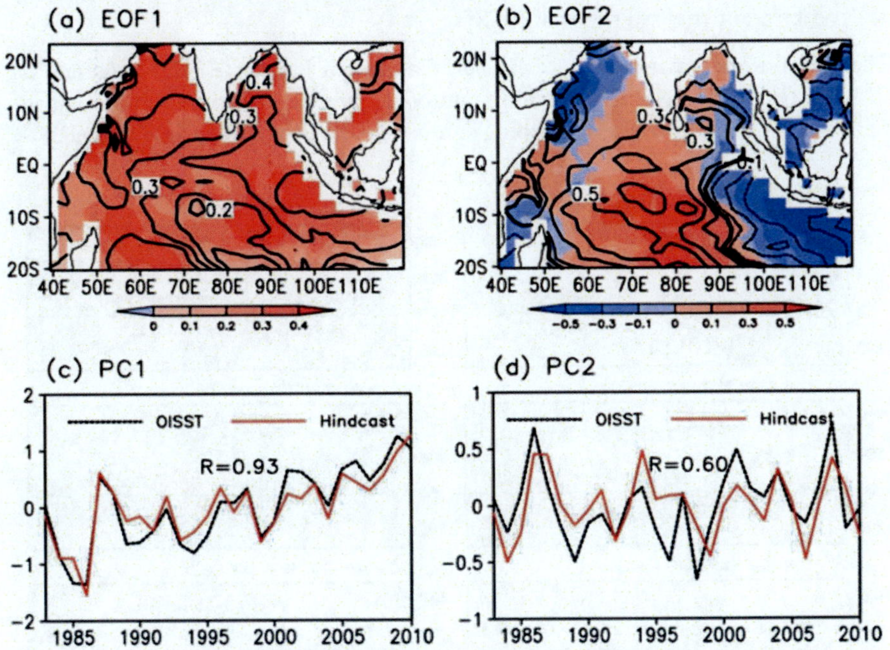

Fig. 8.4 **a** Spatial patterns and **c** principal components of the first mode of an EOF analysis of the temporal covariance matrix of JJA-averaged SST from the OI SST (shadings for spatial pattern and dashed lines for principal component) and from 0-month lead hindcast (contours for spatial pattern and solid lines for principal component). **b** and **d** are the same as (**a**) and (**c**), except for the second mode. After Jiang et al. (2013c)

Besides, it overestimates the IOBW magnitude in winter and spring due to its overestimation of ENSO. However, the CFSv2 underestimates the magnitude of IOBW from July to October, which is more obvious as lead time increases.

3. Prediction of the Indian Ocean dipole

The second mode of the EOF analysis of the temporal covariance matrix of observed JJA-averaged SST is characterized by a dipole pattern, with warm SST anomaly (SSTA) in most central and western IO but cold SSTA in the western IO, which is called the IOD and accounts for 11% of the total variance (Fig. 8.4b). The 0-month-lead prediction captures this mode very well in terms of both spatial and temporal variability, with correlation coefficients of 0.74 and 0.60, respectively (Fig. 8.4b, d). The IOD is significantly correlated with ENSO during its developing autumn, which is captured by the 0-month-lead prediction (Fig. 8.3c). As for magnitude, the CFSv2 overestimates the IOD in August and September, but underestimates the IOD in spring. The CFSv2 generally overestimates the relationship between IOD and ENSO. These biases become larger with increasing lead time. Note that the CFSv2 cannot capture the anomalies of wind and rainfall associated with IOD even for 1-month-lead prediction.

4. Prediction of the two types of ENSO

They are basically two flavors of ENSO: the eastern Pacific (EP) El Niño and CP El Niño (Fig. 8.5). The EP El Niño features a significant warming in the equatorial Pacific Ocean from the developing summer to the decaying spring, although the

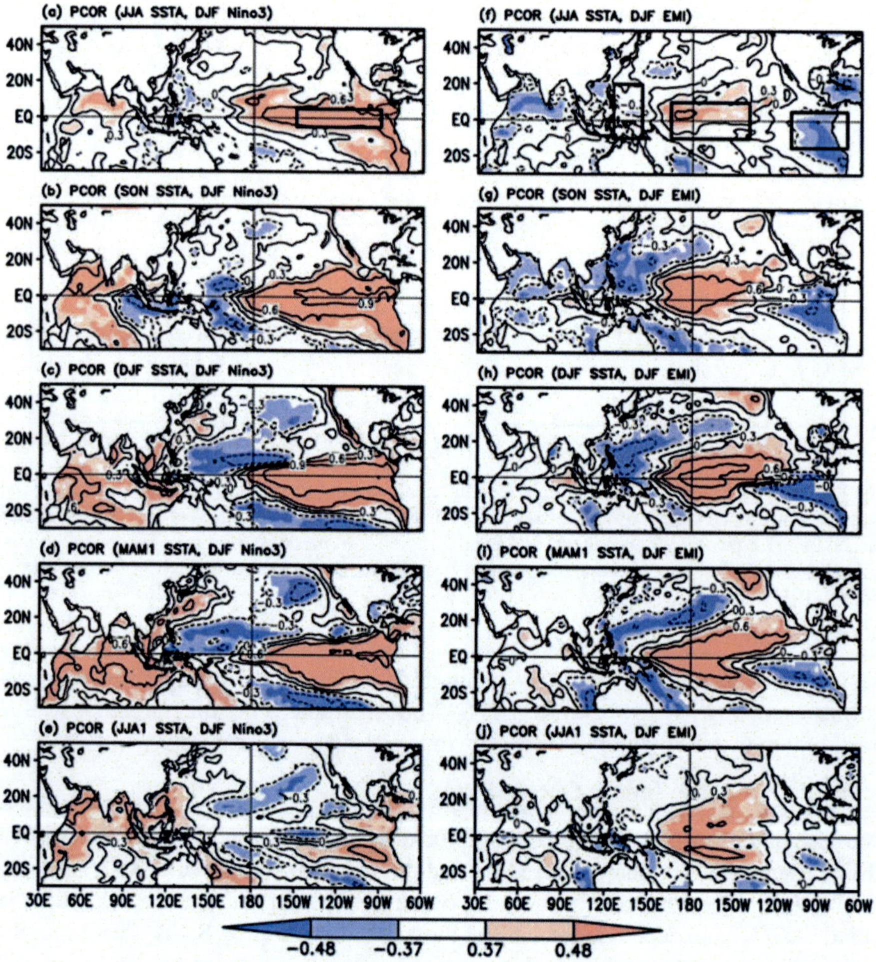

Fig. 8.5 Partial correlations of observed seasonal SSTA with observed normalized DJF Niño3 (left panels) and DJF EMI (right panels). **a** and **f** are for summer; **b** and **g**, for autumn; **c** and **h**, for winter; **d** and **i**, for the following spring; and **e** and **j**, for the next summer. Shadings indicate the significant correlations above the 95% and 99% confidence levels. The domains used to define the Niño3 and EMI are outlined in (**a**) and (**f**), respectively. Niño3 index is defined by the SSTA averaged over (90°–150° W, 5° S–5° S). EMI is defined as [SSTA]C-0.5[SSTA]E-0.5[SSTA]W, where [SSTA]C, [SSTA]E, and [SSTA]W stand for the area-mean SSTA over the central (C: 165° E–140° W, 10° S–10° S), eastern (E: 110°–70° W, 15° S–5° S), and western (W: 125°–145° E, 10° S–20° S) Pacific, respectively (Ashok et al. 2007). After Yang and Jiang (2014)

warming is small and even partly becomes cooling in the decaying summer. On the other hand, a negative SSTA emerges from the western Pacific during the developing summer, forming a cold "boomerang" (Trenberth and Stepaniak 2001), reaches its peak in winter, and persists to the decaying summer. Differently, the CP El Niño is characterized by a significant warming over the central Pacific Ocean from the developing summer to the decaying summer, with cold SSTA to the west and southeast of the warming center (Fig. 8.5).

The cold SSTA related to El Niño in the WNP is important for the maintenance of the anomalous anticyclone over the WNP from the El Niño developing autumn to the following early-summer, which exerts a significant impact on the East Asian climate (e.g., Wang et al. 2000, 2002). The strength and location of SSTA in the WNP are different between the two types of El Niño, resulting in different impacts on the East Asian climate (Weng et al. 2007; Yuan et al. 2012). Note also that the SSTA patterns in the tropical IO are different between the two types of El Niño. The tropical IO SSTA shows a dipole pattern in JJA and SON, and a basin-wide warming pattern from DJF to JJA1 during EP El Niño. Significant negative SSTA is found in part of the IO in JJA and SON, and positive SSTA, in most parts of the IO from DJF to JJA (1) during CP El Niño. However, the positive SSTA over the IO during EP El Niño is larger than those during CP El Niño.

The CFSv2 captures the spatial patterns of SST related to EP and CP El Niño, as well as their differences several months in advance (Yang and Jiang 2014). Prediction skills are different for different phases of El Niño cycle. Figure 8.6 indicates that the CFSv2 has higher skill in predicting the two types of El Niño indices (the EMI and Niño3; see caption of Fig. 8.5 for details) compared to the persistency prediction based on observed SST in JJA and SON. The model has even higher skill in DJF in terms of long lead predictions. The persistency of the EMI index is generally higher than that of the Niño3 index in JJA and SON. The magnitude of SST anomalies related to the DJF EMI is generally underestimated by the CFSv2.

The two types of El Niño have different impacts on the East Asian climate from the developing summer to the decaying summer. The CFSv2 captures the anomalous climate patterns related to the two types of El Niño and their differences, consistent with its skill in predicting SST anomalies. However, it overestimates the areas where the climate is affected by El Niño, but underestimates the magnitude of climate anomalies related to El Niño. The model also shows lower skill in predicting climate impacts of CP El Niño.

8.2.3 Prediction of the Western Pacific Subtropical High

The WPSH plays a key role in the climate variability over East Asia. The ridgeline of the WPSH is located between 18 and 20° N from February to April, when conditions of low temperature, cloudy, and persistent rainfall occur in southeastern China. The ridgeline moves northward to 20°–25° N in June, corresponding to the Meiyu in the Yangtze River and the Changma in Japan. In July, it moves further northward and the

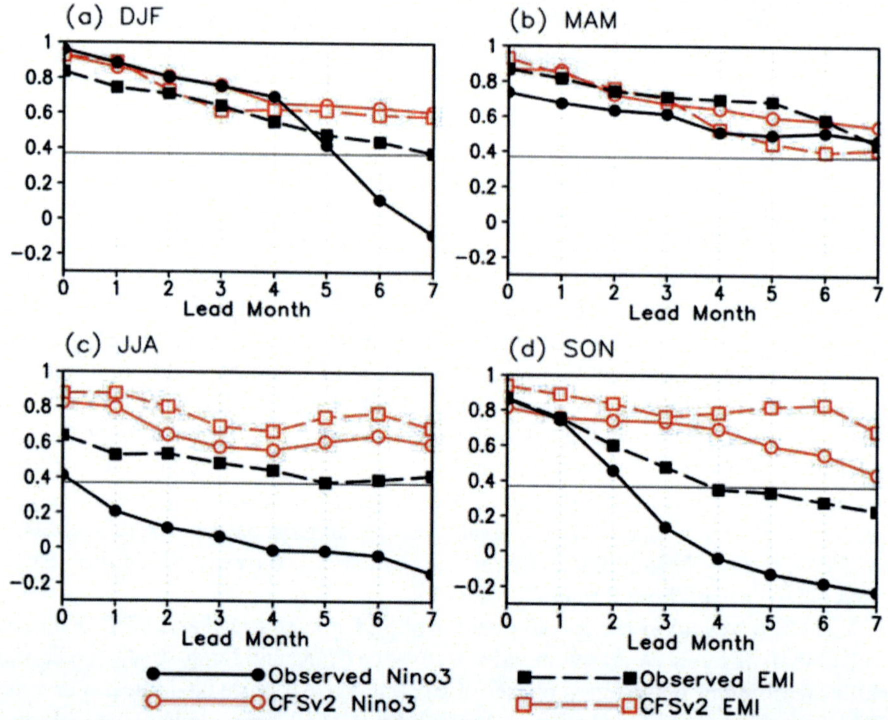

Fig. 8.6 Lagged auto-correlation coefficients between observed seasonal mean SSTA and observed monthly SSTA leading the seasonal mean SSTA (black lines; see text for details), and correlation between observed seasonal mean SSTA indices and CFSv2 hindcast seasonal mean SSTA indices of different leads (red lines). Values are shown for both Niño3 (lines with circles) and EMI (lines with squares). Solid straight line denotes the 95% confidence level. After Yang and Jiang (2014)

rain belt moves from the Yangtze River to the Huanghe and Huaihe. In the meantime, the Yangtze River is under the center of the WPSH, resulted in hot weather and less rainfall in the region. If such conditions persist for a long time, a severe drought will occur. In late-July and early-August, the ridgeline moves to the north of 30° N, and the rain belt is located over northeastern China. In early-September, the ridgeline moves northward quickly, so does the rain belt. Therefore, the study on the variability of the WNPSH received much attention in the past.

The interannual variability of the WPSH is linked to SST anomalies over both the Pacific Ocean and the IO. During the El Niño decaying summer, the WPSH strengthens and shifts southward, resulted in a low-level anomalous anticyclone over the WNP. Note that the location of the anomalous anticyclone changes from June to August, showing a northward movement (Fig. 8.7). In June, cold SSTAs are found over the WNPSH, which are replaced by warm SSTAs in July and August. The upper-tropospheric temperature anomalies are basically zonal symmetric in June, but exhibit a structure of Rossby wave over the western IO and of Kelvin wave over the eastern IO and the WNP in July and August. Thus, the anomalous anticyclone in

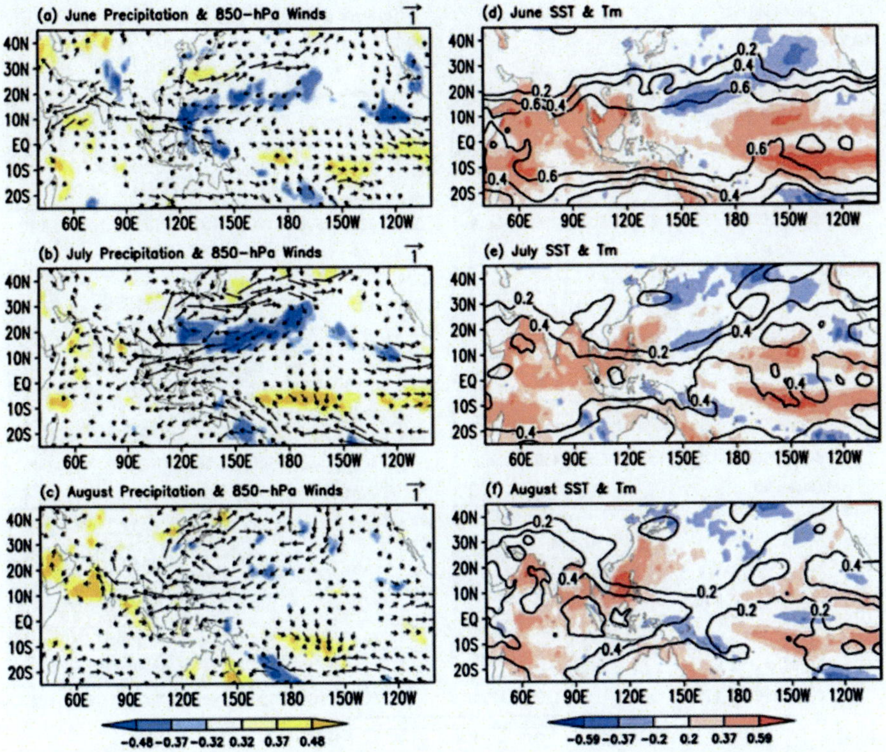

Fig. 8.7 **a** Observed correlation patterns between DJF(0/1) Niño3.4 SST index and June(1) precipitation (shading), and regression of June(1) 850 hPa wind (m s^{-1}; vectors) against DJF(0/1) Niño3.4 SST index; **d** observed correlation patterns of DJF(0/1) Niño3.4 SST index with June(1) SST and mean temperature of 500 and 300 hPa. **b** and **e** are the same as **a** and **d**, except for July(1); **c** and **f** are the same as **a** and **d**, except for August(1). Wind vectors with speed smaller than 0.2 m s^{-1} are omitted. After Jiang et al. (2013c)

June is mostly forced by cold local SST (Wu et al. 2010), while it is mostly forced by SSTAs over the tropical IO (Xie et al. 2009).

The CFSv2 can predict the observed variation of WPSH during El Niño decaying summer (Jiang et al. 2013c). It shows the highest skill for prediction of June anomalous anticyclone, which is forced by the cold local SSTAs via suppressing local rainfall. It can only predict the anomalous anticyclone in July and August when the lead time is less than four months. Thus, CFSv2's skill in predicting the anomalous anticyclone depends on its formation mechanism (Fig. 8.8).

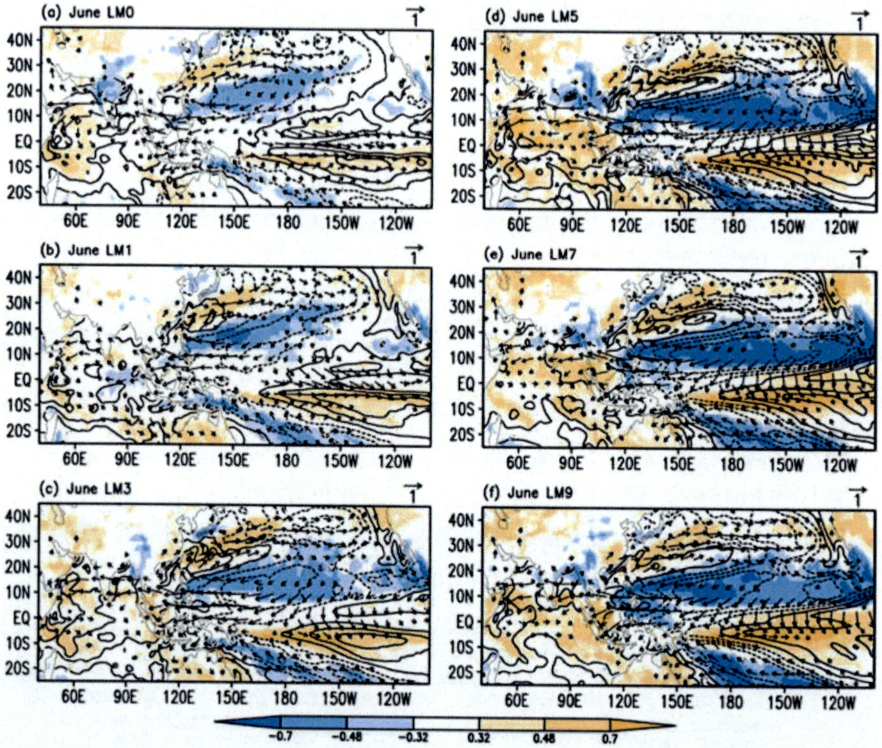

Fig. 8.8 Correlation patterns of observed DJF(0/1) Niño3.4 SST index with June(1) precipitation (shading) and SST (contour; values of the contours are −0.7, −0.48, −0.2, 0.2, 0.48, and 0.7), and regression patterns of June(1) 850 hPa wind (m s^{-1}; vector) against DJF(0/1) Niño3.4 SST index. **a**–**f** are for hindcasts of 0-month lead, 1-month lead, 3-month lead, 5-month lead, 7-month lead, and 9-month lead, respectively. Correlation coefficients of 0.32 and 0.48 are, respectively, corresponding to the 90% and 99% confidence levels for the *t*-test. Wind vectors with speed smaller than 0.2 m s^{-1} are omitted. After Jiang et al. (2013c)

8.2.4 Prediction of MC Rainfall and Related Atmospheric Circulation

1. Variation of MC rainfall and its relationships with large-scale climate anomalies

The MC lies within the Indo-Pacific warm pool, consists of multiple islands of SEA (e.g., Sumatra, Borneo, and New Guinea) and adjacent oceanic areas (Ramage 1968). In climatology, a rain band is located over the MC from November to April, and convergence is observed at low level over the region. In May, rainfall over the MC decreases, with the development of southwesterlies over the northern IO and East Asia. In October, the rain belt gradually returns to the MC, accompanied by the establishment of the Siberian high and reversal of the cross-equatorial flow. Thus, the annual evolution of MC rainfall is clearly characterized by a wet season and a dry season. According to the monthly evolution of rainfall over the MC (Fig. 8.9), we define July–October (December–March) as the dry (wet) season. The wet season of a particular year refers to December of the previous year and January–March of the current year. For example, the wet season of 1984 refers to December 1983 and January–March 1984.

An index is defined as the average rainfall over the MC (95°–145° E, 15° S–10° N), to measure the interannual variability of rainfall over the region. Rainfall over the MC in both dry and wet seasons shows obvious interannual variability, especially in the wet season. In the wet (dry) season, the MC rainfall reaches a maximum of about 9.9 mm day^{-1} in 1999 (7.4 mm day^{-1} in 2010) and a minimum below 5 mm day^{-1} in 1983 (3.2 mm day^{-1} in 1997). Interannual variation of MC rainfall in the dry season is significantly (insignificantly) related to the following (previous) wet season's MC rainfall. The correlation coefficient between the dry season rainfall and the following (previous) wet season rainfall is 0.64 (−0.07).

In the wet season, when the MC rainfall is increased, warm SSTAs appear from the eastern MC to the western North and South Pacific oceans and cold SSTAs appear over the IO, western MC, and the central–eastern tropical Pacific (Fig. 8.10a). As a Rossby response to the enhanced heating over the MC, there are anomalous cyclonic circulations over the southern IO and the northern IO, respectively, accompanied by westerly anomalies from the equatorial IO to the west of the MC (Fig. 8.10). Anomalous westerlies and cyclonic circulation appear over the equatorial Pacific and WNP, respectively (Fig. 8.10a). Low-level winds form an anomalous convergence over the MC, favoring heavy rainfall over the region (Fig. 8.10c). The variation of MC rainfall is significantly related to the variation of rainfall over East Asia and North America (Fig. 8.10c).

In the dry season, positive anomalous MC rainfall is related to positive SSTAs extending from the MC to the western South Pacific and to negative SSTAs over the central–eastern tropical Pacific, and to westerly (easterly) anomalies over the

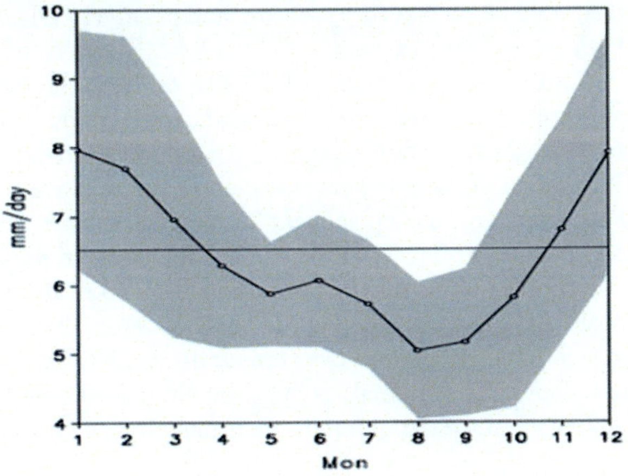

Fig. 8.9 Observed monthly mean MC rainfall (mm day^{-1}; black line) and its standard deviation (shading) from January to December. The horizontal line denotes the annual average MC rainfall. After Zhang et al. (2016a)

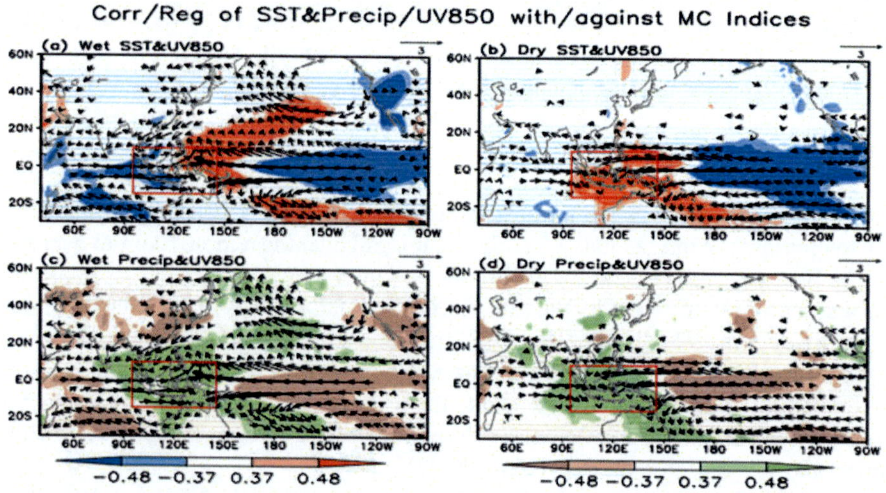

Fig. 8.10 Correlation (shading) of SST (OISST)/rainfall (GPCP) with MC rainfall indices, and regression (vector) of 850 hPa wind (NCEP CFSR) against MC rainfall indices in wet season (left panels) and dry season (right panels). Correlation coefficients exceeding the 95% and 99% confidence levels are shaded, and only the regressed winds significantly exceeding the 95% confidence level are drawn. The domain of the MC is outlined by red box. After Zhang et al. (2016a)

equatorial IO and the eastern Pacific (western–central tropical Pacific) (Fig. 8.10b). The southwesterly flow from India to the Indo-China Peninsula weakens but the cross-equatorial flow over the IO strengthens, accompanied by enhanced rainfall over the Arabian Sea and the Bay of Bengal (Fig. 8.10b). On the other hand, the cross-equatorial flow over 120°–140° E weakens (Fig. 8.10b). These SSTA and low-level circulation patterns lead to positive rainfall anomalies over the MC (Fig. 8.10d). Overall, the interannual variation of the MC rainfall is closely linked to ENSO developing phase in the dry season, and to ENSO maturing-decaying phase in the wet season. Besides, the MC rainfall is negatively (positively) related to rainfall over East Asia in the wet (dry) season, and positively related to rainfall over Australia in both dry and wet seasons.

To separate the effect of ENSO on MC rainfall from that of other ocean-atmospheric anomalies on MC rainfall, we conducted a partial correlation analysis between the MC rainfall index and another variable (SST, rainfall, or 850 hPa wind). When the ENSO influence is excluded, the negative SSTAs over the tropical IO and Pacific Ocean almost disappear for both wet and dry seasons, but the anomalous easterly (westerly) flow over the east (to the west) of the MC still exists, suggesting an independent effect of MC rainfall on the Walker circulation. For the wet season, the positive SSTAs over the Pacific north of 20° N disappear and the strong anomalous anticyclonic circulation over the North Pacific turns to a weak anomalous cyclonic circulation. The anomalous cyclonic circulations over the northern and southern IO and WNP still exist although their magnitudes become much smaller. The positive anomalous rainfall over northeastern Australia becomes much weaker and the associated anomalous winds diminish as well, but the warm SSTAs over the South Pacific become stronger. In addition, the negative rainfall-SST correlation over the western MC disappears, but the rainfall and winds over East Asia are still significantly related to the MC rainfall. These features suggest that in the wet season the rainfall east of Australia is mainly modulated by the ENSO instead of by local SSTAs, and variation of rainfall over the western MC is partly related to the variation of the EAWM (Chang and Lau 1982; Chang et al. 2005a, b). On the other hand, in the dry season, a narrow band of positive SSTA is seen around the MC although with weaker magnitude, and the low-level anomalous winds over East Asia are stronger, suggesting a combined effect of ENSO and local SSTA on MC rainfall, and a close relationship between MC rainfall and the climate anomalies over East Asia.

2. Prediction of MC rainfall variation and related climate anomalies in wet and dry seasonsf

Prediction skills of MC rainfall variation and related climate anomalies in the wet and dry seasons are evaluated using the CFSv2 hindcast outputs. The annual cycle of MC rainfall is generally captured by the CFSv2 in 0-month lead, especially for May and July. However, the model overestimates the MC rainfall in August and September and underestimates the MC rainfall from October to the following June, leading to an overall underestimation in the magnitude of the annual MC rainfall. In the observation, heavy rainfall in the wet season lies within the MC, with maximum centers over the west of Sumatra, between Borneo and Java, and to the south of New Guinea (Fig. 8.11a). The westerly flow is dominant south of the equator from the IO to the MC, while the northeasterly flow prevails north of the equator (Fig. 8.11a). In the dry season, the rain band shifts to the north of the MC, accompanied by southeasterly wind over the southern MC, weaker southwesterlies over the northern MC, and development of the southwesterly monsoon wind over the IO (Fig. 8.11b). The CFSv2 generally captures the major features in both dry and wet seasons, such as the location of the rain band and the reversal of the monsoon circulation (Fig. 8.11c, d). However, the CFSv2 prediction of 0-month lead has dry bias over the region between Borneo and Java, and south of New Guinea in the wet season, and has wet bias near New Guinea in the dry season; the model also predicts weaker winds over the southwestern MC in 0-month lead (Fig. 8.11c, d). The interannual variations of MC rainfall in both dry and wet seasons are well predicted by the CFSv2 in 0-month lead, with correlation coefficient of 0.89 between observed and predicted MC rainfall indices in the two seasons (exceeding the 99% confidence level). However, the wet season MC rainfall is generally underestimated after year 2000. Relationships between MC rainfall variation and climate anomalies are generally well predicted by the CFSv2 in 0-month lead, except for an overestimation of ENSO influence.

The NCEP CFSv2 shows higher prediction skill for MC rainfall from August to October, and lower prediction skill during December–February, April, and June (Fig. 8.12a). As shown in Fig. 8.12b, correlation coefficients between observed and predicted MC rainfall exceeding the 95% confidence level for all leads in both dry

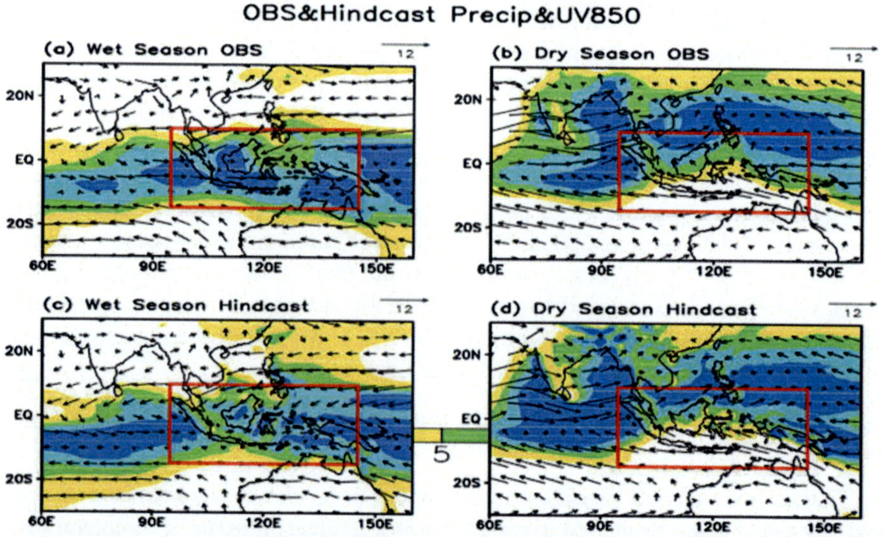

Fig. 8.11 Observed and NCEP CFSv2 (0-month lead) predicted rainfall (mm day^{-1}) and 850 hPa wind (m s^{-1}) in wet and dry seasons. The red box denotes the MC domain. After Zhang et al. (2016a)

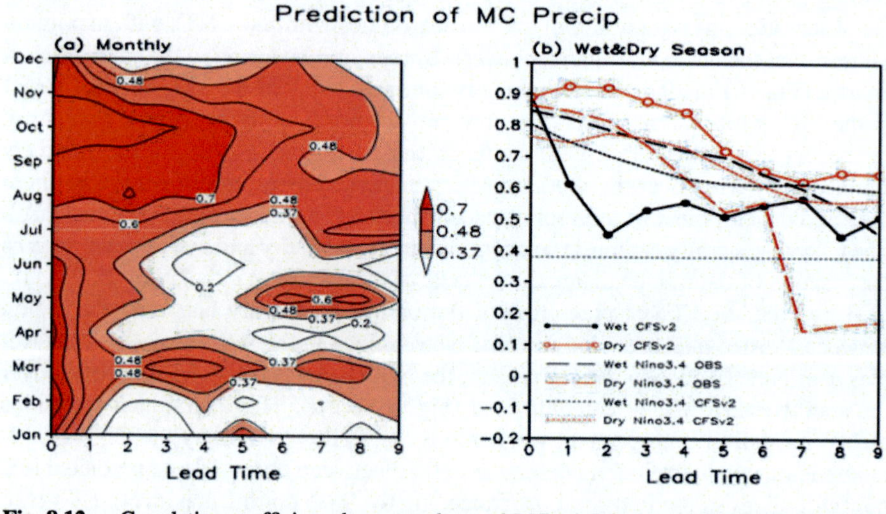

Fig. 8.12 **a** Correlation coefficients between observed MC rainfall and NCEP CFSv2-predicted MC rainfall in different lead times and target months. **b** Correlations of observed MC rainfall with CFSv2-predicted MC rainfall (solid lines with circles), observed Niño-3.4-predicted MC rainfall (dashed lines), and predicted Niño-3.4-predicted MC rainfall (dotted lines) with different lead months in wet season (black lines) and dry season (red lines). The x-coordinate indicates the corresponding lead months. Values significantly exceeding the 95% confidence level are shaded in (**a**), and straight dotted line in (**b**) denotes the 95% confidence level. After Zhang et al. (2016a)

and wet seasons, indicating high prediction skills of CFSv2 for the interannual variability of MC rainfall in the two seasons. In the dry season, correlation coefficient of the model decreases gradually from 1-month lead to 7-month lead, but correlation coefficient exceeds the 99% confidence level for all lead months (Fig. 8.12b). The skills of observed Niño-3.4 index and predicted Niño-3.4 index of the NCEP CFSv2 in predicting the MC rainfall are generally lower than the skills of the model (Fig. 8.12b). On the other hand, the CFSv2 skill in the wet season decreases rapidly from 0-month lead to 2-month lead, with obviously lower prediction skill than that in the dry season (Fig. 8.12b). The skills of both observed Niño-3.4 index and predicted Niño-3.4 index in predicting MC rainfall are generally higher than the skills of the model, suggesting the NCEP CFSv2 has a large bias in predicting the response of wet-season MC rainfall to ENSO (Fig. 8.12b).

Correlations of SST and rainfall (regression of 850 hPa wind) from the hindcast with (against) observed MC rainfall indices are calculated, to further analyze the model's skill in predicting the large-scale climate features that are related to MC rainfall variation. Overall, the prediction patterns do not vary significantly with lead time, and the major features of anomalous rainfall, SST, and 850 hPa wind are generally well captured in all leads by the NCEP CFSv2. However, in the wet season the relationship between MC rainfall index and the rainfall over western MC, especially around Sumatra and Borneo, is unrealistically predicted by the model. Correspondingly, the CFSv2 predicts a weaker anomalous cyclonic circulation over the IO, leading to underestimation of convergence over the western MC and the IO. This feature is consistent with that in Neale and Slingo (2003), indicating that the rainfall bias over the MC could be a driver for other systematic errors over the tropics through the Walker circulation (Fig. 8.13).

The MC rainfall variation is significantly related to ENSO in both observations and CFSv2 prediction. However, previous studies reported large biases of the relationship between the two in climate models (Luo et al. 2005; Jiang et al. 2013b). Therefore, we analyze the relationship between ENSO and predicted rainfall over the MC, focusing on regional features. In the dry season the anomalous patterns related to ENSO, including the anomalous westerly (easterly) wind over the equatorial IO (Pacific Ocean) and the positive rainfall anomalies over the entire MC region, are well predicted by the CFSv2 for all time leads (Fig. 8.12a, c, e, g). However, for the wet season, apparent errors of ENSO-related rainfall are seen around Sumatra, Malay, and Borneo after 1-month lead, and these errors even worsen as lead time increases (Fig. 8.12b, d, f, and h). The relatively lower prediction skill for the wet-season MC rainfall is partly caused by the large bias over Sumatra, Malay, and Borneo, with insignificant correlation between observed and predicted rainfall over Sumatra, Malay, and Borneo after 1-month lead. Correspondingly, the positive correlation between negative Niño-3.4 SST and the rainfall over Sumatra, Malay, and Borneo in the wet season in the observation becomes negative in the predictions. This unrealistically predicted relationship between ENSO and the rainfall over Sumatra, Malay, and Borneo contributes to the lower prediction skill of the rainfall over Sumatra, Malay, and Borneo in the wet season.

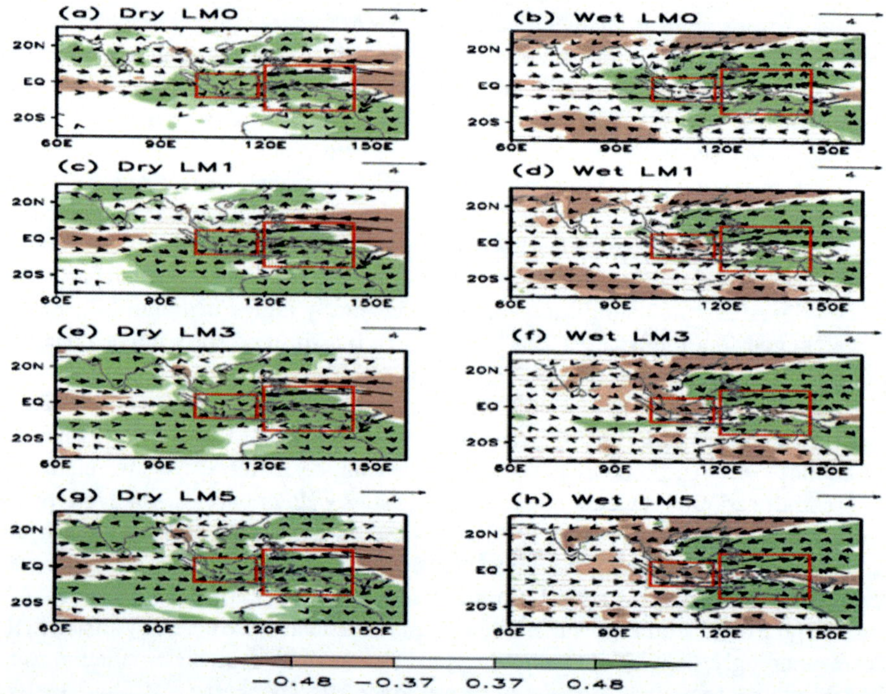

Fig. 8.13 Correlation of rainfall with negative Niño-3.4 index, and regression of 850 hPa wind against negative Niño-3.4 index in dry season (left panels) and wet season (right panels) for the NCEP CFSv2 hindcasts of 0-month lead (**a–b**), 1-month lead (**c–d**), 3-month lead (**e–f**), and 5-month lead (**g–h**). The correlation coefficients between rainfall and MC indices (shading) are plotted at the 95% confidence level, and only the regressed winds exceeding the 95% confidence level are drawn. The domains of the SMB region (100°–118° E, 8° S–5° S) and of the EMC region (120°–145° E, 15° S–10° S) are outlined by red lines. After Zhang et al. (2016a)

3. Impacts of local and remote forcing on the variation of regional MC rainfall

From the above analysis, we have learned that the NCEP CFSv2 has distinctive prediction skills for rainfall over the western MC (WMC) and eastern MC (EMC) with higher skill for the EMC rainfall, especially in the wet season. Here, we consider the WMC (95°–120° E, 13° S–7° S) and the EMC (120°–145° E, 11° S–10° S) separately. From Fig. 8.14a and b, the NCEP CFSv2 has generally high skills for the entire MC rainfall in the dry season, but relatively lower skills in predicting the rainfall over the MC in the wet season, especially for the WMC rainfall. The skill of the WMC rainfall is high from July to November, during which the CFSv2 can well predict the WMC rainfall longer than eight months in advance (Fig. 8.14c). However, the model shows lower skill in predicting the WMC rainfall from December to June (only 1-month lead Fig. 8.14c). The CFSv2 also shows different skills in predicting the EMC rainfall for the dry and wet seasons, but the difference is not as remarkable as for the WMC (Fig. 8.14). In comparison, the model shows slightly higher prediction

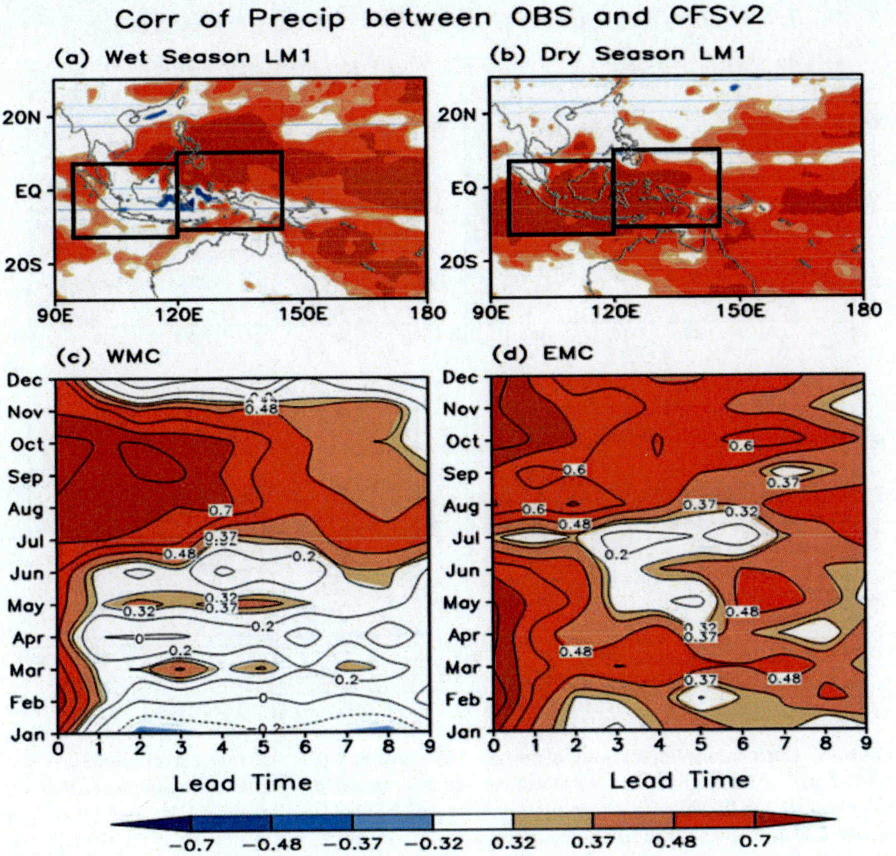

Fig. 8.14 Correlations of rainfall (mm/day) between observations and CFSv2 in 1-month lead for **a** wet season and **b** dry season, and correlations between observations and CFSv2 predictions for area-averaged rainfall for **c** WMC and **d** EMC. Values exceeding the 90%, 95%, and 99% confidence levels are shaded. The domains used to define WMC (95°–120° E, 13° S–7° S) and EMC (120°–145° E, 11° S–10° S) are outlined with black boxes in (**a**) and (**b**), respectively. After Zhang et al. (2016b)

skill for the WMC rainfall than for the EMC rainfall in the dry season, but much lower prediction skill for the WMC rainfall than for the EMC rainfall in the wet season (Fig. 8.14).

Variation of MC rainfall can be significantly modulated by ENSO and local SST (e.g., Hendon 2003; Chang et al. 2004; Zhang et al. 2016a, b, c), but how the ENSO and local SST affect the regional rainfall variation in different seasons remains unclear. As shown in Fig. 8.15a and c, variations of both WMC rainfall and EMC rainfall are closely linked to ENSO, and there is large resemblance between the anomalous patterns related to the WMC rainfall and EMC rainfall variations (Fig. 8.15a, b). However, there are also distinctive regional features, with positive local surface temperature (ST)-rainfall relationship over the EMC while negative

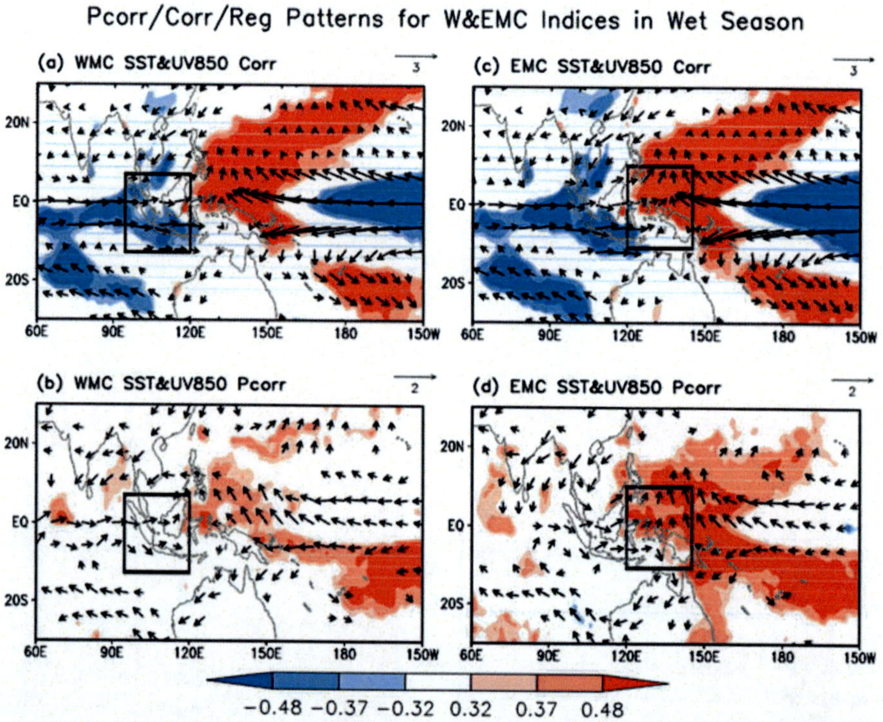

Fig. 8.15 Correlations (regressions) of observed ST (850 hPa wind) with (against) observed **a** WMC rainfall and **c** EMC rainfall indices in wet season. Also shown are partial correlations of observed ST (shading) and 850 hPa wind (vector) with observed **b** WMC rainfall and **d** EMC rainfall indices in wet season. Values (shading) exceeding the 90, 95, and 99% confidence levels are shaded, and only the regressed winds (vector) exceeding the 90% confidence level are plotted. The domains of WMC and EMC are outlined with black boxes. After Zhang et al. (2016b)

local ST-rainfall relationship over the WMC (Fig. 8.15a, b). When ENSO influence is excluded, rainfall is uncorrelated with local ST over the WMC but still significantly correlated with local ST over the EMC (Figs. 8.15a, b). The WMC rainfall is significantly (insignificantly) correlated with the preceding Niño-3.4 indices (local ST), while the EMC rainfall is significantly correlated with both local ST and ENSO. The above features suggest that the WMC rainfall variation is mainly modulated by ENSO and not forced by the local ST, while the EMC rainfall variation is significantly modulated by both ENSO and local ST.

According to the above results, the performance of CFSv2 in predicting the relationships of MC regional rainfall variation with remote (ENSO) and local forcing tends to have an impact on the prediction skill for local rainfall variation. In the wet season, the local ST-rainfall relationship over the WMC in the model is opposite to that in the observation (Fig. 8.16). There is a generally insignificant correlation between the rainfall and local ST over the WMC region, but significant correlation appears over the WMC in the model when the lead time is longer than 0 months

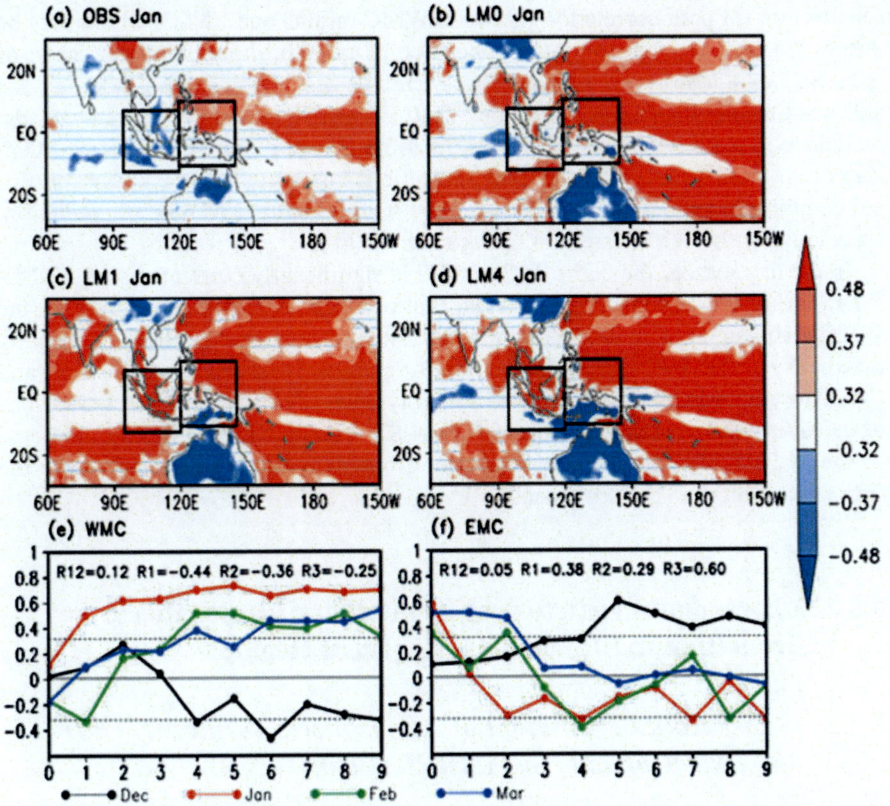

Fig. 8.16 Correlations between ST and rainfall in January for **a** observation, and hindcast ensemble mean of **b** 0-month lead, **c** 1-month lead, and **d** 4-month lead. Correlation coefficients between ST and rainfall over **e** WMC and **f** EMC in wet season months. R12, R1, R2, and R3 represent correlation coefficients in the observations for December, January, February, and March, respectively. Black, red, green, and blue lines represent correlation coefficients in hindcast ensemble mean of different leads for December, January, February, and March, respectively. Values significantly exceeding the 90%, 95%, and 99% confidence levels are shaded in (**a**)–(**d**). Dotted straight line in **e** and **f** denotes the 90% confidence level. The domains of WMC and EMC are outlined with black boxes in (**a**)–(**d**). After Zhang et al. (2016b)

(Fig. 8.16). The CFSv2 also unrealistically predicts the relationship between ENSO and rainfall variation over the WMC. The model overestimates (underestimates) the local ST-rainfall relationship over the EMC in December (in March), and unrealistically predicts the local ST-rainfall relationship over the EMC in January–February (Fig. 8.16). On the other hand, the ENSO-rainfall relationship over the EMC is well predicted by the model, contributing to much higher skills for the EMC rainfall compared to those for the WMC rainfall.

The problem of model's poor performance for the WMC rainfall during the wet season is not limited to the NCEP model (Zhang et al. 2016a, b, c). The complicated relationship (or poor correlation) between WMC rainfall and ENSO, which may be due to the lack of internal spatial coherence of rainfall, the rapid sign change of local SST anomalies once the wet season commences, and the complicated air-sea and wind-terrain interactions over the WMC during wet season, is difficult for the models to capture (Haylock and McBride 2001; Hendon 2003; Chang et al. 2004; Zuo et al. 2013, 2016). Thus, higher resolutions and better representation of air–sea coupling processes in the model are important for better MC rainfall prediction, especially for the WMC rainfall (Zhang et al. 2016b).

In the dry season, the entire MC rainfall is significantly correlated with ENSO. When the ENSO influence is excluded, the local ST-rainfall relationship over the WMC is still significant, but SST signals almost disappear over the EMC, suggesting that the dry-season WMC rainfall variation is significantly related to both local ST and ENSO forcing while the EMC rainfall is mainly affected by ENSO. The relationships of the entire MC rainfall variation with local ST and ENSO are both well predicted by the NCEP CFSv2 in all lead months, contributing to high prediction skills for the entire MC rainfall in the dry season.

8.3 Subseasonal Prediction of Rainfall and Atmospheric Circulation in SEA and Its Adjacent Regions

8.3.1 Subseasonal Prediction of Atmospheric Variability Over the Extratropical Northern Hemisphere

1. The Arctic Oscillation

The AO is the dominant mode of atmospheric variability over the extratropical Northern Hemisphere. It is usually defined as the first mode of an Empirical Orthogonal Function applied to the sea-level pressure north of 20° S. The AO exhibits out-of-phase variation of sea-level pressure between the polar region and the mid and high latitudes, which extends from the surface to the stratosphere (Thompson and Wallace 2000). The AO exerts significant impact on global climate. For example, a positive winter AO is accompanied by above-normal temperature over the high latitude of the Eurasian continent and most of North America, below-normal temperature over eastern Canada, North Africa, and Middle East (Thompson and Wallace 2000). The AO also plays an important role in rainfall variability over East Asia. A positive winter AO is associated with a northward shift of East Asian jet stream, a shallow East Asian trough, above-normal temperature from Siberia to northern China, Korea, and Japan (Chen and Kang 2006). The AO also affects climate in other seasons. Gong et al. (2001) reported that the spring AO could affect summer rainfall anomalies over the middle and lower reaches of the Yangtze River.

The prediction of AO by the Beijing Climate Center Atmospheric General Circulation Model version 2.2 indicates that the prediction skill for monthly AO exhibits strong seasonality (Zuo et al. 2016). The highest skill is found for winter AO (December, January, February, and March), and the lowest for autumn AO (August, September, October, and November). The prediction skill of 1-pentad lead by the model for monthly AO is higher than the persistent skill of observations except for June. However, the prediction of 2-pentad lead only exhibits skill for the AO in December and January. As for the daily AO, the model can predict it skillfully when the lead time is shorter than nine days, compared to the lead time of four days for persist prediction skill.

The variability of AO is associated with winter storm track and related synoptic eddies (Hurrell 1995; Ren et al. 2009), because the feedback of synoptic eddies maintains and enhances low-frequency atmospheric variability (Lorenz and Hartmann 2003). The synoptic eddies simulated by the BCC model are obviously weaker than the observation (Zuo et al. 2016). On the other hand, the downward propagation of stratospheric polar vortex anomalies is an important source for predicting climate anomaly in the troposphere (Sigmond et al. 2013; Tripathi et al. 2015). The prediction by the BCC model indicates that it has a high skill for daily AO prediction when it is initiated on the weak phase of a polar vortex (Zuo et al. 2016). The model simulates a stronger-than-observed coupling between the troposphere and stratosphere. Therefore, improvement in the simulation of storm track activity, synoptic eddy feedback, and troposphere–stratosphere coupling in the Northern Hemisphere may enhance its skill for S2S prediction of winter AO. Besides, some studies reported that ensemble of large size could improve the prediction of AO (Riddle et al. 2013).

2. The Northern-Hemisphere Blocking

The blocking is one of the most important atmospheric phenomena in the extratropics. The formation and collapse of the blocking are accompanied by significant changes in atmospheric circulation over a very large region. The blocking is quasi-stationary and sometimes moves westward. The life cycle of the blocking is normally 5–7 days, but sometimes is more than 20 days. Due to its quasi-stationary and long life cycle, the blocking can cause extremely cold conditions in winter, heat waves in summer, and severe flooding. The blocking affects climate not only over the regions it covers, but downstream and upstream as well.

Jia et al. (2014) reported that the CFSv2 captures the observed features of longitudinal distribution of blocking activity, but underestimates blocking frequency and intensity. The intensity and number of blocking decrease as lead time increases. The frequency of blocking is similar to the climatology when the lead time is shorter than three weeks. The number of blocking over the Pacific sector is slightly higher than that over the Euro-Atlantic sector in the observation. Within 14-day lead time, the prediction skill for blocking over the Euro-Atlantic sector is higher than that over the Pacific. Taking the hit rate of 50% as a criterion (Fig. 8.17), the CFSv2 can predict the blocking over the Euro-Atlantic sector nine days in advance, but it is only seven days over the Pacific sector. The prediction skill for blocking over the Euro-Atlantic sector is lower than that for the Atlantic sector when the lead time is longer than two weeks (Fig. 8.17).

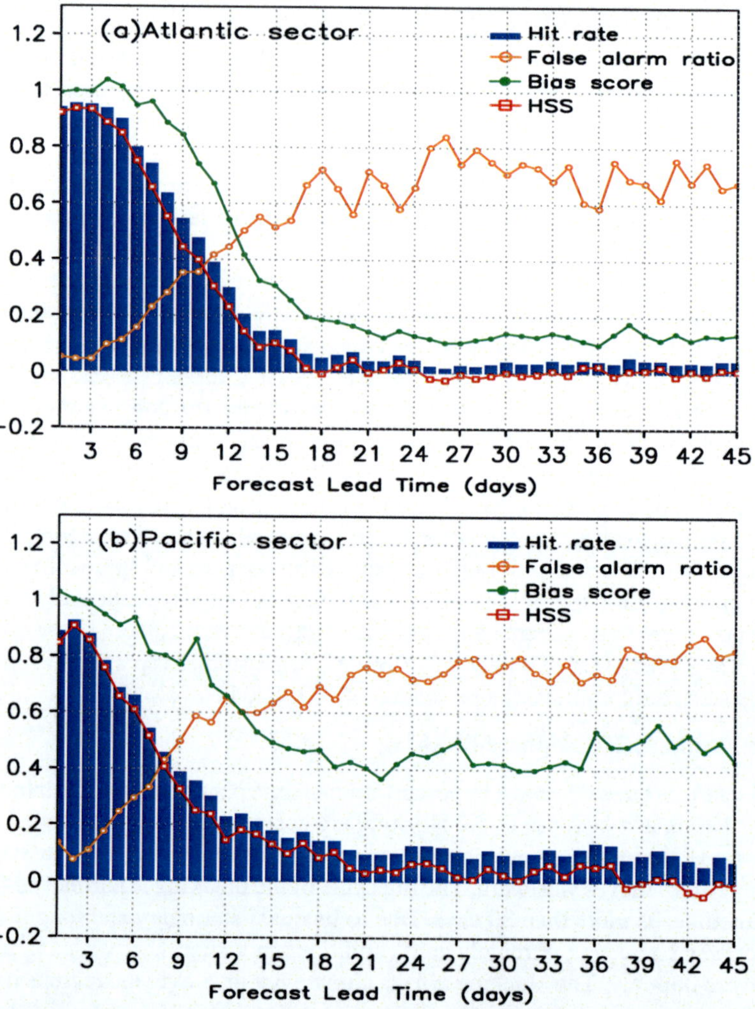

Fig. 8.17 Skill scores for sector-blocking forecast in **a** the Euro-Atlantic sector and **b** the central Pacific sector by CFSv2 with lead times from 1 to 45 days. After Jia et al. (2014)

The predicted sector-blocking episodes decrease as lead time increases. The forecast skill of sector-blocking episodes is slightly higher for the Euro-Atlantic sector than for the Pacific sector. In the Euro-Atlantic sector, more than 50% of sector-blocking episodes can be predicted up to eight days, whereas it is seven days for the Pacific sector. The model has little skill in predicting sector-blocking episodes in both sectors when the lead time is longer than two weeks. The CFSv2 can well predict block onset when the lead time is shorter than five days.

For the Euro-Atlantic sector, when the lead time is shorter than about 10 days, the tropics and subtropics are dominated by negative geopotential height biases, while the extratropical and polar regions are dominated by positive geopotential height biases, which may play an important role in the maintenance of relatively high blocking frequency. On the other hand, the mid-to-high latitudes in the North Pacific sector are dominated by weak negative geopotential height biases, which is unfavorable for maintaining high blocking frequency. Note that CFSv2 shows low skill in predicting wave train patterns associated with the blocking.

3. Cold Surge

Cold surges are the most conspicuous weather events during winter in East Asia. During the occurrence of cold surge, cold air breaks out and moves southward, resulting in severe weather events such as a large temperature drop, strong wind, and heavy freezing rain or snowfall over East Asia. Previous studies investigated the characteristics of cold surges and associated large-scale circulation (Chang and Lau 1980, 1982; Ding and Krishnamurti 1987; Jeong et al. 2006), which focused on the factors that affect the occurrence of cold surge and addressed the influence of large-scale climate phenomena on cold surge. The variability of cold surges is associated with the Siberian high, the AO, ENSO, and the MJO (Jeong and Ho 2005; Takaya and Nakamura 2005; Hong et al. 2008; Park et al. 2010, 2014).

The assessment of winter cold surge prediction by CFSv2 indicates that the model can well predict the frequency, intensity, and location of cold surge at a lead time of about two weeks (Li et al. 2017; Fig. 8.18). When the lead time is longer than two weeks, the model predicts fewer-than-observed cold surge days. The evolution of tropospheric circulation and some surface variables associated with the cold surge occurrence are well captured by CFSv2. However, the model can only skillfully predict precipitation for the lead time less than one week. It has systematic wet biases over the South China Sea, the Philippine Islands, and the Northwest Pacific, and has dry biases over India, the Indo-China Peninsula, and most high-latitude regions. The differences in wave train-like patterns of geopotential height and wind associated with the cold surge occur on both sides of 35° S in the prediction is consistent with the observation. A weaker-than-observed Siberian high and weaker northerly wind over eastern China are found in the predictions, especially with longer lead times.

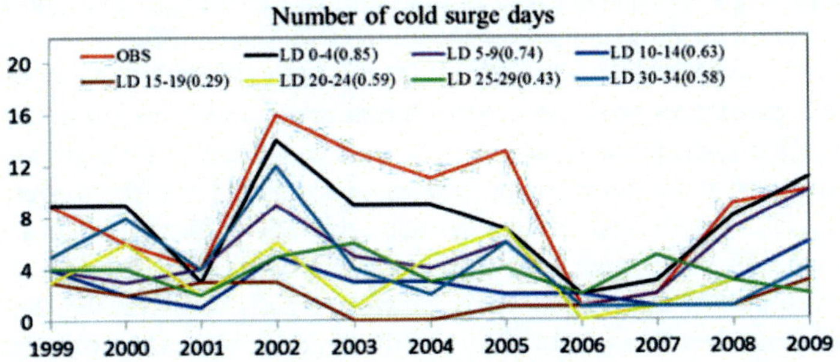

Fig. 8.18 Number of cold surge days in observations and CFSv2 predictions at different lead times. After Li et al. (2017)

Initial conditions are the key factor affecting the model's ability in predicting cold surges.

8.3.2 *Subseasonal Prediction of the Tropical Climate Variability*

The subseasonal prediction of precipitation and circulation in the ASM region is influenced by the intraseasonal evolution of summer monsoon, showing apparent differences in different stages of summer. By taking CFSv2 as an example, we investigated the bias characteristics of the 850 hPa monthly mean wind and precipitation predicted by the model at the beginning of each month (Liu et al. 2013). The results showed that the model basically captured the major evolution features of the ASM, including the eastward retreat of the WPSH and the onset of the ASM in May, the development and prevalence of the ASM in June and July, and the weakening and retreat of the ASM in August and September. These characteristics were similar to the observations in terms of spatial pattern, center location, and precipitation intensity. However, some systematic biases were obviously in model predictions, including the southerly wind bias near Somalia, the westerly wind bias over the eastern tropical IO, the easterly wind bias over the Indian Subcontinent, and the Indo-China Peninsula, the northerly wind bias over SEA, and the cyclonic wind bias over the Northwest Pacific, etc. Corresponding to these wind field biases, the eastern tropical IO had more precipitation while the Bay of Bengal, SEA, and the western edge of the WPSH had less rainfall. Besides, apparent wet biases were found over the south edge of the Tibetan Plateau, the Indian Subcontinent, and the west coast of the Indo-China Peninsula. The causes for these biases might be that, due to limited spatial resolution, the model could not properly reflect the impact of local topography. Note that the prediction biases always changed with time. For example, the easterly wind bias over the eastern tropical IO in April–May was replaced by westerly wind bias in June–August, and the bias center in the Northwest Pacific showed the feature of moving in the north–south direction, similar to that of the WPSH.

Further analysis was carried out, aiming at intraseasonal precipitation and wind fields predicted in different pentads of the summer season at different lead times (Liu et al. 2013). Their findings suggested that for the forecasts with 0–4 day lead times, although the prediction skill was higher, biases were still formed rapidly. Almost in the whole summer, predictions were featured by easterly biases over the South and Southeast Asian monsoon regions, cyclonic wind bias over the western Pacific, and less rainfall over the Indian Subcontinent, the Bay of Bengal, and SEA. When the lead time is increased to 35–40 days, the easterly bias over South Asia and the cyclonic wind bias over the western Pacific were greatly strengthened. At the same time, the wet bias over the equatorial IO and the cyclonic bias center over the western Pacific also developed obviously. However, the dry bias on the west side of the WPSH showed different developing trend, which strengthened in May–June but weakened in July–August with increasing lead time. In general, the model prediction for the ASM showed an increasing trend of wet bias but a decreasing trend of dry bias, although in some areas (such as the Indian subcontinent and the Bay of Bengal), the biases were not seen to increase noticeably.

Liu et al. (2013) evaluated the characteristics of monsoon forecast errors in summer with different lead times by using the ensemble prediction spread and root mean square error (RMSE). The results demonstrated that the growth of the ASM circulation and precipitation forecast biases depended notably on the developing stage of the predicted object and prediction lead time. The period from the end of June to the beginning of July was an obvious time shift point. The prediction biases of monsoon circulation and precipitation changed greatly around this time (Liu et al. 2013). To investigate the distributions of circulation factors for ASM prediction biases and their dependence on prediction lead time, the 850 hPa wind, precipitation, and surface temperature biases predicted in 0–40 day lead times were analyzed with the multivariate empirical orthogonal function (EOF) (Fig. 8.19; Liu et al. 2014). It was found that the first EOF mode embodied the typical distribution characteristics of the ASM biases, that is, the easterly bias over the Indian subcontinent and the Bay of Bengal; the cyclonic wind bias over the northwestern Pacific and the anticyclonic wind bias over the western tropical IO agreed with the situation with less rainfall over the Indian subcontinent and the Bay of Bengal but more rainfall over the western tropical Pacific and eastern IO. Meanwhile, the SSTs near the Bay of Bengal and Northwest Pacific coast were warmer. These features indicated that the predicted WPSH and the South Asian monsoon were weak comparatively. Characteristics of the first principal component (PC1) at each prediction lead time corresponding to this mode indicated that the bias of first EOF mode gradually increased from the end of May to the beginning of July, but decreased slowly from July to September. The evolution process matched with the developing, prevailing, and declining processes of the South Asian summer monsoon and WPSH. Changes of the second EOF mode and its principal component (PC2) denoted that an obvious time shift occurred in the prediction bias field, i.e., there was an anticyclonic wind bias over the Indian subcontinent and equatorial IO, and a tropical cyclonic wind bias and subtropical anticyclonic wind bias over the Northwest Pacific before the end of June. From July onwards, distributions of these biases gradually turned into opposite directions, and their corresponding precipitation bias and surface temperature bias were reversed as well. Moreover, these biases changed exactly at the time when the South Asian monsoon reached its peak, starting to decline, and the WPSH began to jump northward. Although the variance contribution of the second mode was smaller, the PC2 value attained its maximum when the PC1 was quite small in summer, so the prediction bias developing in the whole ASM region can be clearly reflected. In addition, different from the bias from the first mode, which grew rapidly below one-week lead time, the second mode bias basically remained unchanged, but began to increase gradually at the lead time over 10 days. This characteristic, to a certain extent, reflected the impacts of biases from the slowly varying part of the climate system. Overall, the above findings showed that the subseasonal prediction biases of the SAM were closely related to the development of the monsoon climate system,

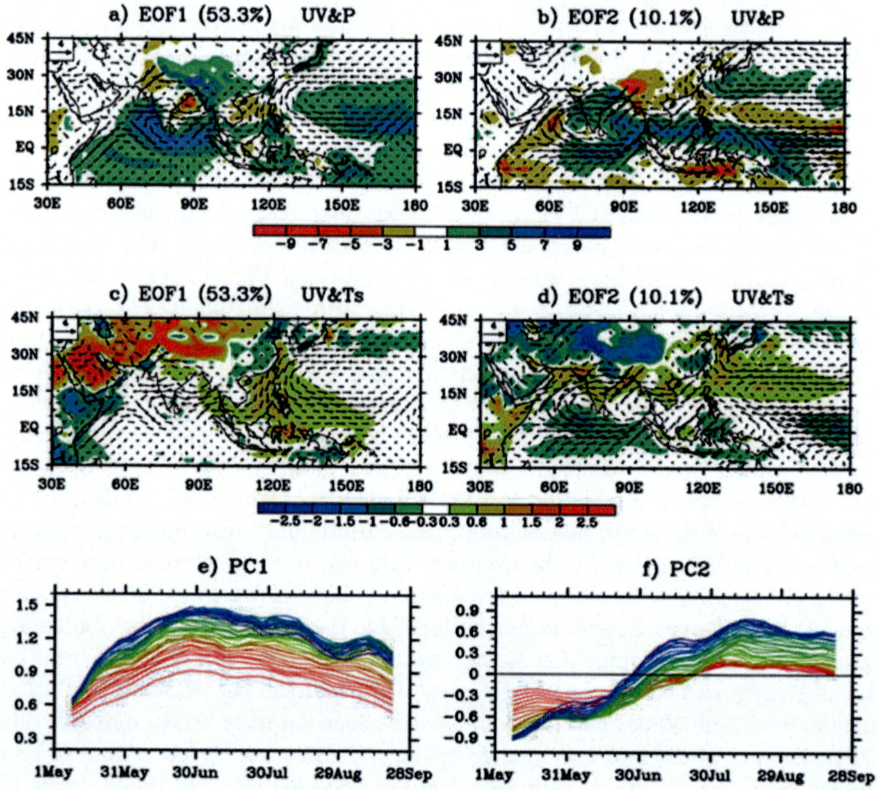

Fig. 8.19 Multivariate EOF analysis of the predicted pentad wind, precipitation, and surface temperature biases at 850 hPa during May–September **a, b** are wind and precipitation characteristics in EOF mode; **c, d** are wind and surface temperature characteristics in EOF mode, in which vectors represent wind, colored area represents precipitation or surface temperature; **e, f** are the principal components corresponding to modes, and the forecast lead times of 0–40 days are marked by the solid line from red to blue in order). After Liu et al. (2014)

causing clearly different bias features in different periods of the summer at different prediction lead times.

In terms of subseasonal variability of monsoon circulation and rainfall, the predictions also showed the features of large spatial differences and sensitive skills to the prediction lead time. Thus, we analyzed the ability of the model in predicting the temporal evolution features of the 850 hPa zonal wind from May to September using the NCEP CFSv2 reforecast outputs (Liu et al. 2014). The multiyear average results showed that the forecast skill of zonal wind at 850 hPa gradually decreased with increasing lead time. When the lead time was shorter than one week, the model showed high skill for prediction of circulation over the entire ASM region, the correlation coefficient of prediction and observation passed the 99% confidence test. When the lead time was longer than two weeks, the high-skill area was reduced apparently, and the highly correlated areas were located mainly in the western tropical IO, South

China Sea, and MC. And when the lead time was longer than three weeks, high correlation skills only existed in scattered areas near the MC. This suggested that the CFSv2 can predict atmospheric circulation over the Southeast Asian monsoon region skillfully with a lead time of about 2–3 weeks. In contrast, the prediction skill of precipitation decreased rapidly with increasing lead time. When rainfall forecast was made one week in advance, the prediction skills were no longer significant over most regions (Liu et al. 2014). This indicated that the precipitation bias grew faster than the circulation bias, so the upper limit of precipitation predictability was reached earlier. The main reason for this result was that the precipitation was restricted by the complex physical processes of cloud and convection in addition to the influence of circulation. Thus, the large uncertainty in rainfall prediction also means faster bias growth rate and lower predictability of precipitation.

The subseasonal predictability of atmospheric circulation and precipitation was further investigated in the key areas of Southeast Asian monsoon (100°–130° E, 5°–25° S). According to the definition by Wang and Fan (1999), the SEA monsoon index was calculated, and the prediction skill for the subseasonal variability of the SEA index was evaluated using the reforecast outputs by the CFSv2. It was found that the multiyear-averaged skill can be more than two weeks, but in some individual years the maximum skill can reach more than five weeks; and the lowest skill is only about 10 days. By analyzing the relationship between the SEA index and the circulation and precipitation, we learned that the enhancement of the SEA index usually corresponds to cyclonic convergence anomalies of local winds over this region, westerly anomalies from the Indian subcontinent to the Philippines, convection enhancement over the western tropical Pacific, and the convection inhibition over the MC. Moreover, as the prediction lead time increased, this relationship would gradually be replaced by the response features of the large-scale circulation, making it more and more difficult to reflect the localized response characteristics. Therefore, for those years when the monsoon index itself was closely correlated with large-scale circulation, the prediction skill was higher, while for the years when the monsoon index was only related to the local circulation, the prediction skill was lower. Thus, we can see that predicting precipitation over the Southeast Asian monsoon region is more difficult than predicting circulation in the region. For the subseasonal variability of regional average precipitation, CFSv2's multiyear-averaged forecast skill was about 10 days, while in individual years it could reach up to three weeks. For the spatial variability of regional precipitation, the average prediction skill was about two weeks, and the highest and the lowest prediction skills were about four weeks and 10 days, respectively (Liu et al. 2014). These results indicated that there are significant interannual differences in subseasonal prediction skills of the Southeast Asian monsoon, which may be affected somewhat by the signals of interannual variability.

A representative feature on the subseasonal scale is the tropical ISO, which is usually referred to as the MJO that propagates from west to east throughout the year in the equatorial region and as the boreal summer intraseasonal oscillation (BSISO) that prevails in summer in the Northern Hemisphere and propagates eastward and northward in the ASM region. Both MJO and BSISO are considered to be important sources of subseasonal predictability. Their predictability in different statistical and dynamic models is a hot issue for many operational centers and research institutions. In recent years, predicting ISO is regarded as one of the major targets of the S2S Prediction Project. For MJO, its forecast skill in most statistical prediction models is usually about two weeks, while that in some other models can exceed three weeks. Compared with the statistical model that focuses on predicting the characteristics of MJO index, the climate model is able to predict the temporal and spatial variation of MJO and to depict the physical relationship between MJO and other climate phenomena, and thus has become a main tool for the current MJO forecasting operation. According to the definition by Wheeler and Hendon (2004), multivariate EOF analysis was performed on the 15° S–15° S averaged outgoing longwave radiation, the 850 hPa zonal wind, and the 200 hPa zonal wind. The obtained first two EOF modes represent the spatial structure of the MJO, and the first two principal components (PCs) represent the time variation characteristics of the MJO. According to the distributions of the two PCs in the quadrant space, the MJO state can be divided into eight phases, and the cyclic evolution of these phases reflects the process of MJO's initiation, eastward propagation, and extinction in the equatorial region. The commonly used index for evaluating the MJO forecast skill of a model is the bivariate correlation coefficient and the RMSEs of the first two PCs between prediction and observation, in which the correlation coefficient decreases with the prolongation of lead time while the RMSE increases with the prolongation of lead time. When the former declines to 0.5 or the latter, to 1.414, the corresponding prediction lead time is considered to be the upper limit of the useful skill for the MJO (Lin et al. 2008). Based on this index, MJO prediction skill in the NCEP CFSv1 was only 10–15 days, while that in the CFSv2 was increased to 20–21 days (Kim et al. 2014; Wang et al. 2014). The MJO forecast skill in the Australian Climate Prediction System was 21 days (Rashid et al. 2011). In comparison, the MJO forecast skill by the ECMWF forecast system kept increasing in the last decade and has now reached 27 days (Kim et al. 2014; Vitart 2014). During the S2S Prediction Project, the prediction system based on the coupled BCC_CSM had an overall forecast skill of 15 days for MJO prediction. After improving initial atmospheric and oceanic conditions, the overall forecast skill of the MJO had been improved to 21 days (Fig. 8.20a; Liu et al. 2017). By analyzing the temporal variation of the improved prediction skills, we can see that the overall prediction skills have significant seasonal differences: the lowest in summer is about 17 days and the highest in autumn and winter can reach 26 days. In some historical periods, the skill even reached above 40 days. In addition, for each phase of the MJO, the forecast skill can be improved somewhat by improving the initial conditions, and especially the forecast skill for initial phase 2 has been improved from 16 to 27 days. This indicates that the improved initial conditions make the predicted propagation characteristics of the MJO over the IO better, although the model still has deficiency in portraying the MJO propagation across the eastern tropical IO and the MC.

The BSISO is a phenomenon active in the tropical to subtropical regions of Asia in summer. It is highly related to the activities of the ASM. According to the definition by Lee et al. (2013), we performed multivariable EOF analysis on the daily 850 hPa zonal wind and outgoing longwave radiation in the (40°–160° E, 10° S–40° S) region from May to October. The obtained 1st and 2nd EOF modes and their corresponding

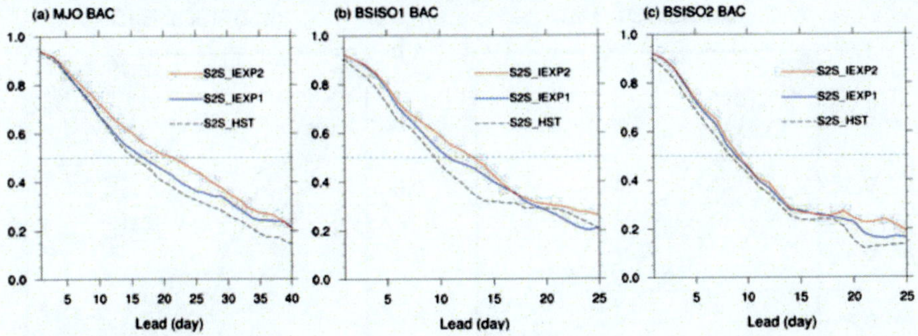

Fig. 8.20 Prediction skills of MJO, BSISO1, and BSISO2 by the BCC_CSM that participates in the S2S Project (S2S_HST, S2S_IEXP1, and S2S_IEXP2, respectively, represent the original hindcast experiment, the experiment to improve atmospheric initial conditions, and the experiment to improve atmospheric and ocean initial conditions. For the hindcast experiments, once snapshot every five days is taken from the period 2000–2013; horizontal dashed line is the upper limit of useful forecast skills, defined as prediction lead time corresponding to skill value equal to 0.5

PCs represent BSISO1, while the 3rd and 4th modes and their corresponding PCs represent BSISO2. The multiple models participating in the S2S Prediction Project show obvious differences in their BSISO predictions. The forecast lead time when the bivariate correlation coefficient of PCs between prediction and observation equal to 0.5 is taken as the maximum useful skill. For the BSISO1 whose oscillation period is about 30–60 days, prediction skills in most models range from one to two weeks, but the prediction skill in the ECMWF model is as high as 24 days, while that in the Russian Hydrometeorological Center model is only six days. For the BSISO2 with oscillation period about 10–30 days, the prediction skills in multiple models range from 6 to 14 days (Fig. 8.21; Jie et al. 2017). Focusing on the strong BSISO events, most models show that the forecast skills for BSISO1 and BSISO2 are relatively low when they are in phases 3–5, which indicates that the BSISO1 is difficult to predict when the convection signal propagates from the IO to the MC and the western Pacific, and that the BSISO2 is also difficult to predict when its convection center lies in the area from the Bay of Bengal to the South China Sea and the Northwest Pacific. For the BCC_CSM model used by the BCC for participating in the S2S Project, the overall prediction skills of BSISO1 and BSISO2 were initially nine and eight days, respectively; but after improving the initial conditions, the skills were improved to 13 and nine days, equivalent to the performance of most models in the S2S project (Fig. 8.20b, c). With the improvement of initialization scheme in the BCC_CSM, the most significant improvement in BSISO1 forecast skill is found in the 8th to the 2nd phase, that is, in the stage of convection weakening over the western Pacific and convection strengthening over the IO. The most significant enhancement of BSISO2 forecast skills is in the 6th to the 7th phase, that is, in the dry phase period when convective inhibition occurs mainly over the IO and the western Pacific.

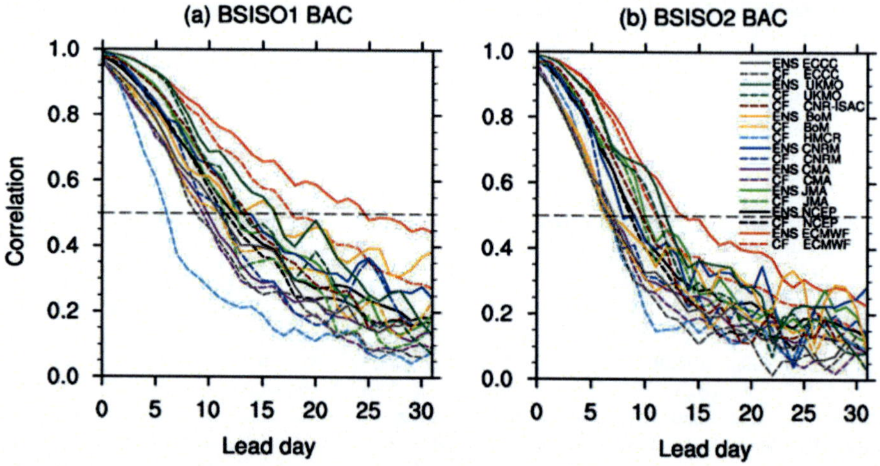

Fig. 8.21 Prediction skills of BSISO1 and BSISO2 by multiple models participating in the S2S Project (the skill is measured by bivariate abnormal correlation coefficient). For the hindcast experiments, twice a week snapshots are taken from May to October each year during 1999–2010. Colored dashed line represents the deterministic forecast; colored solid lines represent ensemble prediction results; horizontal dashed line is the upper limit of useful prediction skills, defined as the prediction lead time corresponding to the bivariate correlation skill equal to 0.5. After Jie et al. (2017)

8.3.3 *Subseasonal Prediction of the Climate Variation Over Southeast Asia*

1. Prediction of MC rainfall variation and associated circulation patterns in the wet-dry transitional seasons

The seasonal evolution of MC rainfall is clearly characterized by a wet season and a dry season, with asymmetric seasonal transitions (Li and Philander 1996; Hung et al. 2004; Chang et al. 2005b; Aldrian et al. 2007). In summer, the maximum convection is located north of the MC, and gradually moves southeastward to the southern MC and northern Australia along the "land bridge" during the boreal fall (Chang et al. 2005a). However, the maximum convection does not return to the north of the MC in the same way during the boreal spring (Chang et al. 2005a). The wet-to-dry seasonal transition is quick but the reversed process is relatively gradual. Chang et al. (2005a) suggested that the asymmetric low-level divergence over the MC, and the mass distributions between the two transitional seasons may contribute to this asymmetry. Matsumoto and Murakami (2000) indicated that this asymmetry was due to the respective annual variations of the equatorial basic flows over the IO and western Pacific, and that the cold surges originated from East Asia during the boreal fall were stronger than those originated from Australia during the boreal spring. On the other hand, although dynamic models showed major improvements in weather and climate forecasts during the past decades, prediction of MC rainfall is still a big challenge due to the complex terrain, land-sea distribution, and interaction over the region (Mori et al. 2004; Ichikawa and Yasunari 2006; Rauniyar and Walsh 2012).

Figure 8.22a shows the time series of climatological (2000–2010) pentad-mean rainfall over the MC (98°–145° E, 9°–1° S; plus 98°–120° E, 1° S–7° S). In the observations, minimum (maximum) value of pentad-mean rainfall over the MC is below 4 mm/day (above 10 mm/day) in the 44th (73rd) pentad, with an annual mean around 6.5 mm/day (Fig. 8.22a). The wet-dry seasonal transitions show asymmetric features, with a sudden decrease in rainfall (about 4.5 mm/day) from the 26th to the 28th pentad and a gradual increase in rainfall (about 2 mm/day) from the 58th to the

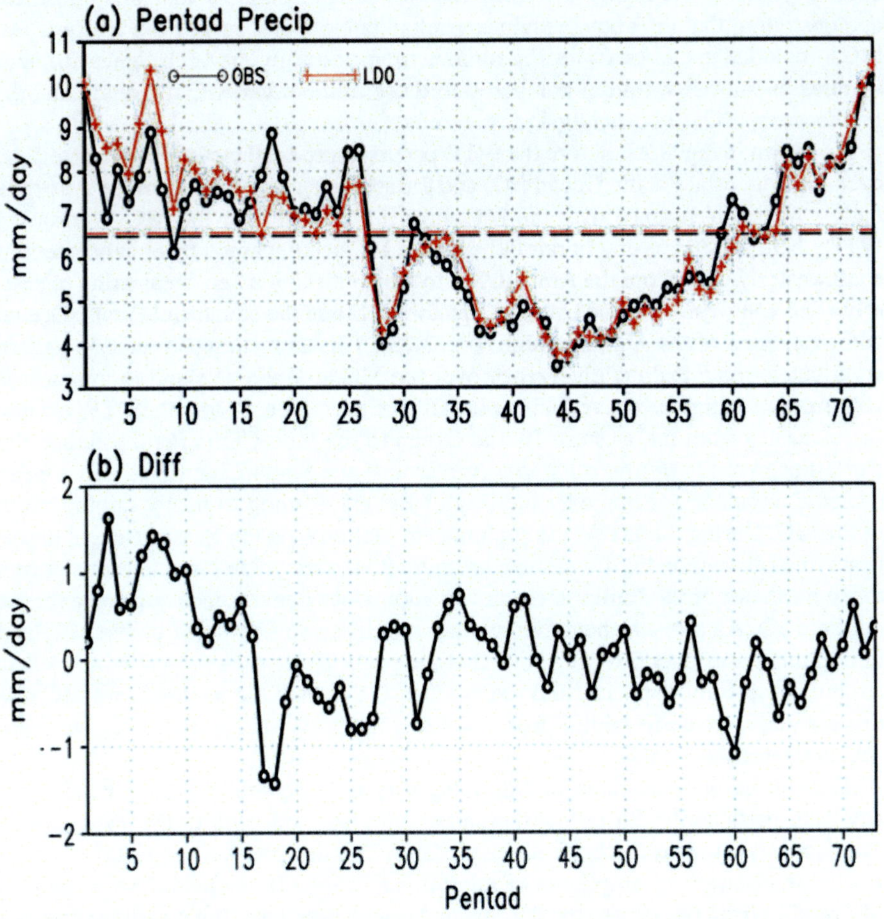

Fig. 8.22 **a** Climatological pentad-mean rainfall (mm day^{-1}) averaged over the MC (98°–145° E 9°–1° S; plus 98°–120° E, 1° S–7° S) for observations (GPCP; black line) and CFSv2 LD0 (red line). **b** Difference of climatological pentad-mean rainfall (mm/day) between CFSv2 LD0 and observation (GPCP). The horizontal black line and red line denote the annual mean values for observation and CFSv2, respectively. After Zhang et al. (2016c)

60th pentad (Fig. 8.22a). To better reflect different features between the two seasons, the transition date from dry to wet season (DTW) is objectively defined as the pentad that satisfies the following criteria: (1) the initial pentad when the above annual mean rainfall occurs during 17th–37th pentads; (2) in three consecutive pentads, rainfall is above the annual mean; (3) rainfall averaged in the initial pentad and the following five pentads is above the annual mean. The pentad that satisfies the opposite criteria is defined as the transition date from wet to dry season (WTD). According to the above definition, the DTW occurs around the 59th pentad and the WTD occurs around the 27th pentad (Fig. 8.22a). Using the pentad data from the Climate Prediction Center Merged Analysis of Rainfall (CMAP), climatological pentad-mean rainfall during 1979–2014 is also calculated, which shows similar results as that in Fig. 8.22a, demonstrating that the above results are reliable. Note that part of the northeastern MC is included in the Asian-Pacific summer monsoon regime, which shows different features of rainfall variation compared to those in the other MC regions; thus, the northeastern MC is not included in this study.

As shown in Fig. 8.23, before the WTD occurs, there are the easterlies (westerlies) over the equatorial Pacific Ocean (IO) and the southeasterly wind from the Australia high, which convergence over the MC, favoring rainfall over the MC. During the WTD, there are decreasing in rainfall over the MC accompanied by enhancement of southwesterly wind from the Arabian Sea to the South China Sea, weakening of easterly wind and westerly wind over the equatorial IO and the equatorial Pacific Ocean, and weakening of the Australia high (Fig. 8.23). This anomalous circulation pattern contributes to anomalous divergence over the MC at low levels and to anomalous convergence at high levels, reducing rainfall over the region. Before the DTW occurs, the easterlies from the western Pacific Ocean to the Indo-China Peninsula and the westerlies over the northern IO converge around the South China Sea, and a weak cyclonic circulation forms over the Bay of Bengal, leading to heavy rainfall north of the MC. During the DTW, the cyclonic circulation over the Bay of Bengal disappears, the Mascarene high weakens, an anomalous anticyclonic circulation appears to the northeast of the Philippines, and an anomalous convergence appears over the northern IO. As a result, negative rainfall anomalies appear north of the MC, and rainfall increases over the MC. Note that the patterns between the DTW and WTD are not completely opposite, and there is late DTW over the southeastern MC due to a gradual increase in rainfall from the WMC to the WMC, and from the northern MC to the southern MC.

The subseasonal variation of MC rainfall is well captured by the CFSv2 of 0-day lead, particularly during the dry season and from the 65th to the 73rd pentad. The model well predicts the features of the WTD, but predicts a late DTW and a weaker-than-observed amplitude of rainfall variation in 0-day lead. The amount of MC rainfall from the 19th to the 73rd pentad is well predicted, but it is overestimated from the 1st to the 16th pentad in 0-day lead. Correspondingly, major features of rainfall and atmospheric circulation pattern related to the WTD and the DTW are generally captured by the CFSv2, including the shifting of rain band, development of the southwesterly monsoon, and weakening of the Australia high during the WTD. However, rainfall over the MC is underestimated (overestimated) from the 25th to

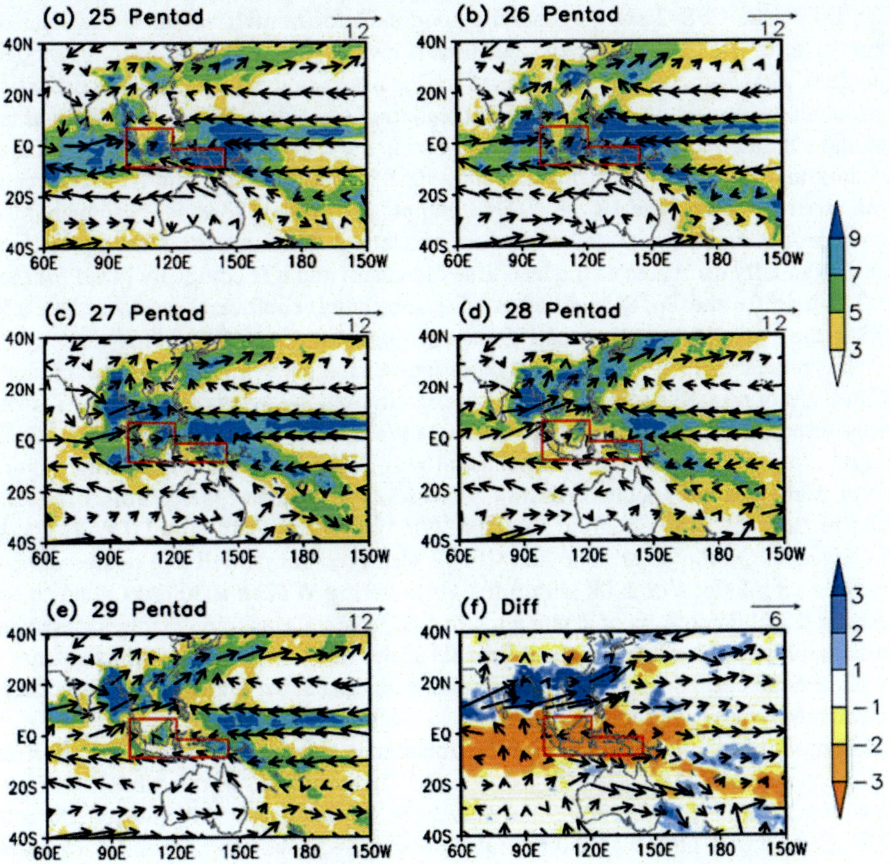

Fig. 8.23 Climatological mean rainfall (GPCP; shading; mm/day) and 850 hPa wind (CFSR; vector; m s^{-1}) for **a** the 25th pentad; **b** the 26th pentad; **c** the 27th pentad; **d** the 28th pentad; and **e** the 29th pentad. **f** Differences between 28th–29th pentad mean and 25th–26th pentad mean. The domain of the MC is outlined by red lines. After Zhang et al. (2016c)

the 28th (from 27th to the 29th) pentads, leading to smaller than the observed change in rainfall during the WTD in 0-day lead. Besides, the NCEP CFSv2 of 0-day lead shows underestimation of rainfall over the MC during the 60th–61st pentads.

The CFSv2 can well predict the major features of rainfall and related atmospheric circulation about two (three) weeks in advance for DTW (WTD). The amplitude of rainfall changes during the two transition seasons, which is underestimated in longer lead days. For the WTD, the related atmospheric circulation in the model shows similar features as those in the observation for lead time of 10 days or shorter. When the lead time is longer than 25 days, rainfall increases over the southeastern MC (but it decreases in the observation), the southwesterly monsoon circulation shifts southward compared to that in the observation, and variation of low-level winds over the western tropical Pacific is nearly opposite to that in the observation. For the DTW, the CFSv2 shows generally good skill for rainfall variation and related atmospheric circulation when the lead time is less than 15 days. However, the model predicts weaker-than-observed changes for the Mascarene high and the atmospheric circulations over the western Pacific for lead time of 15 days of longer, contributing to underestimation of MC rainfall change during the DTW. Pattern correlations of the change in rainfall over (60°–160° E, 30° S–30° S) during the DTW and WTD between the observation and the NCEP CFSv2 outputs of different leads are calculated to provide a quantitative assessment of the model's prediction skills. The prediction skill gradually decreases as the lead time increases, and it is obviously lower for the DTW than for the WTD in all leads. The correlation coefficient is lower than 0.5 when the lead time is longer than 23 days (15 days) for the WTD (DTW).

It is noticeable that the transition dates for wet and dry seasons, as well as prediction skills of rainfall and related atmospheric circulation during the WTD and DTW, vary year to year. The times of the WTD (DTW) are the 27th, 26th, 27th, 27th, 33rd, 21st, 37th, 28th, 27th, and 32nd pentads (64th, 59th, 62nd, 56th, 65th, 58th, 69th, 59th, 59th, and 63rd pentads) during 2000–2009. The CSFv2 well predicts rainfall and related atmospheric circulations in 2006 during the WTD and DTW 10 days in advance. On the other hand, the CFSv2 well predicts rainfall and related atmospheric circulations in 2008 within five days during WTD but 30 days in advance during DTW. Therefore, understanding of MC seasonal transition features and their prediction in different temporal and spatial scales need to be further investigated.

2. Prediction of early-season rainfall over southern China and associated atmospheric circulation pattern

Southern China is a main rice production area in China, which is usually referred to the region of (110°–120° E, 20°–30° S). The early-season rainfall (ESR) over southern China occurs from late-March to early-April and lasts until late-June, which has great impact on the agriculture and economy of southern China. An accurate prediction allows the farmers to choose the best time to plant crops and manage water resources (Yim et al. 2013). In this section, we discuss the hindcast skill of the NECP CFSv2 in predicting ESR. Results show that the CFSv2 can well capture the main features of the ESR at LD0 (LD0, LD1, LD2, … and LD44 represent 0-day lead, 1-day lead, 2-day lead, … and 44-day lead, respectively), including the period of the ESR from pentad 19 to pentad 36, the magnitude of rainfall from 6 to 12 mm day^{-1}, the northward movement and expansion of rainfall center from 20° S in pentad 27 to 30° S in pentad 36, the southwest–northeast pattern, and the transformation of atmospheric circulation associated with the ESR (Fig. 8.24). The model generally overestimates the ESR (Fig. 8.24b).

The CFSv2 at LD0 can well predict the atmospheric circulation during the ESR at both low and high levels from pentad 19 to pentad 36. At the low level, the anticyclonic and cyclonic circulations appear over the tropical-subtropical northwestern Pacific and eastern Tibetan Plateau, respectively. As a result, southern China is under the control of convergent southwesterly wind, which gradually weakens when it reaches higher latitude. At the high level, the strong westerly prevails from Asia to the Northwest Pacific, and high-level divergence can also be found. The model well

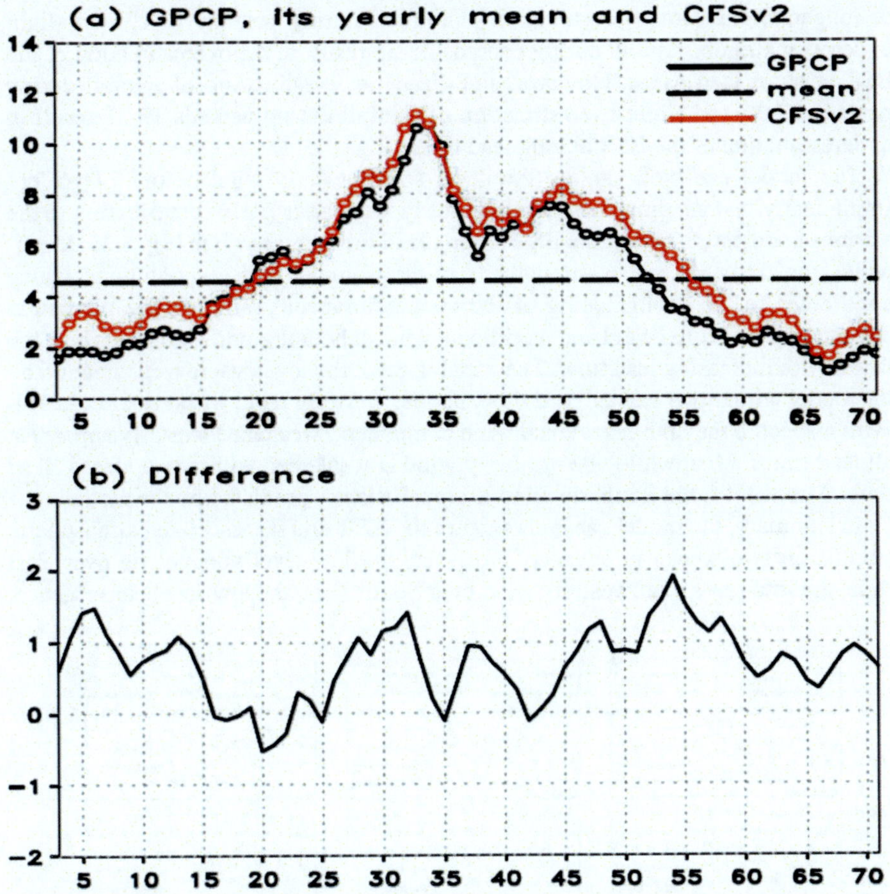

Fig. 8.24 **a** Time series of pentad-mean precipitation (mm/day) averaged over southern China (110°–120° E, 20°–30° S) using GPCP data, its yearly mean, and CFSv2 precipitation in LD0. **b** Differences in precipitation between GPCP and CFSv2 prediction of LD0 (the latter minus the former). After Zhao and Yang (2014)

captures the low-level southwesterly wind, which contributes to the high prediction skill of the outbreak and development of the ESR.

The CFSv2 can well predict the features of rainfall during the ESR by two weeks in advance. The magnitude of rainfall center exceeds 6 mm day^{-1} during the outbreak of the ESR in pentad 19. The range of rainfall gradually extends from south to north after the outbreak of the ESR, especially from pentad 30 to pentad 35. The ESR decays around pentad 36. All the above features are well captured by the CFSv2 at LD0. Compared with the observation, the model also exhibits obvious discrepancy with increasing lead time. The rainfall center exceeding 4 mm day^{-1} occurs earlier in the model. Moreover, the model significantly underestimates the magnitude of the ESR, which drops from 12 to around 10 mm day^{-1} for lead times of two weeks or longer. Significant overestimation of rainfall during pentads 19-27 and slight underestimation of rainfall during pentads 27-36 result in the overestimation of the ESR at short lead times. However, significant underestimation of rainfall during pentads 27-36 and slight overestimation of rainfall during pentads 19-27 result in underestimation of the ESR at long lead times.

The model can well capture low-level southwesterly wind before LD16, but significantly underestimates the southwesterly wind after LD16, contributing to the obvious underestimation of the ESR at long lead times as shown in Fig. 8.25. Meanwhile, the tropical easterly strengthens in late-summer and early-autumn, which contributes to the strengthening of the contemporaneous rainfall. The prediction discrepancies of zonal wind and meridional wind at both low and high levels become larger with increasing lead time. The zonal wind at the low level is well captured by the model within two weeks' lead time, but the westerly wind weakens after LD14. With respect to the high-level zonal wind component, weakened westerly appears at all lead times. Meanwhile, the northerly wind is weakened within two weeks' lead time. After LD14, the northerly wind is unrealistically predicted by the model.

In summary, the model can well capture the ESR and its associated atmospheric circulation two weeks in advance. However, the low-level anticyclone over East Asia, the low-level southwesterly wind over South Asia, the low-level convergence,

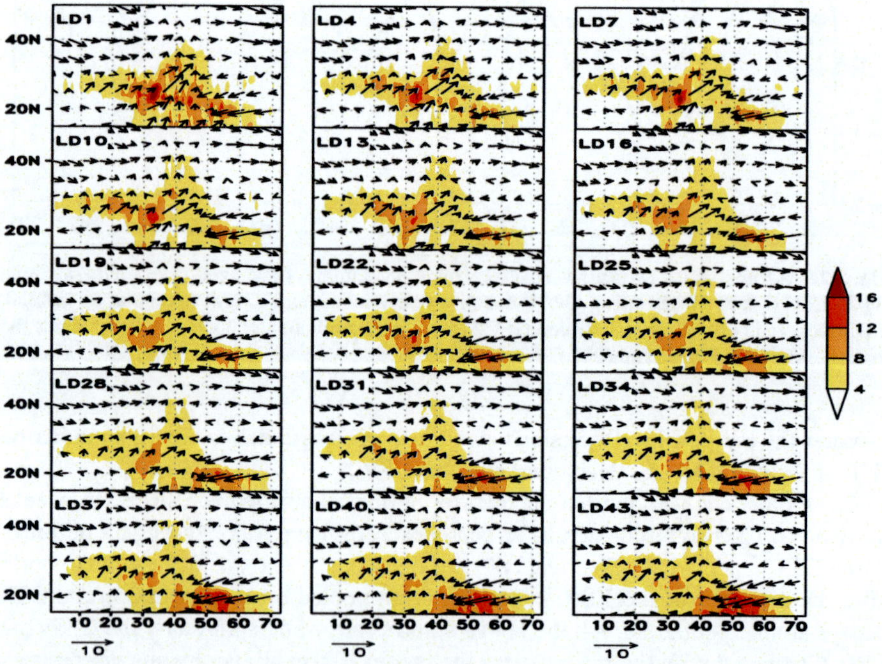

Fig. 8.25 Time–latitude cross-sections of pentad-mean precipitation (shading; mm/day) and 850 hPa wind (vector; m/s) along 110°–120° E from the 3rd to 71st pentad (x-coordinate) predicted by CFSv2 in LD1, LD4, LD7,… LD40, and LD43, respectively. After Zhao et al. (2015)

the high-level divergence over southern China, and the high-level westerly jet are all underestimated by the model after LD14. The above-mentioned underestimation results in underestimated ascending motion over southern China, which contributes to the underestimation of rainfall during the ESR after LD14. In all, the skill of the CFSv2 in predicting the ESR over southern China is higher than that over the other regions in East Asia from April to June.

8.4 Predictability of Climate Variation Over SEA and Its Adjacent Regions

8.4.1 The Most Predictable Patterns of Rainfall and Atmospheric Circulation on Subseasonal and Seasonal Scales

1. Seasonal dependence of the most predictable low-level atmospheric circulation patterns over the Indo-Pacific region

In general, the ensemble mean for ensemble simulations is composed of predictable signals and the residual of unpredictable internal noise. The EOF analysis with maximized signal-to-noise ratio (MSN EOF hereafter) can effectively minimize the effect of internal noise, thus maximizes the signal-to-noise ratio in a moderate ensemble size (Allen and Smith 1997; Venzke et al. 1999; Chang et al. 2000; Sutton et al. 2000; Huang 2004; Liang et al. 2009). When applying to an ensemble hindcast, the leading MSN EOF modes represent the most predictable patterns. The seasonal dependence of the most predictable patterns of monthly 850 hPa zonal wind over the tropical Indo-Pacific region is examined using the ensemble reforecasts by both NCEP CFSv2 and Project Minerva (a collaborative project between the Center for Ocean-Land-Atmosphere Studies and the European Centre for Medium-Range Weather Forecasts).

The spatial patterns of the first MSN EOF (MSN EOF1) modes in different lead times and target months mainly show opposite zonal wind over the equatorial Pacific and IO. As an example, the MSN EOF1 modes of March in 0-month lead, 3-month lead, 6-month lead, and 9-month leads are shown in Fig. 8.26. The percentages of variance of the ensemble mean 850 hPa zonal wind anomalies explained by these MSN EOF1s are around 50–60% in different leads. There are obvious similarities among the MSN EOF1 modes for zonal wind over the equatorial Pacific, but zonal wind over the equatorial IO becomes weaker as the lead time increases, suggesting a relatively lower predictability for the IO region (Fig. 8.26a, c, e, g). The peaks and valleys of MSN PC1s in March generally correspond to El Niño (i.e., 1983 and 1998) and La Niña (i.e., 1999–2001) events, respectively (Fig. 8.26b). To evaluate the prediction skill of these predictable patterns identified by the MSN EOF and to determine whether the highly predictable patterns are robust, the 850 hPa

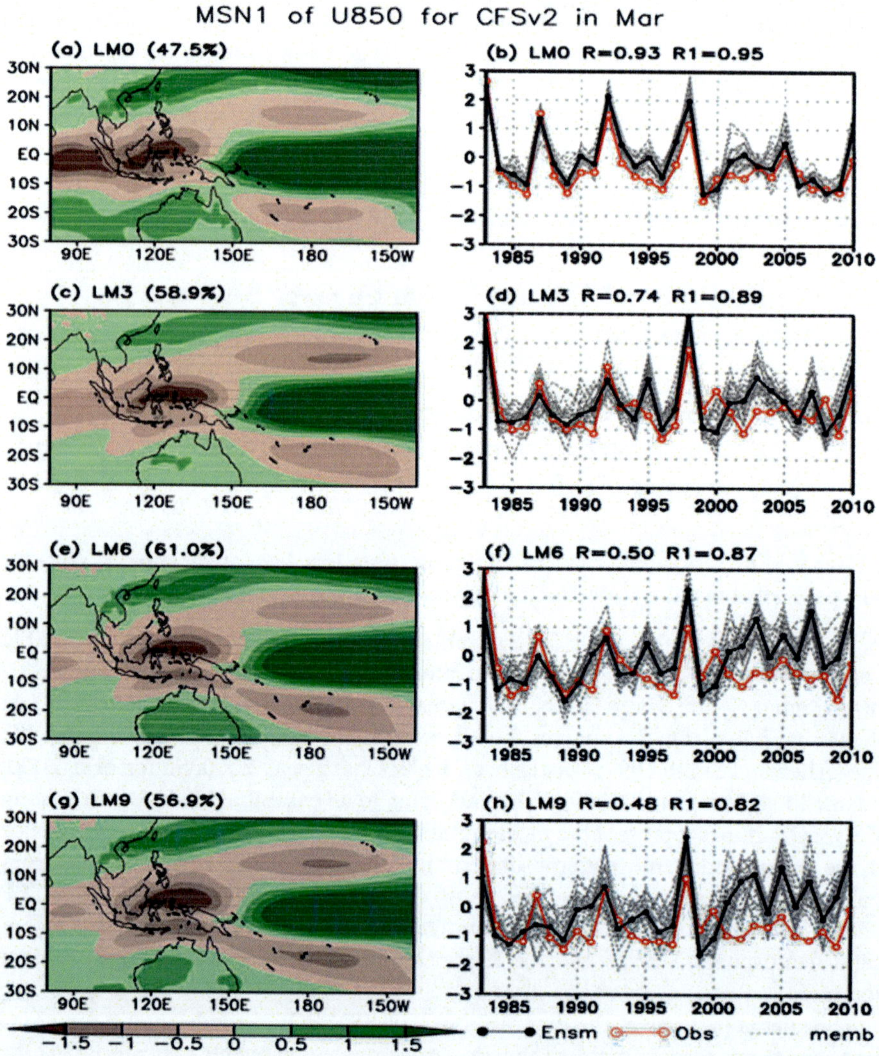

Fig. 8.26 Left: The MSN EOF1 modes of the CFSv2 reforecast 850 hPa zonal wind (m/s) for lead months of 0, 3, 6, and 9 in March. In the right panels, the solid black lines are the PCs of ensemble means. The solid red lines and dashed black lines represent the PCs that are computed by projecting the 850 hPa zonal wind in observation and ensemble members on to the MSN EOF1. R represents the correlation coefficient between ensemble mean PC and projected observational PC. R1 represents averaged correlation coefficient between ensemble mean PC and projected PCs for individual members. SD represents averaged standard deviation of the projected PCs from the ensemble members. After Zhang et al. (2018a)

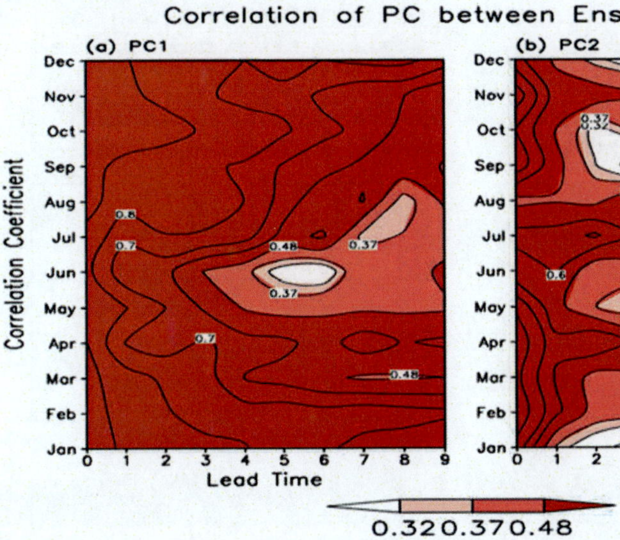

Fig. 8.27 Correlation coefficients between ensemble mean PCs and projected observational PCs for MSN PC1 (left panel) and MSN PC2 (right panel). Values exceeding the 90, 95, and 99% confidence levels are shaded. The x-coordinate indicates the corresponding lead months, and the y-coordinate indicates the corresponding months. After Zhang et al. (2018a)

zonal wind anomalies in observation and individual members are projected onto the MSN EOF1 patterns, respectively. As shown in Fig. 8.26, correlation coefficients (R) between MSN PC1s and their corresponding projected observational PC1s, and averaged correlation coefficients (R1) between MSN PC1s and corresponding projected ensemble member PC1 exceed the 99% confidence level for all leads (R (R1) = 0.93 (0.95), 0.74 (0.89), 0.50 (0.87), and 0.48 (0.82) for LM0, LM3, LM6, and LM9, respectively). The most predictable pattern of 850-zonal wind is well predicted by the NCEP CFSv2 nine months in advance, except for those in June–July, which reflects the spring predictability barrier (Figs. 8.27a, 8.28).

In contrast, the second most predictable atmospheric low-level circulation patterns are more seasonally dependent. The monthly evolution of the predictable patterns during boreal winter and spring is characterized by a southward shift of westerly wind anomalies, which is generated by the interaction between the annual cycle and the ENSO signals (i.e., the combination-mode) and contributes to rapid termination of major El Niño events. The NCEP CFSv2 well predicts the most predictable pattern five months in advance for boreal spring, but shows lower skill for boreal winter (Fig. 8.27b). In boreal summer, the spatial pattern of MSN EOF2 is characterized by opposite variations of anomalous zonal winds around the WNP and its surrounding regions before 4-month lead, and by opposite variations of anomalous zonal winds over the equatorial Pacific and over the regions to its west, north, and south after 4-month lead (i.e., in July) (Fig. 8.27). The NCEP CFSv2 can well capture these predictable patterns in summer (Fig. 8.26b). For example, in July the correlation

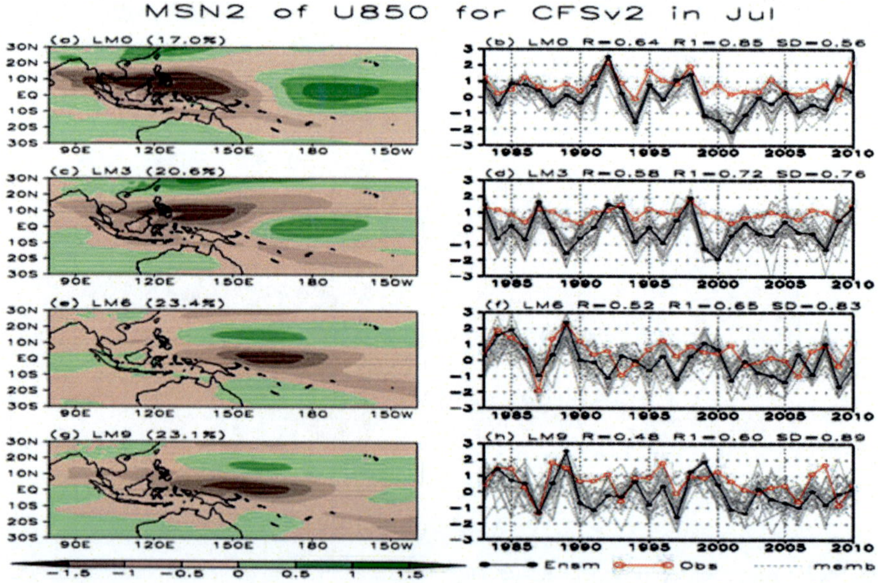

Fig. 8.28 Same as Fig. 8.26, except for MSN EOF2 for July. After Zhang et al. (2018a)

coefficients between ensemble mean PC1 and projected observational PC1 in 0-month lead, 3-month lead, 6-month lead, and 9-month lead are 0.64, 0.58, 0.52, and 0.48, respectively (all exceeding the 99% confidence level). The second most predictable patterns in autumn can be predicted three months in advance (Fig. 8.26b).

Note that there are similar results by the NCEP CFSv2 and Project Minerva. For example, in Project Minerva with T639 resolution, the most predictable patterns during May–November, which are using May initial conditions (ICs), are characterized by anomalous westerlies over the central–eastern equatorial Pacific and anomalous westerlies over the IO to the MC, similar to those by the NCEP CFSv2. The peaks and valleys of MSN PC1s mostly correspond to El Niño and La Niña years, respectively. The centers of easterly loadings over the equatorial IO are generally weaker during summer months. The variances explained by the MSN EOF1 in May–November are 36.2%, 38.9%, 39.2%, 44.5%, 48.5%, 53.2%, and 56.1%, respectively. Correlation coefficients between ensemble mean PC1s and projected observational PC1s, and averaged correlation coefficients between ensemble mean PC1s and projected ensemble member PC1s for May–November all exceed the 99% confidence level, suggesting high predictability of these predictable patterns. For the second most predictable patterns with May ICs in Project Minerva, the predictable signals are mainly located over the WNP, similar to those in the NCEP CFSv2. The spatial patterns of the MSN EOF2 by Project Minerva are characterized by a northward shift of the WNP anomalous anticyclone/cyclone in summer and a southward shift in autumn.

2. The most predictable rainfall patterns on subseasonal time scales over the Indo-Pacific region

The ISO is one of the most important components of the tropical atmospheric variation, which significantly affects the global atmospheric circulation as well as the local weathers and climate. It mainly comprises the MJO with a spectral peak around 30–60 days and the quasi-biweekly oscillation (QBWO) with a characteristic period of 10–20 days. By applying MSN EOF analysis, we investigate the most predictable patterns and predictability of precipitation over the Indo-Pacific region on subseasonal time scales. The reforecast product from the 45-day hindcasts (1999–2014) by the NCEP CFSv2 is used, along with several observational data sets.

The features of most predictable patterns of precipitation over the Indo-Pacific region on 30–60-day time scale exhibit seasonality. The most predictable patterns show tropical zonal dipole during boreal winter and spring, but a meridional dipole during boreal summer and fall. The predictability of predictable patterns is also seasonally dependent, which is the worst in summer but better in the other seasons.

As shown in Fig. 8.29, the most predictable pattern of regional precipitation in boreal winter on the 30–60-day time scale shows a dipole pattern over the northern IO and to the south of the MC. The MSN EOF1 in boreal winter explains 20%–30% of the total variance, 10% higher than the second most predictable pattern. The model shows good prediction of both variability and magnitude of the variation at short leads, and the pattern at LD29 is still predictable with R exceeding certain confidence level (Fig. 8.29f). A sustaining high R1 implies that the most predictable pattern in boreal winter is still quite robust at different leads. The most predictable pattern in boreal spring exhibits similar distribution as that in boreal winter, which shows dipole from the equatorial IO and the main part of the MC to the South China Sea, the eastern Bay of Bengal, and north of Australia. The MSN EOF1s in boreal spring explain at least 20% of the total variance at short leads and around 30% at long leads. Pattern shift can be found after LD30 in boreal spring and after LD25 in boreal winter. For example, the tropical zonal dipole changes to consistent variation after LD25 in boreal winter. Due to the influence of initial memory, the pattern in boreal spring before one week's lead is unstable. After LD7, the pattern in boreal spring remains stable at different leads (Fig. 8.30).

The most predictable pattern of regional precipitation on 30–60-day time scale in summer is quite different from those for boreal winter and spring with a meridional dipole from the eastern Arabian Sea, Bay of Bengal, South China Sea, and Northwest Pacific to the equatorial IO and the MC. The MSN EOF1s explain 40%–50% of the total variance, which is obviously larger than that for boreal winter half of the year and that for the second most predictable pattern. The prediction skill for boreal summer is significantly weaker than that for boreal winter and spring with only 20 days in advance. The most predictable pattern for boreal summer is also robust due to high R1 at different leads. In boreal autumn, the pattern is similar to that in boreal summer but the meridional dipole is weaker over the western Pacific, which is associated with the relatively earlier retreat of monsoon activity over East Asia.

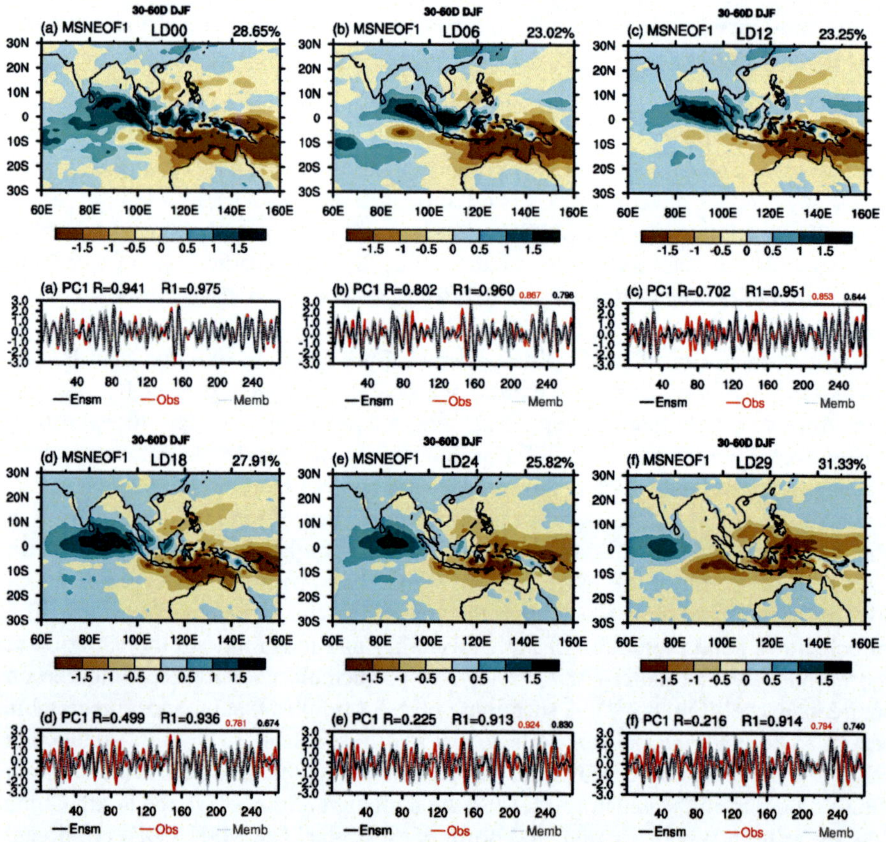

Fig. 8.29 The MSN EOF1 modes and PCs of precipitation (mm/day) predicted by the CFSv2 for lead days of 0, 6, 12, 18, 24, and 29 in boreal winter on 30–60-day time scale. The solid black lines represent the PCs of the ensemble means, the dashed grey lines represent the PCs of different ensemble members, and the solid red lines represent the PCs that are computed by projecting the observed precipitation onto the spatial distribution of the MSN EOF1. R represents correlation coefficient between the solid red line and solid black line. R1 represents averaged correlation coefficient among ensemble members. After Dong et al. (2020)

In summary, the most predictable pattern during boreal winter half of the year with tropical zonal dipole mainly implies the eastward propagation of MJO, while the one during boreal summer half of the year with meridional dipole mainly implies the northward propagation of BSISO. Pattern shift only occurs in boreal winter and spring at long leads when the spread of ensemble members is large.

The prediction skill is the worst in boreal summer but better in boreal winter and transitional seasons. The skill drops with increasing lead time. However, the skill in boreal spring bounces at LD30, which is related to pattern shift in boreal spring. The prediction skill for the second most predictable pattern shows relatively weaker

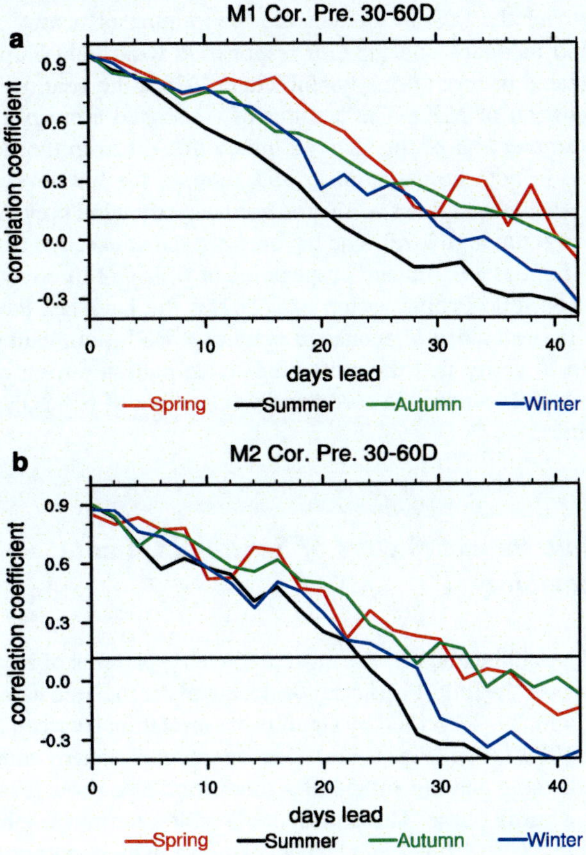

Fig. 8.30 Correlation coefficients between ensemble mean PCs and projecting PCs (computed by projecting the observed precipitation upon the MSN EOF modes) for **a** MSN EOF1 and **b** MSN EOF2 on 30–60-day time scale. The black straight line represents the correlation coefficient at the 95% confidence level. After Dong et al. (2020)

seasonality. In short, the most predictable pattern shows higher predictability than the second one.

The features of associated atmospheric circulation that is related to the most predictable pattern are similar in boreal winter and spring. The regression of OLR shows convective center over the IO. At the low level, the westerly can be found to the west of the convective center while the easterly can be found to the east of the center. What is more, cyclonic circulations appear to the north and south of the westerly. At the high level, the easterly can be found to the west of the convective center, while the westerly can be found to the east of the center; and anticyclonic circulations appear to the north and south of the easterly. The above features are associated with the Gill response. The convective center lags warm SST center by about 1/4 phase, which implies the eastward propagation of MJO. Moreover, the strong easterly over the MC accelerates the evaporation of nearby seas. Both the evaporation-wind feedback and the Gill response in wind field help to verify that the most predictable pattern during boreal winter half of the year mainly shows the eastward propagation of MJO. The features of associated atmospheric circulation during boreal summer half of the year are quite different from those for the winter half of the year. In both boreal summer and autumn, the low-level winds exhibit significant cross-equatorial flows, while the northeasterly wind prevails over the MC and nearby areas at the high level. The lag between the convective center and warm SST center also implies a northward propagation of BSISO. Moreover, the dominant cold SST over the Indo-Pacific region exaggerates the land–sea thermal contrast, which stresses the influence of monsoon activities. Both dominant cold SST and wind fields help to verify that the most predictable pattern during boreal summer half of the year mainly shows the northward propagation of BSISO associated with monsoon activities.

8.4.2 Possible Impact Factors of Seasonal Climate Predictability

The above analyses indicated that the most predictable patterns of atmospheric low-level circulation over the Indo-Pacific region are mainly associated with ENSO during the whole year, which is identified by significant correlation between MSN PC1 and observational ENSO index (Fig. 8.31a). The ocean–atmosphere anomalies related to the most predictable patterns of 850 hPa zonal wind also show features of ENSO developing or maturing phase. These most predictable patterns have high prediction skills for both the NCEP CFSv2 and Project Minerva, which mainly originate from ENSO, suggesting that ENSO is the major source of predictability of low-level atmospheric circulation variation over the tropical Indo-Pacific region. However, the models show relatively lower prediction skills during boreal summer, due to the spring predictability barrier.

In contrast, the second most predictable patterns of the atmospheric low-level circulation over the Indo-Pacific region show strong seasonality. During boreal winter and spring, the most predictable patterns are mainly associated with ENSO decaying phase, with significant correlation between MSN PC2s and Niño-4 index (Fig. 8.31b). Both the NCEP CFSv2 and Project Minerva have high prediction skills for the second most predictable patterns during boreal spring, but show lower prediction skills in winter. In boreal summer, the second most predictable patterns are associated with WNP monsoon in shorter leads, and associated with ENSO in longer leads (Fig. 8.26b, c). Correlation coefficients of MSN PC1 and WNP anticyclone/cyclone (WNP-AC) index exceed the 95% confidence level of the *t*-test for 0-month lead to 4-month lead, and correlation coefficients of MSN PC1 and Niño-4 index exceed the 95% confidence level of the *t*-test for 5-month lead to 9-month lead. Actually, the most predictable pattern and the second most predictable pattern switch order from short

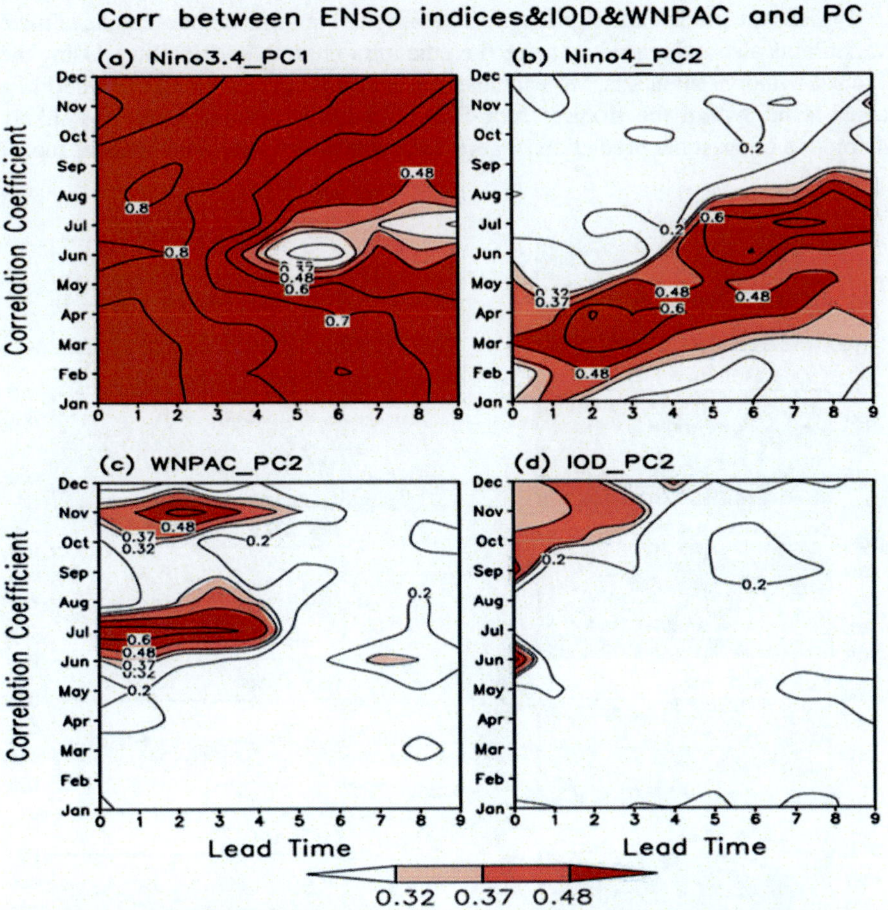

Fig. 8.31 Correlation coefficients between Niño-3.4 index and ensemble mean PC1 (**a**), between Niño-4 index and ensemble mean PC2 (**b**), between WNP-AC index and ensemble mean PC2 (**c**), and between IOD index and ensemble mean PC2 (**d**). Values exceeding the 90, 95, and 99% confidence levels are shaded. The x-coordinate indicates the corresponding lead months, and the y-coordinate indicates the corresponding months. After Zhang et al. (2018a)

to long leads, due to the effect of "spring barrier," which limits the predictability of ENSO. The models generally well capture the second most predictable WNP patterns in summer, but the prediction skill is lower than that for predictable ENSO patterns (Fig. 8.31). The second most predictable patterns are mainly associated with the IOD as well as the WNP-AC in autumn, but can only be well predicted by the models less than three months in advance (Fig. 8.31c, d). The relatively lower prediction skills for the predictable WNP patterns and predictable IOD patterns are partly due to the seasonal nature of the IOD and the WNP monsoon.

Besides the predictable signals, predictability of the tropical Indo-Pacific climate variation is also influenced by noise (i.e., the intra-ensemble variability). Using the Project Minerva hindcasts, we examine the intra-ensemble variability of 850 hPa zonal wind within the tropical Indo-Pacific region, to identify the most likely disrupters of seasonal prediction. Figure 8.32 shows the first EOF modes of model

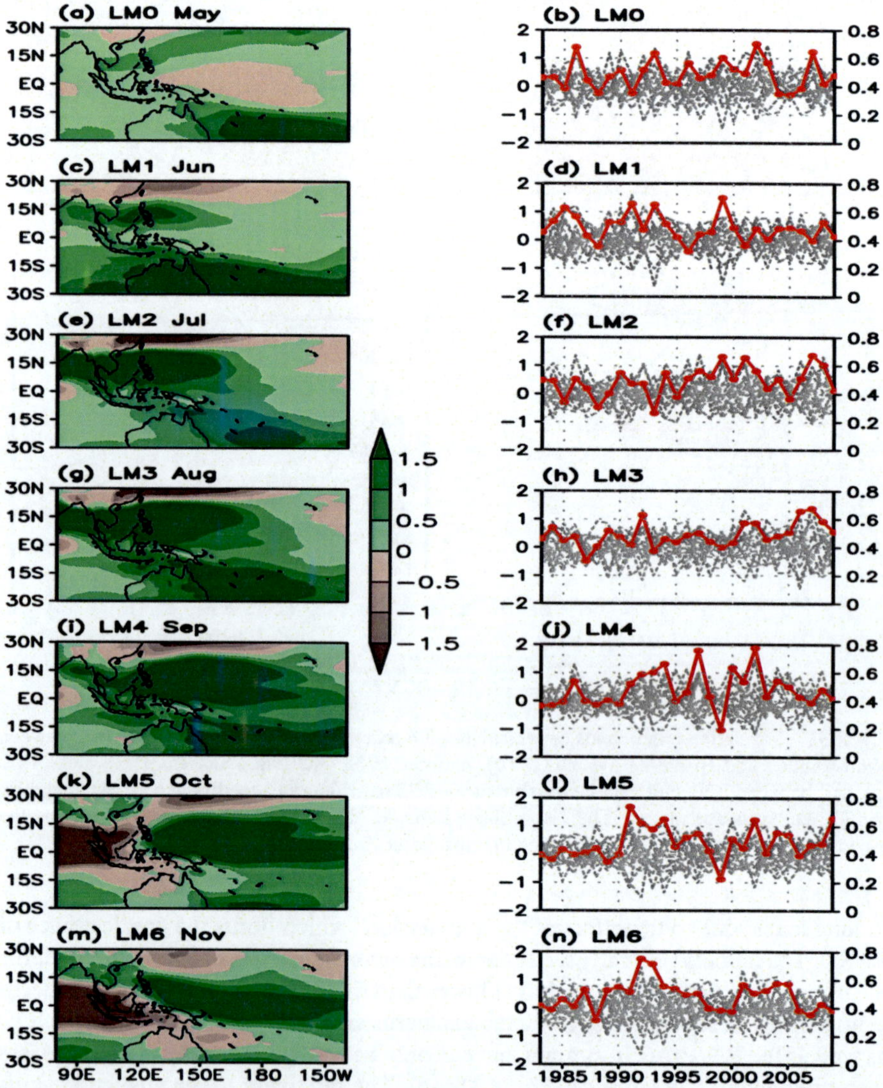

Fig. 8.32 The first EOF modes of model noise data for 850 hPa zonal wind (m/s) in different lead months for T639 with ICs in May. In the right panels, the dashed black lines represent the corresponding PCs for ensemble members, and the red line represents the standard deviation of the ensemble members. After Zhang et al. (2018b)

noise 850 hPa zonal wind from May to November (May ICs) with T639 resolution. In May, the large model spread, which likely originates from the southern extratropics, is mainly located south of 15° S to the east of Australia (Fig. 8.32a). During the boreal summer months, the location of large model spread appears to the north of the equator, with westerly wind anomalies from north of the equator to about 15° S extending from the IO to the WNP region, and with easterly wind anomalies further to the north (Fig. 8.32c, e, and g). During the boreal autumn months, the spatial patterns of noise EOF1 are characterized by the opposite wind anomalies over the equatorial Pacific and the IO (Fig. 8.32i, k, and m). There is large resemblance in spatial structure between the first noise EOF modes and the MSN EOF2 modes in the summer months, and between the first noise EOF modes and the MSN EOF1 modes in autumn, implying that the noise-induced feedback generates the difference in the evolution of the predictable signals among ensemble members during summer and autumn. The interannual variation of the noise level is weak (strong) in late-spring and summer (autumn) (Fig. 8.32).

To identify the mechanisms associated with the intra-ensemble variability during summer and autumn, we calculate correlations (regression) of noise SLP/rainfall (850 hPa wind/SST) with (against) noise PC1s in different lead months. In May, the intra-ensemble variability is mainly related to strong negative SLP anomalies with an anomalous cyclonic circulation south of 20° S to the east of Australia, which seems to have extratropical origin. During the summer months, those correlation and regression patterns are mainly characterized by a dominant WNP cyclone, accompanied by upstream positive SLP anomalies expanding from the IO into the MC and a downstream wave train through Japan. Correspondingly, mildly cold SST anomalies extend from the northern IO to the WNP, and precipitation is generally reduced under the anticyclonic circulation over the IO but increased in the cyclonic circulation over the WNP. These results are consistent with those in Ma et al. (2017). During the boreal autumn months, the intra-ensemble spread of 850 hPa zonal wind is mainly associated with the SST anomalies over the central–eastern equatorial Pacific, and the variation of the Walker circulation. This WNP-IO variation described above can trigger the equatorial wind anomalies in the central Pacific during the boreal autumn season, and then the equatorial wind anomalies seem to evolve via the ENSO-like coupled dynamics (Huang and Shukla 2007; Larson and Kirtman 2015). Consistently, the atmosphere–ocean anomalies related to the intra-ensemble variability are similar to those related to the second most predictable patterns and the most predictable patterns in summer and autumn, respectively.

Figure 8.33 shows the first EOF modes of model noise 850 hPa zonal wind with November ICs (for target months of May–November). In November, the largest model spread is mainly located over the extratropical Northern Hemisphere, with opposite wind anomalies between north and south of 15° S (Fig. 8.33a). The wind anomalies over the North Pacific expand southward in the next two months, triggering wind anomalies over the central equatorial Pacific and the equatorial IO in February (Fig. 8.33c, e, and g). The equatorial zonal wind anomalies are enhanced during the boreal spring season (Fig. 8.33i, k, m). The most striking feature in these correlation and regression patterns for the noise PC1 is the dipole sea-level pressure (SLP) pattern

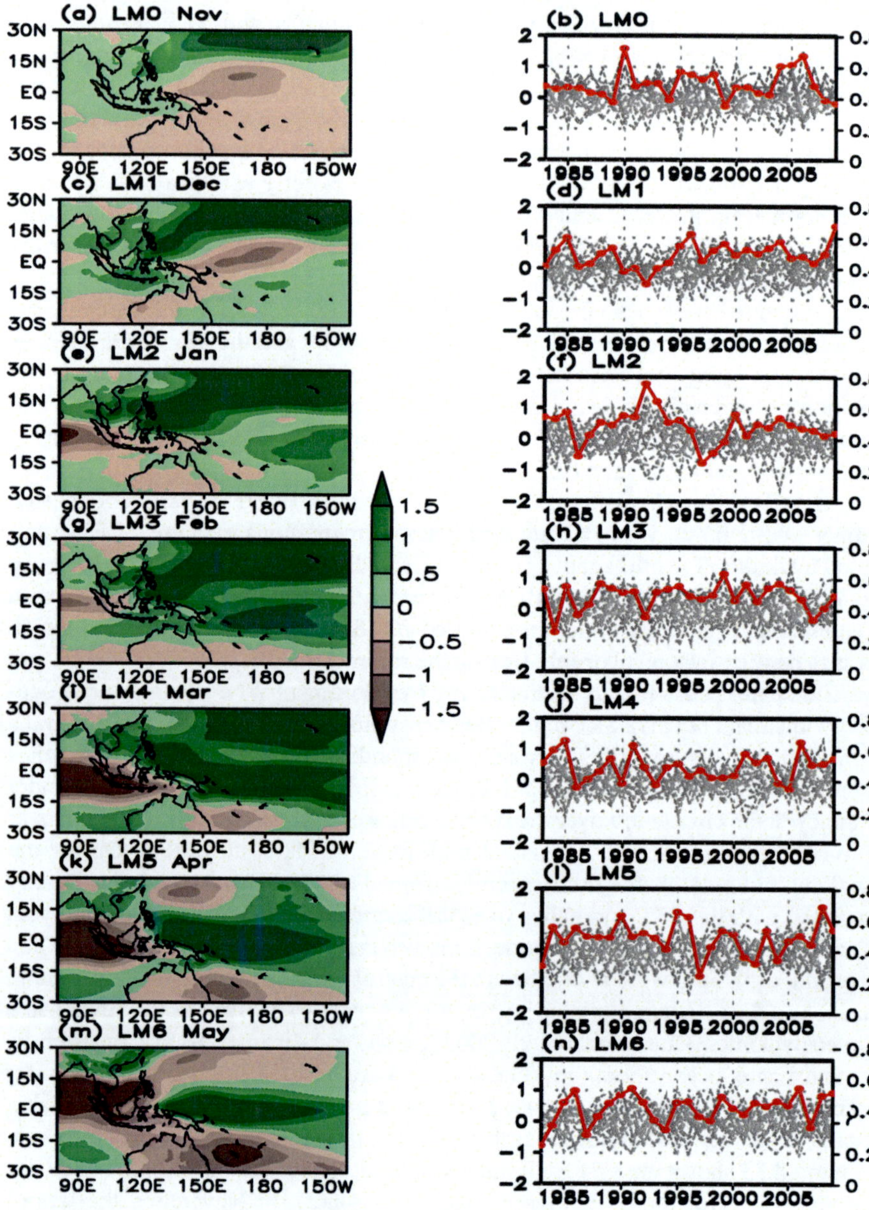

Fig. 8.33 Same as Fig. 8.32, except for November. After Zhang et al. (2018b)

over the North Pacific from November to the following March, which bears a strong resemblance to the North Pacific oscillation (Rogers 1981). The SLP anomalies first expand southwestward in November–December, and positive SLP anomalies establish over the equatorial IO and western Pacific during January–February. The equatorial zonal pressure gradient is established. Weak positive SST anomalies first appear over the central and eastern subtropics of the Northern Hemisphere in winter, and propagate toward the equator. This process is similar to the seasonal footprinting for ENSO triggering (Vimont et al. 2001, 2003). In comparison to those for May ICs, the northern extratropical perturbations affect the tropical circulation more effectively than the southern ones, and may play a major role in prediction uncertainty during the winter half of the year (Zhang et al. 2018a).

The ocean–atmosphere–land interaction is strong over SEA, due to the warmest SST and complex land–sea distribution. The model's skill in predicting the climate variation over this region largely depends on its performance in simulating the ocean–atmosphere–land interaction. Thus, the prediction skill of climate variation over SEA depends on the representation of the air–sea coupling and land surface processes.

Jiang et al. (2013a) analyzed the role of ocean–atmosphere coupling in ASM prediction, indicating that in the absence of the ocean–atmosphere coupling, the model reproduced overall weaker-than-observed ASM. The authors also showed lower skill in predicting annual cycle of rainfall. On the other hand, despite the coupled model showing higher skills in predicting climatological ASM, it did not necessarily lead to better representation of interannual variation of the ASM. This was because the coupled model did not show any advantage in depicting the physical processes associated with the interannual variability of ASM. Jiang et al. (2013b) reported that the coupled model showed better skill in predicting the climatological Asia winter monsoon (EAWM), due to its better representation of rainfall over the western MC. Their results also indicated that the better performance of model in simulating climate variation over SEA led to improved skill for SEA's surrounding regions. Better representation of air–sea coupling processes can contribute to enhanced skill in simulating responses of EAWM to ENSO, which is generally overestimated in the model. However, air–sea coupling not only contributes to enhanced skill for climate variation over several regions but also generates some biases in the model. For example, the NCEP CFSv2, which is a fully coupled model, shows larger biases in simulating rainfall and winds over the western MC and tropical IO compared to the uncoupled model. From the above analyses, ocean–atmosphere coupling apparently contributes to better simulation and prediction of the Asia monsoon, but coupled models need improvement (i.e., the NCEP CFSv2).

Both convection over the MC and heating over the Asia continent are linked to the land surface process; thus, the ability of the model in simulating the land surface process can have an impact on simulation and prediction of climate variation over SEA and its surrounding regions. Yang et al. (2011) pointed out that the model showed improvements in cold bias over the Asia continent, and in simulation of Asia monsoon circulations and rainfall, when the Noah land surface model was replaced by the Oregon State University (OSU) land surface model in the NCEP CFSv1. Actually, the Noah land surface model showed higher skills not only for climatological Asia monsoon but also for its interannual variability, including the onset of the spring monsoon and magnitude of the summer monsoon. Seasonal prediction is more closely related to the land surface model compared to ICs. Moreover, the NCEP CFSv2 shows improvements in cold bias over the Asia continent and simulation of the Asia monsoon compared to the NCEP CFSv1, with the four-layer Noah land surface model in the CFSv2 but the two-layer OSU land surface model in the CFSv1 (Jiang et al. 2013a).

Generally, increase in horizontal resolution contributes to enhanced simulation and prediction skills of climate models, but there is disagreement in results for SEA climate prediction. The atmospheric general circulation models generally produce dry bias over the MC, and there is no improvement in this dry bias even with a threefold increase in horizontal resolution (Neale and Slingo 2003). Neale and Slingo (2003) argued that the diurnal cycle over the islands and the complex circulation patterns generated by land–sea contrast were crucial for the energy and hydrological cycles of the MC and for determining the mean climate. On the other hand, using UK Met Office Unified Model with different horizontal resolutions, Rashid and Hirst (2017) indicated that the model's skill in simulating climate mean rainfall over the MC is enhanced (thus the dry bias is reduced) by increasing horizontal resolution. Better representation and steeper orography at higher resolution lead to enhanced vertical circulation, favoring rainfall over the islands. From above, there are several issues that need to be further investigated: to what extent the prediction of MC rainfall is influenced by the horizontal resolution, and what the associated mechanisms are.

8.4.3 *Possible Impact Factors of Subseasonal Climate Predictability*

The predictability of subseasonal climate variability is, first of all, linked to the nature of climate events themselves. It is shown that the predictability of strong MJO events is higher than that of weak MJO events, and the MJO forecast skills are affected apparently by changing seasons. Usually, forecast skill in winter is superior to that in summer. Such characteristics are closely related to the seasonal change of MJO intensity (Lin et al. 2008; Wang et al. 2014; Liu et al. 2017). Moreover, the predictability of MJO depends highly on the initial or target phase. All the prediction models may have higher skills in forecasting MJO in some phases but lower skills in other phases. For example, both the NCEP CFSv2 and BCC_CSM revealed that the prediction skills are relatively lower for MJO target phases 4–5 with the convection center over the MC. This indicates that it is difficult to predict the propagation of MJO convection from the IO across the MC to the western Pacific (Wang et al. 2014; Liu et al. 2017). However, this common MC prediction barrier does not exist in some climate models (Rashid et al. 2011), which suggests that the dependence of MJO predictability on phase is model-dependent. The predictability of BSISO also relies on the intensity of the event itself, the occurrence time, and the state of the initial

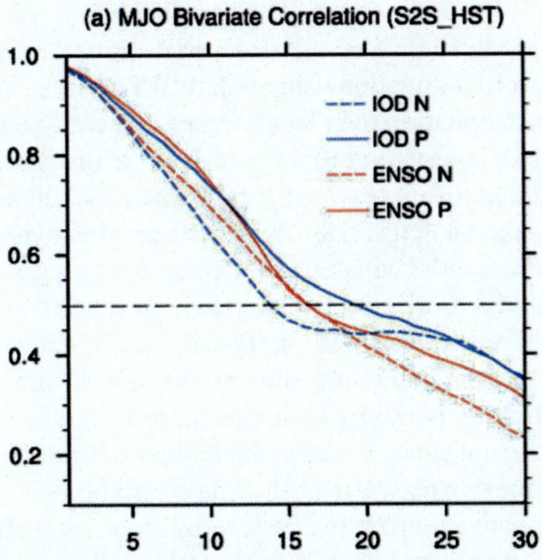

Fig. 8.34 MJO prediction skills in positive and negative phases of ENSO and IOD reforecasted by the BCC_CSM. Skill index is the bivariate correlation coefficient between prediction and observation. After Liu et al. (2017)

and target phase (Liu et al. 2015; Jie et al. 2017). For instance, the multiple models participating in the S2S Project all showed that the forecast skill of strong BSISO is superior to that of weak BSISO; moreover, most of the models performed poor in forecasting the BSISO1 events at phases 4 and 5.

The predictability of subseasonal climate variability is connected to large-scale climate background features. Using the S2S hindcast experiments by BCC_CSM, we investigated the relationship between MJO prediction skills and ENSO/IOD. It was found that the prediction skill of MJO in the ENSO positive phase (El Niño) is higher than that in the ENSO negative phase (La Niña); similarly, it is higher in the IOD positive phase than in the IOD negative phase (Fig. 8.34; Liu et al. 2017). This is because MJO has different characteristics in different phases of ENSO and IOD. When MJO propagates from the eastern IO and the MC to the western Pacific, its intensity in the El Niño stage is much greater than in the La Niña stage; when the MJO convection is initially located over the Western Hemisphere, it propagates much faster and further in the La Niña stage than in the El Niño stage. Also, the convection signal propagates much farther in the negative IOD phase than in the positive IOD phase. In addition, when the MJO propagates from the western Pacific across the Western Hemisphere to Africa, its intensity in the positive IOD phase is greater than in the negative IOD phase, and so on. Such characteristics indicate that MJO activities are very sensitive to the ocean–atmosphere background state. Because a climate model always has some errors in the prediction of MJO itself and also in the prediction of relationship between MJO and climate background, the impact of the large-scale ocean–atmosphere background features on the predictability of subseasonal climate significantly depends on the model itself. In the BCC_CSM, the close correlation between IOD and MJO predictability revealed by the hindcast experiment is related to the overestimation of the relationship between IOD events and MJO intensity in the model.

In addition to the predictability of large-scale climate phenomena, the predictability of climate variability at local scale and its impact factors are also worthy attention. Liu et al. (2015) pointed out that the subseasonal predictability of precipitation over local oceans and continental areas in the ASM region is related to local ocean–atmosphere or land–atmosphere coupling characteristics as well as the relationship between local variability and large-scale circulation. Among them, the subseasonal variability of precipitation over the Arabian Sea, the Bay of Bengal, and the South China Sea is affected by strong local ocean–atmosphere coupling process and is more closely linked to large-scale circulation, resulting in higher predictability of precipitation. In contrast, the land–atmosphere coupling process is relatively weak over the India subcontinent, the Indo-China Peninsula, and southern China, and these links between regional precipitation and large-scale circulation on subseasonal scale are not close; so the predictability of precipitation over these areas is lower. The relationship between local climate variability and large-scale circulation also determines, to some extent, the predictability of local climate, which is bound to be affected by the predictability of large-scale climate events. Figure 8.35 shows the correlations between the prediction skills of subseasonal precipitation over the South China Sea in the NCEP CFSv2 and the average amplitude and occurrence frequency of each phase of BSISO1 and BSISO2. It can be seen that when the BSISO1 events in the 1st and 2nd phases have large amplitude and high frequency of occurrence, the forecast skills of precipitation over the South China Sea are high, and vice versa. This is because the predictability of BSISO1 at phases 1 and 2 is higher in the model, and the South China Sea controlled by the two phases is a central region for convective inhibition, thus making the BSISO event itself closely related to the predictability of South China Sea rainfall (Liu et al. 2015).

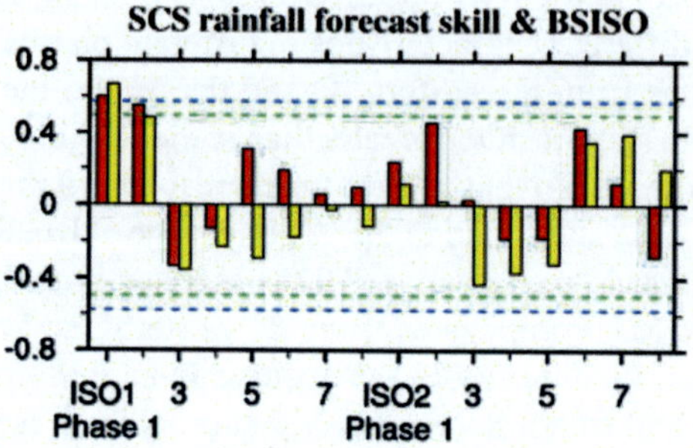

Fig. 8.35 Correlation between forecast skill of the South China Sea rainfall by NCEP CFSv2, and the mean amplitude (red bar) and occurrence frequency (yellow bar) of the observed BSISO1 and BSISO2 in each phase. The regional rainfall forecast skill is defined as the time correlation coefficient of pentad regional average precipitation between prediction and observation in May–September in summer, and the value is the averaged result during the lead times of 0–10 days. Green and blue dashed lines represent the 90% and 95% confidence levels, respectively. After Liu et al. (2015)

Of course, what needs to be emphasized is that the actual predictability of subseasonal climate revealed by a climate model is undoubtedly affected by the performance of the model itself. The ICs of prediction, the model physical parameterization processes, and prediction schemes can determine the subseasonal predictability to a certain degree. For example, the original forecast skill of MJO by BCC_CSM participating in the S2S Prediction Project was only 15 days. After the initial atmospheric condition was improved, the prediction skill improved by 1–2 days. Successively, after the ocean initial condition was further improved, the forecast skills at lead times above one week were improved, and the useful skill was eventually improved by four days to reach 21 days (Fig. 8.20). These results clearly demonstrate the sensitivity of MJO predictability to atmospheric and oceanic ICs. The merits of the model physical parameterization schemes determine the performance of the model in predicting stage, and their impacts on the subseasonal climate prediction have been revealed by many studies. For the BCC_CSM, to reduce the model errors in MJO simulation, we optimized the key physical parameters in the parameterization schemes of cloud and convection processes, including the adjustment time scale and precipitation efficiency for shallow convection, the relative humidity thresholds for low and high clouds, the humidity condition for convection triggering, the precipitation efficiency and precipitation evaporation rate for deep convections, among others. By doing so, we improved the climatology background, variance distribution, spectrum, spatial structure, and propagation characteristics of MJO in the model to some degree. Applying the optimized physical schemes to the prediction model can further increase the MJO prediction skill by 1–2 days, compared to the schemes that only improve the ICs. This demonstrates the contribution of model bias improvement to the enhancement of MJO prediction skill. Ensemble prediction is another important factor that determines the prediction skill of a model. In the multiple models of the S2S Project, the validity of the BCC_CSM ensemble prediction scheme is limited, and there is no apparent difference in the prediction skills of BSISO between the single-member experiments and the ensemble-member experiments. Comparatively, the ensemble forecast of the ECMWF model can improve BSISO skill by about one week compared to the deterministic forecast (Fig. 8.21). These results jointly demonstrate the importance of accurate ICs, reliable physical parameterization schemes, and reasonable prediction schemes for understanding the upper limit of subseasonal climate predictability based on climate models.

8.5 Conclusions and Discussion

The weather and climate over SEA exhibit strong temporal and spatial variability due to its special geographical location and complex distribution of land and sea. The variability of rainfall and atmospheric circulation over SEA has strong interaction with ocean and atmosphere in its neighboring regions. The strong variability of rainfall and atmospheric circulation over SEA always result in flood, drought, and other meteorological disasters. Therefore, accurate prediction of the rainfall and atmospheric circulation over SEA is vital for social and economic development not only over SEA but also over its adjacent regions where weather and climate are significantly affected by convection over SEA. As significant advances in climate system model have been achieved in the recent decade, the outputs of climate prediction system are the most important references for operational climate prediction. Thus, we analyze and review the S2S predictions of rainfall and atmospheric circulation by advanced climate systems, mostly by the CFSv2 from the NCEP and the CSM from the BCC. We also discuss the predictability and its source of S2S prediction of rainfall and atmospheric circulation over SEA.

The climate models have considerable skills for S2S prediction of rainfall and atmospheric circulation over SEA. They have a better skill for atmospheric circulation than for rainfall. The seasonal mean rainfall and atmospheric circulation over the tropics are better predicted than those over the extratropics. The models predict seasonal mean climate more skillfully for oceanic regions than for land regions, with higher skill for the Pacific than for the IO. The model's skill in predicting interannual variability of seasonal mean climate mostly comes from the model's skill in predicting ENSO and associated teleconnections, but the climate impacts of ENSO are overestimated by the models. The ENSO can affect climate over SEA and its adjacent regions by climate anomalies over both the Pacific and IO. The models' prediction skills for seasonal mean rainfall and atmospheric circulation tend to increase as the magnitude of ENSO increases. The prediction skills for the eastern Pacific ENSO and the central Pacific ENSO, as well as their climate impacts, are different, and depend on their phases. The IOBW and its impacts on climate over SEA and its adjacent regions can be well predicted. However, the summer IOD can be predicted only at a very short time lead. The winter climate over SEA is significantly affected by the extratropical climate variability. However, the climate models show low seasonal prediction skill for the most dominant extratropical atmospheric mode, the AO.

The climate models basically can predict most of the dominant subseasonal atmospheric modes two weeks in advance, including the North Atlantic oscillation, blocking, BSISO over the Asian monsoon region, the transition between dry and wet seasons over the MC. However, the MJO can be predicted about 3–4 weeks in advance, which is longer than the prediction for the other subseasonal atmospheric variability. As the MJO has an important role in subseasonal variability of extratropical climate variability, the improvement in simulation of the MJO and its climate impacts is the key to enhance subseasonal prediction skill. Note that the prediction skill for MJO and BSISO over the Asian monsoon region exhibits apparent differences for different phases, which is very low when the signals are located over the MC. Thus, improvement in representation of land–sea–atmosphere interaction over the MC is important for the enhancement of subseasonal prediction by climate models.

A climate model is the most powerful tool for predicting climate variability on various time scales. However, climate model's skill in S2S prediction is low, and is only useful under some specific conditions. Thus, how to improve the S2S prediction is a big challenge for scientific, operational, and user's communities. A WWRP/THORPEX/WCRP joint research project has been conducted to improve forecast skill and understanding on the S2S time scales, and promote its uptake by operational centers and its exploitation by applications community (http://s2sprediction.net). This project has collected the data related to the S2S prediction from major climate centers worldwide, which provides an unprecedented opportunity for studying S2S prediction (Vitart et al. 2017). An improvement in understanding of the prediction and predictability of subseasonal prediction of rainfall and atmospheric circulation over SEA will be achieved by future studies using this database.

Most of the meteorological centers are using different systems for mesoscale prediction, synoptic-scale prediction, and S2S prediction. However, there are strong interactions among the systems with various time scales and spatial scales in the real world, and the MC is just a typical region for studying such interactions. Thus, simulation and prediction of the systems of various temporal and spatial scales are intimated linked, and need to be carried out in a unified framework. Some current studies focus on developing a unified framework for weather and climate prediction. A unified modeling and prediction framework has several clear scientific advantages. For example, having a single numerical model at the heart of the prediction system, we can trace the evolution of error growth across multiple space and time scales to help unravel the key error sources responsible for the failure of the model to adequately capture major modes of climate variability (Brown et al. 2012). Liu et al. (2013) analyzed the daily prediction of the CFSv2, and found that the model-simulated synoptic WPSH was weaker than that in the observation. Jiang et al. (2013a) also found that the CFSv2 simulated a weaker-than-observed seasonal mean WPSH. This is a clear illustration that bias can evolve from synoptic scale to climatological scale. The development of UK Met Office models has exhibited that improvement in the bias at short time scales led to improvement in climatological bias (Brown et al. 2012). Convection is very active over SEA and interacts with the MJO. The large heat source over SEA is also maintained by convection. Therefore, the improvement in the simulation of convection over SEA is essential for the enhancement of S2S prediction. The increase in model horizontal resolution can generally improve the simulation of convection over the MC because of its more realistic description of the complex distribution of land and sea as well as the complex terrain. Therefore, a better S2S prediction will be achieved by a unified model system.

References

Aldrian E, Gates LD, Widodo FH (2007) Seasonal variability of Indonesia rainfall in ECHAM4 simulations and in the reanalyses: The role of ENSO. Theor Appl Climatol 87:41–59

Allen MR, Smith LA (1997) Optimal filtering in singular spectrum analysis. Phys Lett 234:419–428

Ashok K, Guan Z, Yamagata T (2001) Impact of the Indian Ocean dipole on the relationship between the Indian monsoon rainfall and ENSO. Geophys Res Lett 28:4499–4502

Behera SK, Luo JJ, Masson S et al (2005) Paramount Impact of the Indian Ocean Dipole on the East African short rains: a CGCM study. J Clim 18:4514–4530

Chang CP, Lau KM (1980) Northeasterly cold surges and near equatorial disturbances over the winter MONEX area during December 1974. Part II: Planetary-scale aspects. Mon Wea Rev 108:298–312

Chang CP, Lau KM (1982) Short-term planetary-scale interactions over the tropics and midlatitudes during northern winter. Part I: Contrasts between active and inactive periods. Mon Wea Rev 110:933–946

Chang CP, Harr PA, Chen HJ (2005a) Synoptic disturbances over the equatorial South China Sea and western Maritime Continent during boreal winter. Mon Wea Rev 133:489–503

Chang P, Saravanan R, Ji L, Hegerl GC (2000) The effect of local sea surface temperatures on atmospheric circulation over the tropical Atlantic sector. J Clim 13:2195-2216

Chang CP, Wang Z, McBride J et al (2005b) Annual cycle of southeast Asia-Maritime Continent rainfall and the asymmetric monsoon transition. J Clim 18:287–301

Chang CP, Wang Z, Ju J et al (2004) On the relationship between western Maritime Continent monsoon rainfall and ENSO during Northern Winter. J Clim 17:665–672

Chen W, Kang LH (2006) Linkage between the Arctic Oscillation and winter climate over East Asia on the interannual timescale: Roles of quasi-stationary planetary waves (in Chinese). Chin J Atmos Sci 30:863–870

Chen TC, Huang WR, Yoon JH (2004) Interannual variation of the East Asian cold surge activity. J Clim 17:401–413

Chen Y, Ding Y, Xiao Z, Yan H (2006) The impact of water vapor transport on the summer monsoon onset and abnormal rainfall over Yunnan province in May (in Chinese). Chin J Atmos Sci 30: 25–37

Ding YH, Krishnamurti TN (1987) Heat budget of the Siberian high and the winter monsoon. Mon Wea Rev 115:2428–2449

Dong SR, Yang S, Yan X et al (2020) The most predictable patterns and prediction skills of subseasonal prediction of rainfall over the Indo-Pacific regions by the NCEP Climate Forecast System. Clim Dyn 54(5):2759-2775

Gong D, Wang S, Zhu J (2001) East Asian winter monsoon and Arctic Oscillation. Geophys Res Lett 28: 2073–2076

Goswami BN, Krishnamurthy V, Annamalai H (1999) A broad-scale circulation index for the interannual variability of the Indian summer monsoon. Q J R Meteorol Soc 125:611–633

Haylock M, McBride J (2001) Spatial coherence and predictability of Indonesian wet season rainfall. J Clim 14: 3882-3887.

Hendon HH (2003) Indonesian rainfall variability: Impacts of ENSO and local air–sea interaction. J Clim 16:1775–1790

He J, Lin H, Wu Z (2011) Another look at influences of the Madden-Julian Oscillation on the wintertime East Asian weather. J Geophys Res 116:D03109

Holland GJ, Keenan TD (1980) Diurnal variations of convection over the "Maritime Continent". Mon Wea Rev 108: 223–225

Hong CC, Hsu HH, Chia HH et al (2008) Decadal relationship between the North Atlantic Oscillation and cold surge frequency in Taiwan. Geophys Res Lett 32:L24707

Holland GJ (1984) On the climatology and structure of tropical cyclones in the Australian southwest Pacific region. Aust Meteor Mag 32:17–31

Huang B (2004) Remotely forced variability in the tropical Atlantic Ocean. Clim Dyn 23:133–152

Huang B, Shukla J (2007) Mechanisms for the interannual variability in the tropical Indian Ocean. Part II: regional processes. J Clim 20:2937–2960

Hung CW, Liu X, Yanai M (2004) Symmetry and asymmetry of the Asian and Australian summer monsoons. J Clim 17:2413–2426

Hurrell JW (1995) Transient eddy forcing of the rotational flow during northern winter. J. Atmos Sci 52(12):2286-2301

Ichikawa H, Yasunari T (2006) Time-space characteristics of diurnal rainfall over Borneo and surrounding oceans as observed by TRMMPR. J Clim 19:1238–1260

Jia X, Yang S, Song W, He B (2014) Prediction of Wintertime Northern Hemisphere Blocking by the NCEP Climate Forecast System. J Metorol Rea 28: 76-90

Jeong JH, Ho CH (2005) Changes in occurrence of cold surges over East Asia in associated with Arctic Oscillation. Geophys Res Lett 32:85–93

Jeong JH, Kim BM, Ho CH et al (2006) Stratospheric origin of cold surge occurrence in East Asia. Geophys Res Lett 33:289–289

Jie W, Vitart F, Wu T et al (2017) Simulations of Asian Summer Monsoon in Sub-seasonal to Seasonal Prediction Project (S2S) database. Q J R Meteorol Soc 143:2282–2295

Jiang X, Yang S, Li Y et al (2013a) Seasonal-to-interannual prediction of the Asian summer monsoon in the NCEP Climate Forecast System Version 2. J Clim 26:3708–3727

Jiang X, Yang S, Li Y et al (2013b) Dynamical prediction of the East Asian winter monsoon by the NCEP climate forecast system. J Geophys Res 118:1312–1328

Jiang X, Yang S, Li Y et al (2013c) Variability of the Indian Ocean SST and its possible impact on summer western North Pacific anticyclone in the NCEP Climate Forecast System. Clim Dyn 41:2199–2212

Kao H-K, Yu J-Y (2009) Contrasting Eastern-Pacifiic and Central-Pacifiic types of ENSO. J Clim 22: 615–632

Kim H, Webster P, Toma V et al (2014) Predictability and prediction skill of the MJO in two operational forecasting systems. J Clim 27:5364–5378

Kim HM, Webster PJ, Curry JA et al (2012) Asian summer monsoon prediction in ECMWF System 4 and NCEP CFSv2 retrospective seasonal forecasts. Clim Dyn 39:2975–2991

Keenan TD, Morton BR, Manton MJ et al (1989) The island thunderstorm experiment (ITEX) — A study of tropical thunderstorms in the maritime continent. Bull Am Meteorol Soc 70:152–159

Keenan T, May P, Hollandet G et al (2000) The Maritime Continent thunderstorm experiment (MCTEX): Overview and some results. Bull Am Meteorol Soc 81:2433–2455

Kug JS, Jin FF, An SI (2009) Two types of El Niño events: cold tongue El Niño and warm pool El Niño. J Clim 22:1499–1515

Larson SM, Kirtman BP (2015) Revisiting ENSO coupled instability theory and SST error growth in a fully coupled model. J Clim 28:4724–4742

Larkin NK, Harrison DE (2005) Global seasonal temperature and precipitation anomalies during El Niño autumn and winter. Geophys Res Lett 32:L13705

Lee JY, Wang B, Wheeler M et al (2013) Realtime multivariate indices for the boreal summer intraseasonal oscillation over the Asian summer monsoon region. Clim Dyn 40:493–509

Li T, Wang B (2005) A review on the western North Pacific monsoon: synoptic-to-interannual variabilities. Terr Atmos Ocean Sci 16:285–314

Li QP, Yang S, Wu TW et al (2017) Sub-seasonal dynamical prediction of East Asian cold surge. Wea Forecasting 32:1675–1694

Lin H, Brunet G, Derome J (2008) Forecast skill of the Madden–Julian Oscillation in two Canadian atmospheric models. Mon Wea Rev 136:4130–4149

Liu B, Wu G, Mao J et al (2012) Genesis of the South Asian High and Its impact on the Asian summer monsoon onset. J Clim 26(9):2976-2991

Liu X, Yang S, Kumar A et al (2013) Diagnostics of subseasonal prediction biases of the Asian summer monsoon by the NCEP Climate Forecast System. Clim Dyn 41:1453–1474

Liu X, Yang S, Li Q et al (2014) Subseasonal forecast skills and biases of global summer monsoons in the NCEP Climate Forecast System version 2. Clim Dyn 42:1487–1508

Liu X, Yang S, Li J et al (2015) Subseasonal predictions of regional summer monsoon rainfalls over tropical Asian oceans and land. J Clim 28:9583–9605

Liu X, Wu T, Yang S et al (2017) MJO prediction using the sub-seasonal to seasonal forecast model of Beijing Climate Center. Clim Dyn 48:3283–3307

Liang J, Yang S, Hu ZZ et al (2009) Predictable patterns of Asian and Indo-Pacific summer precipitation in the NCEP CFS. Clim Dyn 32:989–1001

Li T, Philander SGH (1996) On the annual cycle of the eastern equatorial Pacific. J Clim 9:2986–2998

Lorenz DJ, Hartmann DL (2003) Eddy–Zonal Flow Feedback in the Northern Hemisphere Winter. J Clim 16:1212–1227
Luo J, Masson S, Behera S et al (2005) Seasonal climate predictability in a coupled OAGCM using a different approach for ensemble forecasts. J Clim 18:4474–4497
Lau KM, Chang CP, Chan PH (1983) Short term planetary scale interactions over the Tropics and mid-latitudes during northern winter. Part II: Winter-MONEX period. Mon Wea Rev 111:1372–1388
Lau NC, Wang B (2006) Interactions between the Asian monsoon and the El Nino/Southern Oscillation. The Asian Monsoon, B Wang, Eds., Praxis, 479-511
Ma J, Xie SP, Xu H (2017) Inter-member variability of the summer northwest Pacific subtropical anticyclone in the ensemble forecast. J Clim 30:3927–3941
Matsumoto J, Murakami T (2000) Annual changes of tropical convective activities as revealed from equatorial symmetric OLR data. J Meteor Soc Jpn 78:543–561
McBride JL (1995) Tropical cyclone formation. Global Perspective on Tropical Cyclones, R L Elsberry, Eds., Tech Doc 693, World Meteorological Organization, 63–105
Mori S, Hamada JI, Tauhid YI et al (2004) Diurnal land-sea rainfall peak migration over Sumatera Island, Indonesian Maritime Continent, observed by TRMM satellite and intensive rawinsonde soundings. Mon Wea Rev 132:2021–2039
Neale R, Slingo J (2003) The Maritime Continent and its role in the global climate: A GCM study. J Clim 16:834–848
Park TW, Ho CH, Deng Y (2014) A synoptic and dynamical characterization of wave-train and blocking cold surge over East Asia. Clim Dyn 43:753–770
Park TW, Ho CH, Yang S (2011) Relationship between the Arctic Oscillation and cold surges over East Asia. J Clim 24:68–83
Park TW, Ho CH, Yang S et al (2010) Influences of Arctic Oscillation and Madden-Julian Oscillation on cold surges and heavy snowfalls over Korea: A case study for the winter of 2009-2010. J Geophys Res 115:D23122
Ramage CS (1968) Role of a tropical "Maritime Continent" in the atmospheric circulation. Mon Wea Rev 96:365–370
Rashid H, Hirst AC (2017) Mechanisms of improved rainfall simulation over the Maritime Continent due to increased horizontal resolution in an AGCM. Clim Dyn 49:1747–1764
Rashid H, Hendon H, Wheeler M et al (2011) Prediction of the Madden-Julian oscillation with the POAMA dynamical prediction system. Clim Dyn 36:649–661
Rauniyar SP, Walsh KJE (2012) Influence of ENSO on the diurnal cycle of rainfall over the Maritime Continent and Australia. J Clim 26:1304–1321
Ren HL, Jin FF, Kug JS, Zhao JX (2009) A kinematic mechanism for positive feedback between synoptic eddies and NAO. Geo Res Let 36(11):269–271
Riddle EE, Butler AH, Furtado JC et al (2013) CFSv2 ensemble prediction of the wintertime arctic oscillation. Clim Dyn 41(3-4):1099-1116
Rogers JC (1981) The North Pacific Oscillation. Int J Climatol 1:39–57
Saha S, Moorthi S, Wu X et al (2013) The NCEP climate forecast system version 2. J Clim 27:2185–2208
Saji NH, Yamagata T (2007) Possible impacts of Indian Ocean dipole mode events on global climate. Clim Res 25(2):151–169
Saji NH, Goswami BN, Vinayachandran PN, Yamagata T (1999) A dipole mode in the tropical Indian Ocean. Nature 401:360-363
Sigmond M, Scinocca JF, Kharin VV, Shepherd (2013) Enhanced seasonal forecast skill following stratospheric sudden warmings. Nature Geos 6(2): 98-102
Shukla J, Paolina DA (1983) The southern oscillation and long range forecasting of the summer monsoon rainfall over India. Mon Weather Rev 111:1830–1837
Sui CH, Lau KM (1992) Multiscale phenomena in the tropical atmosphere over the western Pacific. Mon Wea Rev 120:407–430

Sutton RT, Jewson SP, Rowell DP (2000) The elements of climate variability in the tropical Atlantic region. J Clim 13:3261–3284
Takaya K, Nakamura H (2005) Mechanism of intraseasonal amplification of the cold Siberian high. J Atmos Sci 62:4423–4440
Thompson DWJ, Wallace et al (2000) Annular modes in the extratropical circulation. part I: month-to-month variability. J. Clim 13(5):1018–1036
Tomita T, Yasunari T (1996) Role of the northeast winter monsoon on the biennial oscillation of the ENSO/monsoon system. J Meteor Soc Jpn 74:399–413
Tripathi OP, Charlton-Perez A, Sigmond M, Vitart F (2015) Enhanced long-range forecast skill in boreal winter following stratospheric 10: strong vortex conditions. Environmental research letter 10: 104007
Venzke S, Allen MR, Sutton RT, Rowell DP (1999) The atmospheric response over the North Atlantic to decadal changes in sea surface temperature. J Clim 12:2562–2584
Vimont DJ, Battisti DS, Hirst AC (2001) Footprinting: A seasonal conection between the tropics and mid-latitudes. Geophys Res Lett 28:3923–3926
Vimont DJ, Wallace JM, Battisti DS (2003) The seasonal footprinting mechanism in the Pacific: Implications for ENSO. J Clim 16:2668–2675
Vitart F (2014) Evolution of ECMWF sub-seasonal forecast skill scores. Q J R Meteorol Soc 140:1889–1899
Wang B, Fan Z (1999) Choice of South Asian summer monsoon indices. Bull Am Meteor Soc 80:629–638
Wang B, Lin H (2002) Rainy season of the Asian-Pacific summer monsoon. J Clim 15:386–398
Wang B, Wu R, Xi Fu (2000) Pacific-East Asian teleconnection: How does ENSO affect East Asian climate? J Clim 13:1517–1536
Wang HJ, Xue F, Zhou GQ (2002) The spring monsoon in southern China and its relationship to Large-Scale circulation features. Adv Atmos Sci 19(4):651–664
Wang W, Hung M, Weaver S et al (2014) MJO prediction in the NCEP climate forecast system version 2. Clim Dyn 42:2509–2520
Webster PJ, Magana VO, Palmer TN, Shukla J, Tomas RA, Yanai M, Yasunari T (1998) Monsoons: processes, predictability, and the prospects for prediction. J Geophys Res 103:14451–14510
Webster PJ, Moore A, Loschnigg J, Leban M (1999) Coupled ocean– atmosphere dynamics in the Indian Ocean during 1997–98. Nature 40:356–360
Weng H, Ashok K, Behera SK et al (2007) Impacts of recent El Niño Modoki on dry/wet conditions in the Pacific Rim during boreal summer. Clim Dyn 29:113–129
Wheeler M, Hendon H (2004) An all-season real-time multivariate MJO index: development of an index for monitoring and prediction. Mon Wea Rev 132:1917–1932
Webster PJ, Yang S (1992) Monsoon and ENSO: Selectively interactive systems. Q J R Meteor Soc 118:877–926
Wu B, Li T, Zhou T (2010) Relative contributions of the Indian Ocean and local SST anomalies to the maintenance of the western North Pacific anomalous anticyclone during El Niño decaying summer. J Clim 23:2974–2986
Xie SP, Hu K, Hafner J et al (2009) Indian Ocean capacitor effect on Indo-western Pacific climate during the summer following El Niño. J Clim 22:730–747
Yang G-Y, Slingo J (2001) The diurnal cycle in the tropics. Mon Wea Rev 129:784–801
Yang S, Min W, Yang R et al (2011) Impacts of Land Process on the Onset and Evolution of Asian Summer Monsoon in the NCEP Climate Forecast System. Adv Atmos Sci 28(6):1301–1317
Yang S, Jiang X (2014) Prediction of eastern and central Pacific ENSO events and their impacts on East Asian climate by the NCEP Climate Forecast System. J Clim 27:4451–4472
Yang S, Wen M, Higgins W (2008a) Subseasonal features of the Asian summer monsoon in the NCEP climate forecast system. Acta Oceanol Sin 27(3):88–103
Yang S, Webster PJ (1990) The effect of summer tropical heating on the location and intensity of the extratropical westerly jet streams. J Geophys Res 95:18705–18721

Yang S, Zhang Z, Kousky VE et al (2008b) Simulations and seasonal prediction of the Asian summer monsoon in the NCEP Climate Forecast System. J Clim 21:3755–3775
Yim SY, Wang B, Xing W (2013) Prediction of early summer rainfall over South China by a physical-empirical model. Clim Dyn. https://doi.org/10.1007/s00382-013-2014-3
Yu J-Y, Kao H-K (2007) Decadal changes of ENSO persistence barrier in SST and ocean heat content indices: 1958–2001. J Geophys Res 112:D13106
Yuan Y, Yang S, Zhang Z (2012) Different evolutions of the Philippine Sea anticyclone between eastern and central Pacific El Niño: Possible effect of Indian Ocean SST. J Clim 25:7867–7883
Zhu B, Wang B (1993) The 30-60 day convection seesaw between the tropical Indian and western Pacific Oceans. J Atmos Sci 50:184–199
Zhang R, Sumi A, Kimoto M (1996) Impact of El Niño on the East Asian monsoon: A diagnostic study of the '86/87 and '91/92 events. J Meteor Soc Jpn 74:49–62
Zhang T, Yang S, Jiang X et al (2016a) Seasonal-interannual variation and prediction of wet and dry season rainfall over the Maritime Continent: Roles of ENSO and monsoon circulation. J Clim 29:3675–3695
Zhang T, Yang S, Jiang X et al (2016b) Roles of remote and local forcings in the variation and prediction of regional Maritime Continent rainfall in wet and dry seasons. J Clim 29:8871–8879
Zhang T, Yang S, Jiang X et al (2016c) Sub-seasonal prediction of the Maritime Continent rainfall of wet-dry transitional seasons in the NCEP Climate Forecast Version 2. Atmosphere 7:28
Zhang T, Huang B, Yang S et al (2018a) Seasonal dependence of the predictable low-level circulation patterns over the tropical Indo-Pacific domain. Clim Dyn, 50:4263-4284
Zhang T, Huang B, Yang S (2018b) Predictable patterns of the atmospheric low-level circulation over the Indo-Pacific domain in project Minerva: Seasonal dependence and intra-ensemble variability. J Clim 31:8351–8379
Zhang Y, Sperber KS, Boyle JS (1997) Climatology and interannual variation of the East Asian winter monsoon: Results from the 1979–95 NCEP/NCAR Reanalysis. Mon Wea Rev 125:2605–2619
Zhao P, Sun J, Zhou XJ (2003) Mechanism of formation of low level jets in the southern China Sea during spring and summer of 1998. Chinese Sci Bull 48(12):1265–1270
Zhao SY, Yang S, Deng Y et al (2015) Skills of yearly prediction of the early-season rainfall over southern China by the NCEP climate forecast system. Theor Appl Climatol 122(3–4):743–754
Zhao SY, Yang S (2014) Dynamical prediction of the early season rainfall over southern China by the NCEP climate forecast system. Wea Forecasting 29(6):1391–1401
Zhao P, Zhang RH, Liu JP et al (2007) Onset of southwesterly wind over eastern China and associated atmospheric circulation and rainfall. Clim Dyn 28:797–811
Zuo ZY, Yang S et al (2013) Predictable patterns and predictive skills of monsoon precipitation in Northern Hemisphere summer in NCEP CFSv2 reforecasts. Clim Dyn 40:3071–3088
Zuo JQ, Ren H-L, Wu J et al (2016) Subseasonal variability and predictability of the Arctic Oscillation/North Atlantic Oscillation in BCC_AGCM2.2. Dyn Atmos Oceans 75:33–45